IMMUNE SYSTEM: GENETICS AND REGULATION

ACADEMIC PRESS RAPID MANUSCRIPT REPRODUCTION

ICN–UCLA Symposia on Molecular and Cellular Biology
Vol. VI, 1977

ICN-UCLA Symposia on the Immune System

IMMUNE SYSTEM: GENETICS AND REGULATION

edited by

ELI E. SERCARZ
Department of Bacteriology
University of California, Los Angeles
Los Angeles, California

LEONARD A. HERZENBERG
Department of Genetics
Stanford University School of Medicine
Stanford, California

C. FRED FOX
Department of Bacteriology
and Molecular Biology Institute
University of California, Los Angeles
Los Angeles, California

ACADEMIC PRESS INC. New York San Francisco London 1977
A Subsidiary of Harcourt Brace Jovanovich, Publishers

ACADEMIC PRESS, INC.
111 Fifth Avenue, New York, New York 10003

United Kingdom Edition published by
ACADEMIC PRESS, INC. (LONDON) LTD.
24/28 Oval Road, London NW1

Library of Congress Cataloging in Publication Data

ICN–UCLA Symposia on the Immune System, Park City,
Utah, 1977.
Immune system: genetics and regulation

(ICN–UCLA symposia on molecular & cellular biology ;
V. 6)
1. Immunogenetics—Congresses. I. Sercarz, Eli E.
II. Herzenberg, Leonard A. III. Fox, C. Fred.
IV. ICN Pharmaceuticals, inc. V. California. University. University at Los Angeles. VI. Title.
VII. Series.
QR184.I18 1977 599′.02′0 77-11035
ISBN O-12-637160-1

PRINTED IN THE UNITED STATES OF AMERICA

Contents

V. MHC AND T-CELL RECOGNITION

Preface

This volume presents the proceedings of the ICN–UCLA Symposia on the Immune System held at Park City, Utah in March 1977. Although it was the design of the organizers to restrict the breadth of consideration to genetic aspects of the subject, it is clear that this topic has become all-embracing. We hope that the vigor of current immunological thinking will be evident in these pages. Essentially, ideas were presented concerning the organization and expression of genes relating to antibody structure, to surface markers and receptors on lymphocyte subpopulations, to collaborative cellular interactions, and even to newer systems of somatic cell hybridization.

This meeting was highly participatory, and utilized three formal modes of presentation. The five morning plenary sessions focused on the six major topics indicated in the Table of Contents and involved about 25 conveners and speakers. Approximately 170 poster presentations, the majority of which were presented by younger investigators, were held during the late afternoon under modulated, soft-lighting, ostensibly to foster interaction. Finally, just about everyone contributed to the 19 workshops that were held on four different evenings and covered subjects related to the six major topics. Many of these workshops continued far into the night.

This symposium volume represents each of these modalities: virtually all the plenary papers are here, even our leadoff paper (Chapter 0) a brilliant overview of Ig gene organization, wherein it is shown that being first may be for nought; a summary of each workshop was prepared by the two conveners, with their scribes; about 50 papers selected from the rich repertoire of poster presentations are also included. Luciano Adorini and Keith Krolick are to be commended for the useful Subject Index.

Many different elements have to interweave for a successful meeting. Through a joint venture between the National Cancer Institute, National Institute of Allergy and Infectious Diseases and the Fogarty International Center, Contract No. 263-77-C-0096 was generously awarded in partial support of the present meeting to help defray travel costs for participants. Doctors Earl C. Chamberlayne and William Terry, of the National Institutes of Health, deserve our special thanks for their assistance in facilitating the award. Also, we thank ICN Pharmaceuticals, Inc. for their continued financial sponsorship of these symposia in general. All of us send our appreciation to Accurate Chemical and Scientific Corp., Hicksville, New York, and Becton–Dickenson Electronics Lab, Mountainville, California, who kindly provided the energizing refreshments at the poster sessions.

Fran Stusser accomplished her usual miracle of mothering this enormous conference from point of conception to birth; Carol Parks, serving as the organizers' alter ego, nurtured the scientific program during its growth and development. Robert Williams merits special recognition for competent handling of the financial arrangements. The heroic scribing by Barbara Araneo, Elizabeth Grimm, Michael Katz, Dale Kipp, Keith Krolick, Jean Merrill-Davis, and Robert Yowell from UCLA was of great value in the preparation of the workshop summaries. Joel Sercarz performed flawlessly at slide projection. The Park City hotel management deserves special mention for bringing us close together in a spirit of bunker solidarity; and finally, the beautiful Utah weather and skiing followed the 1974 Squaw Valley precedent for immunological good luck.

THE IMMUNOLOGICAL ORCHESTRA

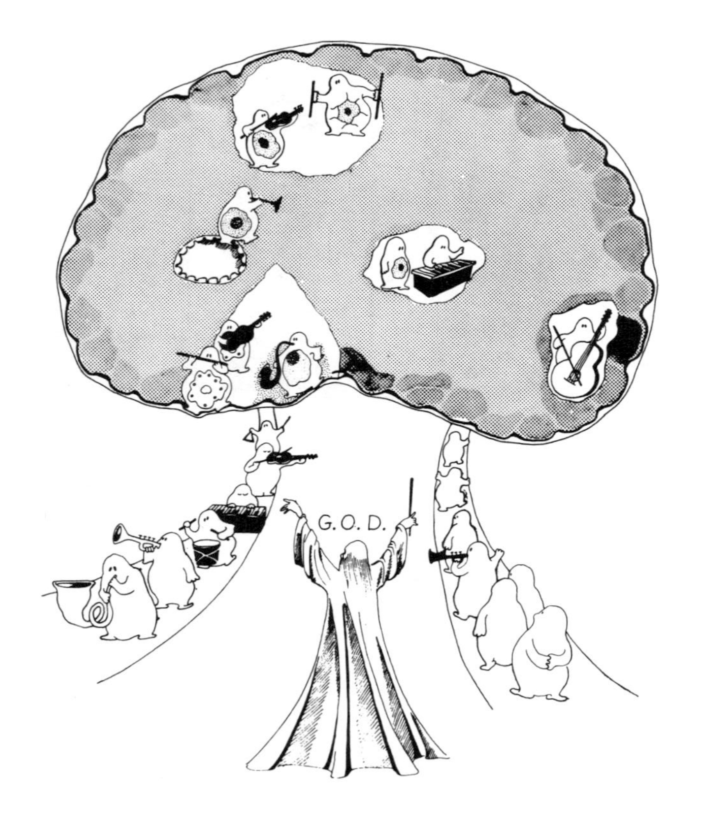
G.O.D.

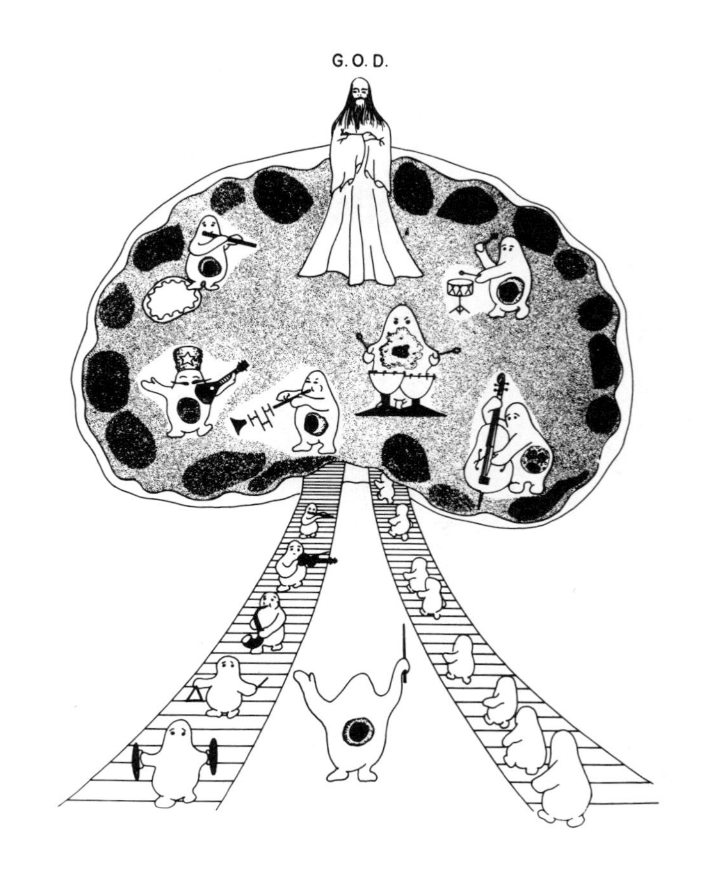
G.O.D.

At the time of the last ICN–UCLA symposium on Molecular Biology dealing with the subject of Immunology (see The Immune System: Genes, Receptors, Signals, *edited by Eli E. Sercarz, Alan R. Williamson, and C. Fred Fox, Academic Press, New York, 1974, p. xxii), which was held at Squaw Valley, California, March, 1974, two versions of the immunological orchestra were presented. These are reprinted here as Figs. 1 and 2. Figure 1 represents the 1968 version of the orchestra and Fig. 2 represents the 1974 version. The main changes made during that six-year period were the enlargement of the cell types involved in the orchestra and also the change in the conductor. The change in the conductor was brought about by the discovery that T cells acted not only to amplify the immune response, as was originally shown in 1968, but also to suppress it, suggesting that some T cells should be thought of as regulatory cells. It was therefore felt that a subpopulation of T cells (probably T_2) should be considered to be the conductor of the immunological orchestra.*

Since that time, regulatory T cells have been divided into three distinct subclasses: an Ly 1 helper cell, an Ly 23 suppressor cell, and an Ly 123 cell (or cells), which amplify or regulate the helper and the suppressor. Thus the notion of a single conductor is no longer tenable, and the 1977 version of the orchestra reflects these changes (Fig. 3). In addition, the complexity and number of the feedback loops between the three regulatory T cell subclasses is reflected by the demeanor of the generator of diversity (G.O.D.).

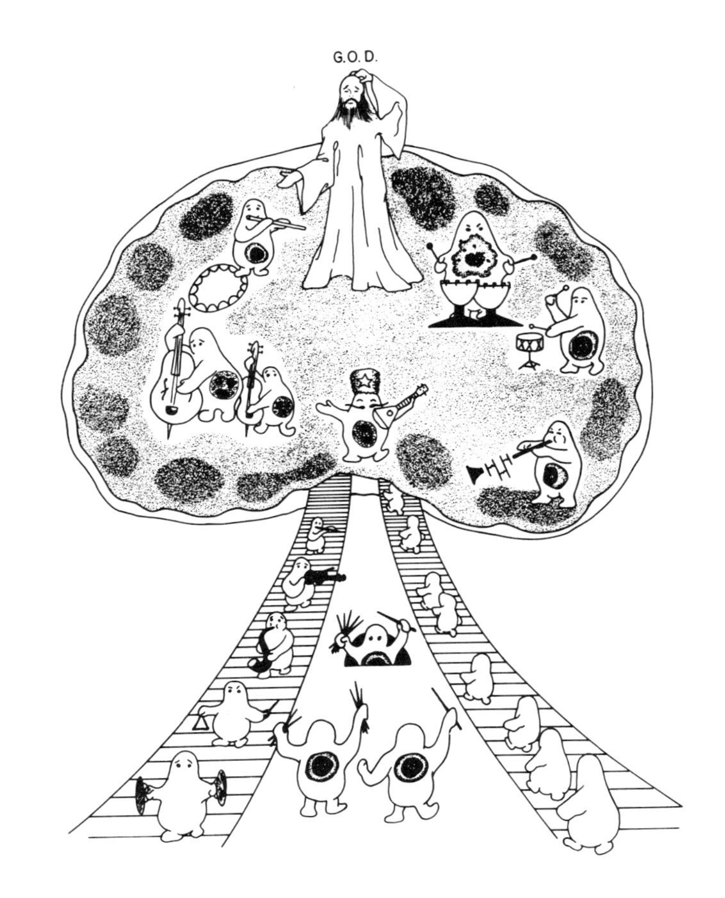

NUCLEIC ACID CHEMISTRY AND THE ANTIBODY PROBLEM

L. Hood, M. Kronenberg, P. Early and N. Johnson

Division of Biology, California Institute of Technology
Pasadena, California 91125

ABSTRACT. Nucleic acid chemistry has and will continue to make important contributions to our understanding of the antibody problem. This paper reviews current knowledge about the organization, expression, diversification and evolution of antibody genes. Particular emphasis is placed on the probable future contributions that will be made by the techniques of nucleic acid chemistry.

INTRODUCTION

Our perception of a discipline is determined by the nature of the tools and methodologies employed to study it. Accordingly, the use of nucleic acid chemistry has had a profound impact on immunology because one can begin to study directly the genes which code for immunity.

Advances in our understanding of immunology have been made in several distinct eras. At the beginning of this century Ehrlich described in a preliminary manner the phenomenology of the immune response. He suggested that preformed antibody molecules could be selectively expressed by interaction with antigen. In the 1920's and 1930's Landsteiner used various immunochemical techniques to elucidate the exquisite specificity of the immune response. These studies led to the view that the antigen must serve as a template to direct the synthesis of complementary antibody molecules. This instructionist view held sway over the next several decades until the early 1960's when classical genetics, sophisticated serology and protein chemistry revealed many general features of antibody molecules and antibody genes. These studies supported the clonal selection hypothesis, led to the formulation of several genetic explanations for antibody diversity, suggested that individual antibody polypeptides were coded by two distinct genes, and gave a general picture of the evolution of antibody genes. The advent of nucleic acid studies of antibody genes in the 1970's has marked the beginnings of yet another era for experimental immunology. Now one can begin to make a direct analysis of the organization of antibody genes, mechanisms for antibody diversity, and strategies for antibody gene regulation.

This paper will summarize the current status of molecular immunology with regard to the organization, expression, and evolution of antibody genes.

I. ANTIBODY GENES AS MULTIGENIC FAMILIES

Antibody molecules are encoded by clusters or families of genes which exhibit close linkage, sequence homology, related or overlapping functions, and multiplicity (Figure 1) (1). The unit of information

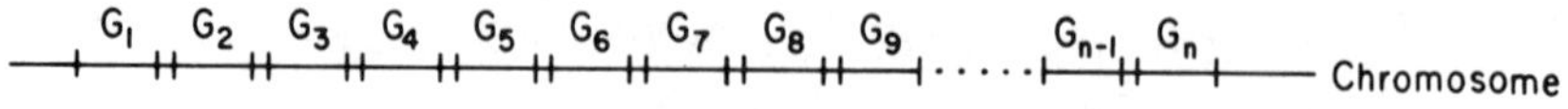

G_1 G_2 G_3 G_4 G_5 G_6 G_7 G_8 G_9 ..... G_{n-1} G_n Chromosome

Figure 1. A model of a multigene family. From reference 1.

in the immune system is the antibody molecule. The immune system has evolved diverse and sophisticated strategies for "information handling." We shall consider several general aspects of the information-handling problem of antibody molecules. i) How is information organized? ii) How is information expression regulated? iii) How is information generated during somatic differentiation? iv) How does information evolve?

One controversial issue should be kept in mind throughout this review. Are the multigenic families of antibody genes relevant as general models for an understanding of other complex eukaryotic systems such as the nervous system? Alternatively, do the information-handling mechanisms that have evolved for vertebrate immunity represent novel adaptations for unique phenotypic problems? We suggest that evolutionary considerations argue that the antibody gene families may be general models for at least certain other complex eukaryotic systems.

Before considering the various categories of information handling, let us describe some of the general techniques of nucleic acid chemistry that are being applied to the analysis of antibody genes.

II. MODERN TECHNIQUES OF NUCLEIC ACID CHEMISTRY

Over the past several years nucleic acid chemistry has made important contributions to our understanding of antibody genes. We will define very simply for subsequent discussion the basic reagents and methodologies of this rapidly progressing technology.

i) The techniques of nucleic acid chemistry are based on the annealing or renaturation reactions of complementary nucleic acid sequences (see 2). Radiolabeled probe sequences are used to follow the reactions of larger amounts of unlabeled DNA or RNA. Commonly these probes are (^{125}I) iodinated immunoglobulin messenger RNA from appropriate myeloma tumors. Radioactive nucleotides may also be incorporated into DNA copies of mRNA by reverse transcription allowing one to use a highly labeled DNA probe.

ii) The rate of reaction of a radiolabeled probe with excess total cellular DNA may be followed (2). This rate is proportional to the number of copies per genome of the probe sequence. Such hybridization kinetics studies have led to the conclusion that there are one or a very few copies of the kappa constant region in the mouse genome

(3). They also indicate a similar low number of V_κ genes within each subgroup tested so far (4-6).

iii) Restriction endonucleases cleave DNA at specific recognition sequences. When genomic DNA is digested with such an enzyme and the resulting fragments are separated by size, a particular single-copy gene will be found only in one or a few fractions. Gene linkage relationships (or the lack thereof) can be determined by examining these fractions with appropriate probes (7). In this manner, Tonegawa has shown that V and C genes for certain light chains are on separate restriction fragments in undifferentiated mouse embryo DNA, but are on the same restriction fragment (and possibly joined) in the appropriate differentiated myeloma tumor DNA (8).

iv) Individual immunoglobulin genes may be amplified by the use of recombinant plasmids (9). DNA for cloning may be obtained directly from genomic DNA fragments by appropriate screening or enrichment procedures, or it may be produced from mRNA by reverse transcriptase. After insertion into a plasmid or bacteria phage λ, large quantities of these genes may be obtained for further characterization. Using these techniques Tonegawa has been the first to isolate a mammalian gene--a mouse V_λ gene from embryonic or undifferentiated DNA (Tonegawa et al., this volume).

v) Rapid new methods for sequencing DNA have been developed by Maxam and Gilbert (10), and in Sanger's laboratory (11). The Maxam and Gilbert procedure uses DNA enzymatically labeled with a 5' terminal ^{32}P. The DNA is randomly cleaved an average of once per molecule under four separate conditions, one preferential for each of the four bases. The products of each reaction are then fractionated on adjacent lanes of a polyacrylamide gel. The sequence from the labeled 5' terminus can be read directly from an autoradiograph of the gel, as successively smaller fragments appear in one of the four lanes. Given an appropriately labeled DNA fragment, one can readily sequence 80-100 nucleotides per day. Certainly the nucleotide sequences of the mouse V_λ gene described above will be of great interest. The nucleotide sequence of mRNAs can also be determined, although by more laborious techniques (Milstein et al., this volume).

Nucleic acid studies promise to tell us a great deal about what genes are present in the germ line and what role the surrounding sequences play in gene joining and the regulation of gene expression.

III. THE ORGANIZATION OF ANTIBODY GENES

1. Three gene families (12). Classical genetic studies have demonstrated that three clusters or families of antibody genes, λ, κ, and H, are present in all mammals studied to date (Figure 2). These gene families are genetically unlinked to one another. The λ and κ gene families code for light chains whereas the H gene family codes for heavy chains.

2. Separate V and C genes (8). The variable (V) and constant

(C) regions of antibody polypeptides appear to be encoded by separate germ line genes (Figure 2). These genes undergo a rearrangement

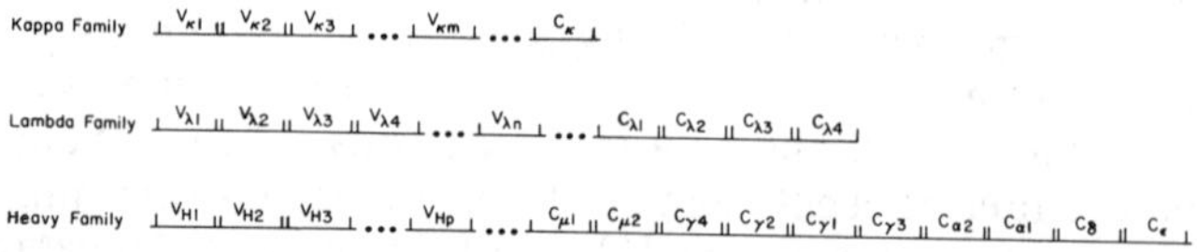

Kappa Family $V_{\kappa 1}$ $V_{\kappa 2}$ $V_{\kappa 3}$ ... $V_{\kappa m}$ ... C_{κ}

Lambda Family $V_{\lambda 1}$ $V_{\lambda 2}$ $V_{\lambda 3}$ $V_{\lambda 4}$ ... $V_{\lambda n}$ ... $C_{\lambda 1}$ $C_{\lambda 2}$ $C_{\lambda 3}$ $C_{\lambda 4}$

Heavy Family V_{H1} V_{H2} V_{H3} ... V_{Hp} ... $C_{\mu 1}$ $C_{\mu 2}$ $C_{\gamma 4}$ $C_{\gamma 2}$ $C_{\gamma 1}$ $C_{\gamma 3}$ $C_{\alpha 2}$ $C_{\alpha 1}$ C_{δ} C_{ϵ}

Figure 2. A model of the genes encoding the three antibody families of man. From reference 1.

during somatic differentiation, presumably to form a contiguous V-C gene (see Tonegawa et al., this volume).

3. Idiotype mapping (13). The idiotypes of at least 12 myeloma proteins or "homogeneous" antibodies appear to be present in the serum of certain inbred mice and lacking in the serum of others. When mice from a positive and a negative strain are crossed and their offspring appropriately tested, these idiotypes appear to segregate in a Mendelian fashion. Moreover, these idiotypes are linked to the C_H genes. The simplest interpretation of these observations is that these idiotypes are coded by germ line V_H genes that are closely linked to their corresponding C_H genes. Recombinational analysis of the idiotypes and C_H gene markers has led to three interesting observations (see Riblet et al., this volume). i) A preliminary ordering of certain of the putative V_H genes with respect to one another and the C_H genes has been possible. ii) One crossing-over event appears to predispose this genetic region to additional crossing-over events. This increased frequency of multiple relative to single crossovers is termed high negative interference. This propensity for extensive crossing-over may be a fundamental feature of multigene families with a large array of closely-linked homologous gene sequences (1). iii) The V_H-C_H chromosomal map distance appears to be at least 7 map units in length. With simplistic genetic assumptions this is sufficient DNA in the H gene family to code for approximately 56,000 V_H genes (14). This calculation does not include the possibility of spacer and/or regulatory DNA sequences. Obviously this map distance, which is calculated from recombinational analyses, may not correlate with chromosomal length because of the high negative interference seen in this chromosomal region. While this map distance appears unreasonably large, it does stress that the V_H segment of the mouse chromosome potentially has sufficient DNA to encode many germ line V_H genes.

Two important qualifications should be noted about the use of idiotypes for genetic mapping. i) The individual idiotype may represent not a single V_H gene, but rather a cluster of closely-related and presumably closely-linked V_H genes. ii) The mapping of idiotypes has been carried out by the genetic analysis of inbred strains of mice expressing or failing to express the idiotype in their serum. It has recently been demonstrated that an inbred strain of mouse lacking a particular idiotype in its serum may express that same idiotype on certain of its lymphocytes (P. Gearhart, personal communication). This raises the possibility that at least some idiotypes may reflect a control mechanism rather than a genetic polymorphism of a structural gene.

4. Future questions. Nucleic acid chemistry should allow us to approach several important questions concerning the organization of antibody genes. Several of these are probably best examined by analyzing recombinant DNA clones derived from the mouse germ line (embryo) or differentiated (myeloma tumor) DNA. i) What are the numbers of V and C genes in the individual families? ii) Are the V (and C) genes separated by untranscribed DNA? iii) What kinds of sequences separate the V and C genes from one another? iv) Are V genes with related sequences adjacent to one another or dispersed throughout the gene family? v) Which mouse chromosomes contain the three antibody gene families? This problem may be approached by the techniques of somatic cell genetics or possibly by special *in situ* hybridization techniques with metaphase chromosomes.

IV. THE EXPRESSION OF ANTIBODY GENES

1. V-C joining may be a fundamental component of the mechanism for lymphocyte differentiation. Each mature lymphocyte expresses one unit of information, that is, one type of antibody molecule. It is attractive to postulate that V-C joining is an important component of the molecular mechanism whereby a single lymphocyte is committed to the expression of one antibody molecule (Figure 3). DNA rearrangements have been shown to regulate gene expression in several other

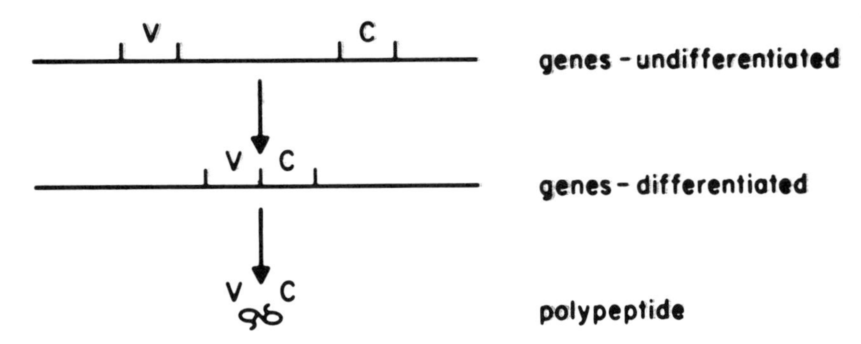

Figure 3. A model for the differentiation of a lymphocyte by the translocation and joining of its V and C genes.

systems. These include the transposable genetic elements studied by McClintock in maize (15) and the flagellar proteins in Salmonella (16).

It should be stressed the V-C joining is only a part of the regulatory machinery that commits individual lymphocytes to the synthesis of one antibody molecule as is obvious from the following observations. i) Within a particular gene family (e.g., H) only a single V sequence may be joined to a C sequence. Most antibody families have multiple C genes (Figure 3), thus the expression of multiple V-C genes is prevented. ii) Only one of the two homologous gene families is expressed, e.g., either the maternal or the paternal H gene family. This observation is designated allelic exclusion. A trans acting control mechanism appears to be required to inform one gene family that its homologue has already joined a V and C gene. The evidence on whether one or both homologous gene families undergo V-C joining is unclear (see Tonegawa et al., this volume). iii) The V-C product from just one light chain family is expressed, e.g., $V_\kappa C_\kappa$ but not $V_\lambda C_\lambda$. Thus within individual lymphocytes, restriction of gene expression must operate within gene families (e.g., only one V_H-C_H), between gene families (e.g., only one $V_L C_L$), and between homologous gene families (e.g., only the paternal V_H-C_H). Accordingly, control mechanisms in addition to V-C joining appear necessary to commit the lymphocyte to the expression of one type of antibody molecule.

2. Future questions. There are a variety of intriguing regulatory phenomena for antibody gene expression than can now be subjected to analysis by nucleic acid techniques.

i) Models of V-C joining. Several alternative models for V-C joining can be proposed (Figure 4). Certain of these models can be tested directly now by the nucleic acid technologies available. For example, in differentiation are V genes lost as would be consistent with the looping-out model? This could be ascertained by hybridization to myeloma DNA with various V gene probes.

ii) The C_H switch. During the differentiation of the immune system, a single clone of antibody-producing cells appears to maintain its V region specificity while shifting from the expression of one class of antibody to another (e.g., IgM to IgG). In molecular terms this shift suggests that the same V_H region and light chain are expressed in the clone throughout this differentiation process, while there is a shift from the expression of a C_μ to a C_γ gene (Figure 5). This hypothesis is supported by a variety of observations including the chemical analysis of biclonal myeloma proteins (17), the idiotypic identity of various classes of antibodies throughout an immune response (18) and an analysis of the progeny of individual lymphocytes in spleen fragment culture (19). Accordingly, a single V_H gene or copies thereof may associate with two or more C_H genes during the differentiation of antibody-producing clones.

Two general models may account for the association of a single V_H gene with two or more C_H genes (Figure 6). First, the simultaneous insertional model suggests that the V_H gene may be copied many times over and separate copies be joined to each C_H gene in the heavy chain family (Figure 6a). The differentiation process would then consist of the successive activation of complete V_H-C_H genes which could occur by transcriptional regulation. Second, the successive insertional model suggests that a single V_H gene could "switch" from one C_H gene to a second or a third during the differentiation process (Figure 6b). Accordingly, each cell could only transcribe a single V_H-C_H gene at a particular point in time. The observation that single myeloma cells

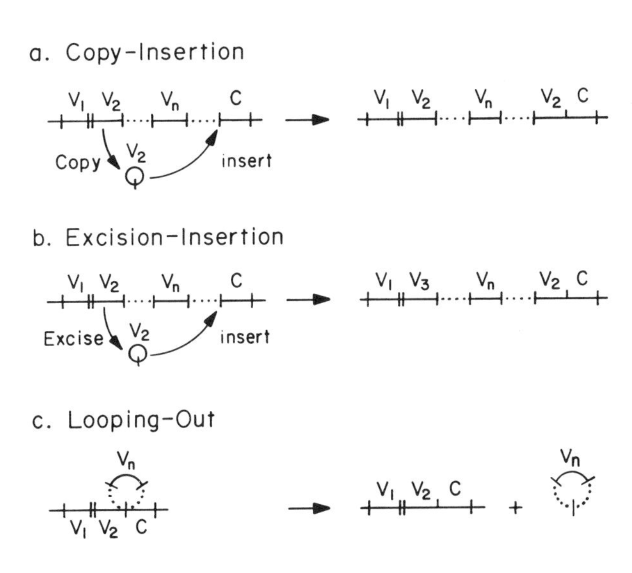

Figure 4. Models for V-C joining. a. Copy-insertion suggests that a V gene is copied and inserted into its corresponding C gene. b. Excision-insertion suggests that a V gene is excised and inserted into its C gene. c. Looping-out suggests that all of the intervening DNA between V_2 and the C gene is deleted as a recombinational event joins these genes.

can synthesize throughout their lifetime two classes of immunoglobulin with identical idiotypes and specificity (H. C. Morris, III, personal communication) tends to favor the simultaneous insertional model. However, the hybridization of a V_H probe to embryonic and myeloma tumor DNA should be able to unambiguously distinguish between a simultaneous insertional model in which copies of the V_H gene are joined to all C_H genes and the successive insertional model. Alternatively, the simultaneous joining of copies of a V_H gene to a more limited number of C_H genes could be detected by the analysis of appropriate restriction enzyme fragments.

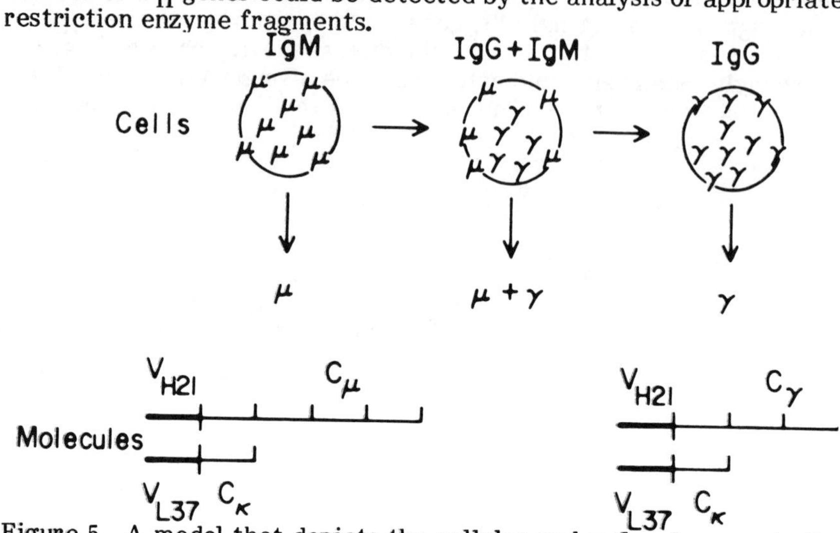

Figure 5. A model that depicts the cellular and molecular events that occur during maturation of the immune response as reflected in the shift from the production of IgM to IgG antibody. From reference 50.

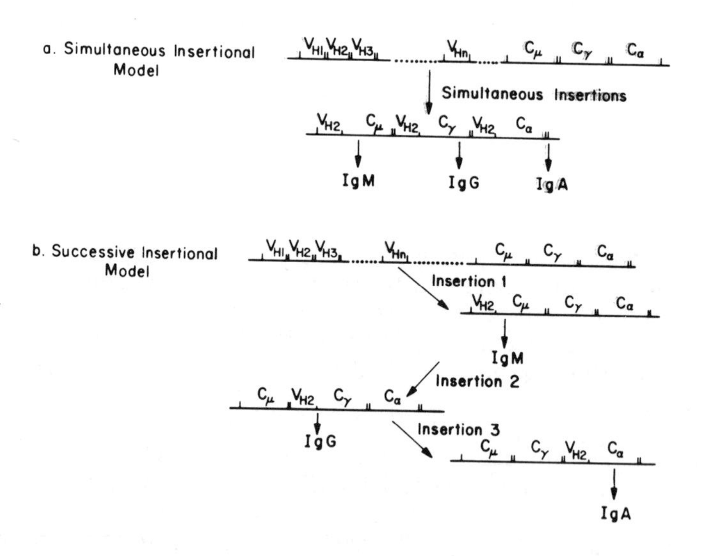

Figure 6. Two models for the association of a V_H gene to multiple C_H genes (see text). From reference 17.

iii) In a single lymphocyte clone there is no indication a single V_L gene is associated with more than one C_L gene. Accordingly, the observations discussed above raise the unattractive possibility that the joining mechanisms for light and heavy chains may not be identical.

iv) During the maturation of the immune response first specific IgM and then specific IgG molecules are expressed (Figure 5). What is the order of C_H genes? Does it show any relationship to the order of C_H gene switching? Nucleic acid techniques may be required to resolve this question in the mouse because to date no genetic recombinants have been detected among the C_H genes tested.

v) It has been suggested that there is a precise developmental order for the ability to express certain types of antibodies (20). Is the order of developmental expression of these genes related to their chromosome map locations?

vi) Do the flanking sequences around the V and C genes give any further clues as to the nature of regulatory events that permit their joining and expression? Are there, for example, short period interspersed repetitive sequences, hypothesized by Britten and Davidson (21) to play a regulatory role in eukaryotes? Or will inverted repeat sequences suggestive of prokaryotic insertion sequences (22) be found?

vii) A variety of studies suggest that individual animals or strains contain V and/or C genes that they do not normally express (23-25). This hypothesis could be tested by hybridization studies with appropriate probes. If latent antibody genes do exist, additional control mechanisms must be postulated to explain their regulation.

V. GENERATION OF ANTIBODY DIVERSITY

1. Mechanisms for information amplification. The vertebrate system can respond specifically to a universe of different antigenic determinants by virtue of molecular interactions with complementary antibody molecules. How then can the gene products from a finite number of antibody genes react with untold numbers of different antigenic determinants? The basic strategies for the amplification of antibody information fall into two broad categories--genetic and molecular (Table 1). Genetic strategies amplify information by producing multiple V genes, whereas molecular strategies amplify information by employing certain fundamental characteristics of antibody molecules themselves. These mechanisms are not mutually exclusive and, indeed, all probably contribute in some degree to the amplification of antibody information. Unfortunately, molecular immunology has focused on the genetic mechanisms almost to the exclusion of the two molecular mechanisms.

2. Molecular mechanisms for the amplification of antibody information.

i) Combinatorial association is possible because antibody molecules are made up of nonidentical subunits (light and heavy chains). This mechanism suggests, for example, that one light chain may associate with many different heavy chains to produce many different antibody

Table 1. Mechanisms for Amplifying Information*

Strategy	Category
1. Multiple germ line genes	Genetic
2. Somatic mutation	
3. Combinatorial association	Molecular
4. Multispecificity	

*From reference 31.

molecules. Thus if 1000 light and 1000 heavy chains can freely associate with one another, 10^6 different antibody molecules would be produced. Accordingly, the amount of information generated may increase as the product of the number of different V_L and V_H genes. X-ray crystallographic and amino acid sequence studies suggest that appropriate residues are highly conserved at the sites of V_L and V_H interaction so that virtually any light chain might in theory associate with virtually any heavy chain (26). In fact, studies have shown that light and heavy chains from sources as diverse as different species, when mixed in solution, can reconstitute intact IgG immunoglobulin molecules (see 27). Clearly combinatorial association may be a powerful strategy for amplifying the information contained in a given number of germ line genes.

ii) Multispecificity is defined as the ability of a single antibody molecule to interact with a variety of different antigens, some presumably related in tertiary structure and others possibly unrelated (28). The antigen-binding crevice is generally a shallow trough formed by V_L and V_H interactions with room in this large site for many different molecular interactions (29, 30). If individual antibody molecules may combine with multiple and different antigenic patterns, antibody molecules derived from a fixed number of germ line V genes may interact with a disproportionately large fraction of the foreign antigenic universe. Thus the inherent degeneracy of the antigen-binding site is an important mechanism for amplifying the information contained in a discrete number of antibody V genes.

Surprisingly few experiments have been directed at assessing the relative contributions of combinatorial association and multispecificity to the total antibody repertoire. Clearly multispecificity and combinatorial association will have to be studied at the level of antibody molecules since their abilities to amplify antibody information arise as a result of the interaction of polypeptide segments with one another--a relationship which does not exist at the nucleic acid level. These molecular strategies appear to offer extremely important mechanisms for amplifying genetic information in multigene families. In contrast, the mechanisms of genetic amplification of antibody information can be approached both at the protein and nucleic acid levels.

3. Genetic mechanisms for antibody diversity. The genetic mechanisms for antibody diversity fall into two general categories (1). The germ line theory postulates that most antibody genes are separately encoded in the zygote or germ line and that these genes arose by gene duplication and changed by mutation and selection during vertebrate evolution. In contrast, the somatic mutation theories postulate that antibody diversity is encoded by a more limited number of germ line genes which diversify by some type of somatic mutational or reassortment process during the differentiation of each individual. There are five general theories of antibody diversity (Table 2). Let us consider the salient observations from protein and nucleic acid chemistry that relate to antibody diversity and then consider the constraints these impose on various theories of antibody diversity.

4. Analysis of V diversity.

i) A caution. A variety of estimates on the sizes of the antibody gene families have come from protein and nucleic acid analyses of genes or gene products from myeloma tumors. Before discussing these data, it is important to stress that these estimates should be viewed with caution. Myeloma proteins may represent only a subset of the total antibody repertoire in the mouse. Myeloma tumors can be artificially induced by injecting mineral oil into the peritoneal cavity of two inbred strains of mice, BALB/c and NZB. The NZB myeloma proteins differ from their BALB/c counterparts by three criteria (31). 1) NZB myeloma proteins are predominately of the IgG class whereas BALB/c tumors are predominately IgA. 2) The profile of simple haptens to which the NZB myeloma proteins bind appears to be distinct from that of their BALB/c counterparts. 3) The V_κ and V_H regions from the NZB and BALB/c myeloma proteins appear to be distinct from one another by amino acid sequence analysis. For example, 25 NZB κ chains have been examined over their N-terminal 23 residues and only one is identical to one of the 43 different V_κ sequences (Figure 7). In contrast, five repeat sequences were found in the first 22 BALB/c sequences examined (32).

On the basis of heavy chain class distribution, antigen-binding properties and sequence analysis, the NZB and BALB/c myeloma proteins appear to be distinct populations. These distinct sets may arise because NZB and BALB/c mice have 1) different V genes, 2) genetic differences outside the V region structural genes which cause different sets of V genes to be expressed, or 3) both of these possibilities. The explanation of different V genes does not account for the fact that myeloma proteins in the two strains bind distinct sets of antigens or that the class distribution is different in the two strains. The more likely explanation is that because of genetic differences outside the

Table 2. Contemporary Theories of Antibody Diversity

Theory	Category	Comments	Definition of germ line V genes	Estimated number of mouse V_κ germ line genes
Classical germ line (43)	Germ line	Most antibody variable regions are encoded by distinct germ line genes	Each distinct V sequence	100's–1000's
Combinatorial mutation (52, 53)	Somatic (footnote a)	Framework and hypervariable sequences are encoded by distinct germ line genes. In one simple formulation of this model, a single framework and three hypervariable genes are joined during differentiation for each V gene	Each distinct framework and each hypervariable region sequence	1000's–10,000's
Ordinary somatic mutation (35, 54)	Somatic	Ordinary spontaneous mutants are selected for clonal expansion	Each distinct framework sequence and each distinct sequence gap (Cohn)	100–200
Somatic recombination (55)	Somatic	Somatic recombination among germ line V genes generates diversity	Uncertain	100's
Special mutation (56, 57, 58)	Somatic	These theories may evoke any one of a variety of special somatic mutational mechanisms which may or may not operate only in hypervariable regions	Depends on the specific form of this theory	Few to 100–200

V genes, the internal environments are distinct in these two strains, and this difference leads to the clonal expansion of different sets of lymphocytes in the two strains (via distinct control genes and/or distinct antigenic environments). Thus the myeloma process may provide

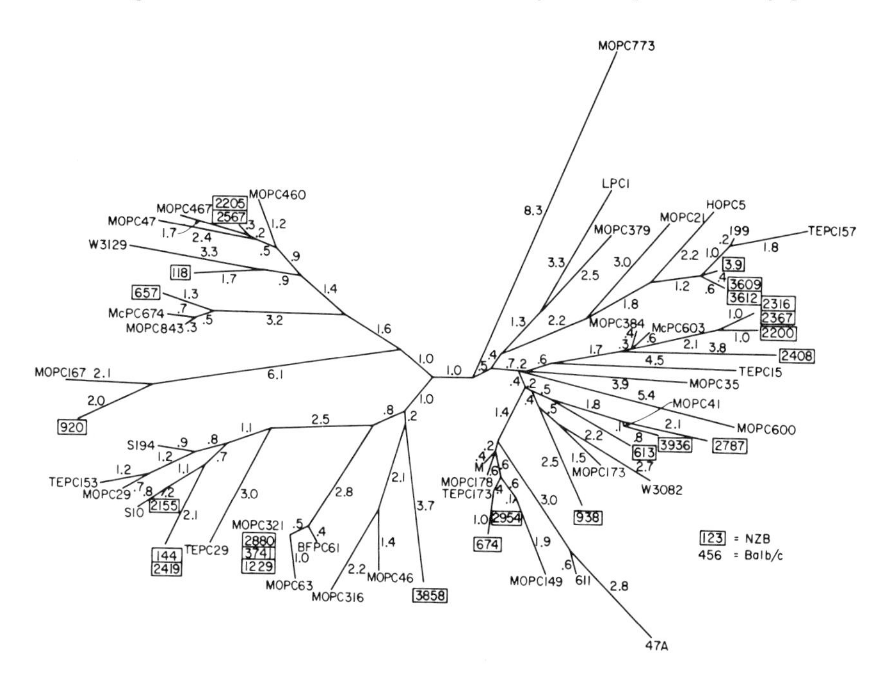

Figure 7. A genealogical tree of the N-terminal 23 residues of BALB/c and NZB V_{κ} regions. The NZB V_{κ} regions are boxed and the BALB/c V_{κ} regions are not. The numbers on the branches indicate the number of nucleotide substitutions separating the nodal points. Fractional numbers designate nucleotide substitutions arising from the averaging process that is used to generate these trees (see reference 2). From reference 31.

windows through which we can glimpse the antibody repertoire, but the windows probably view just a fraction of the total diversity. As it is difficult to say how small this fraction is, it is impossible to estimate what fraction of the antibody repertoire is seen by the myeloma process in each of the inbred strains. Estimates of the number of V region genes in the antibody families, based on the chemical analysis of myeloma proteins, are therefore minimal estimates that could be far too low.

ii) A comparison of the V regions from a particular immunoglobulin family reveals related sets or subgroups of V regions that have similar amino acid sequences and sequence gaps (Figure 8). These analyses have been carried out primarily on homogeneous immuno-

```
Subgroup V_κ region 1       5          10         15         20         25    a b c d e f    30         35

I        M70     D I V L T Q S P A S L A V S L G Q R A T I S C R A S Q S V B B[- -]S G I S F M N W
         M321  -----------------------------------------------------------K-----N T[- -]Y--N------Z--
         T124  -----------------------------------------------------------Q---------W[- -]Y--N------Z--
         M63   -----------------------------------------------------------------S[- -]---------------

II       T15   ---------M---------T F-------T A S K K V---------T----Z--L Y S S K H K V H Y L A--
         H8    ---------M---------T F-------T A S K K V---------T----Z--L Y S S K H K V H Y L A--
         S107  ---------M---------T F-------T A S K K V---------T----Z--L Y S S K H K V H Y L A--
         S63   ---------M---------T F-------T A S K K V---------T----Z--L Y S S K H K V H Y L A--

III      M167  ---------I-----D E L--D P--T S--E S V S--T-----S--K--L L Y K[-]B--K T Y L B--
         M511  ---------I-----D E L--K P--T S--E S V S--T-----S--K--L L Y K[-]D--K T Y L----

IV       U10   -------Q M----T T S----S A------D--V------------Z [-------]B I S B Y L B--
         Y5476 -------Q M----T T S----S A------D--V------------Z [-------]B I S B Y L B--
```

Figure 8. The N-terminal sequences from BALB/c κ chains that fall into four distinct subgroups. Lines indicate where these sequences are identical to M70. Differences are indicated by the one letter amino acid code. Brackets indicate that a deletion is required to make these V_{κ} regions homologous to those of subgroup II. From reference 31.

globulins derived from myeloma tumors in man, mouse, rat, and dog, and on restricted antibodies from rabbits (33). The historical importance of V region subgroups is that most immunologists concede that each subgroup is encoded by at least one germ line V gene. Thus two V gene-counting problems are of interest. First, is each subgroup coded by a single V gene? Several V_κ subgroups from the BALB/c mouse have been studied by nucleic acid hybridization kinetics (4-6). The results seem to indicate that each subgroup is specified by one or a very few germ line genes. However, questions have been raised as to the ability of this technique to count genes with the precision that is claimed (34). Second, how many V genes (subgroups) are present in each antibody family? The difficulty in currently defining a subgroup raises serious questions about the use of this means for V gene counting in a given antibody family. As more amino acid sequence data have accumulated, the sharp divisions of the subgroups as depicted in Figure 8 have broken down. V regions that differ from a particular V sequence, for example, by 1 residue, 5 residues, 10 residues, 15 residues, 20 residues, etc. have been found. Thus the definition of a subgroup has become imprecise and, indeed, dependent upon one's theory of antibody diversity.

iii) Amino acid sequence analysis has shown that the mouse κ family is more heterogeneous than the H family, which in turn is more heterogeneous than the λ family. The N-terminal sequence of 81 BALB/c kappa chains has been examined over their N-terminal 23 residues. Sixty-two of the 81 V regions have distinct sequences in this region and a majority differ from one another by multiple amino acid substitutions (33). Moreover, the first 20 sequences examined yielded approximately as many new sequences as did later groups of twenty. Clearly the diversity in the mouse κ family is extensive (Figure 7). These V regions appear to represent more than 30 different V subgroups. Accordingly, V region sequence counting and various statistical calculations (35) suggest that V_κ regions are encoded by 100's to 1000's of germ line V genes (Table 2). The N-terminal sequence of 45 BALB/c heavy chains has been examined over their N-terminal 20 residues. Twenty-seven of the 45 sequences differ by one residue or more (33). This family falls into less than 10 subgroups and, accordingly, the diversity in this family is clearly less striking than in the κ family. The mouse λ family exhibits only two subgroups (33).

iv) Several sets of closely-related V regions have been studied in the mouse κ, λ, and H families. Certain models of somatic mutation suggest the diversity within each set should be encoded by a single germ line V gene. If so, the diversity patterns seen in these sets should reflect the nature of the somatic mutational mechanism which produces the variant V sequences.

The V_λ set. Twelve of 18 complete V regions studied are identical and the 6 variant sequences differ by 1 to 3 residues from the most common sequence (Figure 9). The striking observation in this set of V_λ regions is that all of the substitutions fall within the three hypervariable regions, those extremely variable polypeptide segments that comprise the walls of the antigen-binding crevice. The numbers of V_λ genes required to explain this diversity according to the theories of antibody diversity are quite clear. The germ line theory would require 7 V_λ germ line genes, one for each sequence, whereas the somatic mutational theories would require but one. A careful effort has

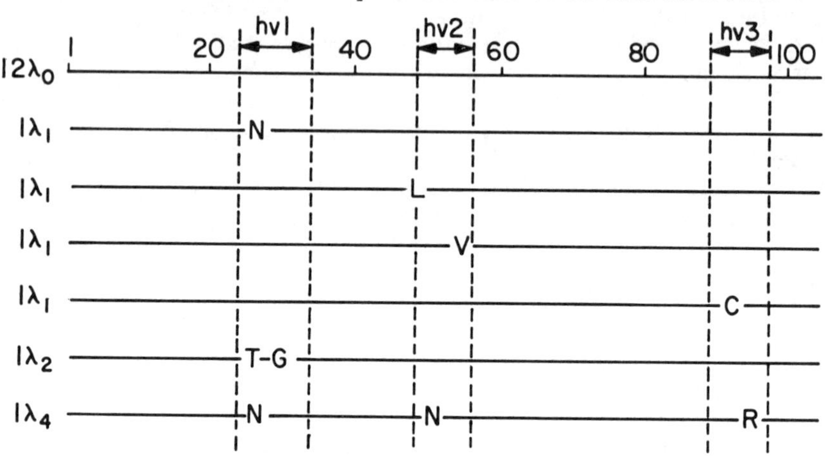

Figure 9. The amino acid sequences of 18 mouse V_λ regions. The three binding-site associated hypervariable regions are indicated by hv1, hv2, and hv3. All V regions are compared against the prototype λ_0 sequence. λ_1, λ_2, and λ_4 indicate the number of nucleotide substitutions that separate these sequences from the prototype. Amino acid differences are indicated by the one letter code. From reference 31.

been made to measure the numbers of V_λ genes by hybridization kinetics (4). The results indicate there are relatively few V_λ germ line genes, certainly fewer than would be predicted on a germ line analysis of V_λ sequence diversity. Accordingly, these data suggest that the V_λ variants are produced by somatic mutation.

The PC V_H subset. Four of 9 V_H regions derived from BALB/c myeloma proteins binding phosphorylcholine are identical (Figure 10). The variants differ by 1 (H8) to 11 (M167) residues from the prototype V_H region (T15). Four of six different V_H sequences have sequence gaps (insertions or deletions). If one contends this diversity can arise from a single V_H gene, these observations place several new constraints on a somatic mutational mechanism. First, mutation can occur outside of hypervariable regions in what is termed the framework regions. Second, the mutational mechanism must be capable of generating sequence gaps. Third, as many as 11 mutations must occur during the differentiation of a single lymphocyte line.

The M70 V_κ subset. This is perhaps the most promising set of closely-related V sequences for examining diversity patterns because approximately 10% of NZB myeloma V_κ regions appear to fall into this subset (M. Weigert, personal communication). These V_κ sequences have been designated the M70 subset because of the similarity of these NZB V_κ sequences to a BALB/c κ chain, M70, that had been previously sequenced. Moreover, specific serological reagents have been prepared

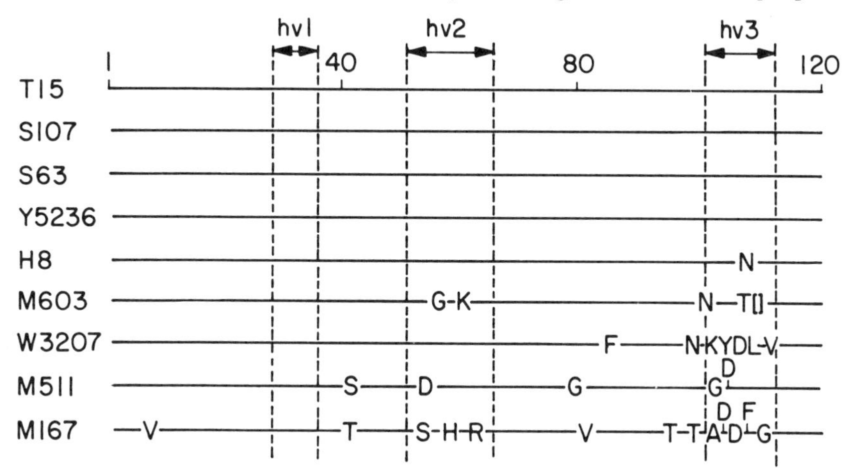

Figure 10. The complete amino acid sequences of 9 mouse V_H regions from myeloma proteins binding phosphorylcholine. Sequence gaps are indicated by ( ). Sequence insertions are designated by a vertical bar which lifts the residue above the main sequence. See legend to Figure 9. From reference 31.

to facilitate the identification of members of this subset (M. Weigert, personal communication). Thus very large numbers of V_κ regions are available for protein or nucleic acid analysis.

Two of nine V_κ sequences appear identical (2880 and 1229) while the remainder differ by 1-14 residues from one another (Figure 11). Thus almost every V_κ sequence differs from all of the others. If a somatic mutation theory contends these nine V_κ sequences are derived from a single V_κ gene, then somatic mutation must be capable of generating at least 14 V_κ mutations during the differentiation of a single lymphocyte line. Moreover, there are three sub-subsets (2880-1229-7132, 7175-2485 and 7043-6308) which share identical amino acid substitutions in framework as well as hypervariable regions. This suggests that the mutation mechanism must be capable of producing multiple identical (parallel) substitutions in separate individuals. Alternatively, the need for parallel mutations in independent lymphocyte lines may be avoided by postulating a separate germ line V_κ gene for each sub-subset. The mutational data for these three subsets are summarized in Table 3. Clearly one must ask of the various somatic mechanisms: 1) how many substitutions can they produce in a single lymphocyte line and

2) are they capable of producing extensive parallel mutations? From the number of different substitutions and their location, one can calculate the number of germ line genes the various theories need to explain these data (Table 4). The general guidelines for counting germ line V genes are given in Table 2. The combinatorial mutation, classical germ line, and ordinary somatic mutation theories require, respectively, 49, 21, and 12 germ line genes for 21 different V sequences. Since

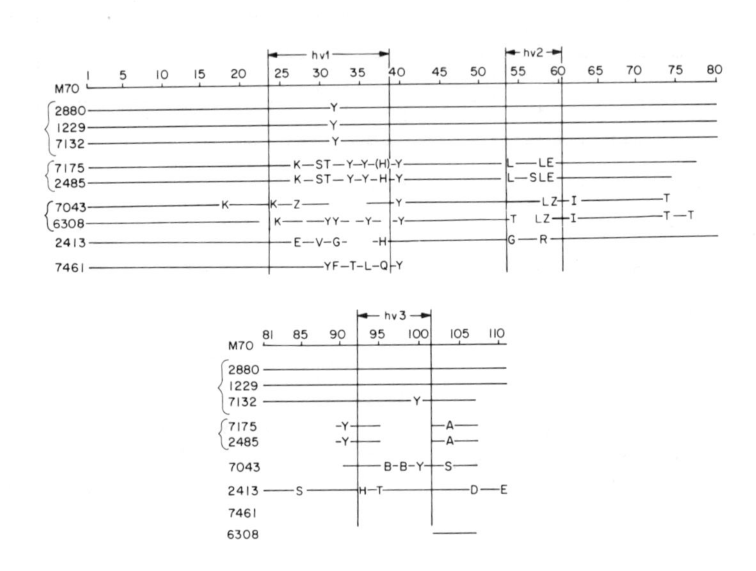

Figure 11. The partial amino acid sequences of M70-like NZB V_κ regions. Brackets indicate sub-subsets. From E. Loh, B. Black, J. Schilling, M. Weigert and L. Hood, unpublished results.

Table 3. Nucleotide Substitutions in Closely-Related V-Region Sets

	V_H	V_κ	V_λ	Total
Total residue substitutions	27	67	9	103
Hypervariable residue substitutions	19	48	9	76
Framework residue substitutions	8	19	0	27
Two-base substitutions	7	15	1	23
Sequence gaps	4	0	0	4

these subsets presumably constitute a very small fraction of the V gene repertoire in the mouse, each of these theories suggest that there are large numbers of V_κ genes. Special somatic mutational mechanisms will require fewer genes; how few depends on the specific model.

v) Future experiments. The modern techniques of nucleic acid chemistry appear to offer general fruitful approaches toward the analysis of mechanisms of antibody diversity. 1) Does somatic mutation

Table 4. Gene Counting Among Closely Related V-Region Sets

Set	Number of distinct V-regions	Theory		
		Classical germ line	Combinatorial mutation	Somatic mutation (Cohn)
PC-V_H	6/9	6	15 (11+4)	5
M70-V_κ	8/9	8	22 (17+5)	6
V_λ	7/18	7	12 (11+1)	1
Total	21/36	21	49 (39+10)	12

occur in V genes as the antibody-producing cell differentiates? This could be tested directly by independently cloning several embryonic V genes of a particular subgroup. These clones could then be compared to their somatic counterparts by DNA sequence analysis. 2) If somatic mutation occurs within a subgroup, can it produce deletions and insertions, and is there an upper limit to the number of base changes that can be made? This will require comparison of an extensive group of related sequences such as the M70-subset from myeloma tumors with their counterparts in the germ line. 3) What is the size of the individual antibody gene families? Various novel and sophisticated approaches can be taken toward this problem (36).

5. Antibody repertoire size. Estimates for the size of the antibody repertoire for an individual mouse range from 10^5 to 10^9. In our view there is no compelling data to support any particular number. With combinatorial association and $\sim$1000 V_L and 1000 V_H genes, a germ line theory could produce diversity consistent with a repertoire size of 10^5 to 10^6. To explain larger repertoire sizes either additional germ line genes or somatic mutation must be postulated.

6. Where do we stand now? Most immunologists would agree that both germ line V genes and somatic mutation contribute to antibody diversity. The immune system has at least four basic strategies for amplifying antibody information-multiple germ line genes, somatic mutation, combinatorial association and multispecificity. Thus the

unresolved problem for the future is, What is the relative contribution of each of the four mechanisms to the total functional diversity of vertebrate antibody molecules? This problem will require a spectrum of sophisticated immunologic, chemical, genetical, and biological experimental approaches.

VII. EVOLUTION OF ANTIBODY GENES

1. General scheme. Antibody polypeptides are linearly differentiated into homology units of about 110 amino acid residues (Figure 12) (37). The constant homology units demonstrate significant amino acid

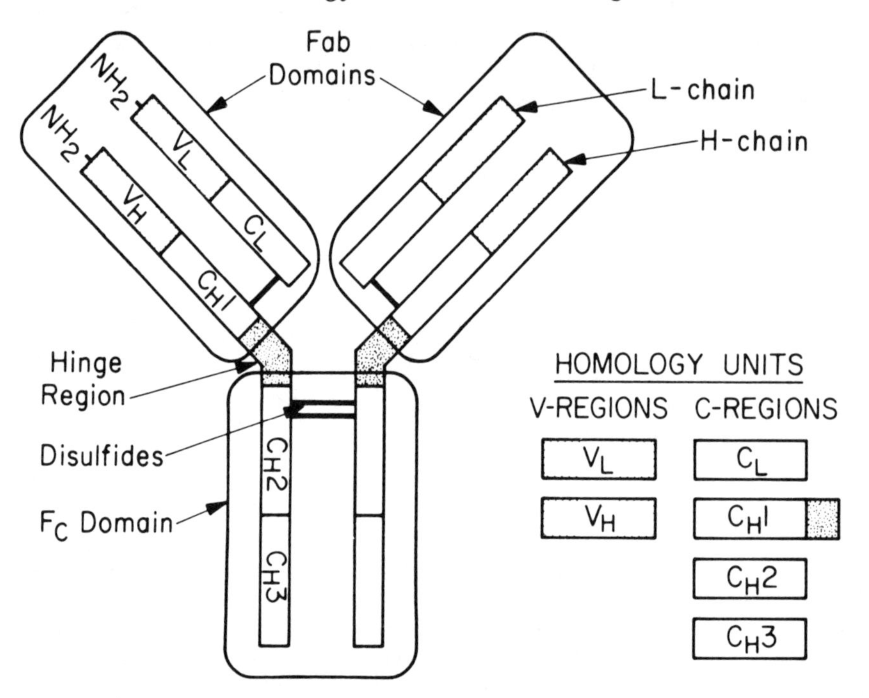

Figure 12. A model depicting the basic structure of the IgG immunoglobulin molecule.

sequence homology to one another and so do the variable homology units. The existence of variable- and constant-region homology units and the observation that the tertiary structures for V and C homology units are very similar (38) indicate that the antibody genes probably evolved from a precursor gene coding for a single ancestral homology unit. A hypothetical evolutionary scheme is diagrammed in Figure 13. The hypothetical precursor gene duplicated at a very early time to produce the ancestral V and C genes. Subsequently, the V gene duplicated many times over to generate a primordial multigenic family.

This original family may have coded for primitive membrane receptor molecules. This multigene family in turn was duplicated either by polyploidization or by duplication and translocation of a chromosomal segment to produce a primitive antibody gene family. Subsequent duplication of this primitive antibody family may have produced the three families that evolved to become contemporary λ, κ, and H families. Contiguous or fused gene duplication led to C_H genes comprised of three or four homology units in the heavy chain family. Accordingly,

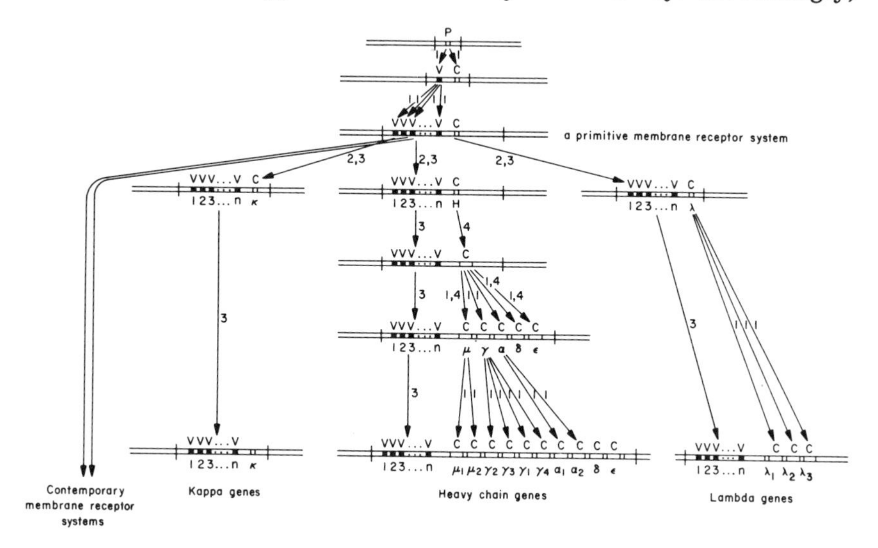

Figure 13. A hypothetical scheme for the evolution of the antibody gene families. The order of gene duplication events is unknown. Adapted from 51.

the evolution of antibody genes employed all of the major mechanisms of gene evolution–point mutation, discrete duplication, contiguous duplication, polyploidization, and/or translocation (39).

2. Unusual evolutionary features. Antibody V genes exhibit two evolutionary features that are not generally expected in gene evolution. The first feature is that V gene families can expand or contract their gene numbers rapidly in terms of evolutionary time. Some mammalian species express many V_κ and few V_λ genes, whereas others appear to express many V_λ and few V_κ genes (40). This suggests that the information required to encode the light chains of diverse antibodies can be contained either in the λ or κ light-chain families and that dramatic shifts in the information content of each family can occur over the 75 million years of divergence of the mammalian evolutionary lines.

The second unusual evolutionary feature is that the V genes of an antibody family in one species appear to evolve structural changes together or in parallel. We designate this process coincidental evolution (Figure 14). In portions of the V region that are highly conserved, coincidental evolution is reflected by species-specific residues at certain positions that distinguish most of the immunoglobulin chains of one species from those of a second (41, 42). For example, most V_κ regions of rabbit are distinguished from their counterparts in the mouse by the presence of a valine and glycine at positions 11 and 17 respectively rather than a leucine and glutamic acid (41). In highly diverse antibody

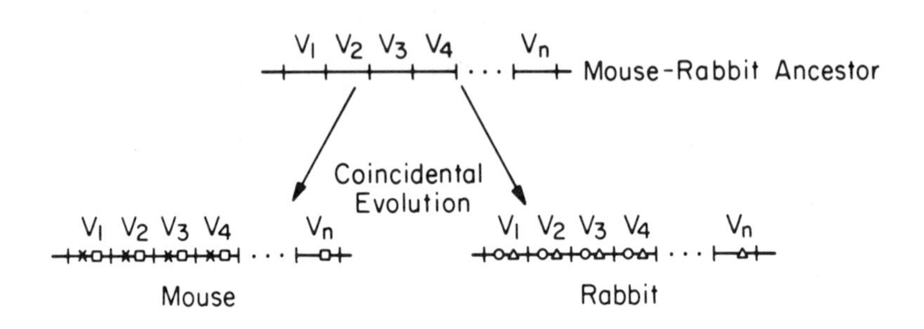

Figure 14. A diagrammatic representation of coincidental evolution during the divergence of the mouse and rabbit evolutionary lines. X, □, Q, and Δ represent coincidental changes in the respective evolutionary lines. From reference 1.

families, coincidental evolution is illustrated by the presence of species-specific branches on a genealogical tree constructed from the V regions of two or more species (43). Thus V genes or sets of V genes from a particular family appear to evolve together or in a coincidental fashion.

A number of multigene families including the histones and ribosomal RNA genes appear to exhibit these same evolutionary features of gene expansion and contraction and coincidental evolution [see (1) for review]. Thus these features appear to be a fundamental feature of known multigene families. A variety of evolutionary mechanisms have been proposed to explain gene expansion and contraction and coincidental evolution (1). Homologous but unequal crossing-over appears to be a particularly attractive mechanism in that it can explain both of the phenomena described above (1, 44).

3. Evolution of multigene families. New multigene families can arise in two ways: 1) gene duplication from a single gene, or 2) duplication and translocation of all or part of a preexisting multigene family. It obviously requires a long period of evolutionary time to produce a large multigene family starting with a single gene. In contrast, the duplication of an entire multigene family generates new gene families by a single genetic event. Structural genes as well as

their attendant (cis) control mechanisms are duplicated simultaneously. Accordingly, one unit of evolution becomes the entire multigene family. Obviously the evolutionary ancestry of multigene families may be traced by looking for homologous control mechanisms as well as for homologous gene products. For example, it will be fascinating to determine whether other complex and multigenic systems, such as those presumably associated with cell recognition or the nervous system, exhibit V and C regions and a joining mechanism for differentiation.

4. Functions of newly-arising multigene families. Multigene families arising by duplication of an ancestral family can assume a variety of functions. They can evolve to interact intimately with related multigene families, e.g., the interactions of the gene products from the heavy and light gene families. In addition, some duplicated gene families can evolve to perform totally new and unique functions. In this regard the recent report that the human transplantation antigens (HLA antigens) show a partial homology to immunoglobulins is particularly interesting (C. Terhorst and S. Strominger, personal communication). This apparent homology could be explained by convergent evolution because of similar structural constraints in the two molecules. This hypothesis will be difficult to rule out unless more extensive homologies are found upon subsequent amino acid sequence analysis. However, if this report is confirmed, the implication is that immunoglobulins and transplantation antigens may share a common ancestor.

5. Other potential multigene families. The antibody genes are the first complex eukaryotic system that has been studied in detail at a variety of different levels. Strategies that the immune system has employed for information handling are summarized in Table 5. Do other complex eukaryotic systems have similar informational requirements? Certainly the nervous system exhibits exquisite specificity in its wiring requirements (45, 46). Various organs and tissues appear to interact through highly specific cell surface molecules (47) and, of course, there are a variety of multigenic families encoding cell surface molecules that are located on chromosome 17 of the mouse (see reference 14). The T/t locus has at least six linked complementation groups and appears to regulate early neuroectodermal development (48). The H-2 complex appears to encode at least two multigenic families--the immune response genes (I region) and the complement genes (S region) (49). Indeed, there are suggestions that the genes coding for transplantation antigens may be more complex in nature than previously suspected and possibly even constitute a third, perhaps rather simple, type of multigenic family in the H-2 region. In this regard, the possible homology relationship between antibodies and transplantation antigens formally raises the possibility these two gene families could have descended from a common ancestral multigene family.

Nature is basically conservative. Once an effective system for handling information has evolved, it will probably be duplicated and employed in other roles. In our view it is likely that the antibody gene

Table 5. Strategies of the Immune System for Information Handling

Strategy
Information Generation
Multiple different germ lines
Somatic mutation and selection
Combinatorial association
Multispecificity
Information Expression
V-C joining
Allelic exclusion
C_H switch
Expression of subsets of genes (?)
Information Evolution
Homology units and domains
Duplication of multigene families to assume new functions or interact with the old family
Special evolutionary mechanism(s) leading to coincidental evolution and change in family size

families will share a common multigenic ancestor with other contemporary multigene families that have come to assume a variety of different roles (Figure 13). Accordingly, we believe the antibody gene systems may well serve as a general model for understanding other complex eukaryotic systems.

6. Future experiments. Nucleic acid chemistry can shed a variety of insights into the evolution of multigene families.

i) How rapidly can multigene families change significantly in size? One could examine antibody family sizes in appropriately selected species.

ii) What genetic mechanisms do the gene families employ for gene expansion and contraction and coincidental evolution? The various mechanisms make different predictions as to the organization of the antibody V genes (see 1).

iii) Do other multigenic systems share structural genes, control mechanisms or organizational features that are homologous to those of the antibody gene families? Certainly the very difficult task of isolating and characterizing genes from the T/t locus, the H-2 complex, and other similarly complex systems will provide important insights into the organization, expression and evolution of these fascinating multigenic systems.

SUMMARY

The study of complex and multigenic families is a common denominator which ties together such diverse disciplines as modern genetics, molecular biology, immunology and cell biology. We have seen that studies of the nucleic acid level can shed light on how diversity is maintained or generated within a multigene family. These methods promise also to tell us how expression of such genes are regulated and how gene families evolve. They may also shed light into the evolutionary relationships of diverse multigene families such as those encoded by chromosome 17 of the mouse. It appears that many intriguing problems of modern immunology can be effectively approached using the rapidly advancing technologies of nucleic acid chemistry.

ACKNOWLEDGEMENTS

This work was supported by grants from National Institutes of Health (AI 10781 and GM 06965) and National Science Foundation (PCM 76-81542). Thanks to Bernita Larsh for her usual superb job of typing.

FOOTNOTE

(a) The proponents of the combinatorial theories of antibody diversity have defined them as germ line (52, 53). However, in these theories information is rearranged during somatic differentiation. Thus the major component of antibody diversification has a somatic and not a germ line basis. Indeed, genetically it appears illogical to call somatic recombination (which also somatically rearranges V genes) a somatic theory and the combinatorial models germ line theories. Admittedly this discussion is not unlike the story about the three blind men attempting to describe an elephant.

REFERENCES

1. Hood, L., Campbell, J. H. and Elgin, S. C. R. (1975) Ann. Rev. Genetics 9, 305.
2. Hood, L. E., Wilson, J. H. and Wood, W. B. (1975) Molecular Biology of Eucaryotic Cells, W. A. Benjamin, Menlo Park. See the appendices to Chapter 2 (DNA-DNA reassociation) and Chapter 5 (RNA-DNA hybridization).
3. Honjo, T., Packman, S., Swan, D., Nau, M. and Leder, P. (1974) Proc. Nat. Acad. Sci. USA 71, 3659.
4. Tonegawa, S. (1976) Proc. Nat. Acad. Sci. USA 73, 203.
5. Rabbitts, T. H., Jarvis, J. M. and Milstein, C. (1975) Cell 6, 5.
6. Leder, P., Honjo, T., Swan, D., Packman, S., Nau, M. and Norman, B. (1975) In: Molecular Approaches to Immunology, eds. E. E. Smith and D. W. Ribbons, Academic Press, New York, p. 173.

7. Roberts, R. J. (1976) CRC Critical Rev. Biochem. 3, 123.
8. Hozumi, N. and Tonegawa, S. (1976) Proc. Nat. Acad. Sci. USA 73, 3628.
9. Abelson, J. (1977) Science 196, 159.
10. Maxam, A. M. and Gilbert, W. (1977) Proc. Nat. Acad. Sci. USA 74, 560.
11. Sanger, F., Air, G. M., Barrell, B. G., Brown, N. L., Coulson, A. R., Fiddes, J. C., Hutchison, C. A. III, Slocombe, P. M. and Smith, M. (1977) Nature 265, 687.
12. Mage, R., Lieberman, R., Potter, M. and Terry, W. D. (1973) In: The Antigens, ed. M. Sela, Academic Press, New York, p. 300.
13. Weigert, M., Potter, M. and Sachs, D. (1975) Immunogenetics 1, 511.
14. Klein, J. (1975) Biology of the Mouse Histocompatibility-2 Complex, Springer-Verlag, New York, p. 218.
15. McClintock, B. (1965) In: Brookhaven Symp. Biol. 18, 162.
16. Zieg, J., Silverman, M., Hilmen, M. and Simon, M. (1977) Science 196, 170.
17. Sledge, C., Fair, D. S., Black, B., Krueger, R. G. and Hood, L. (1976) Proc. Nat. Acad. Sci. USA 73, 923.
18. Oudin, J. and Michel, M. (1969) J. Exp. Med. 130, 619.
19. Gearhart, P. J., Sigal, N. H. and Klinman, N. R. (1975) Proc. Nat. Acad. Sci. USA 72, 1707.
20. Klinman, N.R. and Press, J. E. (1975) J. Exp. Med. 141, 1133.
21. Davidson, E. H. and Britten, R. J. (1973) Quart. Rev. Biol. 48, 565.
22. Saedler, H., Reif, H. J., Hu, S. and Davidson, N. (1974) Molec. Gen. Genet. 132, 265.
23. Rivat, L., Gilbert, D. and Ropartz, C. (1973) Immunology 24, 1041.
24. Bosma, M. J. and Bosma, G. (1974) J.Exp. Med. 139, 512.
25. Strosberg, A. D., Hamers-Casterman, C., Van der Loo, W. and Hamers, R. (1974) J. Immunol. 113, 1313.
26. Poljak, R. J., Amzel, L. M., Chen, D. L., Phizackerley, R. P. and Saul, F. (1975) Immunogenetics 2, 393.
27. Olander, J. and Little, J. R. (1975) Immunochemistry 12, 383.
28. Inman, J. K. (1974) In: The Immune System, Genes, Receptors, Signals, eds. E. E. Sercarz, A. R. Williamson and C. F. Fox, Academic Press, New York, p. 37.
29. Amzel, L., Poljak, R., Saul, F., Varga, J. and Richards, F. (1974) Proc. Nat. Acad. Sci. USA 71, 1427.
30. Padlam, E. A., Segal, D. M., Cohen, G. H. and Davies, D. R. (1974) In: The Immune System: Genes, Receptors, Signals, eds. E. Sercarz, A. Williamson and C. F. Fox, Academic Press, New York, p. 7.
31. Hood, L., Loh, E., Hubert, J., Barstad, P., Eaton, B., Early, P., Fuhrman, J., Johnson, N., Kronenberg, M. and Schilling, J. (1977) Cold Spring Harbor Symp. Quant. Biol. 41 (in press).
32. Hood, L., Potter, M. and McKean, D. (1970) Science 170, 1207.

33. Kabat, E. A., Wu, T. T. and Bilofsky, H. (1976) Variable Regions of Immunoglobulin Chains.
34. Smith, G. (1977) Cold Spring Harbor Symp. Quant. Biol. 41 (in press).
35. Cohn, M., Blomberg, B., Geckeler, W., Raschke, W., Riblet, R. and Weigert, M. (1974) In: The Immune System: Genes, Receptors, Signals, eds. E. E. Sercarz, A. R. Williamson and C. F. Fox, Academic Press, New York, p. 89.
36. Stavnetzer, J. and Bishop, M. (1977) Biochemistry (in press).
37. Edelman, G. M., Cunningham, B. A., Gall, W., Gottlieb, P., Rutishauser, U. and Waxdal, M. (1969) Proc. Nat. Acad. Sci. USA 63, 78.
38. Poljak, R. J., Amzel, L., Avey, H., Chen, B., Phizackerley, R. and Saul, F. (1973) Proc. Nat. Acad. Sci. USA 70, 3305.
39. Ohno, S. (1970) Evolution by Gene Duplication, Springer-Verlag, New York.
40. Hood, L., Grant, J. A. and Sox, H. C. (1970) In: Developmental Aspects of Antibody Structure and Formation, ed. J. Sterzl, Academia, Prague, vol. 1, p. 283.
41. Hood, L., Eichman, K., Lackland, H., Krause, R. and Ohms, J. (1970) Nature 228, 1040.
42. Capra, J. D., Wasserman, R. W. and Kehoe, J. M. (1973) J. Exp. Med. 138, 410.
43. Hood, L. (1973) In: Stadler Genet. Symp. 5, 73.
44. Smith, G. P. (1973) Cold Spring Harbor Symp. Quant. Biol. 38, 507.
45. Sperry, R. W. (1963) Proc. Nat. Acad. Sci. USA 50, 703.
46. Changeux, J.-P. and Danchin, A. (1976) Nature 264, 705.
47. Moscona, A. A. (1975) In: Developmental Biology, eds. D. McMahan and C. F. Fox, W. A. Benjamin, Menlo Park, p. 19.
48. Bennett, D. (1975) Cell 6, 441.
49. McDevitt, H. O. (1976) Fed. Proc. 35, 2168.
50. Hood, L. (1972) Fed. Proc. 31, 177.
51. Gally, J. A. and Edelman, G. M. (1972) Ann. Rev. Genetics 6, 1.
52. Capra, J. D. and Kindt, T. J. (1975) Immunogenetics 1, 417.
53. Klinman, N. and Press, J. (1975) Transplant. Rev. 24, 41.
54. Jerne, N. K. (1971) J. Immunol. 1, 1.
55. Gally, J. A. and Edelman, G. M. (1970) Nature 227, 341.
56. Brenner, S. and Milstein, C. (1966) Nature 211, 242.
57. Baltimore, D. (1974) Nature 248, 409.
58. Leder, P., Swan, D., Honjo, T. and Seidman, J. (1977) Cold Spring Harbor Symp. Quant. Biol. 41 (in press).

LIGHT CHAIN mRNA

Cesar Milstein, Kirk C.S. Chen*, Pamela H. Hamlyn, Terence H. Rabbitts and George G. Brownlee

Medical Research Council Laboratory of Molecular Biology
Hills Road, Cambridge CB2 2QH, England

ABSTRACT. Early structural studies of L-chain mRNA have been based on enzymatic degradation and fractionation of [^{32}P]mRNA. The description of the fingerprints produced in this way have been useful in genetic studies. Comparative studies between two different kappa chain [^{32}P]mRNAs have now permitted us to extend at the RNA level the notion of variability and constancy of the two halves of the light chains. In addition they proved that the 3' untranslated region is also constant. More recent sequencing methods have been based on preparation of complementary cDNA. The analysis of the products was at first carried out by degradative procedures. Much better results were obtained by rapid sequencing methods based on gel separation of the cDNA. These required specific primers and two of them are described. Priming with d(pT$_{10}$-C-A) permitted the elucidation of 88 bases at the 3' untranslated region. The oligonucleotide T-T-G-G-G-T primes at the end of the C-region and although it is only a hexanucleotide it displays an unexpected specificity. The specificity of short primers is further discussed. Light chain cDNA has been fractionated and components of defined size isolated. The largest is thought to include the whole of the V-region. Its hybridisation kinetics show that very few genes can cross-hybridise with the MOPC 21 light chain cDNA sequence. The use of these probes to measure the total pool of V-genes is discussed.

INTRODUCTION

The understanding of the molecular and genetic basis of the immune response has been, and remains, to a considerable extent dependent on the description of the amino acid and nucleotide sequences of the molecules involved. Advances have been triggered off largely by the introduction and development of appropriate techniques. In addition the technology

*Present address: Dept. of Pathobiology, School of Public Health and Community Medicine, University of Washington, Seattle, Washington 98195, USA.

required for the description of molecular structures has often opened up new perspectives. For instance, protein fingerprint techniques - a basic technique for amino acid sequence determination - became a tool for the definition of structural markers useful in genetic studies (1,2,3).

In recent years we have devoted considerable effort to studies of the nucleotide sequence of the light chain mRNA of the mouse myeloma MOPC 21. Progress has not been as fast as we had originally hoped. But we were dealing with a problem for which the technological approaches were being tested. It is worth remembering that the earliest light chain oligonucleotide sequences published (4) were indeed the first for any eukaryotic mRNA other than viral proteins. These studies were performed in parallel with attempts to define the number and arrangement of Ig genes by molecular hybridisation techniques.

Here we would like to give a progress report on some of our recent studies and give some examples of how the development of methods for sequence analysis has provided useful tools for genetic analysis.

STUDIES WITH [^{32}P]mRNA

The earlier structural characterisation of light (4) and heavy (5) chain mRNA utilised established procedures for [^{32}P] RNA. This involved digestion with T_1 ribonuclease fractionation of the products by two-dimensional fingerprints, and further analysis of T_1 oligonucleotides by digestion with pancreatic ribonuclease and U_2 ribonuclease. The critical limitation of this approach was the amount of radioactivity available, and this emphasised the need for new methods of sequence analysis. Nonetheless fingerprint analysis of [^{32}P] mRNA has remained the method of choice for a simple characterisation of mRNA as, for instance, in comparing light chain mRNA from different tumours (see below). This procedure was used for the detection of inactive light chain mRNA in a mutant line (6) and for the characterisation of a point mutation in the heavy chain (7). It has been invaluable in assessing the chemical purity of the mRNA preparations. This has been measured by calculating the molar yields of specific oligonucleotides isolated from fingerprints (8). Another application of the method has been the demonstration that the molecular species that hybridise to non-repetitive DNA contain light chain mRNA sequences from both C- and V-regions (8).

STUDIES OF cDNA TRANSCRIPTS OF mRNA

(a) Priming from the poly(A) tract

Complementary transcripts of mRNA can be made by reverse

transcription in the presence of a DNA primer. A commonly used primer is oligo(dT)$_{10}$ which is complementary to the poly (A) sequence present at the 3' end of the mRNA. Radioactive cDNA has been made in this way and used for both molecular hybridisation and sequence analysis (9). For the latter, the radioactive cDNA can be digested with endonuclease IV which splits at C residues. Unfortunately the splits are only partial and fingerprints very variable. Large fragments are also produced and considerable portions of the sequences do not show up in the fingerprints, or occur in very low yields. In consequence when all the products from a given cDNA digest were analysed they did not overlap each other as if they did not correspond to a continuous stretch. In addition minor impurities in the cDNA preparation introduced considerable confusion because of the relative high yield of the sequences next to the primer.

The introduction of the rapid DNA sequencing methods (10) based on gel separation has opened up much better perspectives. These methods have been applied to mRNA sequencing via cDNA transcription (11). The procedure requires phased primers; oligo(dT) can start at a number of points since the poly(A) sequence is 200 bases long (4). For a single starting point a more elaborate primer was needed. To construct it, preliminary experiments were done to establish the sequence adjacent to poly(A). The experiments established the sequence 5'--- C-U-U-G-poly(A)$_{(OH)}$. Based on this observation, an oligonucleotide d(pT$_{10}$-C-A$_{OH}$) was synthesised by Gillam and Smith to be used as a primer in the following way:

$$\text{---- C-U-U-G-A}_{10}\text{-A}_{190}\text{-OH}$$

$$\text{A-C-T}_{10}\text{-p}$$

With this primer, cDNA was prepared and analysed by the methods described by Sanger and Coulson (10) and adapted to sequencing mRNA by Brownlee and Cartwright (11). A sequence of 88 bases was thus established and is presented in Fig. 1. This sequence includes a number of T$_1$ oligonucleotides that were characterised (but not overlapped) by direct analysis of [^{32}P]mRNA, as described in the previous section. In particular it includes part of an oligonucleotide (t$_0$, see below and Fig. 7) which, until this study, remained unassigned to any region of the mRNA. The reason why we had failed to place it previously was its unusual size - almost 50 residues long. This increases our estimate of the 3' untranslated region to 200-250 bases long - an increase in the minimum from 150 to 200. The length of this region was unexpectedly long when first reported (12) and its function remains a puzzle. It includes a hexanucleotide A-A-U-A-A-A, about 17 residues

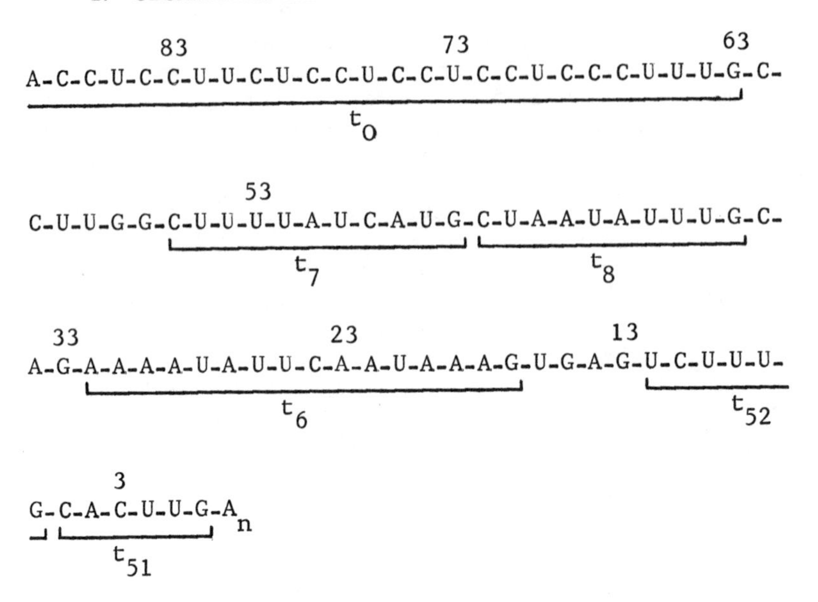

Fig. 1. Nucleotide sequence of part of the 3' untranslated region of MOPC 21 L-chain mRNA. Arrows indicate the corresponding oligonucleotides that have been identified in fingerprints of [^{32}P]mRNA (e.g. Fig. 7).

removed from the poly(A) that is also found in rabbit and human α- and β-globin mRNA and in ovalbumin mRNA. It may represent a signal required by mRNAs for an unidentified function (13). The light chain mRNA sequence shown in Fig. 1 contains other unusual features. In the sequence from 60 to 88 the trinucleotide C-C-U is repeated seven times. This may represent an internal duplication resembling the type of "spacer" duplication observed in Xenopus 5S DNA (21). However, this similarity should not be taken as evidence for lack of selective pressure on the 3' non-coding region.

(b) Priming in the C-region

The availability of oligonucleotide sequences derived from the fingerprints of [^{32}P]mRNA permitted the design of a number of synthetic primers complementary to different regions of the light chain mRNA. But, though we were convinced of the correctness of both the sequence of the mRNA and of the synthetic deoxynucleotides, no priming activity could be detected with some oligonucleotides - up to 9 residues long. On the other hand, a pentanucleotide had surprisingly good priming activity (9). This activity was shown to originate at two priming sites in the C-region (9) (sites 1 and 2, Fig. 2). The pentanucleotide (synthesised by C.C. Cheng) (pT-T-G-G-G_{OH}) was extended by Gillam and Smith by one residue,

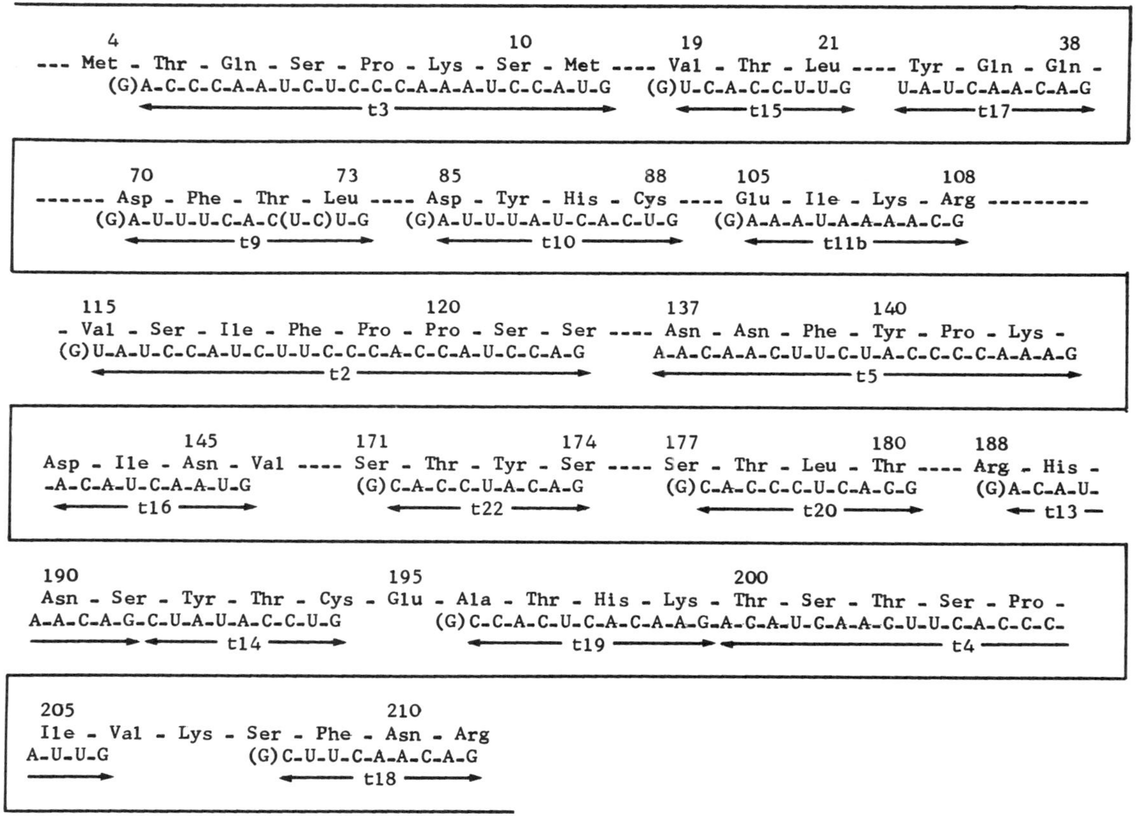

```
 4                                        10        19          21                  38
--- Met - Thr - Gln - Ser - Pro - Lys - Ser - Met ---- Val - Thr - Leu ---- Tyr - Gln - Gln -
    (G)A-C-C-C-A-A-U-C-U-C-C-C-A-A-A-U-C-C-A-U-G   (G)U-C-A-C-C-U-U-G    U-A-U-C-A-A-C-A-G
     <--------------------t3------------------->    <------t15------>    <-----t17------>

       70                73       85                88       105                108
------ Asp - Phe - Thr - Leu ---- Asp - Tyr - His - Cys ---- Glu - Ile - Lys - Arg ---------
    (G)A-U-U-U-C-A-C(U-C)U-G   (G)A-U-U-U-A-U-C-A-C-U-G   (G)A-A-A-U-A-A-A-A-C-G
     <--------t9-------->        <--------t10-------->       <-------t11b------->

  115                          120                    137               140
- Val - Ser - Ile - Phe - Pro - Pro - Ser - Ser ---- Asn - Asn - Phe - Tyr - Pro - Lys -
(G)U-A-U-C-C-A-U-C-U-U-C-C-C-A-C-C-A-U-C-C-A-G      A-A-C-A-A-C-U-U-C-U-A-C-C-C-C-A-A-A-G
 <-------------------t2-------------------->        <-----------------t5----------------->

         145                 171                174      177                180      188
Asp - Ile - Asn - Val ---- Ser - Thr - Tyr - Ser ---- Ser - Thr - Leu - Thr ---- Arg - His -
-A-C-A-U-C-A-A-U-G      (G)C-A-C-C-U-A-C-A-G      (G)C-A-C-C-C-U-C-A-C-G     (G)A-C-A-U-
 <------t16------>        <-------t22------>         <--------t20-------->       <-- t13 --

190                             195                           200
Asn - Ser - Tyr - Thr - Cys - Glu - Ala - Thr - His - Lys - Thr - Ser - Thr - Ser - Pro -
A-A-C-A-G-C-U-A-U-A-C-C-U-G       (G)C-C-A-C-U-C-A-C-A-A-G-A-C-A-U-C-A-A-C-U-U-C-A-C-C-C-
------->  <------t14------>          <--------t19-------> <----------------t4-------------

205                            210
Ile - Val - Lys - Ser - Phe - Asn - Arg
A-U-U-G                 (G)C-U-U-C-A-A-C-A-G
------->                  <-------t18------->
```

Fig. 2. Oligonucleotide sequence coding for C-region of L-chain mRNA of MOPC 21. The oligonucleotides identified as components of [^{32}P]mRNA fingerprints (e.g. Fig. 7) are indicated as ←— t2 —→, ←— t25 —→, etc. Taken from ref. 9.

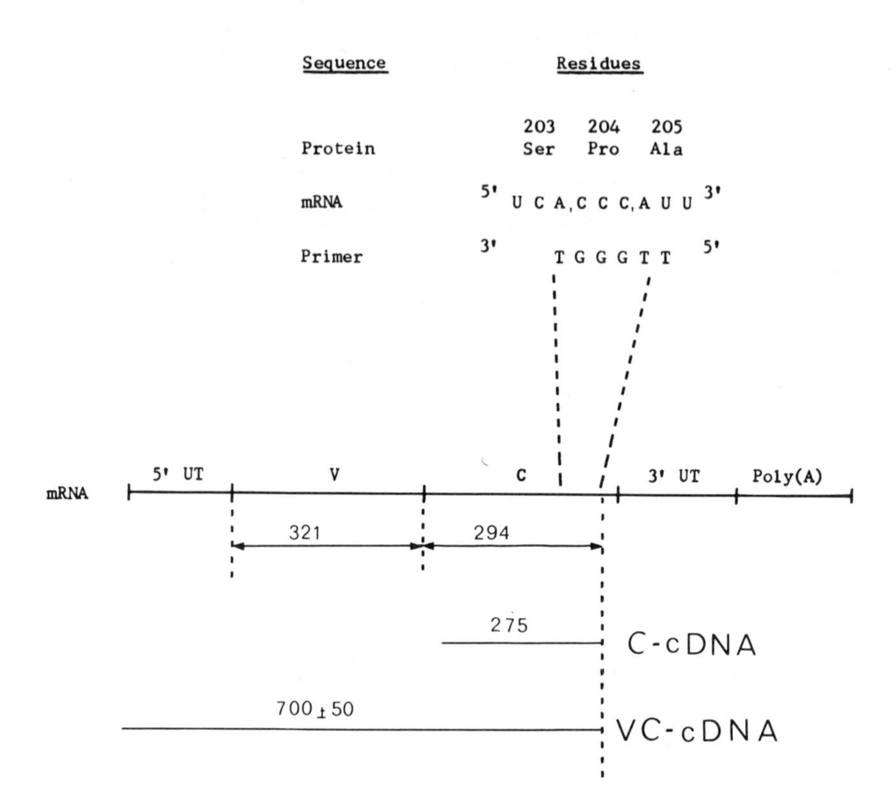

Fig. 3. Specific priming by synthetic oligonucleotides. The length of cDNA transcripts correspond to the indicated bands of Fig. 4.

(pT-T-G-G-G-T_{OH}) to restrict the priming activity and preclude initiation at site 2. Priming with the hexanucleotide was then tested by a variety of means. A single continuous sequence could be derived from the cDNA prepared in this way. This is indicated by a thick line starting at amino acid residue 203 and extending for over 60 bases to amino acid residue 185. The results obtained with this primer indicated a high degree of purity of the cDNA, in spite of our use of only a partially purified (about 50%) preparation. This observation was quantitated by measuring the back-hybridisation of cDNA (made by priming the transcription with T-T-G-G-G-T) to other mRNA preparations. Over 90% of the cDNA was hybridised in a single kinetic transition (14). All these results suggested to us that priming was not only dependent on correct complementarity, but (no less important) accessibility of the primer site. There are now many lines of evidence that indicate that mRNA contains a high degree of secondary structure. Indeed we should not regard mRNA as only containing a secondary structure characterised by hairpin loops stabilised by Watson-Crick base pairs. As demonstrated with tRNA (15), more complicated tertiary interactions are likely to be present. This would lead to a rather restricted number of accessible sites. It is only at these accessible sites (where there is no competition with the internal folding) that short primers are successful. This combination of sequence and accessibility gives a very unexpected specificity to a rather short sequence.

Preparation of cDNA by priming with T-T-G-G-G-T (specific primer cDNA) has several advantages. It starts at a very specific point in the sequence and totally excludes the whole of the 3' untranslated region (Fig. 3). Transcripts longer than 300 bases are enough to include V-region sequences. Indeed a complete transcript, including the whole of the V-region and the untranslated region beyond (5' untranslated region), should be shorter than 800 residues. We have analysed these cDNA transcripts and the results are shown in Fig. 4. This shows a heterogeneous population, but including well-defined bands representing an accumulation of material at specific points along the transcript. These bands are thought to be due to a combination of factors, the principal ones being the folding of the chain and local peculiarities in base sequence. The largest transcript we can detect in Fig. 4 is about 700 bases long. This must include the whole of the mRNA region translated into a complete light chain, with the exception of a few residues corresponding to the C-terminus of the chain (615 bases). This leaves about 100 bases for the precursor part and 5' untranslated region. This estimate is of course subject to error in estimation of the total length (700 bases). The sharp cut-off in the transcription beyond the 700 bases-long band (Fig. 4) may be due to transcription having reached the end of the mRNA. In this case we can estimate that the 5' untranslated portion is quite short, probably about 50 residues. This length would agree with estimates of the length of the mRNA and of the poly(A) stretch (4) as shown in Fig. 5. Fractionation of the T-T-G-G-G-T primed cDNA gave specific bands that could be eluted from the gel. One such band includes the whole of the C-region; the other both V- and C-regions. These components are very important tools as molecular probes. The longest cDNA, including the whole of the V-region, has been used for measurement of the repetitive frequency of the V- and C-genes (Fig. 6). Similar measurement (Cot kinetics) have been made previously using different probes (e.g. 16,17,18). What is new in this

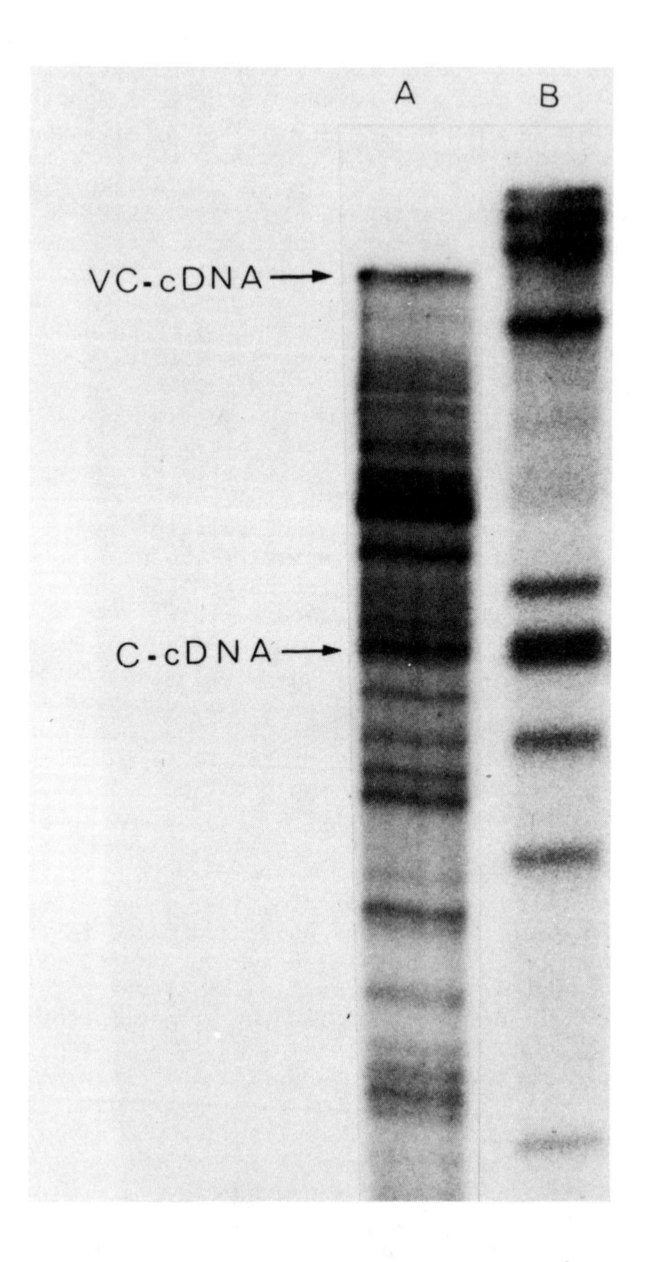

Fig. 4. Acrylamide gel electrophoretic analysis of [^{32}P] cDNA. The cDNA (slot A) was made using MOPC 21 L-chain as template, T-T-G-G-G-T as primer and [^{32}P] TTP as labelled input. Electrophoresis in 4% acrylamide 7 M urea. M.W. of cDNA calibrated using restriction fragments (Hae III)f rom ØX174 (slot B).

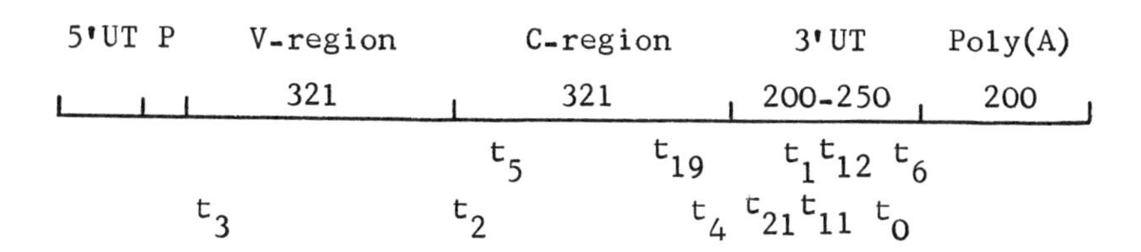

Fig. 5. Schematic diagram of L-chain mRNA. Location of T_1 oligonucleotides (see Fig. 7) is taken from Fig. 1 and from ref. 4.

curve is that for the first time a homogeneous cDNA encompassing the whole of the V-region has been used. In addition the kinetic rate of hybridisation has been measured by the hydroxyapatite procedure. The advantage of this method is its better capacity to detect cross-hybridising genes. The results shown in Fig. 6 show that very few (possibly only one) V-genes can cross-hybridise with the MOPC 21 sequence (19).

We hope to use these probes in our attempts to measure the total pool of V-genes. For this purpose the basic idea is to utilise the cross-hybridisation properties of the V-regions. We have argued (9,19) that cross-hybridisation under stringent conditions is unlikely to include V-genes differing by more than 10% in terms of amino acid sequence. On the other hand, by lowering the stringency it may be possible to cross-hybridise V-genes differing by perhaps 20% or 30%, i.e. belonging to different subgroups. These estimates are based on the extent of cross-hybridisation of C_χ genes of different species (9,19).

SEQUENCE HOMOLOGY BETWEEN DIFFERENT L-CHAIN mRNAs

In the first part of this paper we have explained the fingerprint analysis of [^{32}P]mRNA and its usefulness in defining selected sequences along the light chain mRNA. We have now used fingerprint procedures to compare the nucleotide sequences of two different kappa light chains. The comparison is shown in Fig. 7, which shows that all the oligonucleotides described in the kappa chain of MOPC 21 mouse myeloma, and previously located at the C-region or at the 3' untranslated end of the light chain mRNA, are also present in the fingerprint of the kappa chain mRNA of Sp2 mouse anti-SRBC plasmacell hybrid (20). Identification of the oligonucleotides mentioned has been carried one step farther than direct fingerprint comparison. This was done by digesting individual T_1 oligo-

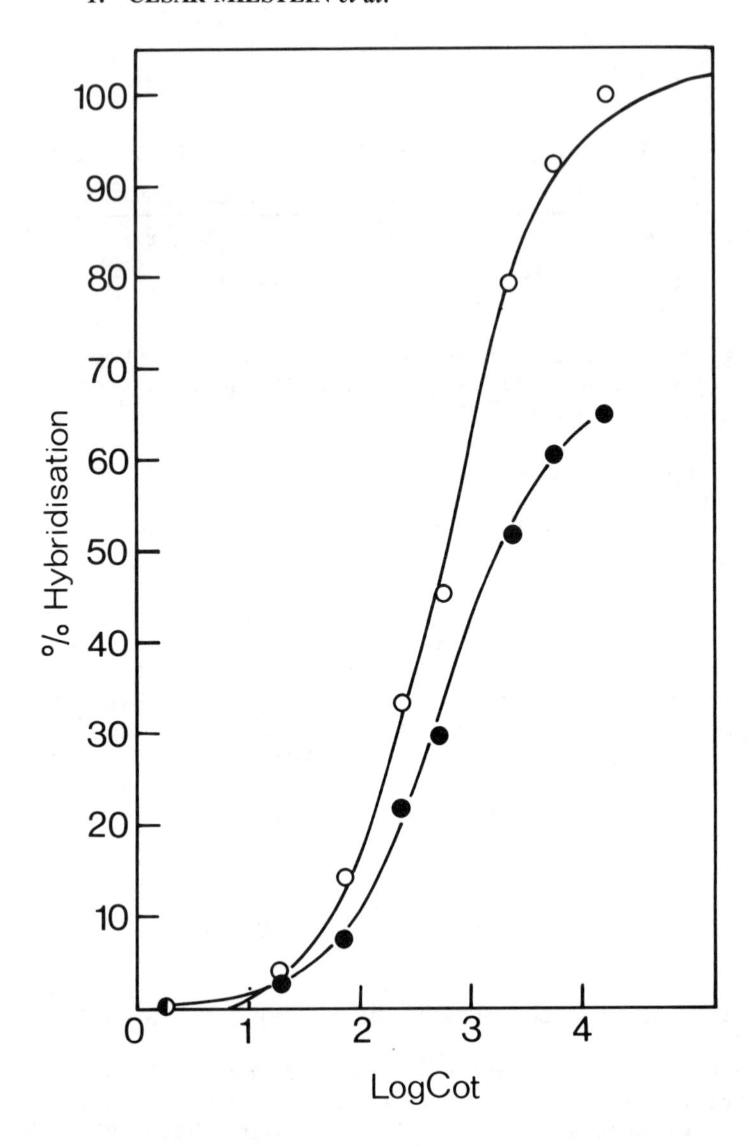

Fig. 6. Hybridisation of specific VC-cDNA to myeloma DNA as measured by binding to hydroxyapatite. The [^{32}P]VC-cDNA utilised in this experiment was eluted from a gel band as in Fig. 4. Hybridisation was at 70° in 0.24 M sodium phosphate buffer pH 6.8. Experimental points (●) have been normalised (o) and a theoretical second order reaction curve fitted. The points follow the theoretical hybridisation of a homogeneous cDNA population with a nominal reiteration frequency of 3.

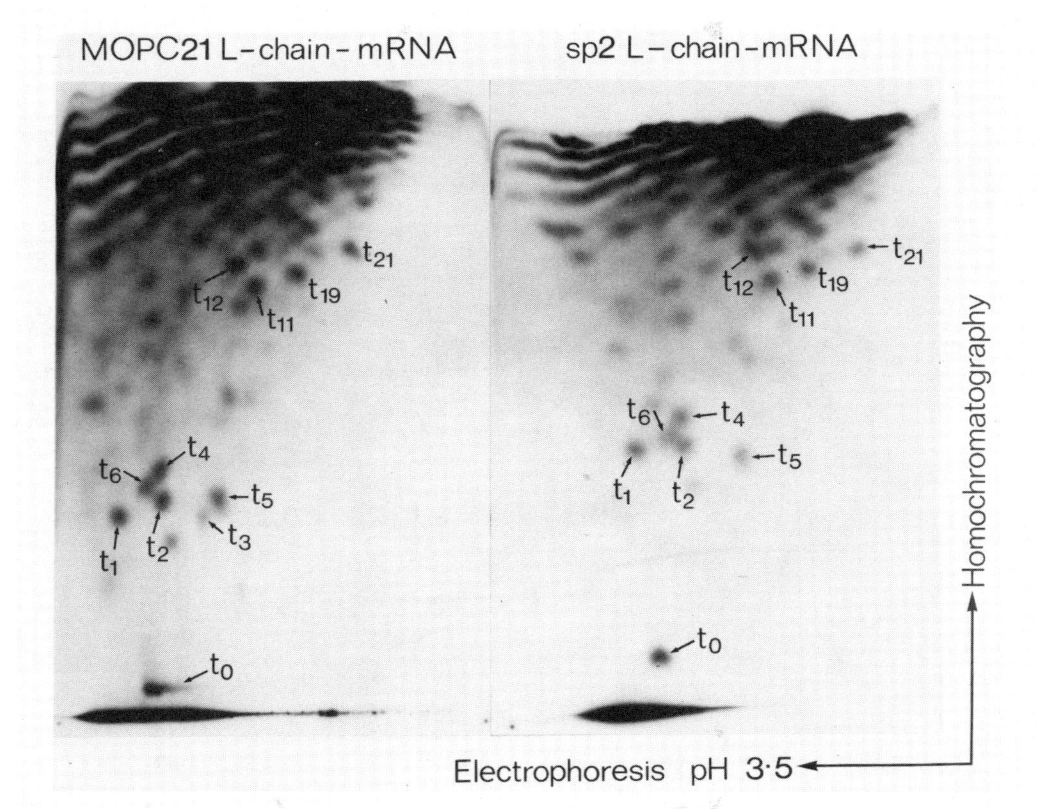

Fig. 7. Comparative fingerprints of the kappa chain mRNAs from the myeloma protein MOPC 21 and the anti-SRBC hybridocytoma Sp2/HL. The numbered oligonucleotides have been located in the regions of the mRNA as shown in Fig. 5.

nucleotides with pancreatic and U_2 ribonucleases. Fig. 8 shows a comparison of the U_2 digestion patterns of the longest of the oligonucleotides shown in Fig. 7. While no difference could be demonstrated with the C-region and 3' untranslated region oligonucleotides, the oligonucleotide t_3, which is part of the V-region of MOPC 21, was absent in Sp2 (Fig. 7). The results extend at the RNA level the notion of the variability and constancy of the two halves of the light chains. But in addition they prove that the 3' untranslated region is also constant. They suggest that this region of the mRNA belongs to the C-region gene.

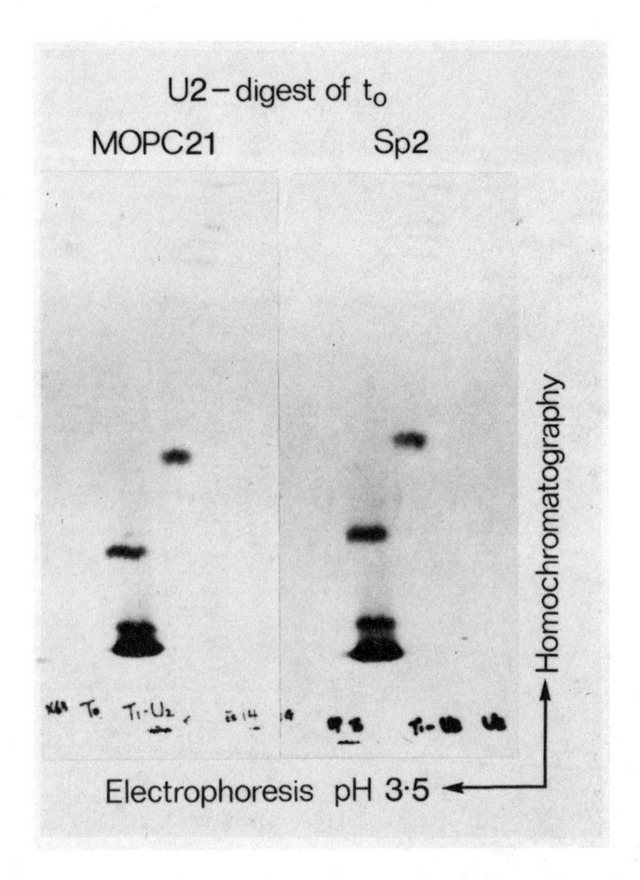

Fig. 8. Comparison of oligonucleotide t_0 from MOPC 21 and Sp2/HL taken from Fig. 7.

ACKNOWLEDGEMENTS

P.H.H. is the recipient of an MRC Training Fellowship and K.C.S.C. of a Helen Hay Whitney Foundation Fellowship.

REFERENCES

1. Milstein, C., Milstein, Celia P. and Feinstein, A. (1969) Nature 221, 151.
2. Edelman, G.M. and Gottlieb, P.D. (1970) Proc. Nat. Acad. Sci. U.S.A. 67, 1192.
3. Mole, L.E. (1975) Biochem. J. 151, 351.
4. Brownlee, G.G., Cartwright, E.M., Cowan, N.J., Jarvis, J.M. and Milstein, C. (1973) Nature 244, 236.
5. Cowan, N.J., Secher, D.S. and Milstein, C. (1976) Eur. J. Biochem. 61, 355.
6. Cowan, N.J., Secher, D.S. and Milstein, C. (1974) J. Mol. Biol. 90, 691.
7. Milstein, C., Adetugbo, K., Cowan, N.J., Köhler, G., Secher, D.S. and Wilde, D.C. (1976) Cold Spring Harbor Symp. Quant. Biol., in press.
8. Rabbitts, T.H., Jarvis, J.M. and Milstein, C. (1975) Cell 6, 5.
9. Milstein, C., Brownlee, G.G., Cheng, C.C., Hamlyn, P.H., Proudfoot, N.J. and Rabbitts, T.H. (1976) 27th Mosbacher Colloquium. Ed. by F. Melchers and K. Rajewsky, Springer-Verlag, Berlin, Heidelberg, pp.75-85.
10. Sanger, F. and Coulson, A.R. (1975) J. Mol. Biol. 94, 441.
11. Brownlee, G.G. and Cartwright, E.M. (1977) J. Mol. Biol., submitted.
12. Milstein, C., Brownlee, G.G., Cartwright, E.M., Jarvis, J.M. and Proudfoot, N.J. (1974) Nature 252, 354.
13. Proudfoot, N.J. and Brownlee, G.G. (1976) Nature 263, 211.
14. Rabbitts, T.H., Forster, A., Smith, M. and Gillam, S. (1977) Eur. J. Immunol., in press.
15. Jack, A., Ladner, J.E. and Klug, A. (1976) J. Mol. Biol. 108, 619.
16. Honjo, T., Packman, S., Swan, D. and Leder, P. (1976) Biochemistry 15, 2780.
17. Tonegawa, S. (1976) Proc. Nat. Acad. Sci. U.S.A. 73, 203.
18. Farace, M.G., Aellen, M.F., Briand, P.A., Faust, C.H., Vassalli, P. and Mach, B. (1976) Proc. Nat. Acad. Sci. U.S.A. 73, 727.
19. Rabbitts, T.H. and Milstein, C. (1977) Contemporary Topics in Molecular Biology. Ed. by R.R. Porter and G. Ado, in press.
20. Köhler, G. and Milstein, C. (1976) Eur. J. Immunol. 6, 511. 511.
21. Brownlee, G.G. (1976) Dahlem Konferenzen. Life Sciences Research Reports Vol. 4. Ed. by V.G. Allfrey, E.K.F. Bautz, B.J. McCarthy, R.T. Schimke and A. Tissières, pp.180-186.

ARRANGEMENT AND REARRANGEMENT OF IMMUNOGLOBULIN GENES

Susumu Tonegawa, Nobumichi Hozumi, Christine Brack and Rita Schuller

Basel Institute for Immunology, 487 Grenzacherstrasse, Postfach 5, Basel 5, Switzerland.

ABSTRACT. Arrangement of Immunoglobulin genes in mouse DNAs of various cellular sources was analyzed by use of Bacterial restriction enzymes. A high-molecular weight DNA was digested to completion and resulting DNA fragments were fractionated according to size in preparative agarose gel electrophoresis. DNA fragments carrying gene sequences coding for the variable or constant region of light chains were detected by hybridization with purified ^{125}I-labeled, whole light chain mRNAs and with their 3'end halves. The corresponding patterns of hybridization were completely different in the genomes of embryo cells and of plasmacytomas. The results, together with those of control experiments, lead us to conclude that the V and C genes, which are some distance away from each other in the embryo cells, are joined to form a contiguous polynucleotide stretch during differentiation of lymphocytes. There seems to be a strict correlation between such V-C joining and expression of the joined V gene. Relevance of these findings with respect to activation of a specific V gene and allelic exclusion in immunoglobulin gene loci is discussed.

In addition, isolation of a phage λ which carries an embryonic mouse DNA fragment as an insert in its genome is described. The DNA fragment is 3 kilobase long and contains a V_λ gene.

INTRODUCTION

Both light and heavy chains of immunoglobulin molecules consist of two regions: the variable region (V region) and the constant region (C region) (6,19). This led Dreyer and Bennett to put forward the hypothesis that two separate DNA segments, one each for V and C regions, may be involved in the synthesis of a single immunoglobulin chain (2). This attractive concept remained speculative until recently, partly because of its unorthodox nature, but mostly because of lack of direct experimental evidence.

NUCLEOTIDE POSITION

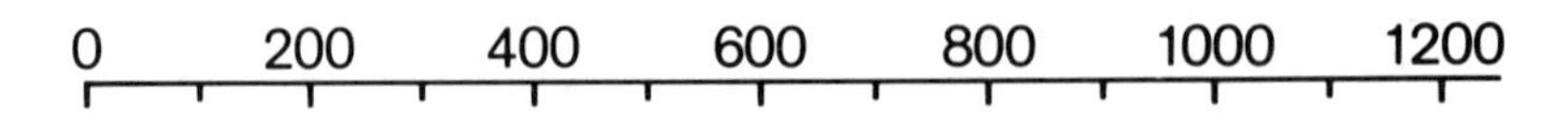

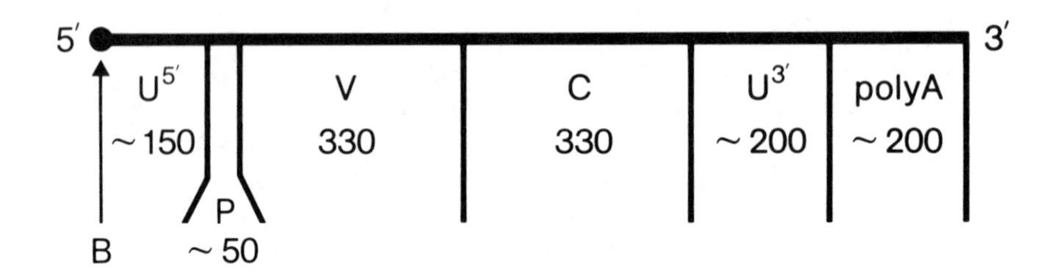

B = Blocked 5′-end
U$^{5'}$ = Untranslated region at the 5′-end
P = Region coding for extra peptide in the precursor
U$^{3'}$ = Untranslated region toward the 3′-end

Figure 1. Schematic illustration of light chain mRNA.

Use of bacterial restriction enzymes allowed us to provide direct experimental evidence for such a concept. How, then, is the genetic information in two separate DNA segments integrated to generate a single polypeptide chain? We recently reported evidence for the hypothesis that such integration takes place directly at the DNA level (6).

In this article we shall review our experimental results pertinent to these two concepts and discuss some implications drawn from them.

EXPERIMENTAL APPROACH

Bacterial restriction endonucleases recognize and cleave specific sequences of base pairs within DNA duplexes. High molecular weight DNA purified from twelve to fourteen day old Balb/c mouse embryos, or from various myeloma tumors, was digested to completion with restriction enzyme Bam H-I or Eco RI. The resulting DNA fragments were fractionated according to size by electrophoresis in 0.9% agarose gels. DNA eluted from gel slices was assayed for V and C gene sequences by hybridization with ^{125}I-labelled whole κ or λ light chain mRNA, or half RNA fragments containing the 3' terminal. The assay is based on the fact that the sequences corresponding to V and C genes are on the 5' end and 3' end halves of the mRNA, respectively (Figure 1). Thus, the 3' end half is an RNA probe for C gene sequences, and the V gene sequences are determined indirectly from the difference in the two hybridization levels obtained with the whole RNA molecule and the 3' end half. The details and basis of this assay are described elsewhere (6).

JOINING OF V - C GENES AT THE DNA LEVEL

We recently reported evidence for V-C joining at the DNA level (6). The evidence was based on analysis of embryonic and myeloma DNA by procedures similar to those described in the previous section. Here we present results of such an analysis, together with some additional controls. In Figures 2A and 2B, DNAs from twelve day old Balb/c embryos and MOPC 321 myeloma were separately digested with a restriction enzyme, *Bam* H-I. Both MOPC 321 κ mRNA and MOPC 104E λ mRNA were used as the sequence probes. Of the two major embryonic DNA components which hybridized with the κ mRNA, only the larger (6.0 million M.W.) hybridized with κ 3' end half (data not shown here, see reference 6). The patterns of hybridization of the κ mRNA with the embryonic DNA and the myeloma DNA are dramatically different. Both the whole κ mRNA and its 3' end half (latter data not shown) hybridized with a single major component of 2.4 million M.W. in the myeloma DNA. In contrast, hybridization patterns of the λ mRNA in the two DNAs are virtually indistinguishable.

As we argued elsewhere, these results are best interpreted as follows. The V_κ and the C_κ genes which are some distance away in the embryo genome are brought together, during differentiation of lymphocytes, in order to form a continuous DNA stretch. An alternative explanation of the results, namely, that accumulation of mutations leading to either loss or gain of *Bam* H-I sites generated the observed pattern difference, is not impossible. On this view, there would have to be a *Bam* H-I site close to the V-C junction in embryo DNA. This *Bam* H-I site would have to be lost by mutation in the MOPC 321 tumor. By itself, such a mutation would cause the appearance of a single 9.9 million component in the tumor. To achieve the M.W. of the single component actually observed in the tumor (2.4 million), there would have to be new *Bam* H-I sites created by mutation between the V_κ gene and the nearest site on either side. Since there is

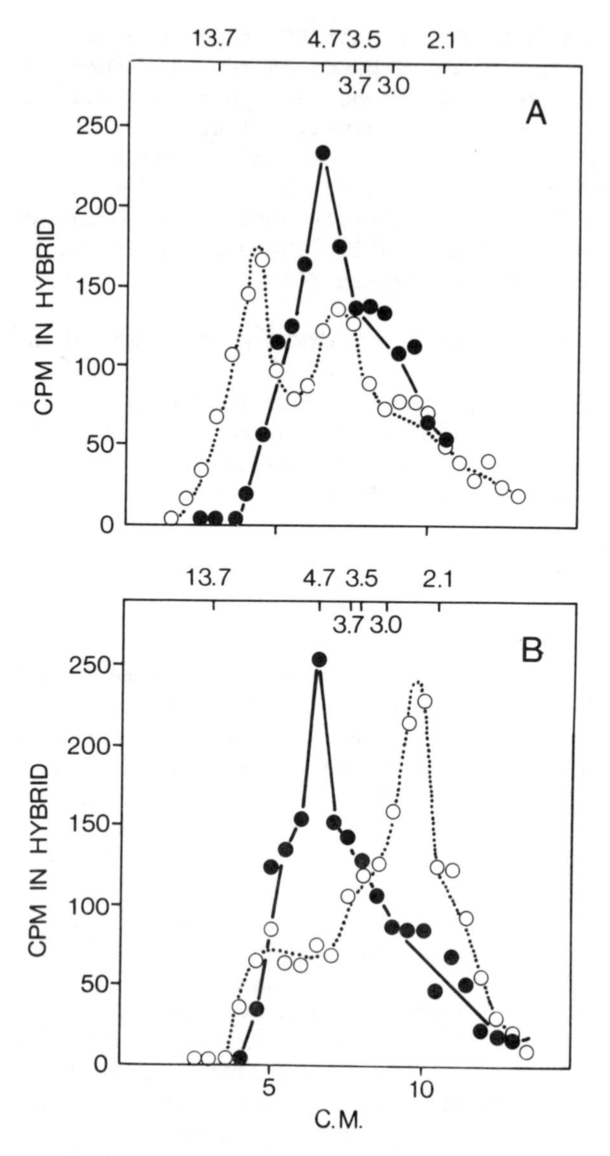

Figure 2. Gel electrophoresis patterns of Bam H-I digested embryo DNA (A) and MOPC 321 DNA (B). Bam H-I digestion, DNA fractionation and hybridization were carried out as described (6). Whole MOPC 321 κ mRNA (o----o, 1250 cpm, 5×10^7 cpm/μg) or MOPC 104E λ mRNA (●——●, 1250 cpm, 7×10^7 cpm/μg was annealed with DNA. In this and Figures 3, 4 and 5 numbers on top of the panel indicate the molecular weights (in millions) of DNA fragments used as migration marks.

no reason why there should be any selective pressures involving Bam H-I sites, the occurrence of three mutations would seem to be quite unlikely.

In addition, constancy of λ mRNA hybridization patterns suggest that there is no massive scrambling of DNA sequences throughout the chromosomes during generation and propagation of this myeloma. This view is also supported by the experiments involving the converse combination, namely, the analysis of DNA from λ chain-producing myeloma with a κ mRNA (Figure 3A). The hybridization pattern was indistinguishable from that of embryonic DNA. The case can be extended to DNAs from normal adult tissues as representatively shown in Figure 3B with kidney cells. Thus there seems a good correlation between the V-C joining event and expression of the joined immunoglobulin gene.

We have also digested embryo and MOPC 321 tumor DNAs with another restriction enzyme, Hind III (data not shown). With embryo DNA, two major components of 6.0 and 2.8 millions hybridized with the whole MOPC 321 κ mRNA, of which only the latter hybridized with the 3' end half. Thus, V and C gene sequences are in 6.0 and 2.8 million components, respectively. With the MOPC 321 DNA, there was only a single major component of about 2.8 millions observed, with which both RNA probes hybridized. No 6.0 million component existed in the tumor DNA pattern. Thus, it seems that unfortunately in this particular case, the embryonic, C gene fragment was not resolved in the electrophoresis from the myeloma, V-C fragment. The results, nevertheless, provide another case where hybridization patterns are clearly different between embryo and myeloma DNAs. Since Bam H-I and Hind III recognize different base sequences, the alternative interpretation mentioned above becomes even more unlikely.

SUBGROUPS AND V-C JOINING

Our earlier hybridization studies, as well as those of others, indicated that a group of closely related V regions are somatically generated from a few, probably single, germ line genes (7,14,18-21). While there still remains some ambiguity, such a V region group is best approximated to the subgroup as defined by Cohn and his co-workers (1). A direct demonstration of separate germ line V genes for two V regions of different subgroups are presented in Figure 4A. Here embryonic DNA digested with Bam H-I enzyme was analyzed with two mRNAs coding for two κ chains of different subgroups, MOPC 321 and MOPC 21. These two κ chains show little homology in their V regions whereas they have identical sequences in

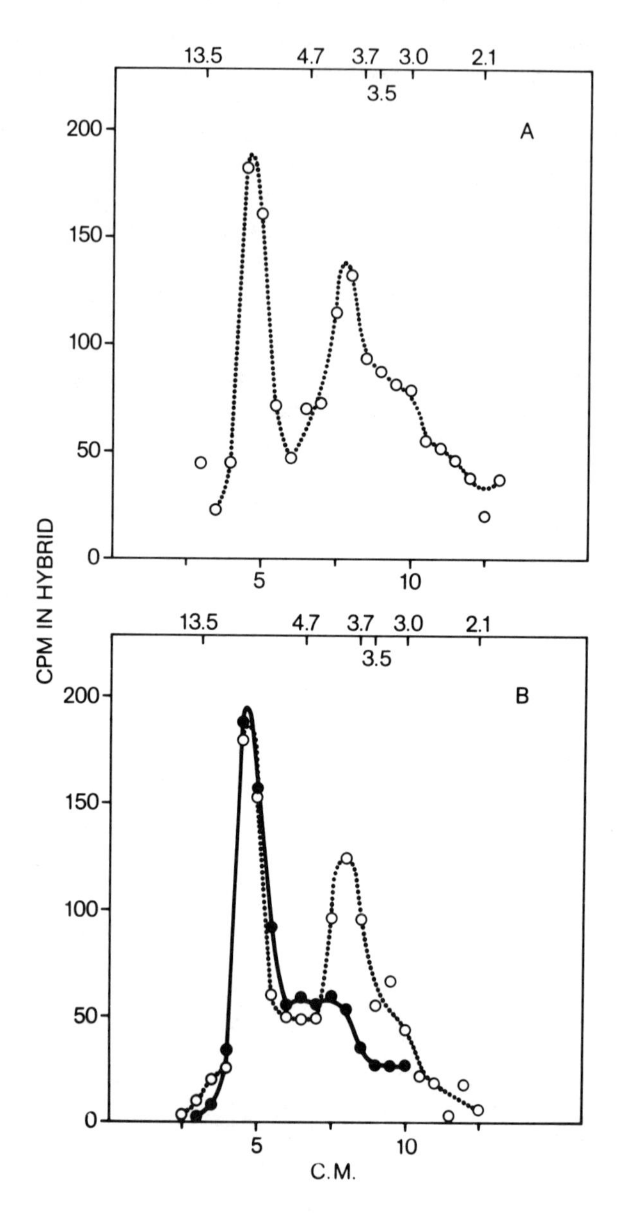

Figure 3. Gel electrophoresis patterns of λ J558 DNA (A) and kidney DNA (B) digested with Bam H-I. Whole MOPC 321 RNA (o----o, 1250 cpm, 7 x 10^7 cpm/μg) or its 3' end half fragment (●——●, 600 cpm, 7 x 10^7 cpm/μg) was hybridized to fractionated DNA.

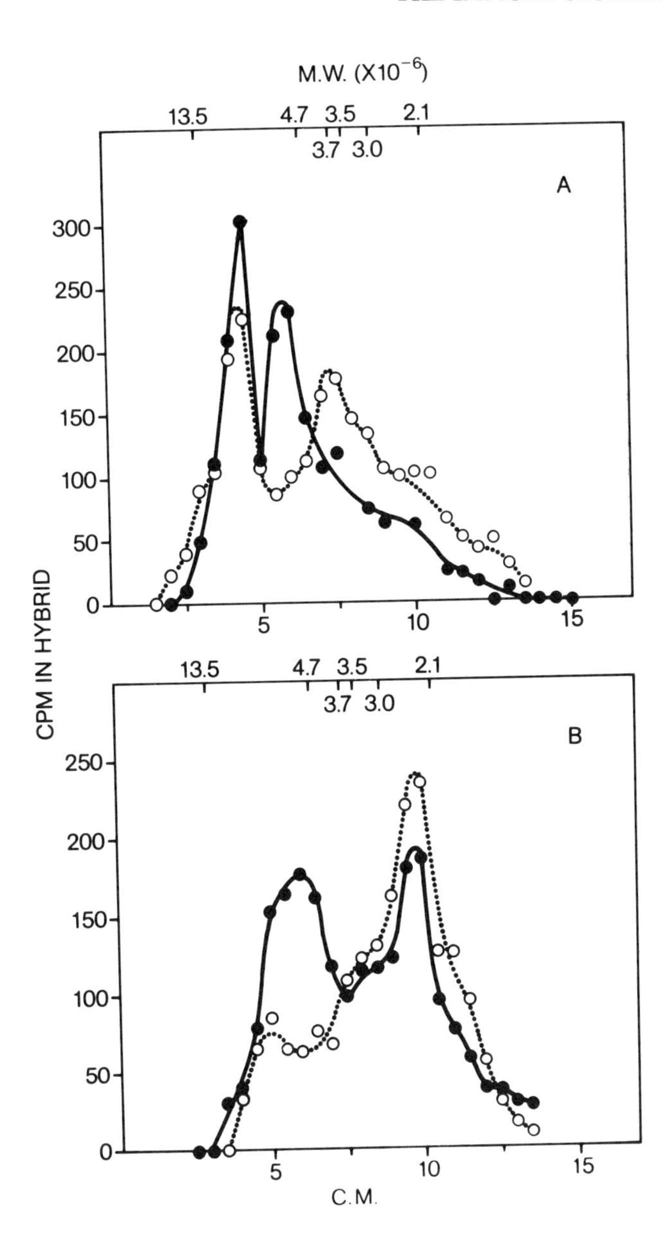

Figure 4. Gel electrophoresis patterns of embryo DNA (A) and MOPC 321 DNA (B) digested with Bam H-I. Whole MOPC 321 mRNA (o----o, 1250 cpm, 7 x 10^7 cpm/μg) or MOPC 21 mRNA (●——●, 1220 cpm, 8 x 10^7 cpm/μg was annealed with extracted DNA.

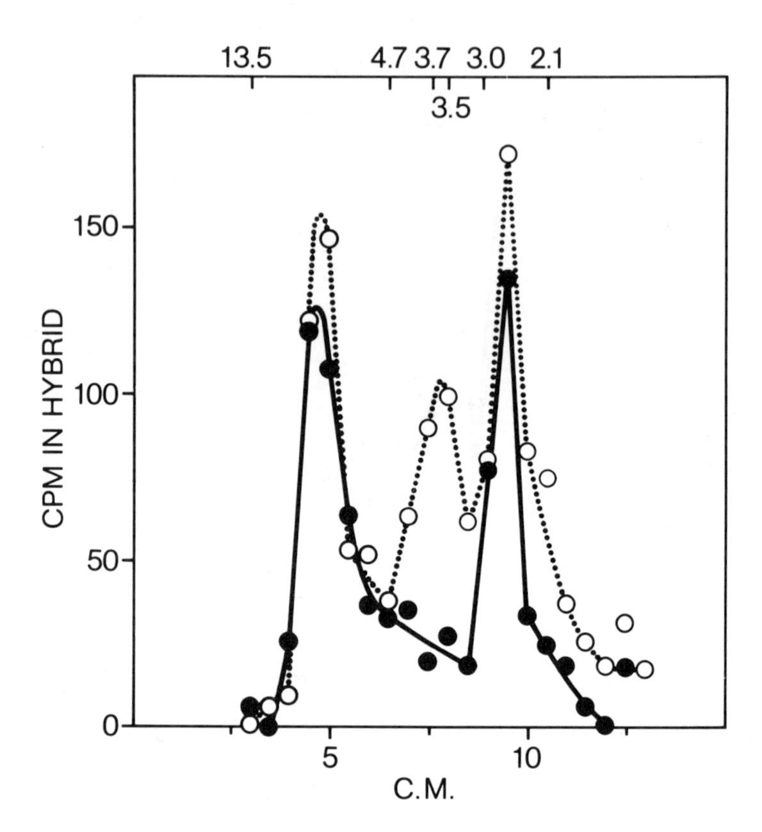

Figure 5. Gel electrophoresis patterns of TEPC 124 DNA digested with Bam H-I. Whole 321 mRNA (o----o) or its 3' end half fragment (●——●) was annealed with extracted DNA as described in the legend to Figure 4.

the C region (8,10). As expected, both RNAs hybridized with the 6.0 million component which carries the C_κ gene (see above). In addition, each of the two RNAs hybridized with a second, but mutually different, DNA component. These DNA components of 5.0 and 3.9 millions M.W. should carry MOPC 21 and MOPC 321 V gene sequences respectively.

Is a V_κ gene for a given subgroup joined with a C_κ gene in the myeloma which synthesizes a κ chain carrying a V region of another subgroup? That this is not the case is shown in Figure 4B, where MOPC 321 DNA was analyzed with the homologous (MOPC 321) and heterologous (MOPC 21) κ mRNAs. As already pointed out, the homologous κ mRNA hybridized with a major DNA component of 2.4 million M.W. While the 6.0 million M.W. C_κ gene component disappears, the 5.0 million M.W., $V_{\kappa,MOPC\ 21}$ gene fragment remains at the embryonic position. These results provide another example for strict correlation between the V-C joining event and expression of the joined immunoglobulin gene.

What about a V_κ gene in the myeloma synthesizing another κ chain of the same subgroup? Analysis of TEPC 124 DNA with MOPC 321 mRNA is shown in Figure 5. The two κ chains synthesized by these myelomas are different in three amino acids in the V regions, and belong to a single subgroup (8). Three major DNA components hybridized with the whole κ mRNA, two of which hybridized also with the 3' half fragments. The size and hybridization properties of these DNA components suggest that the overall pattern is a composite of the two patterns obtained when embryonic and MOPC 321 DNA were analyzed by the same RNA probes (Figures 2A and 2B). Thus, the principal difference in the two hybridization patterns, one of embryonic DNA (Figure 2A) and the other of TEPC 124 tumor (Figure 5), is the addition of the 2.4 million component in the latter. This component hybridized with both whole and 3' end half RNA probes. Each of the two components which hybridized with the 3' end half ought to contain a complete C gene sequence, for nucleotide sequence studies of a κ mRNA indicate that there is no Bam H-I cleavage site in the C gene (9).

We have previously shown that two nucleotide sequences coding for two V regions of a single subgroup are extensively homologous (18,20,21). Therefore, in interpreting these results, we assume that this is the case for the nucleotide sequences coding for MOPC 321 and TEPC 124 κ chains are extensively homologous. Given this assumption, two likely explanations can be given. First, there are two separate germ line genes for the two κ chains. While $V_{\kappa M\ 321}$ remains at the embryonic position, $V_{\kappa T\ 124}$ is joined with one of the C_κ genes. The alternative explanation is that there is a common germ line V gene from which two V regions are somatically derived. This V gene is joined with the C_κ gene only in one of the multiple homologous chromosomes. While it is fortuitous in the first instance that the size of the DNA fragment carrying the joined V-C gene is identical in MOPC 321 and TEPC 124 myelomas, it is a logical consequence in the second.

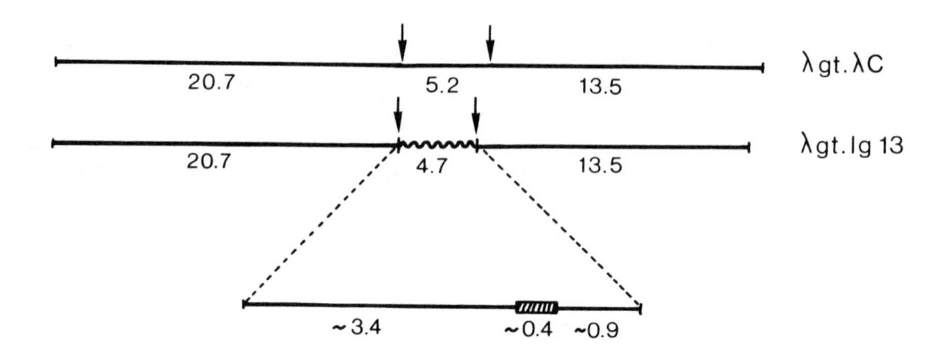

Figure 6. Schematic illustration of λgt-λC WES and λgt-Ig 13 DNAs. Arrows indicate the positions of *Eco* RI cleavage sites. Length of DNA is given in kilobase. The shaded box at the bottom indicates the DNA segment homologous to V_λ gene sequences.

NEW APPROACHES TO THE STUDY OF THE ARRANGEMENT OF IMMUNOGLOBULIN GENES

We are applying the *in vitro* DNA recombination technique (11,16) to the immunoglobulin genes. Two kinds of approaches are being made in our laboratory. In one, we are isolating plasmids carrying DNA inserts of κ or λ chain genes, which were enzymatically synthesized from purified mRNA. We intend to use these hybrid plasmids as hybridization probes in the analysis of DNAs from natural lymphocyte clones.

In the other approach, we are isolating DNA fragments carrying immunoglobulin genes directly from chromosomal DNA. One such DNA fragment carrying a V_λ gene was successfully isolated from embryonic DNA as an insert in the genome of a phage λ (λgt-Ig13). Electronmicroscopy and hybridization studies demonstrated that the fragment carries no $C_{\lambda I}$ gene sequence and that the V_λ gene sequence is internally located within the DNA fragment (Figure 6). We are currently determining the nucleotide sequence of this V gene.

The above work was carried out in a P-3 physical containment facility in combination with EK-2 biological containments in accordance with the N.I.H. guideline on recombinant DNA work issued in June 1976. As the EK-2 biological containments, we used phage λgt.λC *Wam* 403, *Eam* 1100, *Sam* 100 (3) or *E.coli* X1776.

DISCUSSION

We consider that our interpretation of the difference in the hybridization patterns being due to V-C joining is well justified, given the various control experiments described here. Details of V-C joining, however, are still to be studied. We intend to do so by the use of cloned immunoglobulin genes.

A committed B-lymphocyte or plasma cell produces antibody of only one specificity (7,13,15). In particular, it expresses only one light chain V gene. Since there are not only V_λ and V_κ genes but also multiple V_κ genes, there must exist a mechanism for the activation of a V gene which is directly coupled to its becoming joined to a C gene. For instance, in the "excision-insertion" model or in the "inversion" model, (see reference 6), a promotor site might be created by the insertion of the V gene fragment. This would activate that particular V gene for transcription.

Alternatively, all V genes and even the C genes might be activated for transcription early in lymphocyte differentiation. The RNAs synthesized would be rapidly degraded in the nucleus by a 5'-specific nucleotidase. The sequence created by the joining event at the insertion site could serve, directly or indirectly, as a signal for preventing the exonuclease from proceeding further down toward the 3'-end. This model looks rather wasteful at first sight, but it should not be discarded solely for this reason. It is, in fact, in agreement with the fact that in eucaryotic nuclei a large proportion of DNA is constitutively transcribed into large RNA molecules (HnRNA), most of which turn over rapidly before reaching the ribosomes. Only a small proportion (a few per cent) of these large RNA molecules are "processed" to become mRNA.

In any case, in view of the above considerations, the current terminology of "V genes" and "C genes" is somewhat inappropriate. Rather there are two types of segments of DNA, one specifying the V region and the other specifying the C region. The "gene" is created by joining.

For immunoglobulin loci, only one allele is expressed in any given lymphocyte (13). This is not the case for any other autosomal genes studied until now. Our results of MOPC 321 DNA (Figures 2A and 2B) suggest an interesting explanation for this phenomenon of allelic exclusion - that the two homologous chromosomes are, in any given plasma cell, homozygous. Homozygocity could result from the loss of one homologue followed by reduplication of the other, or it could result from somatic recombination (presumably mediated by a specialized recombinase) between the centromere and the immunoglobulin locus (12). This explanation of allelic exclusion is independent of whether the joining takes place only in one of the two homologous chromosomes or both, in a single lymphocyte. On the other hand, our results on TEPC 124 DNA suggest that the joining takes place only in one of the two homologous chromosomes and no homozygosis is to take place. While this, too, can conveniently explain allelic exclusion, there is an apparent discrepancy between the two cases. The discrepancy could arise from the known abnormality in caryotypes of myeloma cells (23). MOPC 321 myeloma might have lost the homologous chromosomes on which V and C genes lie separate. Conversely, TEPC 124 myeloma might have acquired an additional chromosome of non-lymphoid origin during generation or propagation of the tumor (22).

Resolution of the discrepancy and elucidation of the mechanism of allelic exclusion would require detailed analysis of DNA from Ig synthesizing cells, in particular DNA from natural lymphocyte clones.

ACKNOWLEDGMENTS

We thank Dr. M.O.Potter and Dr.M.Cohn for myeloma cells, Dr.R.Curtis for E.coli strain X1776, and Dr.R.Weisberg for λgt-λC WES.

The technical assistance of Mr.G.R.Dastoornikoo, Ms.P.Riegert and Mr.A.Traunecker is highly appreciated. We thank Ms. K. Perret-Thurston for excellent secretarial help.

REFERENCES

1. Cohn, M., Blomberg, B., Geckeler, W., Raschke, W., Riblet, R., and Weigert, M. (1974) In *The Immune System, genes, receptors, signals*. (eds.) E.E. Sercarz, A.R. Williamson and C.F. Fox, p. 89, Academic Press,New York.
2. Dreyer, W.J. and Bennett, J.C. (1965) *Proc. Natl. Acad. Sci., U.S.A.*, 54, 864. 3.
3. Enquist, L., Tiemeier, D., Leder, P., Weisberg, R., Sternberg, N. (1976) *Nature* 259, 596.
4. Hilschmann, N. and Craig, L.C. (1965) *Proc. Natl. Acad. Sci. U.S.A.* 53,1403.
5. Honjo, T., Packman, S., Swan, D., and Leder, P. (1976) *Biochemistry* 15, 2780.
6. Hozumi, N., and Tonegawa, S. (1976) *Proc. Natl. Acad. Sci. U.S.A.* 73, 3628.

7 Mäkelä, O. (1967) *Cold Spring Harbor Symp. Quant. Biol.* 32, 423.
8. McKean, D., Potter, M., and Hood, L. (1973) *Biochemistry* 12, 760.
9. Milstein, C., Brownlee, G.G., Cheng, C.C. Hamlyn, P.H., Proudfoot, N.J., and Rabbitts, T.H. (1976) In *The Immune System.,* (eds.) F. Melchers and K. Rajewsky, p. 75., Springer-Verlag, Heidelberg.
10. Milstein, C. and Svasti, J. (1971) In *Progr. Immunol.* (ed.) E. Amos, p. 33., Academic Press, New York.
11. Morrow, J.F., Cohen, S.N., Chang, A.C.Y., Boyer, H., Goodman, H.M. and Helling, R.B. (1974) In *Proc. Natl. Acad. Sci., U.S.A.*, 71, 1743.
12. Ohno, S. (1974) In *Chromosomes and Cancer.,* (ed.) J. German, p. 77. John Wiley and Sons, New York.
13. Pernis, B., Chiappino, G., Kelus, A.S. and Gell, P.G.H. (1965) *J. Exp. Med.* 122, 853.
14. Rabbitts, T.H., Javis, J.M. and Milstein, C. (1975) *Cell* 6, 5.
15. Raff, M.C., Feldmann, M. and De Petris, S. (1973) *J. Exp. Med.* 137, 1024.
16. Thomas, M., White, R.L. and Davis, R.W. (1976) *Proc. Natl.Acad. Sci.U.S.A.,* 73, 2294.
17. Titani, K., Whitley, E., Avogardo, L, and Putnam, F.W. (1965) *Science* 149, 1090.
18. Tonegawa, S. (1976) *Proc. Natl. Acad. Sci. USA* 73, 203.
19. Tonegawa, S., Bernardini, A., Weimann, B.J. and Steinberg, C. (1974) *FEBS Lett.,40,* 92.
20. Tonegawa, S., Hozumi, N., Matthyssens, G. and Schuller,R. (1976) *Cold Spring Harbor Symp. Quant. Biol.* 41, in press.
21. Tonegawa, S., Steinberg, C., Dube, S. and Bernardini, A. (1974) *Proc. Natl. Acad. Sci. USA* 71, 4027.
22. Warner, T.F.C.S. and Krueger, R.G. (1975) *J. Theor. Biol.* 54,175.
23. Yoshida, T.H., Imai, H.T. and Potter, M. (1968) *J. Nat. Cancer Inst.* 41, 1083.

RECOMBINANT DNA PROBES INTO THE EXPRESSION OF IMMUNOGLOBULIN GENES

Maureen Gilmore-Hebert and Randolph Wall
Dept. of Microbiology and Immunology
and The Molecular Biology Institute
University of California
Los Angeles, California 90024

ABSTRACT

Using a procedure developed for the cloning of poly(A)-containing mRNA, we have established recombinant DNA clones containing MOPC 21 immunoglobulin κ light chain mRNA sequences. Purified κ light chain recombinant plasmid DNA hybridized exclusively with a 13S cytoplasmic RNA species shown to be immunoglobulin κ mRNA. Recombinant plasmid DNA hybridization with nuclear RNA revealed two classes of κ specific nuclear RNA sedimenting, under denaturing conditions, at approximately 34S and 16S. These findings suggest that 13S κ light chain mRNA may be derived from larger nuclear RNA precursors.

INTRODUCTION

Recently, several laboratories have established procedures for the construction of recombinant DNA clones from complementary DNA (cDNA) synthesized on polyadenylic acid-containing messenger RNA (1 – 4). Such recombinant DNA clones containing specific eukaryotic mRNA sequences represent absolutely pure probes for the isolation and mapping of cellular genes and for resolution of the nuclear RNA species and processing events in gene expression. We have constructed recombinant DNA clones from purified, murine myeloma MOPC 21 immunoglobulin kappa (κ) light chain mRNA (Wall, Gilmore-Hebert, Higuchi, Komaromy, Paddock, Salser and Toth; manuscript in preparation). In this report, we present initial results which establish that these cloned κ light chain recombinant DNA provide highly specific hybridization probes for analysis of the events in the biogenesis of immunoglobulin κ light chain mRNA. Using such recombinant cDNA probes, we have detected discrete classes of nuclear RNA containing κ mRNA sequences which, under stringently denaturing conditions, appear to be larger than the cytoplasmic κ light chain mRNA.

TABLE 1

HYBRIDIZATION OF NUCLEAR AND CYTOPLASMIC RNA WITH PLASMID DNA.

RNA SAMPLE	INPUT CPM	CPM HYBRIDIZED TO FILTERS BLANK	pκ4	pMB9	pHb72
P3 nuclear RNA	1.7×10^8	800	85,000	1,000	2,900
P3 cytoplasmic RNA	2×10^8	430	130,000	590	530

Nuclear and Cytoplasmic RNA were isolated from approximately 10^8 P3 cells labelled for 3 hr with 20 mCi ^{3}H-uridine (14, 15). RNA samples were hybridized with nitrocellulose filters containing 20 μg of the various DNA at 65 C for 24 hr in 0.3 M NaCl, 0.001 M EDTA, 0.01 M TES (pH 7.4), 0.2% SDS. The DNA used are: pκ4, an immunoglobulin κ light chain cDNA recombinant clone; pHb72, a rabbit β globin cDNA recombinant clone; pMB9, the plasmid vector used in the cloning of pκ4 and pHb72. Blank filters lacking DNA were included in the hybridizations. Hybridization was scored following digestion with pancreatic and T1 RNAse.

RESULTS

Recombinant DNA clones containing immunoglobulin κ light chain mRNA sequences were prepared using procedures previously established in the generation of recombinant clones from rabbit globin mRNA (2). Basically, cDNA synthesized from purified MOPC 21 κ light chain mRNA by reverse transcriptase, was made double-stranded, further modified by the addition of terminal poly (dA) sequences to facilitate annealing with the plasmid vector, pMB9 (tet^r, 5) containing complementary poly(T) homopolymer "tails". Transformed, tetracycline resistent _E. coli_ colonies were screened by the Grunstein-Hogness colony hybridization procedure (6) to identify recombinant colonies containing κ light chain mRNA sequences. Five κ light chain recombinant clones showing the most definitive hybridization in the colony hybridization assay were selected and have now been extensively characterized. Recombinant clones containing only κ constant region sequences as well as clones containing both κ constant and variable region sequences are represented in this collection. Purified DNA from one of the characterized κ light chain clones, designated pκ4, was used in the studies presented here.

For ease of labelling, the P3 line of MOPC 21 cells

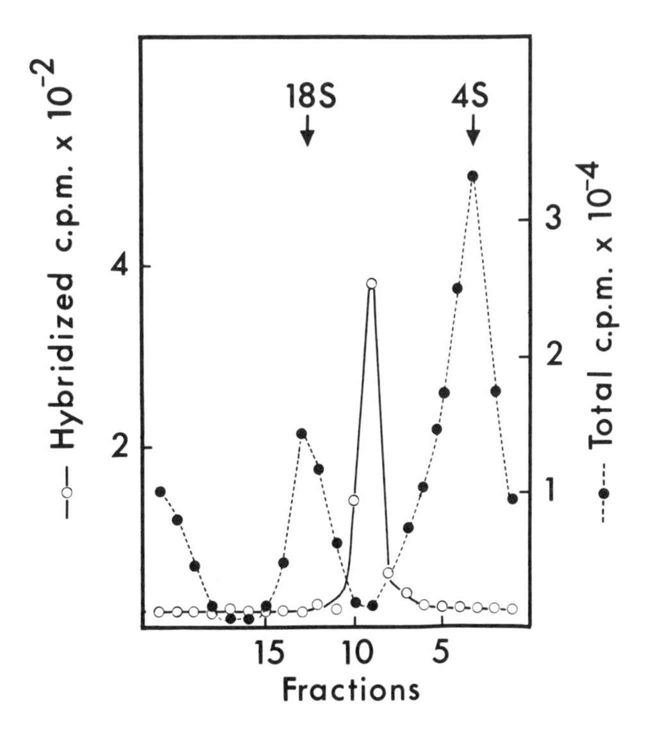

Figure 1. Sucrose gradient analysis of P3 cytoplasmic RNA hybridizing with κ light chain recombinant clone pκ4 DNA.

Total cytoplasmic RNA was prepared from 1 X 10^8 P3 cells labelled for 1 hr with 20 mCi ^{3}H-uridine (15). The RNA sample was denatured with 75% dimethyl sulfoxide (DMSO) at 65 C for 3 min and fractionated on a 15 - 30% sucrose gradient (0.1 M NaCl, 0.01 M Tris-HCl (pH 7.5), 0.001 M EDTA) spun 16.5 hr at 35,000 rpm in a Beckman SW41 rotor. The direction of sedimentation is right to left. Individual gradient fractions were exhaustively hybridized to nitrocellulose filters containing 20 μg of denatured pκ4 DNA as described in Table 1. The hybridization results shown are corrected for binding to blank filters without DNA.

---

adapted to continuous culture was used in these studies (7). The results in Table 1 show that both P3 total nuclear and cytoplasmic RNA exhibit significant hybridization with pκ4 DNA, while the binding of P3 RNA to pMB9 plasmid vector DNA, or to a rabbit β globin mRNA recombinant clone (pHb72) also constructed using the (dA:T) homopolymer annealing procedure (2), is not significantly greater than the binding to control filters lacking DNA. The control using the rabbit β globin recombinant plasmid DNA establishes that the P3 RNA hybridi-

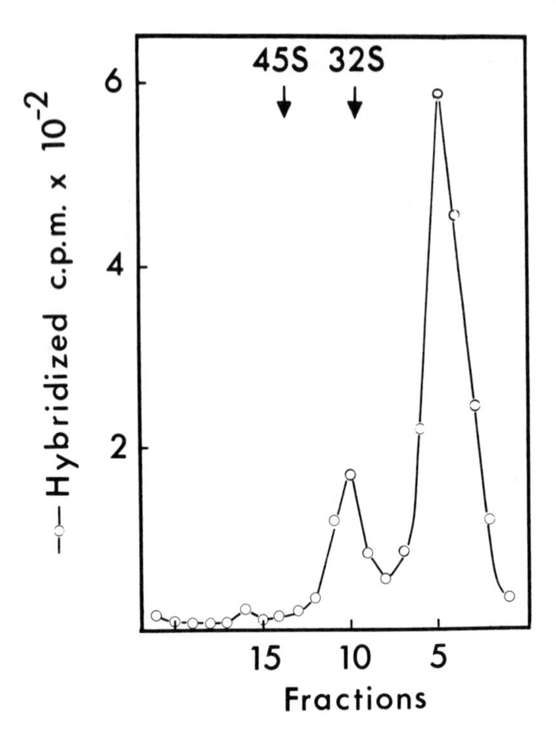

Figure 2. Sucrose gradient analysis of P3 nuclear RNA hybridizing with κ light chain recombinant clone pκ4 DNA.

Nuclear RNA from 1 X 10^8 P3 cells labelled for 1 hr with 20 mCi ^{3}H-uridine was denatured in DMSO and fractionated on sucrose gradients as described in Figure 1 except centrifugation was for 4 hr at 40,000 rpm. Gradient fractions were hybridized with pκ4 DNA filters as noted in Figure 1.

---

zation with pκ4 DNA is not mediated through the poly(T) sequences present in recombinant plasmids constructed by the (dA:T) homopolymer annealing procedure.

When κ recombinant plasmid pκ4 DNA is hybridized with the isolated sucrose gradient fractions of labelled P3 total cytoplasmic RNA, hybridization occurs exclusively with RNA sedimenting at 13S. This corresponds to the published value for immunoglobulin κ light chain mRNA (8 - 11). Further evidence for the identity of this 13S RNA species as κ light chain mRNA comes from results (to be presented elsewhere) which indicate that the 13S RNA isolated by hybrid selection and elution from pκ4 DNA-sepharose columns is efficiently translated in the cell-free wheat germ translation system to produce MOPC 21 light chain precursor (9, 10, 12, 13).

When recombinant pκ4 plasmid DNA is similarly hybridized with sucrose gradient fractions of nuclear RNA from P3 cells labelled for 60 min, two well defined classes of κ specific nuclear RNA are detected (Figure 2). The more rapidly sedimenting species appears to be approximately 34S (calculated from the 32S and 45S cellular pre-rRNA species whose positions are denoted by arrows in Figure 2), while the remaining class of κ specific nuclear RNA sediments broadly with a peak at about 16S. The more slowly sedimenting class of κ specific nuclear RNA may represent more than one RNA species, possibly including 13S κ mRNA sized molecules. These two classes of κ specific nuclear RNA larger than 13S κ mRNA are not aggregated RNA complexes because they were analyzed under stringent denaturing conditions which eliminate such complexes (14). The more rapidly sedimenting class of κ specific nuclear RNA detected under these conditions calculates to be some 6 - 7 X larger than cytoplasmic 13S κ light chain mRNA.

DISCUSSION

These findings suggest that κ light chain mRNA may be derived from a much larger nuclear RNA precursor in stages possibly involving discrete intermediate RNA species. Pulse labelling experiments using very brief periods (1 - 10 min incorporation), as well as structural studies on the nature of the sequences in the κ specific classes of nuclear RNA, are now underway to resolve the events in the biogenesis of κ mRNA. Such studies are made feasible by the availability of pure recombinant DNA probes containing immunoglobulin κ light chain mRNA sequences.

BIOHAZARD CONSIDERATIONS

The cloned recombinant kappa light chain DNA probes used in the studies reported here were constructed from purified MOPC 21 kappa light chain mRNA in a single cloning experiment carried out in December, 1975. This cloning experiment was conducted in full compliance with the Asilomar Guidelines then in effect, using biocontainment conditions designated P3, EK1 under the NIH Guidelines for Research Involving Recombinant DNA Molecules issued in June, 1976. On the basis of our subsequent characterization of these immunoglobulin κ light chain DNA clones, the NIH Recombinant DNA Molecule Program Advisory Committee approved the characterized light chain clones for P3, EK1 levels of biocontainment in January, 1977.

ACKNOWLEDGMENTS

These studies were supported by NIH Grant, CA 12800. Maureen Gilmore-Hebert was the recipient of an NIH Postdoctoral Fellowship. We thank Ms. Kathleen Toth for excellent technical assistance.

REFERENCES

1. Rougeon, F., Kourilsky, P., Mach, B. (1975) Nucleic Acids Res. 2, 2365.
2. Higuchi, R., Paddock, G.V., Wall, R. and Salser, W. (1976) Proc. Nat. Acad. Sci. USA 73, 3146.
3. Maniatis, T., Kee, S.G., Efstratiadis, A. and Kafatos, F. (1976) Cell 8, 164.
4. Rabbits, T.H. (1976) Nature 260, 221.
5. Rodriquez, R.L., Bolivar, F., Goodman, H., Boyer, H.W. and Betlach, M. (1976). ICN-UCLA Symposia on Molecular and Cellular Biology, Vol. 5, p. 471, Nierlich, D.P., Rutter, W.J. and Fox, C.F. (eds.), Academic Press, New York.
6. Grunstein, M. and Hogness, D. (1975) Proc. Nat. Acad. Sci. USA 72, 3961.
7. Horibata, R. and Harris, A.W. (1970) Exptl. Cell Res. 60, 61.
8. Swan, D., Aviv, H. and Leder, P. (1972) Proc. Nat. Acad. Sci. USA 69, 1967.
9. Tonegawa, S. and Baldi, I. (1973) Biochem. Biophys. Res. Commun. 51, 81.
10. Mach, B., Faust, C. and Vassali, P. (1973) Proc. Nat. Acad. Sci. USA 70, 451.
11. Brownlee, G.G., Cartwright, E.M., Cowan, N.J., Jarvis, J.M. and Milstein, C.(1973) Nature New Biol. 244, 236.
12. Milstein, C., Brownlee, G.G., Harrison, T.M. and Mathews, M.B. (1972) Nature New Biol. 239, 117.
13. Schmeckpeper, B.J., Cory, S. and Adams, J.M. (1974) Mol. Biol. Reports 1, 355.
14. Fedoroff, N., Wellauer, P.K. and Wall, R. (1977) Cell (in press).
15. Penman, S. (1969) Fundamental Techniques in Virology, p. 35, Habel, K. and Salzman, N. (eds.), Academic Press, New York.

STUDIES ON CONTROL OF IMMUNOGLOBULIN BIOSYNTHESIS

Williamson, A.R., Bennett, J.N., Fitzmaurice, L.C., Laidlaw, S.A., Mosmann, T.R., Schuch, W. Singer, H.H. and Singer, P.A.

Department of Biochemistry, University of Glasgow, Glasgow G12 8QQ, Scotland.

ABSTRACT. The methods being used to study immunoglobulin biosynthesis are summarised. 1. Messenger RNA coding for L and H chains can be isolated as a suitable starting point for molecular cloning. 2. Restriction endonuclease analysis of ribosomal DNA from myeloma tumour cells reveals fragments characteristic of the particular cell line. 3. The efficiency of isolation of hybrid cell lines formed by fusion of two myeloma cell lines has been greatly increased. 4. A hybrid cell line has been used to analyse a possible defect in L chain secretion. 5. Analysis of the products of the myeloma line SAMM368 suggests that it may have arisen by a spontaneous fusion of two immunoglobulin producing cells *in vivo*. 6. The surface IgM of the Burkitt lymphoma line DAUDI can be isolated linked to another surface protein

INTRODUCTION

Our aim is an understanding of the arrangement of genes coding for antibodies and the control of expression of those genes. To those ends we are studying methods for:
1. isolation and purification of immunoglobulin (Ig) mRNA; 2. cell-free translation of Ig mRNA; 3. use of restriction endonucleases in analysis of genome DNA from Ig-producing cells; 4. characterising the synthetic capacities of cultured Ig-producing cells; 5. isolation of hybrid lines obtained by fusion of such cells. These are the starting points for determining the genetic basis of the various controls on Ig synthesis.

mRNA FROM MOUSE MYELOMA AND HUMAN LYMPHOID CELLS

Isolation of mRNA from immunoglobulin producing cells is a starting point for studies of the translation of specific mRNA *in vitro* and for the purification of specific mRNA molecules. Pure nucleic acid probes for complete and partial sequences of a variety of H and L chains are needed for direct analyses of immunoglobulin genes and their expression. Molecular cloning is the method of choice, and has already proven successful for complete globin sequence (1,2) and immunoglobulin L chain sequences (3). For the preparation of cDNA to be inserted into bacterial plasmids, preservation of intact immunoglobulin mRNA is of prime importance; physical purification of specific mRNA need only be partial since cloning affords purification. Selection of isolated clones by hybridization procedures (4) requires that small amounts of more highly purified specific mRNA be prepared. There have been several reports of the isolation of mRNA from myeloma tumours with preservation of biologically active mRNA-L and in certain cases considerable purification of mRNA-L has been achieved (e.g. 5-8). Isolation of mRNA-H is apparently more difficult but preservation of activity and partial purification of mRNA-H have been reported (9,10). We are currently isolating mRNA from the polyribosomes of mouse myeloma cells and preparing cDNA from that mRNA for cloning in plasmids. Translation of mouse myeloma cell mRNA and mRNA prepared by similar methods from human lymphoid cells is being studied in cell-free systems.

We have obtained good yields of total polyribosomes by lysis of cells with the detergents DOC and Triton X-100 in the presence of cycloheximide which prevents run-off of ribosomes from mRNA during isolation (11) and heparin as a nuclease inhibitor. Polyribosomal RNA was then fractionated using oligo dT cellulose or poly U sepharose to yield poly A-containing mRNA. Profiles of isolated RNA are compared by analytical polyacrylamide gel electrophoresis (PAGE) in 99% formamide (12). By this sensitive criterion ribosomal RNA (rRNA) can be shown to be intact, while poly A-containing mRNA displays a heterogeneous profile with sizes ranging from about 10S to > 28S. Preparative fractionation of mRNA is being carried out on sucrose density gradients (SDG) and by formamide-PAGE. Individual fractions of mRNA from both methods are assayed by cell free translation. Formamide-PAGE affords much higher resolution and hence better purification of individual mRNA molecules than does fractionation on SDG, although the latter method presently affords better recovery of RNA.

CELL-FREE TRANSLATION PRODUCTS OF mRNA

Translational capacity of mRNA is assayed in cell-free systems prepared either from a wheatgerm (WG) extract (13,14) or from messenger RNA dependent reticulocyte lysate (MDL) (15). Analysis by SDS-PAGE of the products of translation of mRNA in the MDL system showed a good correspondence of band pattern with the products labelled in the cultured cells from which the mRNA was extracted. These patterns are good evidence for the general preservation and translational capacity of isolated mRNA. The principal exceptions to the correspondence of the patterns were the L and H chain bands. For each of the cells examined (5563, MOPC70, LPC1, P1, 1788) the products of cell free synthesis putatively identified as immunoglobulin chains by the criteria of immune precipitation and SDS-PAGE analysis differed significantly from the immunoglobulin chains synthesised *in vivo*. In each case the L chain band was replaced by a band of apparently higher molecular weight. That this band is a precursor of L chain, similar to the pre L chains described for other myeloma proteins (5-8,19), is suggested by the fact that this component is precipitable by anti-immunoglobulin antisera. Identification of the H chains synthesised *in vivo* has proved more complicated. The products in the MDL system include polypeptides of apparent molecular weight (by SDS-PAGE) similar to H chains homologous to the source of mRNA. Immune precipitation reveals putative H chains of apparent molecular weight lower than or similar to homologous H chains. If H chains are synthesised with a precursor sequence, as might be expected by comparison with L chains, then the increase in molecular weight of the cell-free product might be countered by the failure to add carbohydrate side chains to the polypeptide synthesised in the cell-free system.

The products of translation of total mRNA from a given cell line in the wheatgerm system failed to show a general correspondence with the pattern of polypeptides synthesised by those cells in culture. The complex pattern of products in the WG system is apparently due to the generation of many incomplete polypeptide chains. We cound find no evidence that these incomplete chains are the result of proteolytic action in the WG system. There was however evidence of nuclease activity in the system. Nuclease activity in the WG system was assayed by formamide-PAGE analysis of ribosomal RNA incubated in the system. The pattern of degradation was consistent with a mixture of exonuclease and endonuclease activity. Rat liver ribonuclease inhibitor did not block the activity appreciably. Despite these problems translation of mRNA-L to yield pre-L chains similar in size to those observed in the MDL system occurs efficiently in the WG system. A polypeptide similar in size to H chain is also synthesised in the WG system. However, specific immune precipitation reveals a spectrum of polypeptide chains intermediate in size between H and L chains as well as polypeptides shorter than L chain. It has been suggested that H chain translation is less efficient than that of L chain in the WG system and that this might reflect specificity of chain initiation (16). The patterns of incomplete immunoglobulin chains, the larger of which appear to be H chain fragments, caution that efficiency of initiation can not be directly inferred from estimates of only complete products of translation in the WG system.

RESTRICTION ENDONUCLEASE ANALYSIS OF DNA CODING FOR RIBOSOMAL RNA IN IMMUNOGLOBULIN PRODUCING CELLS

Restriction endonuclease digestion of complex DNA followed by electrophoretic fractionation of the restriction fragments and identification of specific fragments by hybridisation procedures has been described recently (17). The first results of the application of similar procedures to the analysis of immunoglobulin genes (18) are consistent with the evidence for the inheritance of separate V and C genes which become joined in an Ig producing cell. At present the restriction maps of Ig specific DNA sequences can not be unambiguously interpreted in terms of a mechanism for the joining of V and C genes. The use of pure nucleic acid probes in hybridisation analysis and the study of DNA from cells with distinct Ig producing capacities are required. While such systems were under development we have studied the restriction mapping of ribosomal DNA sequences in the DNA of murine myeloma cells as a model system.

High molecular weight DNA (over 150 kilobase pairs as judged by analytical centrifugation and agarose gel electrophoresis) was extracted from mouse myeloma tissue culture cells or solid tumours, digested with restriction endonucleases, and separated on 0.8% agarose gels. DNA was then denatured and neutralised in the gels, transferred to membrane filters by diffusion or electrophoresis and rDNA was detected by autoradiography after hybridisation to (^{32}P) or (^{125}I) labelled rRNA.

For a given BALB/c myeloma cell line the pattern of rDNA restriction fragments is specific for the restriction endonuclease used. However, the fragments produced by one enzyme are not identical from cell line to cell line.

Minor heterogeneity was observed in Hind III digests of 47A, LPC1 and 104E DNA. Major differences were observed in Eco RI digests of DNA from 47A, 104E and P1 and LPC1 cell lines.

Previously minor heterogeneity in repeat length of rDNA of various sources has been demonstrated (17). These

differences arise out of length differences of the non-transcribed spacer DNAs. The major differences in the Eco RI digests might have three explanations:
a) they might be artifactual; however, the different specific banding patterns are observed reproducibly,
b) they might be due to mutations at the endonuclease recognition sites,
c) they might be due to different methylation patterns in different myeloma DNA's.

In cases b) and c) these changes must have been established throughout the whole family of repeated genes. Which of these possibilities might prevail is unknown.

However, it suggests that caution is needed in analysing multigene families and their restriction patterns.

Ig PRODUCTION IN CULTURED CELL LINES

Isolation of hybrid cell lines. The control of synthesis of immunoglobulins by a single cell appears to be complex, since several discrete patterns of immunoglobulin synthesis can be recognised. Some of these patterns are also found in various *in vitro* cell lines, which should allow investigation of the nature of the controls by biochemical and genetic techniques. Cell fusion is a useful tool for analysing genetic controls in such cell lines, and we have been investigating the requirements for cell fusion of myeloma cells. Other workers have reported (20,21) that mouse myeloma cells show a low spontaneous rate of fusion (e.g. 1 hybrid cell recovered from 10^6 parent cells) and Sendai virus has little or no effect on this low rate. Our early experiments confirmed this difficulty, and in addition, experiments using polyethylene glycol did not result in isolation of viable hybrid cell lines, even when obvious cytoplasmic fusion could be observed microscopically.

We then examined the possibility that although hybrid cells were being formed, they could not grow under the conditions used. Both parental lines used in this work had no minimum cell density requirement for growth, but this behaviour is unusual amongst myeloma cell lines newly established in tissue culture. Either a high cell density or the presence of a low amount of whole mouse blood (22) would supply the feeder effect required for most lines, and so these two factors were investigated in the recovery of hybrid cells. Using either mouse blood (1% final concentration) or a continuous line of human lymphoblast cells to supply feeder functions, we found that hybrid cell lines could be obtained at high frequency even without using a fusion agent such as Sendai virus or polyethylene glycol. Spontaneous fusion frequencies ranged from 300 to 1600 viable hybrid clones per 10^6 parent cells, which represents an increase of more than 100-fold over previous reported values. All hybrids so far isolated by this method are unable to grow unassisted at low cell densities, confirming the necessity of providing feeder functions during the initial selection and until the hybrid clone is grown to sufficient numbers. Reconstruction experiments with known numbers of hybrid cells showed that the efficiency of isolation of such hybrids was approximately 30%. Other cell lines are now being examined to determine whether or not these growth properties are a general property of hybrid myeloma cell lines.

The secretion of immunoglobulin is being examined in a cell line derived from MOPC315. A variant cell line was selected whichsecreted only light chains, and then a further selection was carried out to obtain a cell line synthesising but not secreting λ_2 light chains. The secretion defect in this cell line was examined by cell fusion. Since the defective cell line synthesised λ_2 light chains, a second cell line was constructed which synthesised and secreted κ light chains, and was resistant to bromodeoxyuridine, and unable to grow in hypoxanthine-aminopterin-thymidine(HAT) medium. The λ_2 non-secreting cell line was resistant to 6-thioguanine and unable to grow in HAT medium. Fusion and selection were carried out as described above, using HAT medium in which only a hybrid cell line would be able to grow. A hybrid clone was isolated and the synthesis of light chains examined by a pulse-chase experiment. The hybrid clone synthesised both κ and λ_2 light chains, but secreted only κ chains. This strongly suggests that the λ_2 light chain in the variant cell line has a structural alteration which prevents it from being secreted. A less likely possibility is that separate κ and λ_2 light chain secretion mechanisms exist, and that the λ_2 chain mechanism is not operative in either the κ-secreting or λ non-secreting cell line.

Cells producing two classes of Ig. A spontaneously-occurring double-producing myeloma cell line is being examined and compared with hybrid lines. SAMM368 has been reported to produce γ_{2b}, α and κ chains, but the α and γ_{2b} chains apparently may not share the same idiotype (22). We have established SAMM368 in tissue culture and isolated several clones. All clones tested continued to synthesise both α and γ_{2b} heavy chains, as shown by biosynthetic incorporation of radioactivity, immune precipitation and analysis on three different electrophoretic systems. In addition, two closely-migrating light chain bands were seen. Although we have not yet ruled out post-translational modification as a cause of these two bands, our results so far are consistent with the hypothesis that SAMM368 is a natural fusion product of two myelomas or a myeloma and a normal plasma cell, one synthesising α and κ chains, and the other synthesising γ_{2b} and a separate κ chain. Such a fusion event is not unlikely, given the high spontaneous rates of fusion described above, and the high incidence of plasmacytomas induced in BALB/c mice.

Surface immunoglobulin. To study the differences in synthesis of Ig that determine whether it will be incorporated into the plasma membrane or will be secreted, we are comparing various human Ig-producing cell lines. The Burkitt lymphoma line DAUDI is a good model for surface IgM production (24). Several IgM secreting lymphoid cell lines are being used as comparisons. Analysis by SDS-PAGE showed the chain of DAUDI IgM to have a higher apparent molecular weight than that found in secreted IgM from the lines 1788, BEC-11 and Tay. Molecular weight differences could be due to either the size of the polypeptide chain or to the amount of carbohydrate added. Cell-free synthesis is being used to look directly at the polypeptide chains.

Study of DAUDI surface IgM has also shown that it can be isolated as a complex in disulphide linkage with a protein of approximately 30,000 molecular weight. This complex could be instrumental in adapting IgM to functioning as a surface receptor.

REFERENCES

1. Rougeon, F., Kourilsky, P. and Mach, B. (1975) *Nucleic Acids Res.* 2, 2365.
2. Maniatis, T., Gek Kee, S., Efstratiadis, A. and Kafatos, F.C. (1976) *Cell* 8, 163.
3. Mach, B., Rougeon, F., Aellen, M-F., Farace, M-G. and Vassalli, P. (1976) *Ann. Immunol.* 127C, 419.
4. Grunstein, M. and Hogness, D.S. (1975) *Proc. Nat. Acad. Sci. USA* 72, 3961.
5. Swan, D., Aviv, M. and Leder, P. (1972) *Proc. Nat. Acad. Sci. USA* 69, 1967
6. Milstein, C., Brownlee, G.G., Harrison, T.M. and Mathews, M.B. (1972) *Nature New Biol.* 239, 117.

7. Mach, B., Faust, C. and Vassali, P. (1973) *Proc.Nat. Acad. Sci. USA* 70, 451.
8. Tonegawa, S., Steinberg, C., Dube, S. and Bernadini, A. (1974) *Proc. Nat. Acad. Sci. USA* 71, 4027.
9. Bernadini, A. and Tonegawa, S. (1974) *FEBS Lett.* 41, 73.
10. Cowan, N.J., Secher, D.S. and Milstein, C. (1976) *Eur. J. Biochem.* 61, 355.
11. Buckingham, M.E., Caput, D., Cohen, A., Whalen, R.G. and Gros, F. (1974) *Proc. Nat. Acad. Sci. USA* 71, 1466.
12. Duesberg, P.H. and Vogt, P.K. (1973) *J. Virol.* 12, 594.
13. Roberts, B.E. and Paterson, N.M. (1973) *Proc. Nat. Acad. Sci. USA* 70, 2330.
14. Marcu, K. and Dudock, B. (1974) *Nucleic Acids Res.* 1, 1385.
15. Pelham, H.R.B. and Jackson, R.J. (1976) *Eur. J. Biochem.* 67, 247.
16. Sonensheim, G.E. and Brawerman, G. (1976) *Biochemistry* 15, 5501
17. Southern, E.M. (1975) *J. Mol. Biol.* 98, 503.
18. Hozumi, N. and Tonegawa, S. (1976) *Proc. Nat. Acad. Sci. USA* 73, 3628.
19. Schechter, I. (1973) *Proc. Nat. Acad. Sci. USA* 70,2256.
20. Kohler, G. and Milstein, C. (1975) *Nature* 256, 495.
21. Margulies, D.H., Keuhl, W.M. and Scharff, M.D. (1976)
22. Metcalf, D. (1973) *J. Cell Physiol.* 81, 397.
23. Morse, H.C., Pumphrey, J.G., Potter, M. and Asofsky, R. (1976) *J. Immunol.* 117, 541.
24, Klein, E., Klein, G., Nadkarni, J.S., Nadkarni, J.J., Wigzell, H.,and Clifford, P. (1967) *Lancet ii*, 1068.

SYNTHESIS OF KAPPA CHAINS FROM THYMUS RNA BY CELL FREE TRANSLATION

David Putnam*, Ursula Storb* and James Clagett*†

Department of Microbiology and Immunology*
and Department of Periodontics†
University of Washington, Seattle, Washington 98195

ABSTRACT. Cell-free translation of murine thymocyte poly A RNA and immunoprecipitation of the products with anti-Kappa mouse light chain antibody has revealed that T cell mRNA bears the coding capacity for immunoglobulin light chain. Whole cell mRNA was prepared from both mouse thymus and spleen, as well as Kappa producing myelomas. Wheat germ extract translation of these RNA's was monitored by ^{35}S-methionine incorporation. The products were immunoprecipitated to specifically isolate Kappa chain, and the precipitates resolved by polyacrylamide slab gel electrophoresis. Fluorographic detection of labeled protein indicated that the thymic mRNA directed the translation of a heterogeneous population of complete intact Kappa chains similar in size distribution to that of spleen cells. The possibility that this result is due to B cell contamination is extremely remote. Immunofluorescence checks on each sample at the time of thymocyte cell preparation reveal 99.8% theta positive cells, and $\leq$ 0.2% B cells or 0.05% plasma cells.

INTRODUCTION

The identity of the T cell antigen receptor still remains controversial. There is increasing evidence that the putative receptor molecule has antibody like characteristics (1,3,4,5,7); but in general skepticism abounds primarily for two reasons: 1. The findings have not been easily or widely reproduced 2. There is not convincing demonstration that the T cell itself synthesizes the detected moiety.

To approach the latter question, we chose to determine whether T cells have the potential capacity to synthesize an antibody molecule by examining the T cell mRNA pool for immunoglobulin coding sequences. That such RNAs are present for at least the light chain portion was first demonstrated by nucleic acid hybridization using Kappa cDNA probes (12, 13). The next step was to ascertain if these sequences are translatable. Evidence is presented herein which confirms that thymic mRNA will translate in a cell free system a significant amount of complete Kappa light chains.

MATERIALS AND METHODS

RNA PREPARATION

Thymi were removed from 4-6 week old Swiss or Balb/c mice and a single cell suspension prepared by pressing through nylon screen. After pelleting, a representative aliquot of cells was removed for viability and immunofluorescence staining. Whole cell RNA was prepared from the remaining cell pellet using 7 M guanidine hydrochloride (GuHCl) by adaption of the Zsindeley et al. method (15). Splenic whole cell RNA was prepared by the direct homogenization of intact spleens in 7 M GuHCl.

Kappa positive and negative control RNA's were prepared by phenol extraction from a variety of myeloma and lymphoma cell lines. All RNA's were fractionated for poly A RNA on Oligo dT cellulose and translated by cell free wheat germ extract as previously described (11).

IMMUNOPRECIPITATION

Translational products were prepared and immunoprecipitated according to Palmiter's method (9). The anti-mouse Kappa light chain antiserum was produced in New Zealand rabbits using DEAE purified MOPC-41 Bence-Jones protein (10). Direct immunoprecipitation was achieved using the same purified Kappa chain as carrier antigen (11).

PAGE AND FLUOROGRAPHY

Total products and immunoprecipitates were analyzed by SDS-15% polyacrylamide gel electrophoresis (PAGE) by the Laemmli method (6) using a Biorad Model 220 slab gel apparatus. Stained gels were DMSO-PPO impregnated for fluorography (2) and exposed to Kodak X-Omat R X-ray film at -70°C.

IMMUNOFLUORESCENCE

Thymic cell samples were screened for the numbers of cells bearing surface theta (T cells) or Ig markers (B cells) and cytoplasmic Ig (plasma cells) as described previously (12).

RESULTS

Figure 1 illustrates that the anti-Kappa antibody employed in this direct immunoprecipitation scheme provides a clean and highly specific assay when resolved by PAGE and fluorography. The antibody demonstrates specific cross reactivity with all Kappa producing myelomas tested thus far, as well as with serum, and spleen cell products from a variety of different mouse strains (data not shown). Moreover, the antibody is reactive with not only the secreted (sec) and cytoplasmic (cyto) ^{35}S-met labeled products of these lines, but with their mRNA directed light chain precursor translational products (trans) as well. No binding activity has been detected with the products of Kappa negative controls: mastocytoma P-815 (not shown), lambda producer S-178A (Fig. 1 and 2), leukemia L-1210 (Fig. 2), and yeast (Fig. 3) or endogenous wheat germ (Fig. 2 and 3) translates.

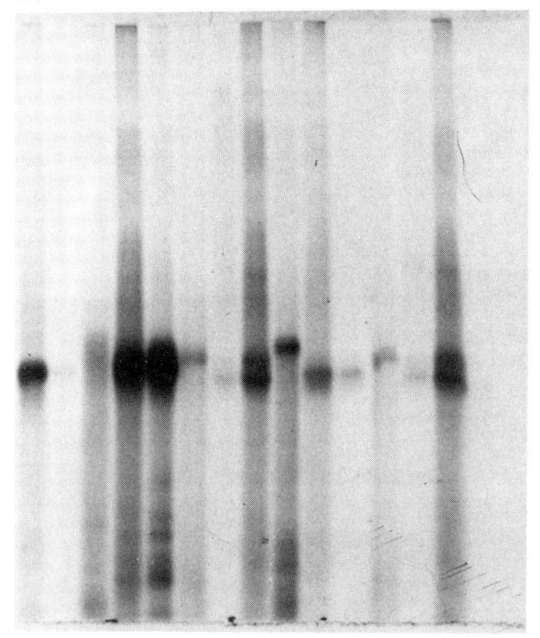

Fig. 1. Anti-Kappa immunoprecipitation of myeloma light chains: Messenger RNA from myelomas P3K, MOPC 45.6, MCPC 774, MOPC 41 and S178A was translated cell free and the precursor light chains (trans) compared with the respective secretory (sec) or cytoplasmic (cyto) products of internally labeled cells. S178A trans is presented in Fig. 2.

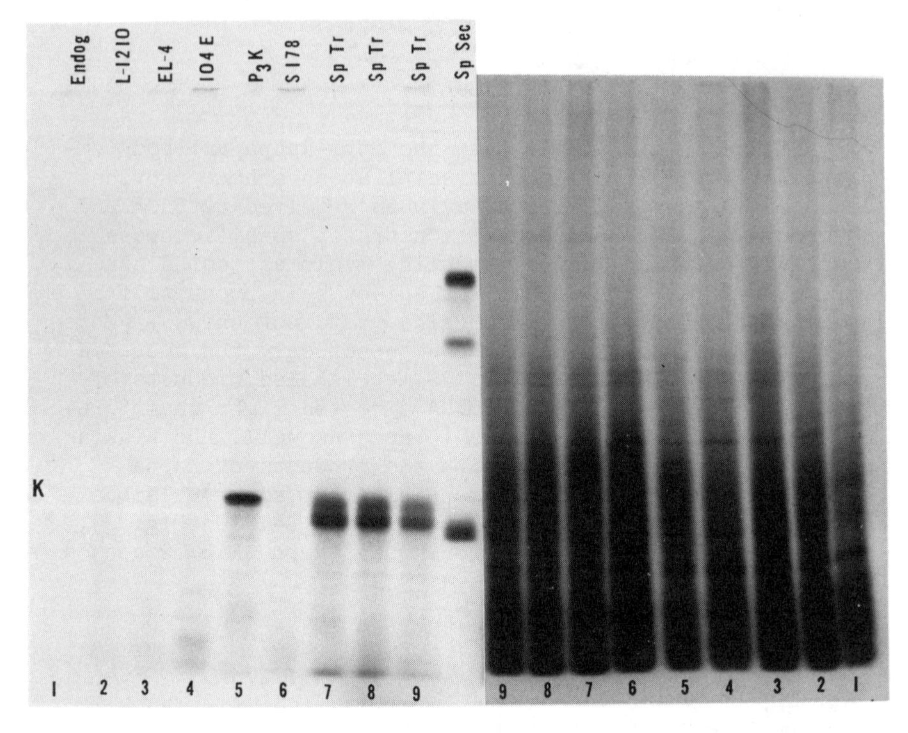

Fig. 2. Total translational products (right) and respective anti-Kappa immunoprecipitated samples (left) of Swiss spleen cell (Sp Tr) mRNA are compared to positive and negative control translates.

Spleen mRNA codes for a heterogeneous population of translatable precursor Kappa chains (Sp Tr) as demonstrated in Figure 2. These molecules are slightly larger than the secreted spleen cell products labeled metabolically in vitro (Sp Sec). The Kappa producing P_3K myeloma mRNA translates a precursor larger than secreted molecules, yet falling in the range encompassed by the spleen precursor pool, as expected. A small amount of immunoprecipitable Kappa is perhaps detectable in the 104E sample, but this lambda secreting myeloma also contains Kappa RNA (8, 13). It is not known if the faint bands appearing in the region smaller than secreted Kappa represent specifically immunoprecipitated degraded product or incompletely translated chains, or unspecific label precipitation.

Immunoprecipitation of thymus mRNA translate (Figure 3) reveal Kappa chain precursor products with the same heterogeneous size distribution as that of the spleen. As noted before, these molecules are larger than secreted Kappa chain, yet encompass the region wherein MCPC 774 mRNA translated precursor migrates. Negative control translates of endogenous wheat germ mRNA (End TP), yeast mRNA (Yea TP) and translates of spleen RNA which is larger than 18S (Sp h TP) do not yield detectable products in this region upon immunoprecipitation. As in Figure 2 the specificity of the immunoprecipitation is supported by the fact that it is not possible to delineate the Kappa bands in the total translational products, whereas other more predominant bands which are readily visible, are not present in the precipitate.

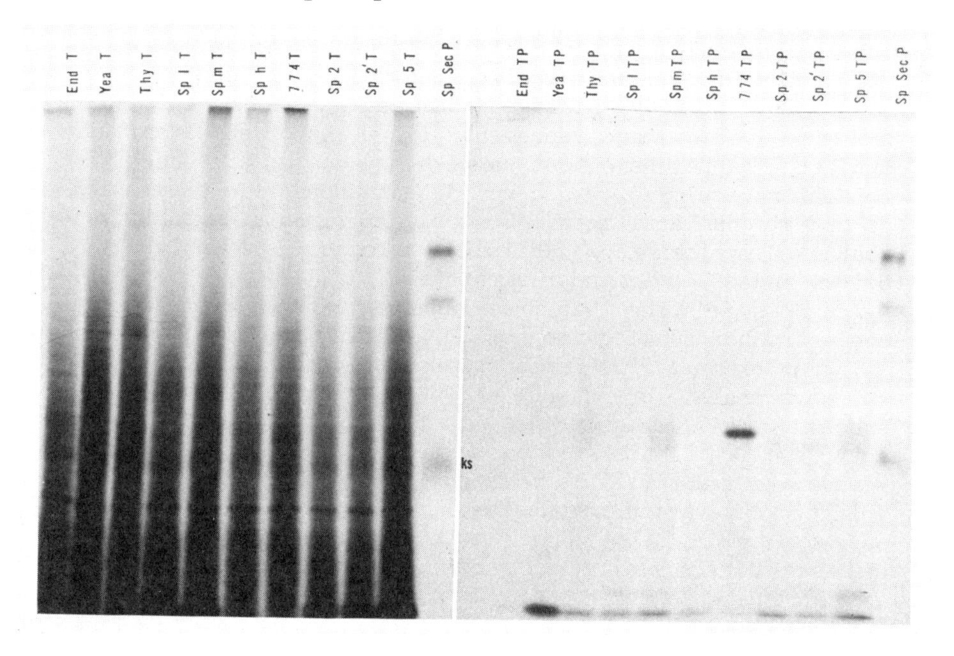

Fig. 3. Total translation (left) and immunoprecipitation (right) of samples from thymus and spleen are compared relative to controls, similar to Fig. 2. Immunoprecipitated splenic secretory Kappa (Sp Sec P) serves as a marker of mature light chains (Ks). MCPC 774 mRNA translate immunoprecipitate (774 TP) is the precursor control. Thymus mRNA translate immunoprecipitate (Thy TP) exhibits Kappa chain precursors with a size distribution similar to spleen samples (Sp L TP from spleen mRNA ≤ 9S; Sp m TP from spleen mRNA 10-18S; Sp 2, 2' and 5 TP are spleen RNAs translated under varied cell-free conditions).

DISCUSSION

We feel that the immunoprecipitation assay employed is a highly specific and clean system for the following reaons: 1. The antibody is specific for Kappa light chains, either in the free secreted monomeric state, in association with heavy chain, or as precursor molecules. 2. The direct immunoprecipitation reaction has a double specificity control, i.e. the specificity of the anti-Kappa antibody, and the purity of the cold carrier antigen, Kappa. 3. The immunoprecipitation is carried out in the presence of weak detergent to prevent nonspecific protein interactions, and the complexes are isolated by centrifugation through 1 M sucrose, 10 mM EDTA and detergent to prevent trapping of label. Typically, nonspecific background binding constitutes < 0.01-.1% of negative control label input.

In our opinion, slab gel PAGE and autoradiography is the most critical and representative means of simultaneously analyzing Kappa chain test samples and controls. Under the conditions employed, we do not detect unspecific separation or migrational artifacts. Control experiments demonstrate that free myeloma kappa chains electrophorese identically before and after immunoprecipitation (12, 14). We have no evidence suggesting that free label or labeled proteins coelectrophorese with immunoprecipitating proteins.

For example, the Kappa chain doublet pattern of the spleen translate precursors shows no specific tendency to proceed, trail, or coincide with the migration of the rabbit antibody light chains or the mouse Kappa chain carrier. More significantly, this evidence for accurate electrophoretic separation is substantiated by the fact that the secreted spleen Kappa chains exhibit a similar doublet pattern, but in a different lower molecular weight region, as would be expected for the mature processed product (Fig. 2 and 3).

Therefore, we conclude that immunoprecipitated material prepared from the thymus mRNA translation products which electrophoreses in the light chain region is a heterogenous population of Kappa chain molecules and not an artifact. This finding corroborates the previous report that thymocyte RNA contains Kappa chain sequences as detected by cDNA hybridization (12, 13). Furthermore, it indicates that these sequences are potentially translatable and can code for complete intact Kappa chains, similar to a spleen cell population.

A possible criticism of the method could be that the initial thymocyte preparation is contaminated by infiltrating B cells. To minimize this, only young healthy mice are used before thymic atrophy occurs and infiltration is still low. Such thymi are large and well defined enabling clean extraction. Immunofluorescence analysis of pooled thymocyte cell suspensions typically show 99% theta positive cells, $\leq$ 0.2% surface immunoglobulin positive B cells, and $\leq$ 0.05% plasma cells. We do not believe that these few B lineage cells could possibly contain enough Kappa mRNA to account for the results in the thymus relative to the spleen, considering the large numbers of B cells in the spleen.

ACKNOWLEDGEMENTS

This work was supported by grants from the NIH 5T01 CA-05040, AI 10685, and DE 02600. We wish to thank Lisa Hager and Fred Farin for technical help, and General Mills for wheat germ samples.

REFERENCES

1. Binz, H., and Wigzell, H. (1976). *Scand. J. Immunol.* 5, 559.
2. Bonner, W. M., and Laskey, R. A. (1974) *Eur. J. Biochem.* 46, 83.
3. Cone, R. E., and Brown, W. C. (1976) *Immunochemistry* 13, 571.
4. Eichmann, K., and Rajewsky, K. (1975) *Eur. J. Immunol.* 5, 661.
5. Hammerling, U., Pickel, H. G., Mack, C. and Masters, D. (1976) *Immunochemistry* 13, 533.
6. Laemmli, U. K. (1970) *Nature* 227, 680.
7. Marchalonis, J. J. (1975) *Science* 190, 20.
8. Ono, M., Kawakani, M., Kataoka, T., and Honjo, T. (1977) *Biochem. Biophys. Res. Comm.* 74, 796.
9. Palmiter, R. D. (1973) *J. Biol. Chem.* 248, 2095.
10. Potter, M. (1967) *Methods Cancer Res.* 2, 105.
11. Storb, U., and Marvin, S. (1976) *J. Immunol.* 117, 259.
12. Storb, U., Hager, L., Putnam, D., Buck, L., Farin, F., and Clagett, J. (1976) *Proc. Natl. Acad. Sci. USA* 73, 2467.
13. Storb, U., Hager, L., Wilson, R., and Putnam, D. (1977) submitted.
14. Wilson, R., Putnam, D., and Storb, U. (1977) submitted.
15. Zsindeley, A., Hutal, J., and Tanko, B. (1970) *Acta Biochim. et Biophys. Acad. Sci. Hung.* 5(4), 423.

Workshop No. 1 Gene Organization

A. Williamson, B. Mach, J. Merrill Davis

The workshop discussion centered on the question of how to measure the extent of diversity of genes coding for antibodies. Most of the problems raised by the presently available data relate to the extremely diverse and sometimes paradoxical phenotype of the antibody system contrasted with the apparently restricted genome and the accepted rules for gene expression.

It was suggested that three main questions be addressed by the speakers. (1) Descriptions of approaches to gene counting and the number of genes actually counted using these methods. (2) If there are very few genes, are there enough to accommodate known diversity? (3) If there are not enough genes detected by gene counting, what structural changes occur either in framework or hypervariable regions to generate diversity and how compatible are these changes with accounting for reoccurences of idiotypes within and among strains? Possible models for changes generating diversity were to be proposed at the end of the workshop.

Mach reported on the use of plasmid cloning for isolation and purification of both mRNA and cDNA probes used in counting genes coding for immunoglobulin light and heavy chains. This technique is currently being employed, albeit with restrictions, both in Europe and the United States. Mach stressed the value of such an approach in its use as a tool and not as a goal in itself. Messenger RNA coding for kappa light chain, which need not be totally clean, is reverse-transcribed to cDNA. DNA Polymerase I is used to generate a complete double-stranded DNA (dsDNA) complementary to the original mRNA. Under appropriate conditions, this dsDNA is 850 nucleotides long including V and C region portions; this is fused via homopolymeric tails with the K 38 plasmid and cloned within _E. coli_. Mercurated plasmids are melted to yield positive and negative strands, both of which can be used to isolate hybrids via coupling to SH-Sepharose and _in situ_ elution. The use of DNA titration to maximize any reiteration has still only shown a very low number of kappa light chain genes. Mach finds two genes per haploid genome while Milstein finds four genes for the constant region and two genes for the variable region.

Thus kinetic and excess probe saturation modes of nucleic acid hybridization yield results consistent with a genome containing very few V genes with sequence homology for a single myeloma V region sequence. The simple conclusion is that there are very few V genes coding for each subgroup of V region sequences. This is to redefine a subgroup as that set of V regions sufficiently similar in nucleotide sequence to cross hybridize to a single V gene coding for any one of those V regions. A suggested operational definition of subgroup was that of the upper number of framework genes that can cross hybridize, implying large differences in the frameworks among subgroups. Using refined techniques, there are one to five genes by stringent cross hybridization among sequences of not more than 10% divergence. The problem is then to relate this definition to definitions based on amino acid sequences.

The clear definition that any amino acid sequence difference outside a hypervariable region constitutes a new subgroup has been used to estimate the number of subgroups represented by sequences of the amino terminal 23 amino acids of mouse myeloma kappa chains. Weigert has continued to sequence 18 kappa chains identical through position 23 to ask how many more subgroups are revealed. So far only 4 new subgroups have been seen. One amino acid difference can define a new subgroup but would not lead to a difference detectable by hybridization. Even the maximum difference of 8% between the 4 new subgroup sequences might not allow distinction to be drawn by hybridization. Comparison of the amino acid sequence data and the counting of subgroup V genes by hybridization point to there being several framework sequences arising (by somatic variation?) from a single V gene. Present hypotheses are designed to account solely for selection of somatic variation in hypervariable regions involved in the antibody combining site.

Subgroup specific antiserum revealed a similar frequency of each subgroup in normal immunoglobulin as in the myeloma collection suggesting that the latter is a fair sample. However, Klinman raised the important question as to whether the finding of two identical myeloma protein sequences really represents the coincidence of random events in view of known high frequency expression of certain clonotypes. This calls into question all statistical estimates based on repeat clonotype frequencies; such estimates are in any case minimal values.

Experiments designed to study somatic mutations of immunoglobulin genes in myeloma cells cultured in vitro were described by Secher and by Birshtein. Spontaneous mutation in MOPC21 IgG were observed by Secher five times in 7000 clones of P3 cells. Four mutants were characterized and each involved the C region of γ_1 chain. No L chain or V_H mutants were observed.

Birshtein described mutagenesis of MPC11 using the mutagen ICR191. All characterized mutants of the IgG2b protein involve the C_H region. The most dramatic change detected involved an alteration in the antigenic specificity of the C_H region from γ_{2b} to γ_{2a}. Partial sequence data on parental mutant proteins is consistent with a switch of C_H from γ_{2b} to γ_{2a}. Antigenic and sequence evidence did not reveal any L chain or V_H mutants. These reports suggest that if somatic mutations play a role in generating antibody diversity this must happen at an earlier cell than the plasma cell.

The question of the genotype and its relationship to phenotype in immunoglobulin production was raised in another form by Kindt. An antiidiotypic serum specific for the H chain of a homogeneous antistreptococcal group A carbohydrate (GAC) was used to screen for the inheritance of idiotype and its linkage to V_H and C_H allotypes. The idiotype (HId) was originally defined on an a3, d11, e15 H chain. Generally, inheritance of HId was linked to the genotype a3 d11 e15. However, HId was detected on anti GAC antibody in a rabbit lacking the genes a3 d11 as shown by conventional typing. Nevertheless the purified antibody carrying the HId idiotype was found to be a3 and d11 positive. No satisfactory genetic explanation for the coappearance of three latent markers is currently available; chimerism was suggested as a possible way of accounting for apparent poly-allelism.

The workshop closed with a discussion of models for generation of antibody diversity; the emphasis was on somatic diversity. The question was posed as to whether the demonstration of inherited V_H genes in the mouse is compatible with an apparent need for somatic diversity. Are minor clonotypes repeated? Klinman in answer to this question gave data showing that a very low frequency for B cells have the idiotypes of phosphorylcholine binding myeloma proteins 167 and 603 by contrast with the high frequency for T15 B cells. The low frequencies of B cells specific for influenza virus agglutinin antigens also argued for a total repertoire of at least 10^7 antibodies. Klinman expressed the view that generation of such a repertoire should not rest with chance but that some programmed variation would be more acceptable.

Insertion of hypervariable region genes into a framework V gene was the method of generating diversity espoused by Mach. He noted arguments of previous proponents and added evidence for palindromic recognition sequences which could exist in V region DNA sequences and might have served in insertion-excision processes. These inverted repeats to the right and left of hypervariable regions could give secondary structure which, as an analogy with start/stop signals in mRNA, could be recognition sites for cleavage enzymes or DNA binding proteins which actively generate mutations. Such transposition sequences in plasmids might be effective tools for actually isolating nucleases, mutation producing proteins, or regulator proteins.

Finally Gefter presented a hypothesis describing the generation of extensive diversity from a very few V genes by utilizing the procedure of gene conversion. The latter process can be involved in genetic recombination and is explained by the Holliday model. With successive recombination within codons, occurring between as few as 3 genes, coupled to gene conversion (which involves correction of mismatched base paring) up to 13 different amino acids might be seen in the phenotype. Thus variations in framework and hypervariable regions could be generated by the same mechanism and such variations should appear in an ordered manner. In addition to accounting for the small number of germ line genes being able to generate, within a lifetime, the breadth of diversity required by an organism, Gefter's model helps explain inheritance of idiotype, without inheritance of sequence, in that there is selection at the level of the whole animal for sequences coding for functional antibodies. The cogent case for this hypothesis put by Gefter evoked a mixture of enthusiasm and skepticism which is characteristic of the debate on antibody diversity.

THE ORGANIZATION OF ANTIBODY GENES

Roy J. Riblet

The Institute for Cancer Research, The Fox Chase Cancer Center, Philadelphia, Pennsylvania 19111

The production of antibodies involves unique genetic mechanisms including somatic diversification of antibody genes, allelic exclusion, V-C gene translocation, and possibly V or C gene relocation in the IgM to IgG switch. The study of antibody genes should help to explain these phenomena by determining, first, the number of germ line genes and their relative contribution to antibody diversity, and, second, the structure and organization of these genes. The most revealing system for this analysis has been the heavy chain genes of the mouse. The genes specifying the constant regions of heavy chain classes are identified as the elements controlling allotypic antigens on the constant regions as well as the electrophoretic mobility of their Fc fragments (1, 2). In an analogous manner, variable region genes are identified by interstrain antigenic differences on the V regions. These V region antigens, called idiotypes, are expressed on specific antibody populations. In mice some idiotypes are expressed by every individual of one inbred strain in response to the appropriate immunogen while no mouse of a second strain will make antibodies having these idiotypes. These idiotypes are then equivalent to V region allotypes and can be used for genetic analysis (3-5).

The J558-MOPC-104 idiotype system in the anti-α-1,3 dextran response is the V region marker used in the studies reported here. This V gene was identified and characterized by Weigert and coworkers (6-8). It was discovered through the study of BALB/c myeloma proteins with identified antigen binding ability. Two proteins, J558 and MOPC-104E, were found which bound α-1,3 linked dextrans. The light chains in both antibodies were λ and had the same sequence (7). Antiidiotype sera made against either protein showed cross-reactivity with the other (8). The interpretation of these findings was that these myeloma antibodies were the expression of a heavy chain gene and a light chain gene which were germ line, i.e, genes as they were inherited, without subsequent somatic mutation and alteration of specificity. This suggested that immunization of BALB/c mice with the α-1,3 dextran should elicit antibodies which similarly were expressions of these germ line genes. Indeed, BALB/c anti-dextran antibodies were shown to have λ chains and the J558 idiotype (6). A strain survey showed that some strains were like BALB/c in making this type of antidextran antibody, but other strains made antibodies which contained λ light chains and did not have the J558 idiotype. In the survey there was an obvious association of the idiotype with the Ig-1^a allotype. This suggested that one or more genes specifying the idiotype positive antibody structure was linked to allotype. This conclusion was tested in initial backcross experiments which indicated that the expression of the J558 idiotype in antidextran antibodies was controlled by a single gene which was, in fact, closely linked to allotype (9). Since the allotype genes had been identified as the probable structural genes for the constant regions of antibody heavy chains (1) and since in man and rabbit the heavy chain and light chain genes were not linked (2), it then seemed likely that the idiotype gene was the structural gene for the heavy chain variable region of antidextran antibodies.

The role of light chain was not clear, however, since the J558 idiotype had been shown to require the correct combination of heavy and light chain sequences (8). The structural gene(s) for λ light chain had not been defined in mice, and it was possible that V_λ, at least, was linked to heavy chain allotype and that the response difference was due to a variation in the λ light chain. We have shown that this is not the case. NZB is a typical idiotype negative strain; in crosses with BALB/c the dextran response trait of BALB again segregates as a single gene, linked to allotype (10). In this combination a λ difference was excluded through the sequence analysis of NZB λ chain myeloma proteins. Most of the sequence, including all three hypervariable regions, which form the binding site, was obtained for two NZB λ chains; one was identical to the sequence of the λ chains of J558 and MOPC-104E while the other had a single amino acid substitution (10). Thus NZB mice have λ chains suitable to form idiotype positive antidextran antibodies and the structural difference between BALB and NZB antibodies must be in the heavy chain variable region. Although the gene controlling the expression of this V_H difference could be conceived of as a regulatory gene, the simplest interpretation is that it is a V_H structural gene specifying a sequence like that of J558 or MOPC-104 heavy chains. NZB mice, then do not have this V_HDEX gene. (Although V_HDEX refers here to a single gene, Hansburg _et al_. (11) suggest that this may represent several related genes including V_H558 and V_H104.) In general then an idiotype is taken to define a V_H gene if it is controlled by a single gene which is closely linked to the heavy chain constant region allotype genes, Ig-1 through Ig-6.

To begin the detailed definition of heavy chain genes extensive crosses were done to determine the recombination frequency or genetic distance between V_HDEX and C_H and to generate a panel of recombinant chromosomes for further analysis. The results are presented in Table 1.

TABLE 1. RECOMBINATION BETWEEN V_HDEX AND C_H

Backcross	No. Chromosomes tested	No. Crossovers	Frequency
(BALB/c x A/He)F_1x A/He	2022	8	0.40%
(BAB/14 x C.AL-9)F_1x C.AL-9	882	5	0.57%
(BALB/c x NZB)F_1 x NZB	119	0	
(129 x A/He)F_1 x A/He	179	0	
(BAB/14 x A/He)F_1 x A/He	140	0	
Congenic Strains			
C.B-20, BAB/14, C.AL-20 SJA/9, B.C-9	(62)	1	1.6%
Recombinant Inbred Strains			
CXB, AKXL, NX8	(140)	0	
	3544	14	0.40%

All backcrosses were of the form (Ig-1^x, V_HDEX$^+$ x Ig-1^y, V_HDEX$^-$)F_1 x Ig-1^y, V_HDEX$^-$. Recombinant progeny were detected as Ig-$1^{x/y}$, V_HDEX$^-$ or Ig-$1^{y/y}$, V_HDEX$^+$. All of the 14 crossovers shown here have been confirmed by test crossing. Opportunities for recombination also occur in the breeding of congenic and RI strains. The numbers shown here in parenthesis are calculated backcross equivalents. The CXB RI lines are derived from a cross of BALB/c x C57BL/6 (12). The AKXL RI lines are from AKR x C57L (13). The NX8 lines are from NZB x C58 (Johnson, Riblet and Weigert, unpublished).

The overall recombination frequency was 0.4%. Similar estimates were obtained from the individual crosses although the question of whether the V_HDEX-C_H distance is the same in BALB/c, 129, C57L, and C58 has not been adequately tested.

In other laboratories different V_H genes have been used in recombination studies (summarized in ref. 5), and some of these genes have shown considerably higher recombination with C_H than does V_HDEX. V_HA5A, for example, recombines with C_H at 3.2% (4), and the entire V-C region may extend for 5 map units (% recombination), ten times the V_HDEX-C_H distance.

How many V_H genes are between V_HDEX and C_H? The entire H-2 region is about the same size as V_HDEX-C_H in terms of recombination frequency, and it contains approximately 5 to 10 structural genes, mainly for cell surface antigens. Although this may be an incomplete inventory it does agree with the gene per map unit ratio in Drosophila (5000 bands (= structural genes?)/300 total map units). If the organization of the V_H region is similar there would be 10 V_H genes between V_HDEX and C_H, and since the V_H region seems to be about 5 map units long, the total V_H complement of a mouse would be approximately 100 genes.

To specify the order and organization of heavy chain genes, some of the crossover chromosomes recovered above have been examined at other V_H loci and nearby genes. The V_H genes studied include the V_HT15 gene controlling BALB/c anti-phosphorylcholine antibodies (14), V_HS117, which specifies a minor population of BALB/c anti-group A Streptococcal carbohydrate (15), V_HARS, which determines A anti-phenylarsenate antibodies (16) and V_HA5A, which controls A anti-Streptococcal A carbohydrate (17). Another group of V_H genes does not determine idiotypic antigens but controls the fine specificity of binding of related haptens. These too are single genes, linked to allotype, which control heavy chain variable region structure. Examples of this type of V_H genes are V_HNP and V_HNBrP which determine the responses to the (4-hydroxy-3-nitro phenyl) acetyl and (4-hydroxy-5-bromo-3-nitrophenyl) acetyl haptens (18).

The BAB/14 heavy chain chromosome derived from recombination of the chromosomes of BALB/c and C57BL/Ka (19), as shown:

	DEX	NP	NBrP	S117	T15		Ig
BALB/c	+	-	-	+	+		a
BAB/14	+	-	-	+	+	X	b
C57BL/Ka	-	+	+	-	-		b

(Here the "V_H" is omitted and Ig is used, as is C_H elsewhere, as an abbreviation for the six allotype loci, Ig-1, Ig-2, ... Ig-6, i.e, C_γ2a, C_γA, ..., C_γM. The position of the crossover is marked X.) Similarly the C.B(F5) chromosome derived from the cross BAB/14 x C.AL-9. The bb7 chromosome was recovered by Eichmann from a cross of BALB/c and A/J and was similarly studied (4).

	DEX	ARS		Ig
BAB/14	+	-		b
C.B(F5)	-	+	X	b
C.AL-9	-	+		d

	A5A		DEX	ARS	T15	Ig
BALB/c	-		+	-	+	a
bb7	+	X	+	-	+	a
A	+		-	+	-	e

This information allows the construction of several partial maps in which the V_H genes can be ordered with respect to the crossover sites. Combining the bb7 and BAB/14 data yields this map:

$$---V_HA5A---X^{bb7}---(V_HS117,\ V_HDEX,\ V_HT15)---X^{BAB}---C_H---$$

(Genes within parenthesis are not ordered with respect to each other.) Combining bb7 and C.B(F5) yields the map:

$$---V_HA5A\ ---\ X^{bb7}\ ---\ (V_HDEX,\ V_HARS)\ ---\ X^{F5}\ ---\ C_H---$$

All that can be concluded about the position of V_HNP and V_HNBrP is that they were separated from C_H by the BAB/14 crossover, represented thus:

$$---(V_HNP,\ V_HNBrP)\ ---\ X^{BAB}\ ---\ C_H---$$

Similar partial maps of other combinations of V_H genes and crossover points can be made (4, 5 and Eichmann, this symposium) but a complete heavy chain map awaits the analysis of many more recombinants arising in a variety of strain combinations.

A number of unusual features have become apparent in the genetics of heavy chains. One is the continuing difficulty in assigning the heavy chain genes to a chromosome; despite extensive efforts to map allotype genes by conventional linkage testing (1, 2) and analysis of Recombinant Inbred strains (20 and B.A. Taylor, personal communication), their location remains unknown. The only progress in this regard is Taylor's discovery (20) that the serum prealbumin locus, Pre, is linked to allotype. Pre and C_H recombine with a frequency of 10% and analysis of crossovers indicate the gene order V-C-Pre. The chromosomal locus of Pre is unknown also. Either this group of genes is located far from any of the many genes tested, or an especially high rate of recombination between this group and its neighbors obscures the true linkage.

Another puzzling feature is the distorted recovery of reciprocal classes of recombinants as shown in Table 2. In the CBC series five V_HDEX^- Ig^b chromosomes were recovered and no V_HDEX^+ Ig^d products; the two types should have appeared in equal numbers. Similarly the ACA series yielded five V_HDEX^+ Ig^e products and only one V_HDEX^- Ig^a crossover. This distortion in some as yet unknown way must be the result of the special structures required for antibody expression.

TABLE 2. MULTIPLE RECOMBINATION IN V-C CROSSOVERS

Strain	V_HDEX		C_H		Pre	No. Cross-overs
Parental						
BAB/14	+		b		o	
C.AL-9	-		d		a	
Recombinants						
C.B(F5)	-	X	b	X	a	2X
CBC 300	-	X	b		o	1X
CBC 520	-	X	b		o	1X
CBC 682	-	X	b	X	a	2X
CBC 950	-	X	b	X	a	2X
Parental						
BALB/c	+		a		a	
A/He	-		e		o	
Recombinants						
ACA 356	+	X	e		o	1X
ACA 366	+	X	e	X	a	2X
ACA 556	+	X	e		o	1X
ACA 563	+	X	e		o	1X
ACA 631	-	X	a		a	1X
ACA 704	+	X	e	X	a	2X

The most extraordinary aspect of heavy chain genetics is the very high rate of multiple recombination (Table 2). When chromosomes which had incurred a crossover between V_HDEX and C_H were typed for prealbumin, 5 of 11 had also recombined in the C_H-Pre region. Only a single such double crossover would be expected on the basis of the 10 map unit distance between C_H and Pre. Possibly similar examples of multiple crossing over among V_H genes have been observed by Taylor (reviewed in ref. 5). These high rates of multiple crossing over are indicative of unusual structures and mechanisms in this region of the mouse genome. A possible model is that there are interspersed regularly among the V_H and C_H genes, special recombining sites, perhaps recognition sites for specific nucleases used in the recombination process. These could be for use in the processes of V-C translocation but might also enhance the probability of additional exchanges once the mechanism for an initial crossover is activated in this region.

ACKNOWLEDGEMENTS

This work was supported by National Institutes of Health Grants AI-13797 (RR), CA-06927 and RR-05539 to The Institute for Cancer Research, and an Appropriation from the Commonwealth of Pennsylvania.

REFERENCES

1. Herzenberg, L.A., McDevitt, H.O. and Herzenberg, L.A. (1968) Annu. Rev. Genet. 2, 209.
2. Mage, R., Lieberman, R., Potter, M. and Terry W.D. (1973) In Sela, M. (Ed.), The Antigens, vol. 1, Academic Press, N.Y., p. 299.
3. Weigert, M., Potter, M. and Sachs, D. (1975) Immunogenetics 1, 511.
4. Eichmann, K. (1975) Immunogenetics 2, 491.
5. Weigert, M. and Potter, M. (1977) Immunogenetics 4, 401.
6. Blomberg, B., Geckeler, W.R. and Weigert, M. (1972) Science 177, 178.
7. Weigert, M.G., Cesari, I.M., Yonkovich, S.J. and Cohn, M. (1970) Nature 228, 1045.
8. Carson, D. and Weigert, M. (1973) Proc. Nat. Acad. Sci. U.S.A. 70, 235.
9. Riblet, R., Blomberg, B., Weigert, M., Lieberman, R., Taylor, B.A. and Potter, M. (1975) Eur. J. Immunol. 5, 775.
10. Weigert, M. and Riblet, R. (1976) Cold Spring Harbor Symp. Quant. Biol. 41, in press.
11. Hansburg, D., Briles, D.E. and David, J.M. (1976) J. Immunol. 117, 569.
12. Bailey, D.W. (1971) Transplantation 11, 325.
13. Taylor, B.A., Meier, H. and Meyers, D.D. (1971) Proc. Nat. Acad. Sci. U.S.A. 68, 3190.
14. Lieberman, R., Potter, M., Mushinski, E.B., Humphrey, W. and Rudikoff, S. (1974) J. Exp. Med. 139, 983.
15. Berek, C., Taylor, B.A. and Eichmann, K. (1976) J. Exp. Med. 144, 1164.
16. Pawlak, L.L., Mushinski, E.B., Nisonoff, A. and Potter, M. (1973) J. Exp. Med. 137, 22.
17. Eichmann, K. (1972) Eur. J. Immunol. 2, 301.
18. Imanishi, T. and Makela, O. (1973) Eur. J. Immunol. 3, 323.
19. Riblet, R., Weigert, M. and Mäkelä, O. (1975) Eur. J. Immunol. 5, 778.
20. Taylor, B.A., Bailey, D.W., Cherry, M., Riblet, R. and Weigert, M. (1975) Nature 256, 644.

GENETIC IMPLICATIONS OF LINKAGE STUDIES USING CHAIN-SPECIFIC IDIOTYPES

John A. Sogn, Martin L. Yarmush and Thomas J. Kindt

The Rockefeller University
New York, New York 10021

INTRODUCTION

The bulk of available evidence concerning immunoglobulin inheritance suggests that these molecules are encoded by structural genes present at three major loci: H chain, κ L chain and λ L chain. Although there is some controversy concerning precise numbers, it is generally held that each of these loci is composed of a number of V genes closely positioned (linked) to a smaller number of C genes. Even though most data concerned with the organization of genes within these complex units have been taken from studies of H chain genes, it is assumed that each locus is similar in organization. Much of the experimental data to support these concepts has been obtained through studies on immunoglobulin allotypes and idiotypes, especially those of the rabbit (1,2).

Allotypes serve as genetic markers for the C and V regions of antibody H or L chains and many of these have been fully characterized (1). Idiotypes, which serve as markers for antibody binding sites, however, are not as simply classified (3-5). Because H and L chain genes are not linked, an idiotype must be localized to a single chain in order to provide reliable genetic data. This requirement is not often met with homologous rabbit idiotypic antisera, which rarely bind to H or L chains isolated from the immunizing antibodies. In recent studies, an homologous idiotypic antiserum was fortuitously found to contain components which reacted with the idiotypic determinants of an L chain (4539 LId) (6). In addition, an antiserum raised against a recombinant immunoglobulin containing the H chain of interest recognized the idiotype of a specific antibody H chain (4135 HId) (7). Although both chain-specific idiotypes were shown to be inherited, the linkage data obtained from the two studies differed significantly.

These linkage data, as well as recent data concerning the expression of latent idiotypes and allotypes (8,9), point out weaknesses in theories of antibody diversity and suggest that modifications must be made to cover these revealing exceptions.

RESULTS

In the course of studies on homogeneous antistreptococcal antibody (Ab) 4539, two homologous idiotypic antisera were prepared in the usual manner (5). One of these antisera proved to contain components specific for idiotypic determinants limited to the L chain of 4539 Ab. These components (anti-LId) were isolated by immunoadsorbant chromatography on a column of Sepharose 4B to which 4539 L chains had been covalently attached. When the ability of sera to inhibit the reaction between radiolabeled 4539 Ab and insolubilized anti-LId was measured, it was seen (Table 1) that expression of the idiotype was strictly limited to rabbits related to the proband and immunized with the Group C streptococcal vaccine. It was further shown that all LId-positive molecules could be removed from sera by adsorption with the Group C vaccine. Thus, all LId-positive molecules showed the same antigen specificity.

Table 1: Detection of 4539 LId

Source of antiserum	Antigen	Number studied	Crossreaction	
			No. positive	% positive
Related rabbits	Group C	52	29	56
	None*	17	0	0
Unrelated rabbits	Group C	24	0	0
	Group A	20	0	0

* Nonimmune sera from rabbits that produced 4539 LId upon immunization.

It was possible to study linkage of 4539 LId to a C_K gene because 4539 Ab possesses a variant C_K region marker, $b4^{var}$. Although $b4^{var}$ is serologically similar to b4, it can be distinguished from b4 by amino acid sequence analysis (10). As shown below, at least two amino acid interchanges, at positions 121 and 124, distinguish their C regions:

	120					125
b4	- PRO -	ALA -	ALA -	ASP -	GLN -	VAL -
$b4^{var}$	- PRO -	SER -	ALA -	ASP -	LEU -	VAL -

In preliminary breeding experiments, b4 and $b4^{var}$ segregated as alleles. While LId and $b4^{var}$ are inherited within the same family, the chemical typing data on $b4^{var}$ (Fig. 1) suggest that the two traits are unlinked. Note that only 7 of 14 LId-positive rabbits were $b4^{var}$-positive. To confirm these data, it will be necessary to isolate LId-positive molecules and type them for $b4^{var}$.

Because of the obvious usefulness of chain-specific idiotypes as V region genetic markers, methods for eliciting chain-specific idiotypes have been examined. An antiserum specific

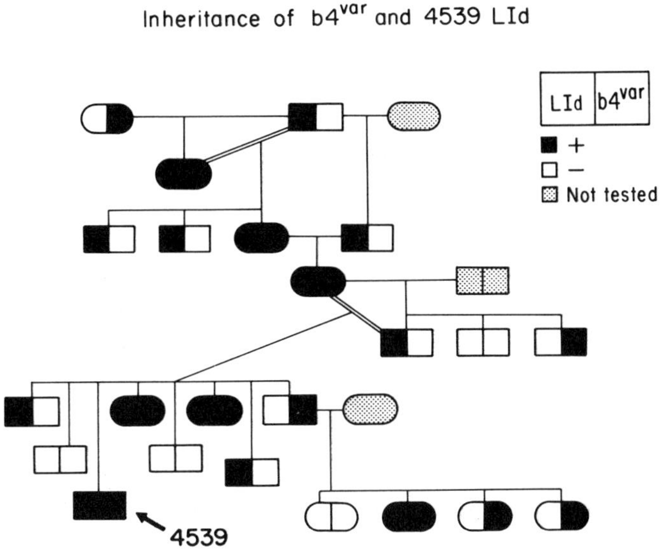

Inheritance of b4^{var} and 4539 LId

LId | b4^{var}

+

−

Not tested

4539

Figure 1

for the H chain of 4135 (anti-4135 HId) was raised by immunizing a rabbit with recombinant immunoglobulin containing 4135 H chains and an L chain pool isolated from the rabbit to be immunized. The resultant antiserum bound equally well to 4135 Ab and to a recombinant molecule containing 4135 H chain. As detected by an inhibition of binding assay, 4135 HId was present in Group C antisera from 30 to 70 related rabbits and only 2 of 24 unrelated rabbits (Fig. 2).

When the distribution of 4135 HId in antisera from related rabbits of different group a allotypes and allogroups was examined (Fig. 2), it could be seen that 4135 HId was present in 74% of rabbits containing the J allogroup (a^3,x^{32},y^-,n^-,f^{72}, g^{74},d^{11},e^{15}) but in only 6% of rabbits lacking this allogroup. Molecular association of 4135 HId with allotypes of the J allogroup was established in several rabbits heterozygous for this allogroup by isolating 4135 HId-positive molecules using an anti-4135 HId immunoadsorbent column.

Because rabbit 4232 ($a^1a^2/b^4b^4/d^{12}d^{12}$) produced a significant amount of HId even though it lacked the J allogroup, serial bleedings from this rabbit were examined in greater detail. A correlation was observed between the presence of HId and the presence of latent a3 allotype; fractionation of serum resulted in coelution of these two markers. HId-positive molecules isolated from an antibody fraction were found to express not only a3 but also d11, a Cγ allotype characteristic of the J allogroup. These data indicate that the idiotype and the two allotypes encoded within the latent allogroup J were present on the same molecules.

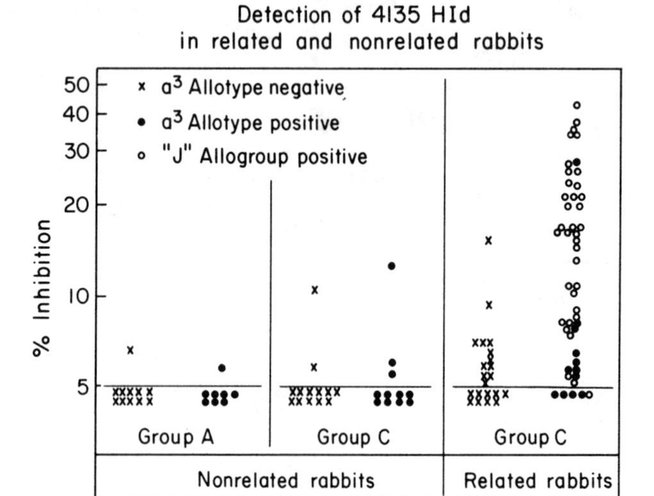

Figure 2. Undiluted sera giving inhibitions less than 10% are considered negative. No preimmune serum gave more than 7% inhibition.

DISCUSSION

Most widely accepted models of antibody genes predict close linkage between V and C region genes. This was strongly confirmed in studies of the H chain idiotype 4135 HId which was shown to be linked to a single H chain allogroup (J) (2). Even though another allogroup (I) in the same family carried the V region allotype a3 of the proband antibody, the idiotype was rarely found in animals of this type. In one branch of the family, a well characterized crossover (11) placed HId and the allotype a3 in linkage with the C_H allotype d12 (allogroup J'). This demonstrates that 4135 HId expression is not dependent upon the presence of allotype d11.

The antibody fractions from a rabbit (4232) expressing HId but lacking the a3 allotype was shown to express, in addition to HId, the a3 allotype and the C region allotype, d11, characteristic of the J allogroup. This provides the first example of coordinate expression of latent markers in a single rabbit and places new constraints on models for control mechanisms regulating the synthesis of immunoglobulin. It also emphasizes that immunoglobulin gene linkage studies are not conclusive in the absence of demonstration of molecular association between the linked markers, because of the possibility of latent allotype expression.

In marked contrast to the results obtained with 4135 HId, a parallel study on the L chain idiotype 4539 LId provided no evidence for linkage of V_L and C_L genes. The L chain from the homogeneous proband antibody carries a C region allotype, $b4^{var}$, that appears from preliminary breeding studies to be an allele of b4, b5, b6 and b9. However, there was no concordance between the presence of LId and of $b4^{var}$. This result was unexpected because of amino acid sequence studies demonstrating clear differences in N-terminal sequence among L chains of different group b allotypes (12,). While the lack of linkage can be rationalized by postulating two crossovers in the 5 generations of the families studied, this number seems excessive and is argued against by data on quantities of $b4^{var}$ present in different animals (10). Further studies will be required to conclusively exclude simple redundancy of 4539 LId on chromosomes carrying different C_κ regions; data on the familial occurrence of this marker do not support this possibility (6).

An increasing number of reports (8,9,14-18) suggest the need to consider genetic regulatory mechanisms in the synthesis of immunoglobulins and the nonrandom expression of immunoglobulin genes. In addition to our earlier studies on latent allotypes and on latent idiotypes presented here, there have been reports of latent allotypes in humans (16,17), mice (15) and rabbits (8,18). Other data obtained in studies of rabbit immunoglobulin which suggest the presence of such regulatory mechanisms include the normal imbalances observed in allotypically heterozygous animals in levels of immunoglobulins bearing different allotypes [the so-called "pecking order" (19)]. In what may be a limiting case of this (20), a rabbit was shown to express only λ-type L chains and no κ chains. Although this could be the result of a defect in structural genes it might also be explained by faulty regulatory genes. In addition, inexact control mechanisms may be responsible for the occurrence of IgG and IgA molecules that result from apparent *trans* synthesis (21,22). These molecules occur at approximately the 1% level in rabbit serum.

It is obvious that conventional models of antibody diversity must be expanded to include the above data. Although an exact formulation is not yet evident certain points may be made. First of all, the phenomena of non-allelic allotype expression must be explained; experimental evidence for the occurrence of latent allotypes is accumulating beyond the point where trivial explanations suffice. The presence of unexpected products argues that genes encoding these molecules must be present in the individual in some form. It is not known whether genetic information for every allotype is present in every individual or whether some subset is present. In the latter case, the subset must necessarily be larger than two genes (or gene complexes) per locus. A second related concept concerns the organization of the immunoglobulin genes in H chain allogroups and L chain linkage groups. No explanation has been put forth to explain the limited number of such groups that have been observed compared to the several thousand that are possible in the rabbit. Any model that attempts to describe the linkage groups is further complicated by expression of latent genetic information. What mechanisms, if any, limit crossovers leading to new allogroups? Is there any relationship between *trans* synthesis that leads to molecules containing combinations of markers in V and C regions not predicted from linkage data, and latent allotype (or allogroup) expression? Is it possible that the allogroup has a functional significance in genetic regulatory mechanisms?

The answers to such questions will not only have an impact on our understanding of immunoglobulin genes, but may also serve to explain some general features of genetic regulation in eucaryotic systems.

ACKNOWLEDGMENTS

This work was supported by USPHS grants from the NIAID AI08429, AI11995 and AI11439. J.A.S. is a Fellow of the Arthritis Foundation. The authors wish to thank Ms. Linda Lee Adams for her excellent secretarial assistance.

REFERENCES

1. Kindt, T.J. (1975) *Adv. Immunol.* 21, 35.
2. Knight, K.L., and Hanly, W.C. (1975) *Cont. Topics Molec. Immunol.* 4, 55.
3. Capra, J.D., and Kehoe, J.M. (1975) *Adv. Immunol.* 20, 1.
4. Oudin, J. (1974) In: *The Antigens*, Vol. 2, M. Sela, ed. Academic Press, New York, p. 277.
5. Sogn, J.A., Coligan, J.E., and Kindt, T.J. (1977) *Fed. Proc.* 36, 214.
6. Sogn, J.A., Yarmush, M.L., and Kindt, T.J. (1976) *Ann. Immunol. (Paris)* 127C, 937.
7. Yarmush, M.L., Sogn, J.A., Mudgett, M., and Kindt, T.J. (1977) *J. Exp. Med.* 145, 916.
8. Mudgett, M., Fraser, B.A., and Kindt, T.J. (1975) *J. Exp. Med.* 141, 1448.
9. Strosberg, A.D. (1977) *Immunogenetics* (in press).
10. Sogn, J.A., and Kindt, T.J. (1976) *J. Exp. Med.* 143, 2475.
11. Kindt, T.J., and Mandy, W.J. (1972) *J. Immunol.* 108,1110.

12. Waterfield, M.D., Morris, J.E., Hood, L.E., and Todd, C. W. (1973) *J. Immunol.* 110, 227.
13. Thunberg, A.L., and Kindt, T.J. (1975) *Scand. J. Immunol.* 4, 197.
14. Farnsworth, V., Goodfleish, R., Rodkey, S., and Hood, L. (1976) *Proc. Nat. Acad. Sci (USA)* 73, 1293.
15. Bosma, M., and Bosma G. (1974) *J. Exp. Med.* 139, 512.
16. Lobb, N., Curtain, C.C., and Kidson, C. (1967) *Nature* 214, 783.
17. Pothier, L., Borel, H., and Adams, R.A. (1974) *J. Immunol.* 113, 1984.
18. Strosberg, A.D., Hamers-Casterman, C., van der Loo, W., and Hamers, R. (1974) *J. Immunol.* 113, 1313.
19. Dubiski, S. (1972) *Med. Clin. N. Amer.* 56, 557.
20. Kelus, A.S., and Weiss, S. (1977) *Nature* 265, 156.
21. Landucci-Tosi, S., and Tosi, R.M. (1973) *Immunochemistry* 10, 65.
22. Knight, K.L., Malek, J.T., and Hanly, W.C. (1974) *Proc. Nat. Acad. Sci (USA)* 71, 1169.

DETECTION AND EXPRESSION OF A V_H SUBGROUP MARKER (U10-173) IN MICE

M. J. Bosma, C. De Witt, M. Potter*, J. Owen and B. Taylor†

The Institute for Cancer Research, Fox Chase Cancer Center, Fox Chase, Philadelphia, Pennsylvania 19111; *National Cancer Institute, Bethesda, Maryland 20014; and †The Jackson Laboratory, Bar Harbor, Maine 04609

ABSTRACT. Similar Heavy (H) chain antigens (U10-173) have been found on BALB/c myeloma proteins known to bind one of four different carbohydrate (CHO) ligands (2,6-levan, 1,6-D-galactan, N-acetyl-glucosamine and N-acetyl-D-mannosamine) and on a few myeloma proteins having unknown binding site specificities. U10-173 determinants are also found on ∿ 1% of normal Igs of most inbred mouse strains. The ability of certain antigens, which contain the above CHO residues, to stimulate increased U10-173 production in normal mice, the association of U10-173 with IgM, IgA and IgG classes, the close linkage between U10-173 and constant region genes for IgG H chains (C_H genes) and the very similar V_H framework sequences of U10-173$^+$ proteins lead us to conclude that U10-173 is a marker for a small number of related V_H subgroups.

INTRODUCTION

Variable (V) region antigens that distinguish V subgroups were first reported for human kappa chains (κI, κII and κIII) by Solomon and McLaughlin (1). This was followed by evidence for subgroup-specific antigens on human lambda (λ) chains (2). And recently, Förre *et al*. (3) reported antisera that distinguish the three major V_H subgroups in humans (V_HI, V_HII and V_HIII). Especially relevant to the latter is the evidence reported here for the serological detection of related V_H subgroups in mice.

METHODS

Ig-congenic mouse strains: BAB-14 (from R. Riblet), C.B-17, -26 and -31 refer to BALB/c mice that carry the allotype of C57BL/Ka mice. The number after the hyphen (*e.g.*, C.B-17) indicates the number of backcrosses that preceded the homozygous derivation of each strain. C3H·SW·Ig^a (CWA) and C3H·SW·Ig^b (CWB) mice came from the laboratory of L. Herzenberg.

Anti-173Fab serum. Rabbit antiserum (prepared by Steve Hausman) to the Fab fragment of an IgG_{2a} myeloma protein (173) was absorbed with myeloma proteins of the IgM (104E), IgA (167), IgG_1 (31c) and IgG_{2b} (195) classes. Subsequently, this absorbed antiserum (anti-173) was found to crossreact with one other IgG_{2a} myeloma protein (U10). We refer to this crossreacting specificity as U10-173.

Quantitation of $U10\text{-}173^+$ molecules. To measure U10-173 determinants exclusively, we constructed U10-173-reactive tubes as follows: the inner surface of 12 x 75 mm polystyrene tubes (Falcon) was sequentially coated with a 1 ml solution of U10 containing 20 μg U10 and 80 μg of bovine serum albumin (BSA), with a 1/2000 dilution of anti-173 and with 6 ml of 0.5% BSA. The manner in which this was done is detailed elsewhere (4). U10-173-reactive tubes were used to carry out antigen competition reactions between a fixed weight of ^{125}I-U10 and varying dilutions of unlabeled protein, H and L chains or mouse sera. The displacement of ^{125}I-U10 by any one of the unlabeled preparations was compared to that obtained with intact U10 protein. Using the U10 competitive curve as a standard, we calculated the relative quantities of U10-173 in the unknowns.

RESULTS AND DISCUSSION

Detection of U10-173 on ∿ 16% of tested mouse myeloma proteins. A screen of 102 myeloma sera revealed 16 $U10\text{-}173^+$ proteins; 3 were of the IgG_{2a} class and 13 were of the IgA class (see Table 1). All but 2 of the proteins in groups B-E were able to displace ^{125}I-U10 completely; X24 and ABE 48 acted as incomplete competitors. Logit transformation (4) of the competitive curves for all $U10\text{-}173^+$ proteins (excepting X24 and ABE 48) gave lines of comparable slope having correlation coefficients close to 1. That U10-173 determinants resided on the H chains alone was clear from other experiments in which isolated H and L chains (of X44, 173 and U10) were used as competitors (unpublished results). Thus it was inferred that similar, if not identical, U10-173 determinants were present on the H chains of most proteins in groups A-F. X24 and ABE 48 apparently lacked some of these determinants.

It can be seen from Table 1 that $U10\text{-}173^+$ myeloma proteins represent at least 5 different antigen-binding specificities: 2,6-levan, 1,6-D-galactan, N-acetyl-glucosamine, N-acetyl-D-mannosamine and 1 or more unknown specificities. Therefore, U10-173 does not likely correspond to determinants near or in the binding site. Further, determinants in the

TABLE 1

DETECTION OF U10-173 DETERMINANTS ON MYELOMA PROTEINS HAVING DIFFERENT LIGAND-BINDING SPECIFICITIES*

Group	$U10\text{-}173^+$ Myeloma Proteins	Ligand-binding Specificities
A	MOEV 48	unknown
B	173, PC 3	
C	U10, Y5476, ABE 48	2,6-levan (5, 6)
D	T601, J539, JPC-1, X44, X24, CBPC 4, S10, T191	1,6-D-galactan (7, 8)
E	S117	N-acetyl-glucosamine (9)
F	M406	N-acetyl-D-mannosamine (10)

*Except for three IgG_{2a} proteins (173, PC 3 and U10), all of the above $U10\text{-}173^+$ proteins belong to the IgA class. References for the ligand-binding specificities are given in parentheses.

C_HI region seem to be excluded because U10-173 was found on some IgG_{2a} and some IgA proteins. We are left to conclude that U10-173 corresponds to subgroup-specific determinants in the V_H framework. Evidence in support of this conclusion comes from a comparison of the available N-terminal V_H sequences of $U10\text{-}173^+$ myeloma proteins. The V_H sequences of S10, T191, X44 (7), S117 and Y5476 (11), and U10 (12) are identical through the first 27 residues; those of JPC-1 (7), 173 (14) and M406 (10) differ by no more than two replacements through the first 20-27 residues; and the nearly completed V_H sequence of J539 (13) differs from that of 173 (14) in only 4 residues of the framework.

Detection of U10-173 on ∿ 1% of normal mouse Igs. U10-173 represented about 1% of the total Ig in most strains examined (15). This is shown in Table 2 for two $U10\text{-}173^+$ strains, BALB/c and C57BL/6. Virtually all detectable U10-173 was in the IgG fraction. Little or no U10-173 was found in the IgM fraction. Also, U10-173 determinants were associated with IgG H chains and not with the L chains. It is important to note that both the IgG and IgM fractions of levan-immunized CWB mice contained U10-173$^+$ molecules whereas CWA mice were U10-173$^-$ before and after immunization with levan.

Evidence for close linkage between genes for U10-173 and C_H allotypes. U10-173 was detected in all but 4 (C3H, CBA, PL/J and AKR) of 25 inbred strains tested (15). Taking advantage of 33 recombinant inbred (RI) strains derived from crosses of the F_2 generation of AKR x C57L (AK x L) and C57BL x C3H (B x H) mice, we were able to compare the segregation of U10-173 and C_H allotypes (the C57L and C57BL strains were U10-173$^+$). The results were very clear. All RI strains with C57L or C57BL allotype markers had detectable U10-173 (16-68 μg/ml) but RI strains having the AKR or C3H allotype markers had no detectable U10-173 (< 1 μg/ml).

TABLE 2

REPRESENTATION OF U10-173 IN DIFFERENT Ig FRACTIONS OF NORMAL AND IMMUNE MICE

Source of Ig	Ig Fraction*	Equivalents of U10/100 ng Protein
Normal BALB/c	IgG	1.3
	IgM	0.02
Normal C57BL/6	IgG	0.4
	IgG H	0.73
	IgG L	< 0.003
Levan-immune CWA	(serum)	< 0.001
Levan-immune CWB	IgG	2.3
	IgM	2.7

*Serum Ig was salt precipitated in 50% $(NH_4)_2SO_4$; the IgM fraction was obtained by filtration (twice) over a 1.5 x 90 cm column of Bio-Gel A-1.5m Agarose (Bio-Rad Laboratories) and the IgG fraction by DEAE-cellulose chromatography at 0.02 M PO_4, pH 7.4. H and L chains of IgG were isolated by the method of Bridges and Little (16).

Several conclusions can be drawn from the preceding results: 1) U10-173 serves as a V_H marker for a small number (5 or more) of related V_H germline genes; this assumes that the minor V_H differences in proteins representative of groups A-E (Table 1) define separate germline genes; 2) the V_H genes for U10-173 are linked to each other and to C_H allotype genes as no recombinants were found in a total of 33 RI strains; and 3) the U10-173 V_H marker can be expressed with C_H genes for μ, γ and α chains (Tables 1 and 2).

Antigen-specific stimulation of U10-173 production. Various antigens containing the CHO ligands indicated in Table 1 were able to stimulate increased production of U10-173 in some but not all U10-173$^+$ strains tested.* The varying stimulatory effects of one such antigen (levan from *Aerobacter levanicum*) are shown in Table 3. The very slight increase in U10-173 production in BALB/c mice was surprising because levan is known to stimulate antibodies mostly specific for 2,6-glycosidic linkages (17) and all BALB/c myeloma proteins known to bind 2,6-levan were U10-173$^+$ (see Table 1). Unlike BALB/c mice, C.B-17,-26,-31, and BAB-14 mice increased their U10-173 production approximately 3- to 4-fold as did C57BL mice. CWA mice lacked detectable U10-173 before and after immunization with levan. But in CWB mice, the U10-173 concentration ([U10-173]) increased $\sim$ 40-fold; absorption with levan-conjugated Sepharose beads (a gift of N. Glaudemans) reduced the elevated [U10-173] back to preimmune levels (not shown). The reason for the 10-fold difference in the U10-173 response of CWB and C57BL/10 mice is not clear. It should be pointed out that anti-levan antibody was not detectable in preimmune sera; however, subsequent to immunization with levan, BALB/c, C57BL and C.B mice made comparable quantities of anti-levan as did CWA and CWB mice (unpublished results).

The results of Table 3 clearly demonstrate allotype-linked control of the U10-173 response to levan. It is interesting to note that BAB-14 mice have given evidence of carrying BALB/c recombinant genes for every V_H marker tested so far (S117, J588, NP, NBrP, T15 and ESE) (as reviewed in 18, 19). The presence of the BALB/c marker for S117 implies that one or more U10-173 genes were also included in the genetic recombination since the markers for the S117 myeloma framework (U10-173) and binding site (S117) presumably define the same V_H germline gene (U10-173/S117). On the other hand, the genetic recombination in BAB-14 mice apparently did not affect the C57BL structural (or regulatory) gene(s) for the U10-173 response to levan because in this respect BAB-14 were quite unlike BALB/c mice and acted more like C57BL mice (Table 3).

---

*Other antigens picked at random (BSA, keyhole limpet hemocyanin, sheep red blood cells and dextran) did not stimulate increased U10-173 production.

TABLE 3

INCREASED PRODUCTION OF U10-173 IN LEVAN-IMMUNIZED MICE*

Mouse Strain	No. Mice Tested	$[U10\text{-}173]_i/[U10\text{-}173]_o$
		av. ratio ± SE
BALB/c	18	1.43 ± 0.11
C57BL/6	25	2.83 ± 0.34
C.B-17	17	2.89 ± 0.25
C.B-26	8	3.21 ± 0.53
C.B-31	8	2.80 ± 0.62
BAB-14	8	3.60 ± 0.60
C57BL/10	6	3.75 ± 0.76
CWB	12	42.0 ± 2.2
CWA	7	0.0

*Preimmune concentrations of U10-173 ($[U10\text{-}173]_o$) were determined 1 day before immunization with 50 μg of levan; $[U10\text{-}173]_i$ refers to the U10-173 concentration 7 days later. $[U10\text{-}173]_o$ in μg/ml ± SE averaged 115 ± 7 in BALB/c, 55 ± 8 in C.B-17, 124 ± 11 in C.B-26, 82 ± 9 in C.B-31, 103 ± 14 in BAB-14, 78 ± 8 in C57BL/10, 33 ± 9 in C57BL/6, and 23 ± 8 in CWB mice. SE indicates the standard error of the mean.

Assuming structural gene control, how can the BAB-14 result be explained? It would seem a coincidence that the genetic recombination in BAB-14 mice just happened to occur between (as opposed to outside of) a small group of tightly-clustered U10-173 genes (U10-173 locus). However, it is easy to visualize a recombination that may have involved the U10-173/S117 gene of BALB/c but not the C57BL gene for U10-173 anti-levan response if we assume that the U10-173 loci in BALB/c and C57BL mice map in nonhomologous positions or that the genes for U10-173 are not clustered, but randomly distributed in the V_H locus. An alternative idea, which cannot be excluded, is that BAB-14 mice do not carry recombinant U10-173 genes of BALB/c origin. This leaves open the possibility that S117 and U10-173 map in different places and that different genes encode for the framework and binding site regions as has been proposed by others (20, 21). Further analyses of U10-173 and associated binding site markers may help resolve this problem.

ACKNOWLEDGMENTS

This work was supported by USPHS grants AI-13323, CA-04946, CA-06927 and RR-05539 from the National Institutes of Health and by an appropriation from the Commonwealth of Pennsylvania.

REFERENCES

1. Solomon, A., and McLaughlin, C.L. (1969) J. Exp. Med. 130, 1295.
2. Tischendorf, F.W., Tischendorf, M.M., and Osserman, E.F. (1970) J. Immunol. 105, 1033.
3. Förre, O., Natvig, J.B., and Kunkel, H.G. (1976) J. Exp. Med. 144, 897.
4. Bosma, M.J., Marks, R., and De Witt, C.L. (1975) J. Immunol. 115, 1381.
5. Cisar, J., Kabat, E.A., Liao, J., and Potter, M. (1974) J. Exp. Med. 139, 159.
6. Lieberman, R., Potter, M., Humphrey, W.Jr., Mushinski, E.B., and Vrana, M. (1975) J. Exp. Med. 142, 106.
7. Rudikoff, S., Mushinski, E.B., Potter, M., Glaudemans, C.P.J., and Jolly, M.E. (1973) J. Exp. Med. 138, 579.
8. Manjula, B.N., Glaudemans, C.P.J., Mushinski, E.B., and Potter, M. (1975) Carbohydrate Res. 40, 137.
9. Vicari, G., Sher, A., Cohn, M., and Kabat, E.A. (1970) Immunochem. 7, 829.
10. Potter, M. (1972) Physiol. Rev. 52, 631.
11. Barstad, P., Hubert, J., Black, B., Eaton, B., Weigert, M., and Hood, L. (unpublished results).
12. Rudikoff, S. (personal communication).
13. Potter, M., Rudikoff, S., Vrana, M., Rao, D.N., and Mushinski, E.B. (1976) Cold Spring Harbor Symp. Quant. Biol., in press.
14. Bourgois, A., Fougereau, M., and de Preval, C. (1972) Eur. J. Biochem. 24, 446.
15. Bosma, M.J., De Witt, C., Hausman, S.J., Marks, R., Potter, M., and Taylor, B. (1976) Immunogenetics, in press.
16. Bridges, S.H., and Little, J.R. (1971) Biochem. 10, 2525.
17. Lieberman, R., Potter, M., Humphrey, W.Jr., and Chien, C.C. (1976) J. Immunol. 117, 2105.
18. Eichmann, K. (1975) Immunogenetics 2, 491.
19. Weigert, M., and Potter, M. (1976) Immunogenetics, in press.
20. Wu, T.T., and Kabat, E.A. (1970) J. Exp. Med. 132, 211.
21. Capra, J.D., and Kindt, T.J. (1975) Immunogenetics 1, 417.

ANALYSIS OF IMMUNOGLOBULIN LIGHT CHAIN GENE EXPRESSION BY ISOELECTRIC FOCUSING

David M. Gibson

Département de Biochimie, Université de Sherbrooke
Sherbrooke, Québec, Canada

ABSTRACT: Light chains isolated from normal immunoglobulin can be resolved into a finite series of discrete "IF-subgroups" by isoelectric focusing in polyacrylamide gels. Analysis of the focusing patterns of light chains from inbred mouse strains has permitted the distinction of four different phenotypes. Strains sharing the different phenotypes are (A) C57Bl/6J, C57Bl/10J, C57L/J, DBA/2J, ST/bJ, C3HeB/FeJ, C57Bl/10Sn, CBA/J, A/J, DBA/1J, Balb/cJ; (B) AKR/J, RF/J, and PL/J; (C) C58/J and (D) NZB/BlnJ and BDP/J. In each case the differences between alternate phenotypes appear as the presence or absence of a distinctive focusing band or bands, suggesting that each band in question may be controlled by a single (v-) gene or a group of very tightly linked (v-) genes.

INTRODUCTION

The number of different immunoglobulin molecules that an animal can produce has been estimated to be of the order of 10^7 and probably no less than 10^6 (1). A significant part of this diversity must result from the fact that the molecule is composed of two different polypeptide chains, heavy and light chains, both of which show extensive diversity in their own right (2). If most combinations of heavy and light chains were able to form functional immunoglobulin molecules, then the total diversity would be equal to the product of the number of light chain sequences multiplied by the number of heavy chain sequences. While it is evident that any attempt to resolve the normal immunoglobulin pool into discrete molecular species would be futile using present methodology, the possibility of reducing the heterogeneity (n) of the normal pool by as much as a factor of $\sqrt{n}$ simply by separating the heavy and light chains makes the resolution problem considerably more interesting. The resolution of light chains by themselves, although certainly a more reasonable proposition than the resolution of total immunoglobulin, still remains a formidable if not insurmountable problem however. Current estimates of the number of different Vκ sequences capable of being produced by a Balb/c mouse range in the order of 700 to 10,000 based on the occurence of 2 identical Bence Jones proteins in a survey of 50 randomly selected proteins (3). Even if the true number of sequences present were the lower of these two estimates, the possibility of resolving such a heterogeneous mixture would be beyond the range of current technology. Although the prospect of achieving complete resolution of even the light chain fraction of normal immunoglobulin at the level of individual species (sequences) seems remote at present, interesting results can now be obtained by partial resolution of the normal light chain pool using gel isoelectric focusing (4). Using this method it is possible to resolve the normal mouse light chain pool into approximately 50-60 major and minor isoelectric focusing bands (IF-subgroups). Some of the more prevalent focusing bands represent as much as 2-3% of the total light chain pattern, but several of the minor bands clearly visible account for less than 0.5% of the total light chain. While it is not certain that even the minor bands visualized represent discrete molecular species of light chain, comparison of various inbred strains of mice has revealed polymorphism involving several of the minor bands and genetic studies indicate that the differences are transmitted in a simple co-dominant fashion (5). In the present work, the nature of polymorphism is examined in greater detail and the analysis of light chain focusing patterns is extended to a wider series of the inbred mouse strains.

METHODS

Normal immunoglobulin was purified from 0.2 ml of mouse serum (6) and was completely reduced and alkylated with ^{14}C-iodoacetamide as described (5). The procedures for separation of heavy and light chains using urea-formate gel electrophoresis as well as the procedure for isoelectric focusing of the separated light chains have also been described (5). The only modification introduced here is that isoelectric focusing was made across the length of the gel giving a longer separation distance (25cm). The maximum voltage applied for the lengthwise separation was 1200 V and this was applied for 5 hours. The final current obtained was usually about 3-4 ma.

RESULTS

Polymorphism in the focusing patterns: In our initial studies we reported that strains AKR/J, C58/J and RF/J possessed at least one extra band visible in the normal light chain focusing pattern compared to the more common (C57Bl/6J-type) light chain pattern. This band was shown to be present in the light chain pattern of (C58 X SWR) Fl hybrid animals indicating that the ability to produce this light chain band was heritable. Based on the observation that a relatively small proportion of the light chain pattern was affected, it was tentatively concluded that the polymorphism concerned the presence of certain v-region sequences unique to strains C58, RF and AKR (5). We have now extended our survey of the light chain focusing patterns of inbred mouse strains and we have found two additional isoelectric focusing phenotypes. The nature of the phenotypic variation is shown in Figure 1, which gives examples of the four patterns which we now distinguish.

The common pattern is that found in the C57Bl/6J and CBA/J strains (Figure 1). The majority of strains so far surveyed could not be distinguished from this pattern (table I). The patterns exhibited by RF/J and AKR/J were identical and shared a number of common features with the C58/J pattern. The AKR and RF patterns could readily be distinguished from the C58 pattern however and this is most clearly illustrated by the presence of band 63. So far no other strains have been found to share all of the features of the C58/J pattern. Interestingly, the fourth phenotype which we distinguish, as illustrated in the light chain patterns of individual NZB/BlnJ and BDP/J mice also shows some common features with the C58/J pattern. This is seen in the absence of band 63 as well as the apparent absence of band 20. The nature of the differences which we observe in the 4 phenotypes are summarized in table I. It is of considerable interest to note that genetic polymorphism of light chain structure has previously been reported to affect the light chains of AKR and C58 strains (7,8) and recent studies have suggested the existence of possible genetic differences in NZB light chain structure (9).

DISCUSSION

Signifiance of the focusing bands: The present results, in agreement with our earlier work (5,10) indicate that despite its extensive heterogeneity, the normal immunoglobulin light chain pool can be resolved into a finite series of discrete focusing "bands" or "IF-subgroups" by gel isoelectric focus-

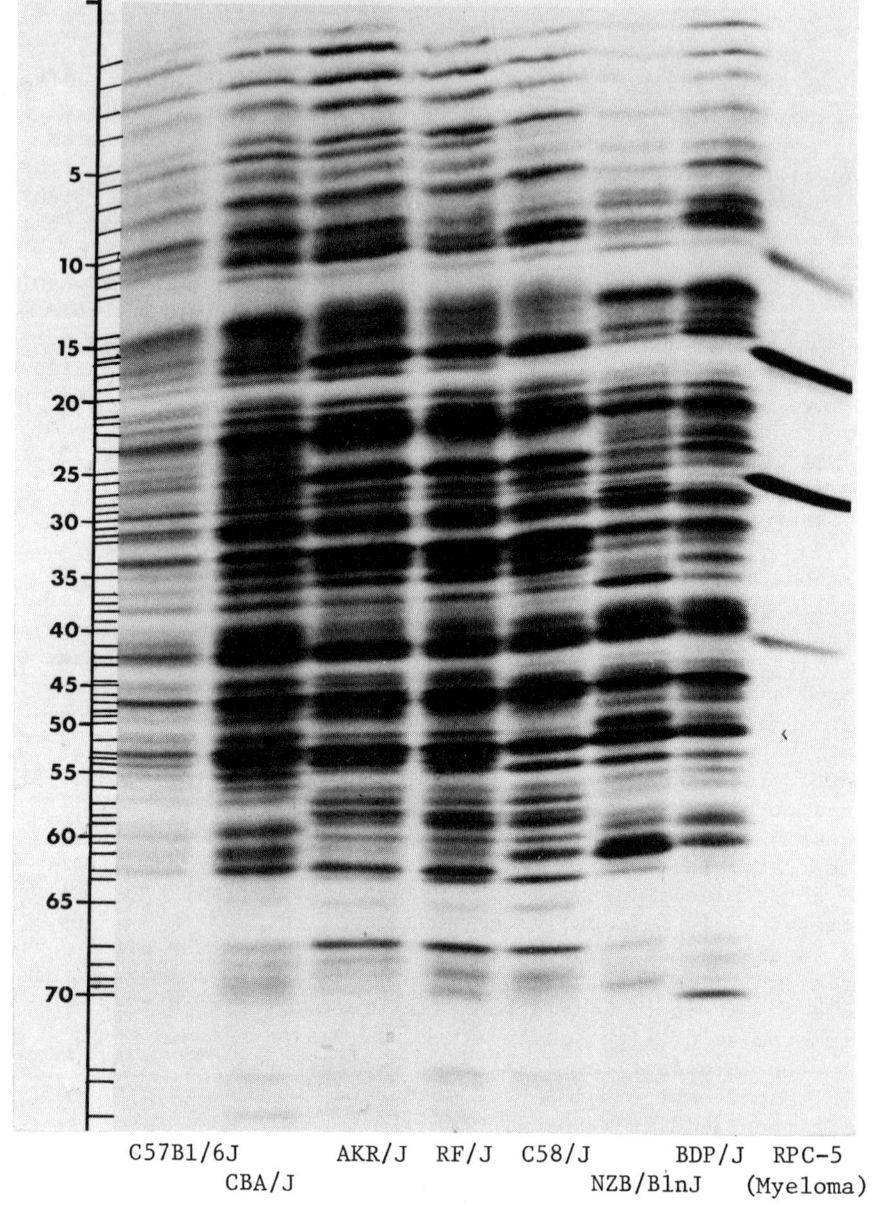

Figure I. Autoradiogram of isoelectric focusing patterns of normal mouse light chains (^{14}C-Carboxymethylamido-derivative). The numbers on the left indicate an arbitrary "band number", used for reference to table I. The gel contained 6,4 M urea and 2% carrier ampholine (LKB), pH range 3.5-10. The anode is at the top.

TABLE I. CHARACTERISTICS OF LIGHT CHAIN PHENOTYPES

Phenotype designation	Strain distribution	Bands elevated compared with phenotype A	Bands reduced compared with phenotype A
A (common)	A/J, A/HeJ, Balb/cJ, BuB/BnJ, CBA/J, CBA/CaJ, CBA/H-T6J,C57B1/6J, C57B1/10J, C57BR/cdJ, C3HeB/FeJ, C57L/J, DBA/2DeJ, DBA/2J, DBA/1J, LP/J, SM/J, ST/bJ	—	—
B	AKR/J, RF/J, PL/J	29,54,58, 59,61,64, 65,66	25,26,55, 56,60,62
C	C58/J	29,58,59, 61,62,64, 65,66	20,25,26, 54,60,63
D	NZB/BlnJ, BDP/J	62	20,22,54, 63

ing. The obvious question raised by these observations concerns the nature of the focusing bands. Although the possibility that normal light chains contain as few as 60-70 predominant sequences cannot be ruled out, it seems unlikely, for it would imply that the myeloma sequence data is totally nonrepresentative of the normal light chain pool. A more likely interpretation of the present results is that the "bands" observed in the isoelectric focusing patterns of normal light chains represent groups of light chains sharing very similar, but not necessarily identical amino acid sequences. Since the method involves separation based on charge, it is to be expected that light chains differing only by neutral substitutions would co-focus at identical positions in the gel. This is in fact observed in the case of the two λ-1 proteins MOPC104E and RPC-20 which we have recently examined. These two proteins differ by a single neutral amino acid substitution (11) and as expected, they focus at identical positions in the gel (corresponding precisely with the position of band 22 in the normal pattern). Co-focusing of several other pairs of closely related but not identical light chains (eg. M167 and M511) (12) has been previously noted (13). This interpretation would favour the suggestion that the focusing bands may represent groups of light chains belonging to the same v-region "subgroup". It is clear that isolation and characterization of individual bands will be required before any definite conclusions can be made regarding their possible relationship to the sequences found in the myeloma light chain data (12).

Nature of the Polymorphism: In order to demonstrate that the differences in focusing patterns observed in various mouse strains are genetically determined, it is necessary to establish that individual animals of the same strain all exhibit the essential features of the pattern. This has been done in the case of the C58, SWR and (C58 X SWR) F1 hybrid patterns (5) and we have now examined the light chain focusing patterns from a larger number of individual C58 and SWR mice. Comparison of the IEF patterns of light chains from individual mice of the same strain reveals significant individual variation, but so far it has been found to be of a quantitative nature and no evidence for the expression of latent phenotypes has been obtained (14). The importance of establishing the penetrance of the phenotype is evident, since individual differences at the level of clonal development would tend to obscure differences existing at the genetic level.

The fact that the differences in the focusing patterns observed in the 4 basic phenotypes affected a relatively small proportion of the light chain pattern suggests that the differences may involve the variable region of the light chain. One possible genetic interpretation for this observation is that certain strains of mice may simply lack the structural (v-) genes for certain "subgroups" of light chain. The fact that the differences in several instances (for example, bands 58,59 and 61 in strain AKR and the absence of band 63 in strain C58) seemed to involve the appearance or disappearance of discrete focusing bands would argue that each band may be under the control of a relatively small number of genes. The possibility that the differences observed may be due to factors regulating the expression of certain subgroups of light chains, as appears to be the case in rabbit heavy chains (15, 16) must also be considered.

ACKNOWLEDGEMENTS

I would like to thank Mme Diane Côté for careful technical assistance and my colleagues Dr. Henri Noël, M. Claude Lazure, Dr. François Lamy and Dr. Gilles Dupuis for encouragement and helpful discussions. I would like to thank Dr. S. Rudikoff, and Dr. M. Potter for supplying mouse myeloma proteins and tumors. This work was supported by the Medical Research Council of Canada, grant no. MT-4317.

REFERENCES

1. Williamson, A.R., Ann. Rev. Biochem. 45, 467 (1976).

2. Smith, G.P., Hood, L. and Fitch, W., Ann. Rev. Biochem 40, 969 (1971).

3. Hood, L., Barstad, P., Loh, E., and Nottenburg, C., in The Immune System: Genes, Receptors, Signals, ed. E.E. Sercarz, A.R. Williamson and C.F. Fox pp 119-39, Academic Press, New York 1974.

4. Awdeh, Z.L., Williamson, A.R. and Askonas, B.A., Nature (Lond.) 219, 66 (1968).

5. Gibson, D., J. Exp. Med. 144, 298 (1976).

6. Noël, H. and Gibson, D., in preparation.

7. Edelman, G.M. and Gottlieb, P.D., Proc. Nat. Acad. Sci. (U.S.) 67, 1192 (1970).

8. Claflin, J.L., Eur. J. Immunol. 6, (1976).

9. Loh, E., Weigert, M., Riblet, R. and Hood, L., Immunogenetics (1976).

10. Gibson, D., J. Immunol 118, 409 (1977).

11. Cohn, M., Blomberg, B., Geckeler, W., Raschke, W., Riblet, R. and Weigert, M. in The Immune System; Genes, Receptors, Signals, ed. E.E. Sercarz, A.R. Williamson and C.F. Fox; pp 89-117, Academic Press, New York 1974.

12. Hood, L., McKean, D., Farnsworth, V. and Potter, M., Biochemistry 12, 741 (1973).

13. Rudikoff, S. and Claflin, J.L., J. Exp. Med. 144, 1294 (1976).

14. Gibson, D., in preparation.

15. Strosberg, A.D., Hamers-Casterman, C., Van der Loo, W. and Hamers, R., J. Immunol 113, 1313 (1974).

16. Mudgett, M., Fraser, B.A. and Kindt, T.J., J. Exp. Med. 141, 1448 (1975).

INHERITED CONTROL MECHANISMS IN MAMMALIAN GENE EXPRESSION

A. Donny Strosberg

Institute of Molecular Biology, Vrije Universiteit, Brussel, B-1640 Sint Genesius Rode, Belgium

ABSTRACT. Recent evidence favors the idea that every member of a species carries a full set of structural genes for all the different alleles in a particular multigene system. A genetically stable control mechanism limits the number of alleles expressed in a normal situation to one gene in a homozygous and to two genes in a heterozygous individual.

Situations reflecting this pluri-allelic inheritance involve mainly highly polymorphic systems such as the immunoglobulin allotypes and the histocompatibility antigens. Serological results indicate striking differences between the supposed genotype in contrast to the found phenotype. Multiple sequence differences between the various alleles may indicate that polymorphism arose by gene duplication and independent mutation. Selection either resulted in the silencing or the elimination of allelic genes.

The expression of unexpected alleles in stress situation suggests the existence in normal circumstances, of a stable control mechanism for repression of silent genes. Various regulatory mechanisms are proposed.

INTRODUCTION

Inherited characteristics, such as blood groups, are generally thought to correspond to multiple forms (alleles) of structural genes. The combination of two alleles, one inherited from each parent, determines the phenotype of the individual. Alternatively structural genes for all allelic variants could be present in every individual, and the expression restricted to only two alleles by a genetically stable control mechanism (2). New evidence in favor of the latter hypothesis was reviewed recently (28) and is discussed further here. In addition, possible control mechanisms for the regulation of multiple gene expression are presented.

EVIDENCE FOR PLURI-ALLELIC INHERITANCE

I. <u>Immunoglobulin genetic markers (allotypes and idiotypes)</u> A high degree of serologic polymorphism has been found to correspond to multiple variants of immunoglobulin molecules (24). These antigenic determinants have been correlated with particular regions on variable and constant domains of the heavy and light chains. Although the polymorphic genes appear to segregate in a Mendelian fashion, a number of situations have been described in which cells or animals display phenotypes which do not correspond to the expected genotypes.

a. Pluri-allelic rabbits and latent allotypes. The discovery of a rabbit with three alleles of the group a and three alleles of the group b allotypes (23) prompted a search for the presence of unexpected alleles in other animals (9, 18, 23, 32). Several laboratories reported the existence of rabbits with more than two alleles of a given allotype and more numerous studies revealed the presence of latent allotypes in large families of rabbits. Recent serological studies reveal the presence of a2 molecules in a1a1 and in a1a3 rabbits (4, 18, 32). These a2 antigenic determinants were also found in cottontail rabbits and in hares, thus confirming the existence of allotypic polymorphism prior to divergence. In rabbits, these "ancient" a2 molecules were detected after hyperimmunization with Micrococcus lysodeikticus. It is likely that repeated massive injections results in a disruption of the normal control mechanisms operative in the immune system and thus leads to the expression of immunoglobulin genes which remain unexpressed in non-immunized animals. The first pluriallelic rabbit to be discovered was producing high concentrations of anti-Micrococcus lysodeikticus antibodies and the unexpected a2 molecules, purified by a specific anti-a2 immunoadsorbent, appeared to be of highly restricted heterogeneity (Strosberg and Van der Loo, unpublished observations).

In addition to the heavy chain variable region a allotypes and the light chain constant region b allotypes, the unexpected expression of the heavy chain constant region d and e markers was also reported (19,34).

b. Inherited idiotypes. Allotypes correspond to the more constant parts of the immunoglobulin molecule, the heavy chain variable region framework or the constant region. These markers are expressed constantly, albeit in sometimes very small concentrations. Idiotypes on the other hand correspond to the antigenic determinants of the immunoglobulin hypervariable regions (6), and their expression is therefore linked to the synthesis of a particular antibody specificity in response to an antigenic stimulus. The inheritance of idiotypic determinants was demonstrated both in the rabbit and the mouse (7, 16, 34). In a recent work, the expression of a heavy chain-specific idiotype was shown to be linked to that of a and d latent allotypes (34).

c. Unexpected immunoglobulin and histocompatibility markers. Unexpected markers may be expressed when animals or cells are subjected to unusual growth conditions. A BALB/c allotype was expressed in a virally infected non-BALB/c strain of mice (3). Various C_H allotypes were detected in human lymphocytes transplanted in neonatal Syrian hamsters (25), cultured in vitro by themselves (17) or with allogeneic cells to induce a mixed lymphocyte reaction (26). A non-self H2 histocompatibility antigen was found in a vaccinia-infected murine tumor cell line (11).

II. Structural variations between allelic forms. The extent of variation between certain allelic forms may in itself constitute evidence for pluri-allelic inheritance. If alleles evolved from each other by mutations and segregation of genes, it is difficult to understand how the b4, b6 and b 9 forms of the rabbit kappa light chain constant region differ by as much as 33% of the 110 residues (8, 29). Similar observations were made for the rat kappa chains (11) and the rabbit heavy chain variable region a markers (15,22). Recent data on the H-2 histicompatibility antigens in the mouse also predict the existence of an unexpectedly large number of amino acid sequence substitutions between the different alleles (5, 31).

If alleles arose at a single gene locus, the observed extent of differences most likely caused heterozygotes to gain a selective advantage over homozygotes. Evolutionary pressure working through recombination, would favor polymorphism in individuals instead of in the species, and would result in the coexistence of the former allelic genes on the same chromosome.

Alternatively, gene duplication may be suggested as a mechanism for generating the various alleles of a given marker. Subsequent independent mutations yield the multiple differences between corresponding polypeptide chains. Close genetic linkage explains the pseudo-allelic behavior in inheritance studies.

The coexistence on the same chromosome of families of closely linked and related genes has been suggested, at least for the a group allotypes, by several studies which describe the simultaneous presence in single rabbits of at least six variants of the a1 allotype (4, 13), and probably as many for the a2 and a3 forms. These serologically defined subspecificities may correspond to the various structural differences found between the framework residues of different a1 molecules (20,33).

EVIDENCE FOR A REGULATORY CONTROL OVER MULTIPLE GENE EXPRESSION

In addition to the observations developed in the

preceding paragraph, the evidence for regulatory mechanisms may be adduced from the following considerations:

I. The existence of a quantitative regulation, as suggested by the unequal synthesis of allelic forms in heterozygotes (pecking order), and of a qualitative regulation, as suggested by suppression experiments in homozygotes, which result in the enhanced expression of otherwise almost undetected markers such as the x and y heavy chain allotypes borne by the blank molecules.

II. The relative high frequency (1 to 5%) of so-called somatic recombinant molecules, displaying the a and d or e allotypic markers in trans rather than in cis configuration.

III. The sequence of expression of the β-like hemoglobin genes during the development of the individual and the occasional activations of silent genes (21).

IV. The selective deficiency of IgA in human disease. This deficiency occurs in ca. 0.1% of all persons. In some families the defect is inherited as an autosomal dominant or recessive gene and may be found associated with two other autosomal genetic abnormalities, ataxia telangiectasia and partial deletion of chromosome 18. In some cases, the defect may be acquired, namely as a consequence of congenital infections such as rubella or toxoplasmosis.

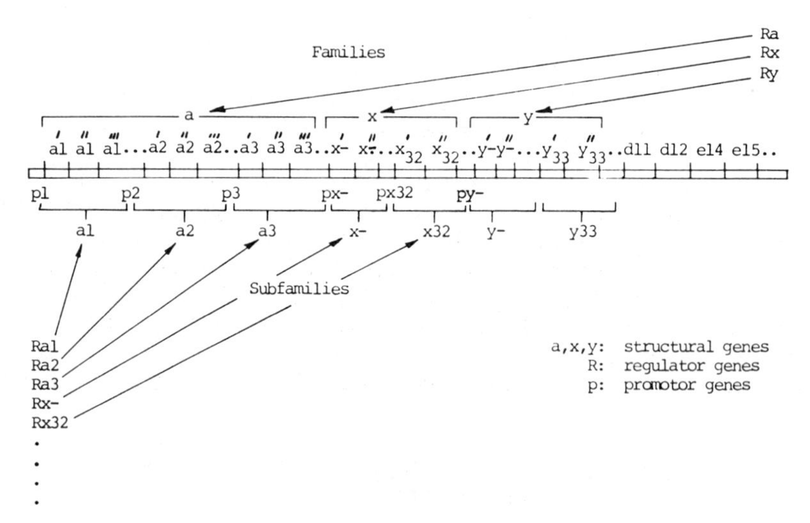

Fig. 1 The rabbit immunoglobulin heavy chain genome as a model for multigene families.

In this model (Fig. 1) the allotypic subspecificities of the a1, a2, and a3 are presented as coexisting on the same chromosome and the existence of similar subspecificities for the x32, x - y33 and y-is predicted. The expression of these various genes is under the control of a series of regulator genes which may be truly allelic. Several levels of control are postulated to account for the fact that the expression of a whole family of genes may be suppressed simultaneously and that the expression of the a1, a2, and a3 subfamilies of genes is interdependent.

The regulator genes act on the respective promotor regions. Deletion of a subfamily promotor may result in "constitutive" pluri-allelism as seen in two related wild rabbits which in addition to complete aI and aI00 sets of genes also displayed incomplete a2 and a3 allotypic specificities (4).

In individual lymphocytes which arrive at a certain degree of maturation, activation of only one type of regulator gene results in apparent allelic exclusion. The switch from the μ to the γ gene expression is also maturation-dependent and probably similarly regulated.

The production of small amounts of alternate gene products leads to the synthesis of autoanti-allotype antibodies which constitute one of the elements of the control circuit.

Similar models may be proposed to represent other multigene families such as the mouse idiotype genes, the histocompatibility antigens or any other family of genes in which whole subfamilies may be permanently or temporarily repressed.

The main advantage of this model is that it eliminates a major objection to the germ-line origin of antibody diversity, namely the potential scrambling of allelism due to multiple and successive crossing-overs between the genes corresponding to the various subspecificities of the a allotypes. If allelism resides only in a small number of control genes, which may be localized so as to prevent their elimination or rearrangement by recombinations, antigenic determinants corresponding to the different alleles may be conserved on multiple genes throughout evolution. The fact that the rabbit allotypic markers are also polymorphic in hares and other lagomorphs suggest that the various alleles arose before speciation and confirms that an *ad hoc* protection mechanism evolved at the same time.

Nature of the regulator genes. The regulator genes might control the fusion of the constant and variable regions of the heavy chains. This could explain the appearance of latent d alleles in conjunction with latent a forms (34) as well as the various findings of linked expression of variable and constant region markers (6).

COMPLEX CONTROL CIRCUITS

Few regulatory mechanisms are well understood at the present time, especially in mammals. Although examples have been found in which bacterial models seem to fit well, it is generally admitted that much more complex regulatory mechanisms operate in multicellular organisms.

Recently (30) a complex control circuit has been proposed to regulate the expression of genes in an inherited fashion in temperate bacteriophages.

Two very different behaviors are displayed by these viruses, the killing of the host bacterium or the establishment of a permanent association with this cell. The choice between a lytic or a lysogenic response depends on complex control interactions between a set of genes and is dependent on environmental factors. Although this model describes monocellular organisms, the ability to choose between two stable pathways reminds one of cell determination and differentiation in higher organisms.

Cell determination for a given character depends on a set of genes whose interactions constitute as many feed-back loops. Additional controls grafted onto these loops will determine which of the possible stationary states of gene expression will be chosen.

One such network of interactions among various components of the immune system was described earlier by Jerne (14), who visualised idiotypes and anti-idiotype antibodies as the interacting signals. The response of the immune system to an antigen constitutes the environmental factor which will perturbate a stationary state and provide the stimulus for a sequence of reactions leading to the multiplication and differentiation of lymphocytes and increased production of antibodies.

More recently an analogous network of interacting components was proposed to be constituted by allotypes, anti-allotype antibodies and allotype suppressor T cells (13). Antigens do modulate the allotype expression and may therefore again constitute one of the external stimuli. An inherited control mechanism may also play an important role. Analogies between the idiotype and the allotype networks strongly suggest common regulatory pathways. These may yet serve as models for other suspected networks operative at the level of the histocompatibility antigens or the interacting components of the nervous system.

REFERENCES

1. Bell, C. and Dray, S., (1973) *Cell Immund.* 6, 375.
2. Bodmer, W.F., (1973) *Transplant. Proc.* 5, 1471.
3. Bosma, M. and Bosma, G. (1974) *J. Exp. Med.* 139, 512.
4. Brézin, C. and Cazenave, P.A. (1976) *Am. Immunol.* (Inst. Pasteur) 127C, 333.
5. Capra, J.D., Vitetta, E.S., Klapper, D.G., Uhr, J.W. and Klein, J. (1976) *Proc. Natl. Acad. Sci.*, U.S.A. 73
6. Capra, J.D. and Kehoe, J.M. (1975) *Adv. Immunol.* 20,40.
7. Eichmann, K. (1975) *Immunogenetics* 2, 491.
8. Farnsworth, V., Goodfliesh, R., Rodkey, S. and Hood, L. (1976) *Proc. Natl. Acad. Sci.* U.S.A. 73, 1293.
9. Francis and Mandy, W.J. (1976) unpublished observations.
10. Garrido, F., Festenstein, H. and Schirrmacher, V. (1976) *Nature* 261, 705.
11. Gutman, G.A., Loh, E. and Hood, L. (1975) *Proc. Natl. Acad. Sci.*, U.S.A. 72, 5046.
12. Horng, W.J., Knight, K.L. and Dray, S. (1976) 116, 117
13. Herzenberg, L.A., Okumura, K., Cantor, H., Sato, V.L., Shen, F. W., Boyse, E.A. and Herzenberg, L.A. (1976) *J. Exptl. Med.* 144, 330.
14. Jerne, N.K. (1971) *Eur. J. Immunol.* 1, 1.
15. Jaton, J.C., Braun, D.E., Strosberg, A.D., Haber, E. and Morris, J.E. (1973) *J. Immunol.* 111, 1838.
16. Kuettner, M.G., Wang, A. and Nisonoff, A. (1972) *J. Exptl. Med.* 135, 573.
17. Lobb, N., Curtain, C.C. and Kidson, C. (1967) *Nature* 214, 783.
18. Mage, R.G., Rejnek, J., Young-Cooper, G.O. and Alexander, C. (1977) *Eur. J. Immunol.* Submitted
19. Mandy, W.J. and Strosberg, A.D. (1977) Manuscript in preparation.
20. Margolies, M.N., Cannon, E., Kindt, T.J. and Fraser, B. A. (1977) *J. Immunol.* In press.
21. Marks, P.A. and Rifkind, R.A. (1972) *Science* 175, 955.
22. Mole, L.E., Jackson, S.A., Porter, R.R. and Wilkinson, J.M. (1971) *Biochem. J.* 124, 301.
23. Mudgett, M., Fraser, B.A. and Kindt, T.J. (1975) *J. Exptl. Med.* 141, 1448.
24. Oudin, J. (1960) *J. Exptl. Med.* 112, 107
25. Pothier, L., Borel, H. and Adams, R.A., (1974) *J. Immunol.* 113, 1984.
26. Rivat, L., Gilbert, D. and Ropartz, C. (1973) *Immunology* 24, 1041
27. Strosberg, A.D., Hamers-Casterman, C., Van der Loo, W. and Hamers, R. (1974) *J. Immunol.* 113, 1313.

28. Strosberg, A.D. (1977) *Immunogenetics*, In press.
29. Strosberg, A.D. and Janssens, L. (1979) *Ann. Immunol.* (Inst. Pasteur) 128 C, 351.
30. Thomas, R., Gathoye, A.M. and Lambert, L. (1976) *Eur. J. Biochem.* 71, 211.
31. Uhr, J.W., Vitetta, E.S., Klein, J., Poulik, M.D., Klapper, D.G. and Capra, J.D. (1977) Cold Spring Harbor Symposium 41
32. Van der Loo, W., De Baetselier, P., Hamers-Casterman, C. and Hamers, R. (1977) *Eur. J. Immunol.*, In press.
33. Van Hoegaerden, M. and Strosberg, A.D. (1976) FEBS Letters 66, 35.
34. Yarmush, M.L., Sogh, J.A., Mudgett, M. and Kindt, T. J. (1977) *J. Exptl. Med.* In press.

Summary of the Workshop on V Region Genetics

Antibody variable regions, either V_HV_L or separate V_H and V_L domains have been classified by at least four methods: serological, isoelectric focusing (spectrotypes), fine-specificity, or primary sequence. Of these, particularly for V_H characterization, the serological method is the one most widely used. The percentage of a serologically defined marker on the total "non-immune" immunoglobulin of an individual varies. At one extreme are the rabbit V_H allotypes which can be detected on 70-90% of the immunoglobulin, and as shown by Margolies (Boston) and R. Mage (Bethesda) these determinants appear to be related to V_H framework sequence differences. Such broadly distributed V_H markers have not been detected in the mouse; instead serologically defined V_H markers associated with a more restricted group of V_H regions such as the U10/173 marker (Bosma, Philadelphia) have been found among myeloma heavy chains with V_H sequence homology, but unrelated specificity. This marker also can be detected in the "non-immune" serum of certain inbred strains. Markers on V_L have been detected by sequencing (Gottlieb, Boston) or I.E.F. (Gibson, Sherbrooke). These are detected in mouse serum at a level of a few percent and are at least in part associated with the framework sequence of a group of related sequences (subgroups?). Finally, there are a number of V region markers that are at low or undetectable levels in unimmunized animals. These examples defined by anti-idiotypic sera, spectrotype or fine specificity are generally dependent on specific V_H and V_L interaction and are generally correlated with antibodies of a particular specificity. Consequently the V region markers detected on specifically induced antibodies fall in this category.

A number of V region markers have been genetically analysed by the use of family studies, congenic strains or recombinant-inbred (RI) strains. These markers ordinarily segregate as single loci and are linked either to the locus controlling immunoglobulin heavy chain allotypes (C_H) or in the case of the V_κ (Claflin, Ann Arbor, Gibson) to Ly-2,3 on chromosome 6. The former result suggests the possibility that these V region markers are coded for by germ-line V_H genes. Alternatively V region markers may be under regulatory gene control.

The possibility of regulatory control has been raised by the observations of latent V region allotypes or idiotypes (Strosberg, Belgium, Kindt, New York, and Sigal, Philadelphia) which can be interpreted to mean that all individuals inherit all V region genes and that an individual or an inbred strain expresses mainly a subset (the nominal allotype) of the total. In certain instances as shown by Mage the latent V_H allotypes of the rabbit represent a minor subpopulation of the spectrum of molecules belonging to the nominal allotype and hence the phenomenon can be explained without invoking regulatory control over allotype expression. On the other hand, as reported by Kindt examples where the latent allotype appears to be equivalent to a nominal allotype have been found suggesting identity between the V_H structural genes of rabbits expressing different nominal allotypes. Thus the issue of structural versus regulatory gene inheritance is not settled and may only be decided by studies at the DNA level and the term V-region marker must be used in an operational sense.

V_H region markers have been analysed in considerable detail with regard to linkage to C_H and other C_H-linked loci (other V_H genes or the locus controlling prealbumin, Pre). The earlier studies on linkage of V_H to C_H in the rabbit that showed a recombination frequency of about 0.4 (summarized in Hamers-Casterman and Hamers, Immunogenetics 2, 597 (1975)) have been extended to the V_H markers in the mouse. Since in contrast to rabbit V_H allotypes, the mouse V_H markers all appear to represent a different subpopulation(s) of the total immunoglobulin, a variety of recombination frequencies between these and C_H have been obtained. They range from 0.41% for V_H-DEX (Riblet, Philadelphia) to 7-11% for V_H-Nase 1 (Pisetsky, Bethesda). Intermediate values are found in other cases: V_H-A5A$^+$, 3%, S117$^+$, 3%, A5A^{Cr}, 1% S117^{Cr}, 2% (Eichmann, Heidelberg), V_H-NP, 0.6%-3%, V_H-NBrP, 2% (Mäkelä, Helsinki). These frequencies are derived from studies on backcross segregants, congenic lines and RI strains and may be subject to a variety of experimental errors particularly in the case of RI strains if expression of a V_H marker is affected by the background genome. This has been indicated by Mäkelä's finding that the quantitative expression of V_H markers can be altered depending on the background genome. Thus recombination frequencies may be over estimated in RI lines as an unexpected phenotype could be observed without a cross-over event especially if the share of the idiotype of the total antibody is low. Such an example of a misleading background effect might be the V_H-NP case for which the recombination frequency can vary from 0.5% to 3% depending on the strains used in the cross. Either the arrangement of the same or similar V_H genes may not be the same in all strains thereby leading to different recombination frequencies as suggested by Eichmann or differential expression due to genetic background of V_H markers may distort the genetic analysis.

Several examples of intra-V_H recombination have been found. These were detected by further analysis of V_H-C_H recombinants that arose in backcross segregants as for example bb7 (Eichmann, Nisonoff) or BAB/14 (Riblet, Pisetsky, Mäkelä, Eichmann). In addition, several examples have been inferred by the distribution of several V_H markers (V_H-ESE, V_H-NP, V_H-A5A^{Cr}, V_H-117^{Cr}) in the BXD RI lines (Taylor, Bar Harbor). It appears that cross-overs initially identified between a given V_H marker and C_H inevitably seem to have occurred intra-V_H. Furthermore in certain instances such as the strain AL/N or BXD RI lines multiple crossing over within V_H has been suggested (Eichmann, Taylor). Another instance of multiple crossing over was described (Riblet) who observed an unusually high frequency of double cross-overs in the region V_H-DEX, C_H, Pre. Thus, barring the above-mentioned reservations regarding the interpretation of V region structural genes, the genetic analysis of V_H markers may indicate that the locus controlling V_H structure is large or that perhaps due to the special multigenic nature of antibody genes, these loci may be subject to unusual genetic effects.

O. Mäkelä
University of Helsinki
Helsinki, Finland

M. Weigert
The Institute for Cancer Research
Philadelphia, Pa.

HEAVY CHAIN VARIABLE REGION IDIOTYPES ON HELPER T CELLS

K. Eichmann

Institute for Immunology and Genetics, German Cancer Research Center, 6900 Heidelberg, F.R.G.

ABSTRACT. The structural basis for the shared idiotypes between the antigen receptors of B and T cells was investigated using anti-idiotypic antisera of defined preferential or exclusive specificity for either the V region of the heavy chain or the V region of the light chain of an antibody to streptococcal Group A carbohydrate. The antisera were tested (i) for their ability to sensitize helper T cells *in vivo* that cooperate *in vitro* with hapten-specific B cells via the recognition of streptococci as carrier and (ii) for their ability to inhibit helper T cells specific for Group A carbohydrate in their cooperation with hapten-specific B cells in the same *in vitro* system. The data show that anti-idiotypic antibodies with V_L reactivity do not interact with T helper cells whereas anti-idiotypic antibodies with V_H reactivity do. It is concluded that the T helper cell carries a receptor that shares with antibodies the V region of the heavy chain. The V region of the light chain is either absent or inaccessable to anti-idiotypic antibody.

INTRODUCTION

T cells possess antigen receptors which carry idiotypic (Id) determinants in common with antibody molecules. This has been brought out by (i) experiments in which anti-idiotypic antibodies (anti-Id) against antibody molecules were analyzed for their reactivity with T helper cells (1,2,3,4) and with T suppressor cells (5), (ii) experiments in which anti-idiotypic antibodies against T cell receptors were analyzed for their reactivity with antibody molecules (6,7). In both types of experiments, idiotypic crossreactivity could be clearly shown to exist using functional assays such as sensitization of helper cells *in vivo* (1), inhibition of helper cells *in vitro* (2,3) and *in vivo* (4), inhibition of graft-versus-host-reactions *in vivo* (6,7). Direct demonstration of idiotypic crossreaction between antibodies and T cells involved hemagglutination assays and the measurement of uptake of antibody onto cells by radioimmune assays or enzyme-tagged detection systems (6,8).

In all of these experiments, the question as to the structural basis of shared idiotypes between B and T cell receptors remained open. Some light was shed on this question by genetic experiments that suggested the linkage between genes that control the responsiveness of T helper cells to anti-idiotypic antibody and genes controlling immunoglobulin (C_H) allotype in the mouse (3,9). These experiments suggested that V_H genes control T cell responsiveness to anti-Id and thus specify the T cell receptor idiotype (3,9). In similar experiments, a genetic linkage was established between the immunoglobulin heavy chain allotype and T cell idiotype expression in the rat (6). In another series of experiments it was shown that anti-idiotypic antisera that preferentially react with the V region of the heavy chain are effective in helper cell sensitization whereas anti-idiotypic antisera that preferentially react with the V region of the light chain are not (9). Preferential reactivity of an anti-idiotypic antiserum in these experiments was demonstrated in two ways: Antisera were tested (i) for linkage or absence of linkage to allotype in the segregation patterns of the idiotypes they detect and (ii) for the binding of radiolabelled artificial antibody molecules recombined from heavy and light chains isolated from the idiotypic antibody molecule and from pooled mouse immunoglobulin (MIg), respectively. Preferential V_H reactivity was concluded when an anti-idiotypic antiserum identified an idiotype that segregates in breeding experiments in an allotypic-linked fashion, and binds quantitatively more of an antibody molecule recombined from the idiotypic heavy chain and MIg light chain than of an antibody molecule recombined from the idiotypic light chain and MIg heavy chain. Preferential V_L reactivity of an anti-idiotypic reagent was concluded in the reverse situation (9).

The previous experiments had revealed an all-or-nothing difference with respect to helper cell sensitization between the two kinds of anti-idiotypic reagents, using *in vivo* adoptive transfers for analysis (9). In the present experiments, dose-response analyses were performed to monitor the relative effectiveness of various anti-idiotypic reagents in helper cell sensitization *in vivo*, using an *in vitro* micro-culture system for analysis. Furthermore, the anti-idiotypic reagents were tested for their relative ability to inhibit helper cell function *in vitro*. The results clearly indicate that only those anti-idiotypic antibodies that react with the V_H region of the idiotypic antibody interact with T cell receptors. Anti-idiotypic antibodies with V_L reactivity are unreactive with T cells.

RESULTS

The experimental system. All of the experiments discussed below employ antibodies and T helper cells specific for Group A streptococcal carbohydrate (A-CHO), generated in strain A/J mice by immunization with Group A streptococcal vaccine (Strep.A) (10). A major portion of the antibodies of strain A mice to A-CHO represents a single antibody species which is defined by its isoelectric focussing spectrum and by idiotype (Id) (11,12). The Id is detected by guinea pig anti-idiotypic antisera (anti-Id) raised against the anti-A-CHO antibody produced by the A/J lymphocyte clone A5A (10), or recently, against the antibody fraction obtained by affinity chromatography from an A/J anti-A-CHO pool, using N-acetyl-glucosamine as the eluting reagent (10). Guinea pig anti-idiotypic antibodies are separated into their IgG1 and IgG2 classes by agarose block electrophoresis (5). The IgG1 fraction thus obtained induces immunity to A-CHO when injected into A/J mice (1). Sensitization expresses itself in the generation of B memory cells and T helper cells specific for A-CHO, which differ from the respective cell populations induced by immunization with Strep.A by their degree of homogeneity. Whereas the latter are idiotypically heterogeneous, virtually all B and T cells that are induced by anti-Id carry the A5A idiotype (2). This experimental system allows the analysis of T helper cell idiotypes by determining the helper activity induced in the spleens of mice injected with anti-Id (IgG1). Helper activity in the spleen is determined _in vitro_ using a microculture system (2,3). The TNP hapten coupled to Strep.A as carrier is added to the cultures and anti-TNP plaque forming cells (PFC) are determined. T helper cell idiotypes can be further analyzed by adding various anti-idiotypic antisera to the cultures and determining the degree of inhibition of carrier-recognition by primed helper cells. The experimental details of the system are described in ref. (2), and have not been altered except for the use of Click's medium as modified by Schwarz (13) instead of Mishell-Dutton medium for cell cultures.

Anti-idiotypic reagents with V_H or V_L specificity. Three anti-idiotypic antisera were used in these studies. Two of these antisera were prepared by conventional immunization of guinea pigs with native A5A antibody as described (10,14). These two antisera were selected from a larger collection of antisera because they showed highly preferential activity towards either the V_L or the V_H region of the A5A antibody (9). The binding of native A5A, native MIg, and of the heterologous polypeptide chain recombinants by these two antisera is represented by the binding curves in Fig.1a and b. Isolation and recombination of immunoglobulin polypeptide chains as well as the binding assay were done as previously described (9,12). A third antiserum was produced using a recombined molecule made from the A5A H chain and MIg L chain as antigen. As shown in Fig.1c, this antiserum has exclusive binding specificity for molecules carrying the A5A H chain, is unreactive with molecules carrying the A5A L chain and the nature of the L chain is seemingly irrelevant for this antiserum.

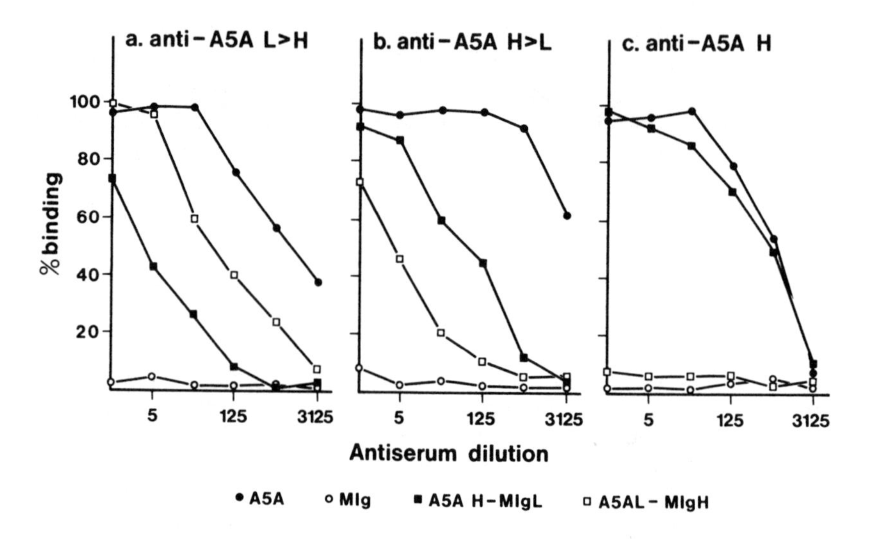

Fig.1

Fig.1: Anti-idiotypic antisera to A5A idiotypic determinants with preferential V_L specificity (frame a), preferential V_H specificity (frame b) and exclusive V_H specificity (frame c). The binding curves were obtained using 20 ng ^{125}J labelled A5A, anti-Id in dilutions given at the abscissa and rabbit anti-guinea pig Ig serum for precipitation (12).

From the 50% points of the binding curves in Fig.1 it can be calculated that the antiserum in frame a binds 26 µg A5AL-MIgH/ml whereas only 2 µg of A5AH-MIgL/ml are bound. Thus, this antiserum has a 13 times greater binding capacity for A5A L chains than for A5A H chains and is referred to as anti-A5A L>H. The antiserum in frame b binds 27 µg A5AH-MIgL/ml and 2 µg A5AL-MIgH/ml, and thus has a 14 times greater binding capacity for A5A H chains than it has for A5A L chains. This antiserum is referred to as anti-A5A H>L. The antiserum in frame c binds 64 µg/ml of native A5A molecules and 58 µg/ml of A5AH-MIgL, whereas the binding for A5AL-MIgH is not above the binding of native MIg molecules. This antiserum will be referred to as anti-A5AH. Unfortunately it was thus far not possible to obtain an antiserum by immunization with A5AL-MIgH that fullfilled the criteria of idiotypic specificity and was essentially V_L specific. Therefore, with respect to the exclusion of V_L determinants on T cells, the experiments reported here rely on the use of the preferentially V_L reactive antiserum represented in Fig.1a, and on the use of another antiserum that bound 4 times more A5AL-MIgH molecules than A5AH-MIgL molecules. The data obtained with this antiserum are not shown because they are in principle similar, though less pronounced, than that obtained with the other preferentially V_L reactive antiserum (Fig.1a).

Differential capacity to sensitize T helper cells of V_L and V_H reactive anti-Id. The IgG1 fractions were obtained from the antisera whose binding curves are represented in Fig.1 by agarose black electrophoresis, as previously described (5). The IgG1 fractions were analyzed for their binding of native and recombinant antibody molecules in a fashion similar to that employed for the analysis of the unfractionated antisera (see Fig.1) and no significant deviations were demonstrated (data not shown). Adult male or female A/J mice were injected with the IgG1 fractions of the three antisera in amounts corresponding to 0.01 µg, 0.1 µg, and 1.0 µg binding capacity for native A5A (IBC). Control mice were injected with 10^9 Strep.A organisms or with saline. Injections were done intraperitoneally as previously described (1).Two-6 weeks later the spleens of two mice of each group were pooled for each experiment and analyzed for Strep.A specific helper activity by determining the generation of plaque forming cells (PFC) against TNP-SRBC after a 4 day culture of 10^6 lymphocytes together with 10^7 Strep.A-TNP organisms in 100 µl of Click's medium (2,13). Of eight identical cultures, 4 were tested for TNP-PFC, 2 for SRBC -PFC and 2 for the number of viable cells. In some experiments, the spleens of mice injected with anti-Id were also analyzed for the induction of A-CHO specific memory B cells in cultures stimulated with uncoupled Strep.A and analyzed for PFC against A-CHO-SRBC (2). The results of three representative experiments are summarized in Table 1. The PFC numbers of each experiment are represented as percent of that of cultures of Strep.A sensitized cells. The helper and precursor activity in such cells in most experiments exceeds that of cells primed with anti-Id and, since the PFC numbers obtained in different experiments are quite variable, a comparison of different experiments is more informative if the PFC numbers are normalized according to the maximum response. The helper cell experiments are done with a high density of TNP groups on the Strep.A organisms because unpublished experiments have shown that in this situation the precursor cell supply is practically unlimited so that the response is limited only by the helper cell number.

TABLE 1

Sensitization of helper and precursor activity by anti-Id sera with V_H or V_L reactivity.

Sensitizing Anti-Id (IgG1)	µg IBC[3]	Exp.1 PFC ā TNP[4]	Exp.1 PFC āA-CHO[5]	Exp.2 PFC ā TNP[4]	Exp.3 PFC ā TNP[4]
-		7[1]	4[1]	11[1]	4[1]
āA5A L>H	0.01	6	24	12	14
	0.1	21	64	19	60
	1.0	42	97	45	118
āA5A H>L	0.01	39	20	61	44
	0.1	59	79	90	80
	1.0	65	114	57	81
āA5A H	0.1	53	84	60	97
Strep.A(10^9)		100 (214)[2]	100 (69)[2]	100 (191)[2]	100 (135)[2]

1) PFC responses in percent of the response obtained with Strep.A primed cells (= 100%)

2) Geometric mean of total PFC number. Standard deviations ranged between 1.1 and 1.5

3) IBC = idiotype binding capacity (see text)

4) TNP-PFC were determined in cultures of 10^6 spleen cells stimulated with 10^7 Strep.A-TNP particles, after 4 days of culture period (2).

5) āA-CHO-PFC were determined in cultures of 10^6 spleen cells stimulated with 3 x 10^6 Strep.A particles, after 4 days of culture period (2).

The results summarized in Table 1 clearly show that anti-A5A H>L is more effective in helper cell sensitization than anti-A5A L>H. All doses of anti-A5A H>L induce helper cells. With increasing doses of anti-A5A H>L injected, the helper effect increases to an optimal response at 0.1 μg and a further 10 fold increase of this anti-Id does either not improve or even impair helper cell induction. In contrast, anti-A5A L>H is inactive at 0.1 μg and an increase of the dose leads to an increase of the helper effect up to the highest dose used. A comparison of the dose-response relationships between thetwo antisera suggest a roughly 10 fold greater effectiveness of anti-A5A H>L than of anti-A5A L>H. This difference reflects the 13 to 14 fold differences in binding activity towards V_H and V_L regions, respectively, between the two antisera. The antiserum with exclusive H chain specificity is as effective in helper cell induction than is anti-A5A H>L.

In order to exclude the possibility that the relative uneffectiveness of anti-A5A L>H was due to a general inability to sensitize cells and was unrelated to its preferential V_L specificity, B cell sensitization was also analyzed in some experiments. It is clear from the anti-A-CHO PFC numbers that there is no significant difference in the number A-CHO specific B memory cells in mice sensitized with anti-A5A L>H and anti-A5A H>L. This result was somewhat unexpected because the A-CHO specific PFC response reflects the priming by anti-Id in both the B and T cell compartments (15). Since all the A-CHO specific PFC are IgM PFC (16), however, it is not unreasonable to assume that this secondary response is less T dependent so that even with poor helper cell activity reasonable responses are achieved. Taken together, the data suggest that with respect to B cell sensitization there is no difference between anti-Id with preferential V_L or V_H activities, and that the differencesin T helper sensitization are therefore related to these specificity differences.

Differential capacity to inhibit T helper cell activity of V_L and V_H reactive anti-Id. The same antisera whose IgG1 fractions were tested for T helper cell induction were analyzed for their capacity to inhibit T helper cells *in vitro*. This was done by including the unfractionated antisera into the culture medium at various dilutions as previously described (2). The spleen cells used for these experiments came from mice preimmunized with the IgG1 fraction of anti-A5A H. It has been previously shown that Strep. A-primed helper cells are poorly inhibited by anti-Id due to the heterogeneity of the antigenic determinants associated with Strep.A (2). Therefore cells primed with anti-Id had to be analyzed.

The results of two representative experiments are summarized in Table 2. The TNP-PFC responses of cultures of anti-Id-primed cells without inhibitor are taken as 100% and the

TABLE 2

Inhibition of helper activity by anti-Id sera with V_H or V_L reactivity.

		Exp.1 PFC $\bar{a}$TNP[5]	Exp.2 PFC $\bar{a}$TNP[5]
Nonprimed cells		4[1]	12[1]
$\bar{a}$Id primed cells		100(104)[2]	100(165)[2]
$\bar{a}$Id primed cells + inhibitory anti-Id			
NGPS[3]	1:1350	79	108
	1: 450	72	85
$\bar{a}$MIg(MIg abs)[4]	1:1500	101	103
	1: 500	89	74
$\bar{a}$A5A L>H	10 ng[6]	89	n.d.
	30 ng	85	88
$\bar{a}$A5A H<L	10 ng	39	45
	30 ng	13	11
$\bar{a}$A5A H	1 ng	50	n.d.
	3 ng	31	34
	10 ng	14	10

1) PFC number in percent of that of cultures without inhibitor

2) Geometric mean of total PFC number. Standard deviations ranged between 1.1 and 1.5

3) Normal guinea pig serum

4) Guinea pig anti-MIg antiserum that has been absorbed on MIg sepharose

5) See footnote 4 Table 1

6) See footnote 3 Table 1

degree of inhibition is reflected by the depression in PFC numbers which are expressed in percent of the results of cultures without inhibitors. Control inhibitors include normal guinea pig serum (NGPS) and a guinea pig anti-MIg antiserum that had been absorbed on MIg-sepharose. These sera were used in concentrations reflecting those at which the anti-Id sera were employed. It is clear from the data that inhibition by unspecific sera does not exceed 30% and that the inhibition by anti-A5A L>H is within that range. In contrast, anti-A5A H>L inhibits in a dose dependent fashion so that with 30 ng IBC/culture the TNP-PFC response is inhibited to background level. Interestingly, the anti-A5A H antiserum is still more effective in helper cell inhibition than the anti-A5A H>L, whereas no significant difference had been seen between the two antisera in the helper cell sensitization experiments. This may be due to the fact that the helper cells used in these experiments were selected by presensitization with this anti-idiotypic antibody. Further experiments will have to be done to investigate the interesting question whether each anti-idiotypic antibody is more effective in the inhibition of helper T cells that have been sensitized by the same anti-Id than by other anti-Id, even though all anti-Id have been raised against the same idiotypic antibody. Apart from this difference, the results are in perfect accord with the sensitization experiments in suggesting that anti-idiotypic antibodies with V_L reactivity do not react with T helper cells whereas those with V_H reactivity do.

DISCUSSION

The results presented in this paper demonstrate, together with data from other laboratories (6,9,17) that the sharing of idiotypic determinants between T cell receptors and antibody molecules is based on idiotypic determinants associated with the variable region of the immunoglobulin heavy chain. The evidence presented here shows that anti-idiotypic antibodies with reactivity towards the heavy chain of the original idiotypic antibody activate T helper cells *in vivo* as well as inhibit the same T helper cells *in vitro*. In previous publications we have reported that antisera with preferential reactivity towards the light chain of the original antibody are poor in sensitizing T cells *in vivo* (9). In this paper it could be shown that such antisera are also able to sensitize T cells effectively but at higher concentrations than antisera with preferential heavy chain reactivity. The differences in the doses needed for optimal sensitization approximately paralelled the differences in binding capacity for recombined antibody molecules consisting of either the idiotypic H chain together with pooled immunoglobulin L chain, or of idiotypic L chain together with pooled immunoglobulin H chain. From these experiments it was concluded that the T helper cell sensitizing antibody population in anti-idiotypic antisera with preferential V_L reactivity is the minor antibody population in these antisera that possesses V_H reactivity. Both kinds of antisera are equally efficient in B cell priming. This shows that their differential effect on T helper cells is not due to a general inability of certain antisera to sensitize lymphocytes. The data were even clearer in the in vitro inhibition experiments in which antisera with preferential V_L reactivity were completely ineffective in helper cell inhibition at concentrations at which antisera with preferential or exclusive V_H reactivity completely inhibited the helper effect. Higher concentrations at which antisera with preferential V_L reactivity are expected to become inhibitory have not been tried because normal guinea pig serum is toxic for the cultured cells at these concentrations (2).

Although data obtained with an antiserum with exclusive V_L reactivity are still lacking it is concluded from the results that anti V_L antibodies are unreactive with T helper cells. Similar, though indirect, results have been obtained by genetic experiments previously (3,9) and by Binz and Wigzell (6, 17). These genetic data showed that an anti-idiotypic antibody only reacts with those T cells that possess in their genome the V genes for the immunoglobulin heavy chain whereas those for the immunoglobulin light chain (as well as the H-2 complex) are irrelevant.

What do these data reveal with respect to the presence or absence of V_H or V_L regions on helper T cells ? No doubt exists that V_H regions are present on T cells and are part of the T cell receptor for antigen. The arguments in favor of this notion rely on serological, genetic, and immunochemical evidence and have been extensively reviewed recently (18). Conclusions regarding the V_L region are more difficult to draw as it is possible that these structures are present but are inaccessable to anti-idiotypic antibody. Independent evidence on isolated T cell receptors, however, also indicates the absence of immunoglobulin light chains (9,19), and a T cell receptor whose combining site is only partially identical with that of antibodies appears to be more in accord with a variety of experimental data on T cell specificity for antigens and on Ir gene control (reviewed in ref. 18). Nevertheless it should be mentioned that in a number of studies light chain analogues have been found on T cell tumors and that some T cell tumors appear to contain K chain mRNA (reviewed in ref. 20). Thus, the question of V_L regions on T cells remains unsettled.

At present, not much more is known about parts of the T

cell receptor other than the V_H region. As has been discussed previously (9,18), none of the constant parts of conventional immunolgobulin molecules appears to be present and a host of evidence suggests the participation of products of the H-2 complex in the construction of the recognition system. The further elucidation of this structure will have to rely on the development of T cell tumors or T cell hybrids with defined specificity that allow the isolation of sufficient quantities for biochemical analysis. It is hoped that many of the puzzling questions such as the H-2 restriction of T cell recognition, the high precursor frequency of H-2 reactive cells,and the discrepancy between the exquisite specificity of T cells on one hand and their genetic defects in antigen recognition on the other hand, will find their answer in the structure of the T cell receptor system.

ACKNOWLEDGEMENT

I thank Miss I. Falk for able technical assistance.

REFERENCES

1. Eichmann, K. and Rajewsky, K. (1975) Eur. J. Immunol. 5, 661.
2. Black, S.J., Hämmerling, G., Berek, C., Rajewsky, K., and Eichmann, K. (1976) J. Exp. Med. 143, 846.
3. Hämmerling, G., Black, S.J., Berek, C., Eichmann, K., and Rajewsky, K. (1976) J. Exp. Med. 143, 861.
4. Cosenza, H., Julius, M.H., and Augustin, A.A. (1977) Transplant. Rev., in press.
5. Eichmann, K. (1975) Eur. J. Immunol. 5, 511.
6. Binz, H. and Wigzell, H. (1977) Cold Spring Harb. Symp. Quant. Biol. 41, in press.
7. McKearn, T. J., Stuart, F. P., and Fitch, F.W. (1974) J. Immunol. 113, 1876.
8. Binz, H., Bächi, T., Wigzell, H., Ramseier, H., and Lindenmann, . (1975) Proc. Nat. Acad. Sci. 72, 3210.
9. Krawinkel, U., Cramer, M., Berek, C., Hämmerling, G., Black, S.J., Rajewsky, K., and Eichmann, K. (1977) Cold Spring Harb. Symp. Quant. Biol. 41, in press.
10. Eichmann, K. (1972) Eur. J. Immunol. 2, 301.
11. Eichmann, K. (1974) Eur. J. Immunol. 4, 296.
12. Eichmann, K. (1973) J. Exp. Med. 137, 603.
13. Click, R. E., Benck, L., Alter, B.J. (1972) Cell. Immunol. 3, 264.
14. Eichmann, K. and Kindt, T.J. (1971) J. Exp. Med. 134, 532.

15. Rajewsky, K., Hämmerling, G., Black, S. J., Berek, C., and Eichmann, K. in "The role of products of the histocompatibility gene complex in immune responses". Eds. Katz, D. and Benacerraf,B. Acad. Press, N.Y. 1976, p.445.
16. Briles, D. and Davie, J. (1975) J. Exp. Med. 141, 1291.
17. Binz, H., Wigzell, H., and Bazin, H. (1977) Nature, in press.
18. Rajewsky, K. and Eichmann, K. in "Contemporary Topics in Immunobiology". (1977) in press.
19. Krawinkel, U. and Rajewsky, K. (1976) Eur. J. Immunol. 6, 529.
20. Marchalonis, J.J. (1976) in "Contemporary Topics in Molecular Immunology" 5, in press.

THE VARIABLE PORTION OF T AND B CELL RECEPTORS FOR ANTIGEN: BINDING SITES FOR THE HAPTEN 4-HYDROXY-3-NITRO-PHENACETYL IN C57BL/6 MICE

M. Reth, T. Imanishi-Kari, R.S. Jack, M. Cramer,
U. Krawinkel, G.J. Hämmerling and K. Rajewsky

Institute for Genetics, University of Cologne,
Weyertal 121, D-5000 Köln 41, F.R.G.

ABSTRACT. The primary antibody response of C57BL/6 mice against the hapten 4-hydroxy-3-nitro-phenacetyl (NP) is restricted to a few antibody species (ref. 2, 3), most or all of which express lambda light-chains. This contrasts with the ability of hyperimmune C57BL/6 mice to produce a large variety of anti-NP antibodies with a wide range of affinities for the hapten and carrying either kappa or lambda light-chains. Nine hybrid cell lines were established by the Köhler-Milstein technique (11), each of which secretes one such antibody in large quantities. The antibodies belong to various immunoglobulin classes and are all idiotypically distinct from the primary anti-NP antibody of C57BL/6 mice. However, the characteristic fine specificity of the primary antibody is shared by three of the nine hybridoma antibodies. Like the primary antibody, these three antibodies carry lambda light chains, whereas all the others possess kappa chains.

A fraction of NP-binding receptor material can be isolated from hapten-sensitized C57BL/6 lymphocytes that lacks class- and type-specific immunoglobulin determinants and appears to originate from T lymphocytes. The majority of this material reacts with anti-idiotypic antiserum against the primary anti-NP antibody, and genetic experiments suggest that the same idiotype is recognized in both cases. The primary anti-NP antibody also shares with the NP-specific receptor material its characteristic fine specificity (ref. 14). The association of lambda light-chains with antibodies expressing this fine specificity makes one wonder whether the similarity between the binding sites of antibodies and receptor molecules of this type might extend beyond the sharing of variable regions of immunoglobulin heavy chains.

INTRODUCTION

In general, a wide variety of humoral antibodies and cell surface receptors participate in the immune response of an animal against a given antigen. There are cases, however, where certain molecular species of antibodies and receptor molecules dominate a particular response. An extreme case is the response of BALB/c mice to phosphorylcholine. It consists under physiological conditions essentially of a single species of antibody molecules (1). A rather similar situation is encountered in C57BL/6 mice upon immunization with the hapten 4-hydroxy-3-nitro-phenacetyl (NP). As Mäkelä and his coworkers have established, only a few antibody species are generated in the primary anti-NP response of these animals as judged by isoelectric focussing analysis (2, 3). Imanishi and Mäkelä have shown that these antibodies bear a strain-specific fine specificity marker in that they have higher affinity for the cross-reacting hapten 4-hydroxy-5-iodo-3-nitro-phenacetyl (NIP) than for the homologous hapten NP (4). This "heteroclicity" is inherited in simple Mendelian fashion and shows close linkage to the immunoglobulin (Ig) heavy-chain allotype (5). It is reasonable to assume that heteroclicity is a marker for a gene or a group of genes encoding variable portions of Ig heavy chains, although the argument is indirect (see also below). The primary anti-NP antibody of C57BL/6 mice can also be defined by anti-idiotypic antisera and the idiotypic marker is again strain-specific and is in breeding experiments transmitted together with heteroclicity (6, 7). The idiotypic marker defined by various anti-idiotypic antisera raised in our laboratory has been found only in mice bearing the $\underline{Ig\text{-}1}^{b}$ allotype. It is present on more than 90% of primary anti-NP antibody of C57BL/6 origin and we have found this to be true in every single individual among some fifty mice analysed (7).

We study this system for two main purposes. The first is the problem of V-gene expression and its regulation: What is the basis of the predominant expression of a particular antibody species in the primary anti-NP response? We show below that hyperimmunized C57BL/6 mice are apparently able to produce a large variety of heteroclitic and non-heteroclitic anti-NP antibodies of various affinities for the hapten. Our approach has been to establish continuous cell lines each of which produces one such antibody in large quantities. The various antibodies can thus be structurally compared.

Our second interest in the system is the analysis of cellular receptors for antigen. We have recently established a method for the specific isolation of hapten-binding receptors from T and B lymphocytes (8, 9). The idiotypic and fine specificity markers in the present system permit a detailed investigation of the variable portions of these receptor molecules and will hopefully enable us to isolate receptor molecules of a high degree of molecular homogeneity.

RESULTS

NP-binding humoral antibodies in C57BL/6 mice. The details of the isolation and characterization of the various anti-NP antibodies of C57BL/6 origin are described in separate publications (7, 10). The protocol can be summarized as follows:

Primary antibodies to NP were induced by injecting C57BL/6 mice intraperitoneally with alum-precipitated NP_{12}-chicken gamma globulin (NP_{12}-CG; 100 μg) and inactivated pertussis vaccine as an adjuvant. The mice were bled 18-28 days later.

In order to establish cell lines producing monoclonal anti-NP antibodies, C57BL/6 mice were hyperimmunized with NP-conjugated group A streptococcal vaccine (Strep.A). Four days after the last of a total of 7 intravenous injections of 10^9 NP-coated Strep.A particles each, spleen cells from the animals were prepared and hybridized with a HGPRTase-negative myeloma cell line as described by Köhler and Milstein (11; the line carries the designation X63Ag8 and was a gift from Dr. C. Milstein. It is a subline of the BALB/c myeloma MOPC 21 and produces an IgG_1 myeloma protein carrying kappa light chains). Hybrid cells were selected in hypoxanthine-aminopterin-thymidine (HAT) medium and analysed for the production of anti-NP antibody in a NP-specific hemolytic plaque assay. A total of 165 clonal cell lines were isolated, eleven of which turned out to produce anti-NP antibody. Some of the positive lines were recloned in soft agar over a fibroblast feeder layer. Finally, nine lines were serially transplanted subcutaneously and/or intraperitoneally in (BALB/c x C57BL/6)F_1 animals. Large quantities of anti-NP antibodies appeared in the serum and/or ascitic fluid of the recipient mice.

The various anti-NP antibodies were isolated by adsorption to NP-conjugated bovine serum albumin (BSA) coupled to Sepharose 4B and subsequent elution from the immunosorbent with free NIP-caproic acid (NIP-cap).

In Table 1 we summarize some properties of the various antibody preparations. The data include (i) chain composition as determined with class- and type-specific antisera in double diffusion and an indirect hemagglutination assay (sen-

TABLE 1

ANTI-NP ANTIBODIES OF C57BL/6 ORIGIN

Antibodies	H-Chain Class, Subclass	L-Chain Type	Affinity to NIP-cap (M) [a]	K_{rel} [b]
C57BL/6 Primary Anti-NP Antibodies	$\gamma 1(\mu)$ [c]	λ [c] [e]	1.6×10^{-7}	4.6
S 8	$\gamma 2a$ [d] [f]	κ [e]	5.7×10^{-9}	0.9
S 10	$\gamma 1$ [d]	κ [d]	1.2×10^{-7}	0.1
S 19	$\gamma 1$ [d] [f]	κ [d]	2.8×10^{-7}	-
S 24	$\gamma 1$ [d] [f]	λ [e]	3.0×10^{-8} [g]	33 [g]
S 34	μ [d]	λ [e]	7.6×10^{-9} [h]	3.5
S 36	$\gamma 1$ [d] [f]	κ [e]	3.5×10^{-7}	0.1
S 43	$\gamma 2a$ [d]	λ [e]	2.4×10^{-8}	8.5
S 74	$\gamma 2b$ [d]	κ [e]	3.0×10^{-7} [g]	0.1 [g]
S 92	$\gamma 2a$ [d]	κ [e]	2.0×10^{-9}	1.6

a) Approximate molar concentration of free NIP-cap giving 50% inhibition in a modified Farr assay.
b) $K_{rel} = \frac{I_{50} \text{ for the homologous hapten}}{I_{50} \text{ for the related hapten}}$, where I_{50} is the hapten concentration for 50% inhibition of hapten binding. c) Determined by indirect radioimmunodiffusion. d) Determined both by double-diffusion analysis and by indirect passive hemagglutination. e) Determined by double-diffusion analysis. f) Except for S 34, the clonal products contained hybrid molecules with H chains of MOPC 21 ($\gamma 1$). The assignment of the C57BL/6-derived H chain to the $\gamma 1$ subclass rests on the failure to detect other classes or subclasses. g) Values determined by HPII. h) The HPII assay yielded a value of 3×10^{-7} M.

sitization of NP-coupled sheep red cells with subagglutinating amounts of anti-NP antibody and subsequent agglutination with defined anti-Ig antisera), (ii) affinity for the NP hapten as determined in a modified Farr assay (12) and by inhibition of haptenated phage inactivation (HPII) (4; the two methods yielded similar results for the hybridoma antibodies except in the case of S 34, an IgM protein; see Table 1) and (iii) the affinity for NIP-cap as compared to that for NP-cap as a measure of heteroclicity.

The primary anti-NP antibody belongs predominantly to the IgG_1 class. This is in accord with previous data of McMichael et al. (2). Some 19S antibody is, however, also present in the preparation (7). Surprisingly, the light chains of the antibody appear to belong in their majority and possibly exclusively to the lambda type. In the cloned antibodies a variety of heavy chain classes are represented. The affinities for NIP cover a range which accommodates also the affinity of the primary anti-NP antibody. Three of the nine hybridoma antibodies are strikingly heteroclitic and these, like the primary antibody, carry lambda light chains. The non-heteroclitic antibodies possess kappa light chains.

Idiotypic analysis revealed that none of the antibodies produced by the hybrid cells carries the idiotypic marker of the primary anti-NP antibody. In addition, an anti-idiotypic antiserum raised against one of the cloned antibodies (S43) revealed that this protein carries an idiotype entirely distinct from that of the primary anti-NP antibody.

These data confirm that the anti-idiotypic antisera used in this study are not merely antisera against constant parts of lambda chains. This possibility had been virtually excluded previously, since the idiotypic specificity detected by the antisera was absent from the lambda-bearing myeloma proteins MOPC 104e and MOPC 315. More importantly, they show that C57BL/6 mice are capable of producing a large variety of distinct, heteroclitic and non-heteroclitic anti-NP antibodies with a range of different affinities for the hapten. This is in good accord with the observation that the secondary anti-NP response of C57BL/6 mice appears strikingly heterogeneous in isoelectric focussing analysis (2, 3 and our own undocumented observations), and also with the fact that our anti-idiotypic antisera react with only part of the anti-NP antibody produced in the secondary response (in the order of 20%). Sequence analysis of our cloned antibodies and of the primary anti-NP antibody might resolve the interesting question of whether the various antibody species are structurally interrelated and possibly derived from each other.

Perhaps the most intriguing aspect of the data in Table 1 is the correlation between heteroclicity and presence of lambda chain. This finding might be taken as an indication that the lambda chain contributes to the characteristic fine specificity fo these antibodies. The third hypervariable region of mouse lambda chains contains two tryptophane residues (13) one or both of which might contribute to binding of the aromatic hapten. This notion is indirectly supported by the observation of a change in the absorption maximum of NIP-caproic acid (NIP-cap) from 430 nm to 480 nm upon binding to two of the lambda bearing monoclonal antibodies (10). This phenomenon is not observed with any of the kappa-bearing antibodies, and we predict that it will also be encountered in the case of antibody S 34 and the primary anti-NP antibody, where we did not so far have enough protein available to carry out the analysis. Present experiments aim at directly demonstrating the participation of V_λ in the binding of hapten to the heteroclitic antibody. If the interpretation is correct, it has important implications for our finding that receptor molecules isolated from C57BL/6 nylon-wool-purified T lymphocytes are as heterclitic as the primary anti-NP antibodies in these animals (14; see also below).

NP-binding lymphocyte receptors in C57BL/6 mice. We have recently developed a method for the specific isolation of hapten-binding receptors from T and B lymphocytes (8, 9). Briefly, T and B lymphocytes bind specifically to hapten-conjugated nylon at $4^{o}C$ (15, 16). When the cells are released from the nylon mesh by temperature shift (17), they leave behind on the immunosorbent hapten-binding material which can be eluted with acidic buffer, 3.5M KSCN or free hapten (8, 14). The material can be detected and titrated on the basis of its ability to inactivate hapten-conjugated bacteriophage T4. When such material, which will be termed "receptor material" from here on, is isolated from hapten-sensitized spleen cells, approximately 2/3 to 3/4 of the phage-inactivating activity can be absorbed on anti-Ig immunosorbents. This fraction of receptor material is called the anti-Ig^+ fraction. The remaining material - the anti-Ig^- fraction - does not detectably carry class-specific determinants of α, γ, μ, δ, k or λ immunoglobulin polypeptide chains. Cell separation experiments have established a clear correlation between anti-Ig^+ fraction and B cells and anti-Ig^- fraction and T cells in the input lymphocyte population (8, 9, 14). These results make us think that the anti-Ig^- fraction represents antigen receptors of T cells, although we have no evidence so far that these molecules are in fact produced by T lymphocytes.

The occurrence of a major idiotype in the anti-NP response of C57BL/6 mice has led us to investigate the expression of this idiotype in the nylon-eluted receptor material, in particular in the anti-Ig^- fraction. For this purpose NP-binding receptor material was isolated from NP_{12}-CG-sensitized spleen cells derived from C57BL/6 and CBA/J mice. In addition, a (C57BL/6 x CBA/J)F_1 generation and from this a F_2 generation and a backcross to CBA were established. The F_2 and backcross animals were typed for their Ig-1 allotype, grouped accordingly, and NP-binding receptors were isolated from each of the various groups of animals. The material was first absorbed on insolubilized polyspecific anti-Ig serum which bound 65 to 75% of the hapten-binding activity. The remaining activity, i.e. the anti-Ig^- fraction, was then absorbed with insolubilized anti-idiotypic antiserum or its IgG fraction. A summary of the results appears in Table 2. It can be seen that the anti-idiotype always binds more than 50% of the anti-Ig^- fractions if - and only if - the material originates from animals carrying the $Ig\text{-}1^b$ allotype. The anti-idiotypic immunosorbent exhibits specificity also in the sense that it does not bind the anti-Ig^- fraction of C57BL/6 2,4-dinitro-phenyl (DNP)-specific receptor material (Table 2). In addition, undocumented specificity controls have shown that in contrast to the primary humoral anti-NP antibodies, the anti-Ig^- fraction of C57BL/6-derived NP-binding receptors is not absorbed by insolubilized anti-lambda antiserum (raised against MOPC 104e light chain, a gift from Dr. H.C. Morse). These results clearly suggest that in the NP system in C57BL/6 mice the major part of the anti-Ig^- receptor fraction carries the same idiotypic marker as the major part of the humoral antibody produced in the primary response to that hapten, whereas it lacks determinants of the constant part of lambda light chains. (For detailed results and discussion see ref 14.)

The anti-Ig^- receptor fraction is thus serologically defined by (i) its property to bind antigen, (ii) its lack of class- and type-specific immunoglobulin determinants, and (iii) the presence of an idiotypic marker which is presumably - but not necessarily - located on the heavy chain. In addition, experiments documented elsewhere (14) demonstrate that the molecules in the anti-Ig^- receptor fraction bind the NP hapten with an affinity similar to that of humoral antibody and express the same kind of heteroclicity which, as in the case of humoral antibody, appears to be allotype-linked.

CONCLUSIONS

In the primary response of C57BL/6 mice to the NP hapten a few species of antibody molecules are regularly and selectively expressed. They mostly belong to the IgG_1 class, carry

TABLE 2

ABSORPTION OF HAPTEN-BINDING ANTI-Ig^- RECEPTOR MATERIAL BY INSOLUBILIZED ANTI-IDIOTYPIC SERUM

Origin and Specificity of Hapten-Binding Material	Ig-1 Allotype of Lymphocyte Donors[a]	% Absorption of Haptenated Phage Inactivating Activity from anti-Ig^- Fraction[b]
C57BL/6J, anti-NP	b/b	54
C57BL/6J, anti-NP[c]	b/b	85
(C57BL/6JxCBA/J)F_1, anti-NP	a/b	74
(")F_1xCBA, anti-NP	a/a	6
(") " , anti-NP	a/b	57
(")F_2, anti-NP	a/a	6
(")" , anti-NP	a/b	57
(")" , anti-NP	b/b	68
CBA/J, anti-NP	a/a	6
C57BL/6J, anti-DNP	b/b	7

a) Determined in double diffusion with serum of lymphocyte donors.
b) Fraction of hapten-binding receptor material not binding to insolubilized polyspecific anti-Ig serum.
c) Receptor material isolated from splenic lymphocytes passaged over nylon wool columns. 75% of the cells were negative for surface Ig in fluorescence analysis.

For experimental details see ref. 14.

a characteristic fine specificity marker and can also be idiotypically defined. As shown here, most, and possibly all of these antibodies carry lambda light chains. The restricted antibody pattern in the primary response contrasts strikingly with the potential of hyperimmune C57BL/6 mice to express a large variety of anti-NP antibodies of various classes, carrying kappa or lambda light chains and covering a large range of affinities for the hapten. We here report the construction, by the Köhler-Milstein technique (11), of nine continuous cell lines each producing one particular species of such anti-NP antibodies of C57BL/6 origin. The secretion of large amounts of these antibodies by the cell lines in vivo opens the way for a structural comparison of the various molecules at the level of amino acid sequence analysis. So far our results indicate that all the hybridoma antibodies are idiotypically distinct from the primary anti-NP antibodies. This is also true for those antibodies that express a fine specificity similar to that of the primary response antibody, i.e. bind the crossreacting hapten NIP-cap with a higher affinity than the homologous hapten NP-cap (heteroclicity). However, a striking correlation of heteroclicity with the expression of lambda light chains is found. This and other reasons (see Results) make us think that the lambda light chain might contribute to the heteroclitic property of the antibody molecule. Lambda chains have been found to show little variability in their variable portions, and it may well turn out that all our lambda bearing (heteroclitic) anti-NP antibodies share the same or very similar lambda chains. It is on the basis of these considerations that we consider it likely that our anti-idiotypic antisera (which do not cross react with the other lambda bearing molecules) recognize determinants on the heavy chain of the primary anti-NP antibody or that the idiotypic determinants are characteristic for certain combinations of heavy and light chains.

In the C57BL/6 strain, the idiotypic marker of the primary anti-NP antibody appears to be also present on the majority of a fraction of NP-binding receptor molecules which we consider to be T-cell-derived. This is suggested by the observation that the anti-idiotypic antisera specifically detect on these molecules determinants which show the same strain distribution and genetic linkage to the Ig-1 allotype as the idiotypic marker on the serum antibody. This result is in excellent agreement with previous work of our own group and that of others (see 9, 18, 20), suggesting the expression of heavy-chain variable regions in T-cell receptors for antigen. Our previous suggestion along the same line, namely that receptor molecules isolated from rabbit T cells might bear a-locus allotypic determinants (9) has been further confirmed in recent experiments (21) and is again in agreement with recent experiments from another laboratory (22).

The data on the a-locus allotype and the various idiotypic markers strongly indicate that the entire V_H region is expressed in T-cell receptors. The problem of light chain variable regions on T lymphocytes is complicated by the lack of suitable genetic markers of the V_L region. Indirect evidence argues against the expression of V_L regions in T-cell receptors (9, 19, 20). However, our findings of similar affinities for hapten and of the heteroclitic property in both humoral antibodies and receptors of putative T-cell origin (14) make us feel that this interpretation should for the present be considered with caution. This is particularly so, since as discussed above, heteroclicity appears to be associated with lambda light chains at the level of humoral antibody.

A striking result of the receptor analysis is the suggestion of predominant expression of the primary anti-NP idiotype also at the level of T-cell receptors. A similar situation is encountered in other systems of "major" idiotypes (18, 20, 23), and the phenomenon may therefore be general. The problem of the predominant expression of certain idiotypes in the immune system is still unresolved, but in systems like the one described here it can now be approached by both functional and structural analysis.

ACKNOWLEDGEMENTS

We wish to thank Dr. C. Milstein for giving to us the X63Ag8 cell line, Dr. H.C. Morse for a gift of antiserum against murine λ_1 light chains, Ms. Sigrid Irlenbusch for competent technical help, Ms. Christa Müller for help in typing the various antibody preparations and Ms. Åsa Böhm for her expert help in the preparation of this manuscript.

This work was generously supported by the Deutsche Forschungsgemeinschaft through SFB 74.

REFERENCES

1. Consenza, H. and Köhler, H. (1972) Science 176, 1027.
2. McMichael, A.J., Phillips, J.M., Williamson, A.R., Imanishi, T. and Mäkelä, O. (1975) Immunogenetics 2, 161.
3. Mäkelä, O. and Karjalainen, K. (1976) Cold Spring Harbor Symp. Quant. Biol.41, in press.

4. Imanishi, T. and Mäkelä, O. (1973) Eur. J. Immunol. 3, 323.
5. Imanishi, T. and Mäkelä, O. (1974). J. Exp. Med. 140, 1498.
6. Mäkelä, O. and Karjalainen, K. (1977) Transplant. Rev. 34, in press.
7. Jack, R.S., Imanishi-Kari, T. and Rajewsky, K., submitted for publication.
8. Krawinkel, U. and Rajewsky, K. (1976) Eur. J. Immunol. 6, 529.
9. Krawinkel, U., Cramer, M., Berek, C., Hämmerling, G.J., Black, S.J., Rajewsky, K. and Eichmann, K. (1976) Cold Spring Harbor Symp. Quant. Biol. 41, in press.
10. Reth, M., Hämmerling, G.J. and Rajewsky, K., in preparation.
11. Köhler, G. and Milstein, C. (1975) Nature 256, 495.
12. Mitchison, N.A. (1971) Eur. J. Immunol. 1, 10.
13. Cesari, I.M. and Weigert, M.G. (1973) Proc. Nat. Acad. Sci. US 70, 2112.
14. Krawinkel, U., Cramer, M., Imanishi-Kari, T., Jack, R.S., Rajewsky, K. and Mäkelä, O., submitted for publication.
15. Rutishauser, U. and Edelman, G.M. (1972) Proc. Nat. Acad. Sci. US 69, 3774.
16. Kiefer, H. (1975) Eur. J. Immunol. 5, 624.
17. Kiefer, H. (1973) Eur. J. Immunol. 3, 181.
18. Rajewsky, K. and Eichmann, K. (1977) Contemp. Top. Immunobiol. 7, 69.
19. Binz, H. and Wigzell, H. (1977) Contemp. Top. Immunobiol., in press.
20. Eichmann, K., this volume.
21. Krawinkel, U. Cramer, M., Mage, R.G., Kelus, A.S. and Rajewsky, K., submitted for publication.
22. Cazenave, P.-A., Cavaillon, J.M. and Bona, C. (1977) Immunol. Rev., in press.
23. Cosenza, H., Julius, M.H. and Augustin, A.A., this volume.

4. [illegible] T. and McNeil[illegible] O. (19[illegible]) [illegible] J. Immunol. [illegible]
5. [illegible], T. and McNeil, O. (1974) J. Exp. Med. 140, 140[illegible]
6. [illegible] and [illegible] (197[illegible]) Transplant. Rev. [illegible] in press.
7. [illegible] and [illegible] submitted for publication.
8. Krawinkel, U. and Rajewsky, K. (1976) Eur. J. Immunol. [illegible]
[illegible]
9. [illegible] (1976) Cold Spring Harbor Symp. Quant. Biol. 41, in press.
10. [illegible] M., Hämmerling, [illegible] and [illegible], in preparation.
11. [illegible] (197[illegible]) [illegible]
12. [illegible] Immunol. [illegible]
13. [illegible] and [illegible] (19[illegible]) [illegible]
[illegible]
14. [illegible]
[illegible]
15. [illegible]
[illegible]
16. [illegible] (197[illegible]) [illegible] J. Immunol. [illegible]
17. [illegible] (197[illegible]) [illegible] J. Immunol. [illegible]
18. [illegible] and [illegible] (197[illegible]) Contemp. Top. [illegible]
[illegible]
19. [illegible]
[illegible]
20. [illegible]
[illegible]
21. [illegible]
[illegible]
22. [illegible] volume.

T LYMPHOCYTE RECEPTORS FOR ALLOANTIGENS

Hans Binz and Hans Wigzell

Department of Medical Microbiology, Division of Experimental Microbiology, University of Zürich, POB, 8028 Zürich, Switzerland, and Department of Immunology, Biomedical Center, University of Uppsala, Uppsala, Sweden.

ABSTRACT. Anti-idiotypic antibodies raised against alloantibodies or against purified peripheral T lymphocytes can be shown to react with normal immunocompetent B and T lymphocytes. Such anti-idiotypic antibodies suppress in a specific way T lymphocyte functions such as MLC, GvH reactions and CML. Idiotypic, alloantigen-reactive T and B cell derived molecules can be detected in normal sera and can be purified by the use of anti-idiotypic immunoabsorbents. B cell derived molecules consist of 7 to 8s IgM. T cell derived molecules can be found in three different forms, namely with a molecular weight of 150,000 daltons, 75,000 daltons and with a molecular weight between 40,000 and 30,000 daltons as analyzed on SDS gels. The molecules forming the 150,000 daltons peak can be split under reducing conditions into single polypeptide chains with a mocular weight of about 70,000 daltons. The molecules with a molecular weight between 30,000 and 40,000 daltons are degradation products of the 70,000 daltons molecules. Neither serological determinants of conventional immunoglobulins nor determinants coded for by the genes of the major histocompatibility locus could be found on the T cell derived molecules.

INTRODUCTION

A great deal of evidence has been accumulated during the past few years that T and B lymphocytes which have the information to react against the same antigens share idiotypic determinants (1,2,3). Workers using two entirely different antigenic systems came to similar conclusions (1,2). First it could be shown that alloantibodies and T lymphocytes mediating GvH-reactions, MLC and CML to the same alloantigen share very similar or even identical idiotypes (4). Secondly, it was found that anti-idiotypic antibodies could specifically sensitize idiotype-bearing B and T lymphocytes, indicating that antibody molecules and functional receptors on T helper or suppressor cells express similar or identical idiotypic determinants (2). In both systems, the idiotype V genes can be shown to be linked to heavy chain region genes and are not linked to light chain genes (5,6). This would strongly suggest that B and T lymphocytes reacting against the same antigenic determinants use, in order to create the antigen-binding receptor, the same set or subset of V_H genes.

Using antisera directed against idiotypic determinants on normal T lymphocytes with specificity for major histocompatibility antigens, it has been shown that substantial amounts of naturally occurring, idiotypic and antigen-binding molecules can be found in the serum of normal rats (7). We have studied the cellular nature of these naturally occurring idiotypic molecules and have compared them with receptors that are shed from normal lymphocyte cultures (8). We would like to present data giving partial characterization of a T cell receptor for alloantigens and we will discuss these findings with respect to the possibility, reported by others, that products of genes located within the major histocompatibility complex are serious candidates for the T-cell receptor for antigens (9,10).

MATERIAL AND METHODS

The relevant procedures and the animals used have been described in detail in a series of articles (1,4,5,7,8).

RESULTS

The naturally occurring, idiotypic, alloantigen-binding molecules of normal Lewis serum (7) consists of four distinct groups of molecules.

Fig. 1.

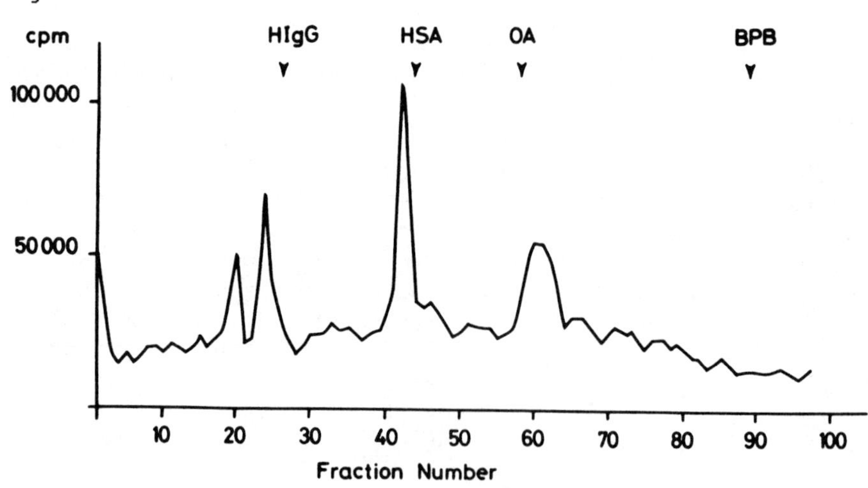

Normal Lewis serum was absorbed on anti-(Lewis anti-DA) immunoabsorbent. Bound material was eluted with low pH and high ionic strength, concentrated, dialyzed and radiolabelled with 125-Iodine (8). Radiolabelled material was analyzed on 5% SDS gels and Fig. 1 shows such a profile. A strikingly similar pattern was obtained when supernatants from normal Lewis spleen cell cultures were similarly isolated. Further analysis of the four peaks revealed the following (8):

Peak I. This peak has a molecular weight of about 180,000 daltons and is derived from B lymphocytes. These molecules will dissociate under reducing conditions into two types of molecules: 68,000 and 24,000 daltons in size. For serological markers see Table I. This peak represents 7 to 8s IgM molecules most likely shed from the correspdoning DA alloantigen reactive Lewis B lymphocytes.

Peak II. Molecules with a molecular weight of about 150,000 daltons form this peak. They derive from T lymphocytes. Purified molecules can be split under reducing conditions into type III-peak molecules with a molecular weight of slightly above 70,000 daltons. Most likely peak II consists of dimers of peak III chains but there is no direct evidence for that. For serological markers see Table I.

Peak III. The molecular weight of the molecules forming peak III is around 70,000 daltons. They represent single polypeptide chains as treatment with 1 M 2-ME, 6 M urea or 6 M guanidine-HCL failed to split peak III molecules. Upon incubation with normal Lewis serum at room temperature overnight in the absence of proteolytic inhibitors these single chain molecules will be degraded into peak IV size molecules. Peak III molecules also derive from T cells. For serological markers see Table I.

Peak IV. The size of these molecules varies between 30,000 and 40,000 daltons. They cannot be further split by any reducing procedure, but will be completely destroyed if subjected to prolonged standing at room temperature in the absence of proteolytic inhibitors. Peak IV molecules can be found in normal Lewis urine (7).

We then looked for serological markers on peak I and peak III molecules (as representatives of the peak II-IV group). These findings are summarized in Table I.

TABLE I. Absorption of T and B Lymphocate Derived Material on Different Immunoabsorbents

Immunoabsorbent	125-I labelled material absorbed	
	Peak I	Peak III
Rabgit anti-rat Ig No. 1 (polyvalent)	+	-
Rabbit anti-rat Ig No. 2 (polyvalent)	+	-
Rabbit anti-rat Ig No. 3 (polyvalent)	+	-
Rabbit anti-rat Ig No. 4 (polyvalent)	+	-
Rabbit anti-rat IgM* No. 1	+	-
Rabbit anti-rat IgM* No. 2	+	-
Goat anti-rat IgM*	+	-
Rabbit anti-rat gamma 2c*	ND	-
Rabbit anti-rat gamma 2a* No. 1	ND	-
Rabbit anti-rat gamma 2a* No. 2	ND	-
Rabbit anti-rat gamma 1*	ND	-
Rabbit anti-rat epsilon* No. 1	ND	-
Rabbit anti-rat epsilon* No. 2	ND	-
Rabbit anti-rat IgD**	-	-
Chicken anti-human IgD***	ND	-
Rabbit anti-mouse IgD***	ND	-
DA anti-Lewis No. 1	ND	-
DA anti-Lewis No. 2	ND	-
DA anti-Lewis No. 3	ND	-
DA anti-Lewis No. 4	ND	-
Anti-Lewis anti-DA IgG	+	+
Protein-A	ND	-
Lentin***	ND	-
Normal (Lewis x DA)F_1 serum IgG	-	-
DA spleen cells	+	+
(Lewis x DA)F_1 spleen cells	+	+
Lewis spleen cells	-	-

* Gift from Dr. H. Bazin, ** Gift from Dr. Ch. Leslie, *** Gift from Dr. P.A. Peterson

ND denotes not done. Peak I and III molecules were purified from a first 5% polyacrylamide gel, dialyzed against PBS and absorbed on the different immunoabsorbents (8).

Immunosorbents made with several different polyvalent rabbit anti-rat Ig antisera and with antisera directed against myeloma proteins of different classes and subclasses failed to remove purified peak III type molecules. In addition three antisera against different IgD preparations, two of which were known to cross react strongly with rat IgD also failed. Antisera directed against rat IgM and the polyvalent anti-Ig antisera, on the other hand, removed purified peak I molecules, suggesting again strongly the 7-8s IgM nature of these molecules. Four different preparations of DA anti-Lewis antibodies known to contain antibodies against Ia-like structures did not remove molecules from peak III. However, anti-Lewis anti-DA anti-idiotypic antibodies as well as DA bearing lymphocytes could fully absorb peak III type molecules. In conclusion we can state that T cell derived molecules as represented by molecules from peak III neither bear antigenic structures of conventional immunoglobulins nor determinants coded for by genes located within the major histocompatibility locus.

DISCUSSION

We were able to isolate from normal Lewis serum and from supernatants from normal Lewis lymphocyte cultures idiotypic, antigen-binding molecules by the use of anti-(Lewis anti-DA) anti-idiotypic immunoabsorbents. These express antigen-binding ability as well as idiotypic determinants whilst being single polypeptide chains. Furthermore, we have so far failed to demonstrate any polypeptide chain of light chain Ig type in the T cell receptor material. Finally, as reported elsewhere (4), all T cell receptor idiotypes seem to exist also on isolated heavy chains from conventional antibodies with the same antigen-binding specificity as the T cell receptors. Our data are thus fully consistent with the assumption that T cells express only those idiotopes that are coded for by single heavy chains of Ig type. The T cell receptors fail to express those idiotypic determinants which are introduced via the variable region of the light chains in the B cell receptors molecules (4), the classical immunoglobulins. Exactly similar conclusions have been reached by Klaus Eichmann in his fine analysis of T cell idiotypes in an anti-streptococcal system. Further support for the above statement can be drawn from genetic analysis of inheritance of T cell idiotypes, where complete positive linkage to heavy chain Ig constant genes have been found both in mouse (6) and rat (5).

We were so far not able to demonstrate any serological markers of constant immunoglobulin type using a battery of allo- or xenogeneic antisera (8). We must therefore conclude that the T cell derived molecules constitute a new immunoglobulin heavy chain class. Why? 1. The idiotypes on immunoglobulin heavy chains and the T cell chains are the same. 2. These idiotypes are coded for by variable genes intimately linked with the genes coding for the heavy chain Ig constant regions (5). We therefore suggest in accordance with previous nomenclature the greak T or tau for our T cell chain. The locations of the genes coding for the constant portions of this tau chain should be in the same gene cluster as the genes that code for M, A, D, G or E.

How can T lymphocytes achieve sufficient diversity, since we have failed so far to find additional chains participating in the build-up of the antigen-binding receptors? B cell receptors can take advantage of the light chain variable regions in their build-up of diversity of antigen-binding sites. The following explanations can be considered: 1. It is possible that all lymphocytes have the capacity of expressing two variable gene products. B lymphocytes express one Ig heavy chain V gene and one light chain V gene. T cells could be postulated to express two Ig heavy chain V genes. Several experimental results point in that direction (8) without telling whether a single polypeptide chain (= two V gene products on the same tau chain), or two distinct tau chains with different idiotypes are involved. 2. One could argue that T cell receptors do in fact contain an additional chain yet to be found and so loosely attached to the tau chain that even mild isolation procedures result in its loss. 4. Our isolated receptor only represents the forms in which the receptors are present on immunologically virginal T lymphocytes. Weak support for such an hypothesis stems from very preliminary results where anti-idiotypic antisera produced against T lymphoblasts receptors seem to react against idiotypes present on blast cells but absent from virginal T lymphocytes of the same individual (11).

Another difficulty stems from the findings of several groups of workers that antigen-specific molecules supposedly produced by immune T cells display serological markers indicating these molecules to be products of immune response genes located within the major histocompatibility complex (9,10). Biochemical data suggest that these molecules are distinct from the receptors described above (9,10). No markers of constant Ig type have been described on these Ia-Ir gene factors.

However, a recent preliminary finding suggests the existence of idiotypic determinants of the expected Ig type on one such factor (Mozes, T., personal communication). If confirmed, this could mean that the immune T cell receptor consists of one chain from the Ia-IR gene cluster and of another (our tau chain) from the heavy chain Ig cluster. It should be realized, however, that the molecular size of the Ia-Ir factors described is smaller than our intact T chain, thus implying that such hypothetical hybrid molecules must be made up of a partially degraded tau chain. Only further analysis of the respective T derived, antigen-binding idiotype-specific molecules from immune cells can solve these problems.

ACKNOWLEDGEMENTS

This work was supported by the Swiss National Science Foundation (Grant No. 3.688-0.76), Swedish Cancer Society, and NIH grant AI.CA. 13485-01 and NIH contract NO-CB-64033.

REFERENCES

1. Binz, H. and Wigzell, H. (1975) J. Exp. Med. 142, 197.
2. Eichmann, K. and Rajewsky, K. (1975) Eur. J. Immunol. 5, 661.
3. McKearn, T.J. (1974) Science (Wash.) 183, 94.
4. Binz, H. and Wigzell, H. (1977) Contemp. Top. Immunobiol. 7, 111.
5. Binz, H. and Wigzell, H. (1976) Nature 264, 639.
6. Eichmann, K. and Berek, M.C. (1973) Eur. J. Immunol. 3, 599.
7. Binz, H. and Wigzell, H. (1975) Scand. J. Immunol. 4, 591.
8. Binz, H. and Wigzell, H. (1976) Scand. J. Immunol. 5, 559.
9. Taussig, M.J., Munro, A.J., Cambell, R., David, C.S. and Stains, N.A. (1976) J. Exp. Med. 142, 694.
10. Tada, T., Taniguchi, M. and Takemori, T. (1975) Transpl. Rev. 26, 106.
11. Aguet, M., Wight, E., Binz, H., Andersson, L.C. and Wigzell, H. (1977) in preparation.

SPECIFICALLY INDUCED RESISTANCE TO SYSTEMIC GVH DISEASE IN RATS

Donald Bellgrau and Darcy B. Wilson
Immunobiology Research Unit
Department of Pathology, University of Pennsylvania School of Medicine
Philadelphia, Pennsylvania 19074

ABSTRACT. These studies demonstrate that a radioinsensitive, specific resistance to lethal systemic graft vs. host (GVH) disease is induced in F_1 rats that have been inoculated with subclinical dosages of parental strain lymphocytes. This resistance can be demonstrated not only by protection from GVH mortality, but also by a specific inhibition of parental blast cells which normally occur in the thoracic duct lymph of F_1 rats undergoing systemic GVH reactions. Requirements of the "immunizing" parental lymphocyte population include T cells which have immunocompetence for host strain alloantigens and which presumably, therefore, bear anti-host T cell receptors. The specificity of this resistance in A/B F_1's, while directed at cells from the immunizing strain (A) and not from B nor C, extends to F_1 lymphocytes sharing the relevant, immunizing parental genome (A/C).

The implications of these findings are (a) that a specific host (A/B) anti-receptor anti-(A anti-B) immunity affords a significant protective effect for irradiated animals against lethal GVH disease and (b) that this induced immunity may detect T cell receptors A anti-B, C anti-B on third party (A/C) F_1 lymphocyte populations which are either extensively cross reactive or are represented on the same lymphocyte subpopulation. It is clear that this model of GVH resistance will serve as a useful probe for studies of the nature of the anti-MHC T cell receptor.

A. INTRODUCTION

Systemic graft vs. host (GVH) disease is a complex, sometimes lethal, syndrome caused by immunocompetent parental lymphocytes inoculated into F_1 animals which are usually considered to be unable to reject them for genetic reasons. Nevertheless, it is a long-standing observation that high, weight-dependent doses of parental cells are needed to kill intact F_1 animals (6). A possible explanation for this resistance, suggested by Ramseier and Lindenmann (8), involves the potential immunogenicity in F_1 animals of T cell receptors present in the GVH inducing parental cell inoculum having specificity for host alloantigens. Such a possibility has received support recently from the studies of Binz and Wigzell (1), demonstrating the presence of an anti-idiotypic antibody in the serum of F_1 animals immunized with parental T cells or with specific alloantibody.

These considerations of anti-idiotypy raise the possibility that a state of specific resistance to systemic GVH disease can be induced in F_1 animals by prior "immunization" with receptor bearing parental lymphocyte populations.

B. SPECIFICALLY INDUCED PROTECTION AGAINST LETHAL GVH DISEASE

To avoid the complications of an apparent naturally occuring resistance in normal animals, F_1 rats were given sublethal whole body irradiation (450r) to render them acutely sensitive to GVH disease (3).

Table 1 shows that irradiated L/BN F_1 rats routinely die within three weeks following inoculation with as few as 10×10^6 Lewis thoracic duct lymphocytes (TDL) or 20×10^6 BN TDL.

TABLE 1

SUSCEPTIBILITY OF SUBLETHALLY IRRADIATED ADULT L/BN HOSTS TO GRADED DOSES OF PARENTAL TDL

Parental cell dose ($\times 10^6$)	Mortality (# dead/total) of L/BN rats following inoculation with lymphocytes from parental strain donors:	
	L	BN
10	10/10	3/5
20	1/1	3/3
30-100	4/4	4/4
> 100	24/24	N.D.

Any protective, specific resistance to GVH disease caused by inoculation with parental cells prior to irradiation would be reflected in the survival of F_1 recipients following reinjection with ordinarily lethal dosages of parental lymphocytes. L/BN rats were "preimmunized" with 30 million Lewis TDL 7 days prior to irradiation (450r). These recipients proved to be resistant to systemic GVH disease caused by as many as 500 million L TDL (50 fold lethal dose) (Table 2). The resistance is quite specific for the immunizing population; i.e. injection with lymphocytes of the other parental strain (BN) does not protect hosts from mortality.

TABLE 2

SPECIFIC PROTECTION OF "IMMUNIZED" IRRADIATED L/BN HOSTS FROM GVH INDUCED MORTALITY

Parental cell dose ($x\ 10^6$)	Mortality (# dead/total) of L/BN rats following inoculation with lymphocytes from parental strain donors:	
	L	BN
30-100	2/18	3/3
100-250	4/32	8/8
250-500	1/6	N.D.

C. INHIBITION OF PARENTAL CELLS IN HOST LYMPH GVH

On of the characteristic features of systemic GVH disease, caused by the transfer of large numbers of parental cells into irradiated F_1 rats, is a massive and sustained outpouring of activated donor blast T cells into the recipient lymph 3-4 days later (4,10). Presumably these cells are the mitotic products of the subpopulation of parental T cells specific for host alloantigens, and which were sequestered early after transfer in host tissues. It was, therefore, of some interest to determine whether this parental blast cell population was affected in F_1 animals known to be protected from lethal GVH by prior immunization with parental cells.

This experiment involved four groups of L/BN rats, three immunized with L TDL (30 m), and one left untreated. These animals received 450r seven days later and, two days after this, thoracic duct fistulae were established. L or BN TDL (200m) were injected iv. 24 hours later to cause systemic GVH disease, and the cellular output in the lymph was monitored over the subsequent five day period. Fig. 1 shows the parental cell recovery in the F_1 lymph at various times plotted as percent per hour of cells injected. The control group of F_1's, which were irradiated and given GVH inducing inocula of L TDL, shows the bi-modal recovery of parental cells in the host lymph, characteristic of systemic GVH reactions. The first peak at ~20 hours consists of small lymphocytes specifically depleted of cells reactive to host alloantigen (4), and the second peak (~72 hours) consists mainly of donor blast cells. The characteristic pattern was also obtained in the F_1 group immunized with TDL, but given GVH inducing inocula of the same parental strain failed to display the second peak, but instead showed only the first peak which reflects the normal blood-lymph recirculation seen also in syngeneic lymphocyte transfers. The final group of immune F_1's, not given GVH inducing inocula, showed only the low level TDL output typical of uninoculated animals. This experiment demonstrates that GVH resistance involves a marked and specific inhibition of the appearance of parental blast cells in host lymph. Whether inhibition of donor cell proliferation in host tissues also occurs has yet to be established.

Fig. 1

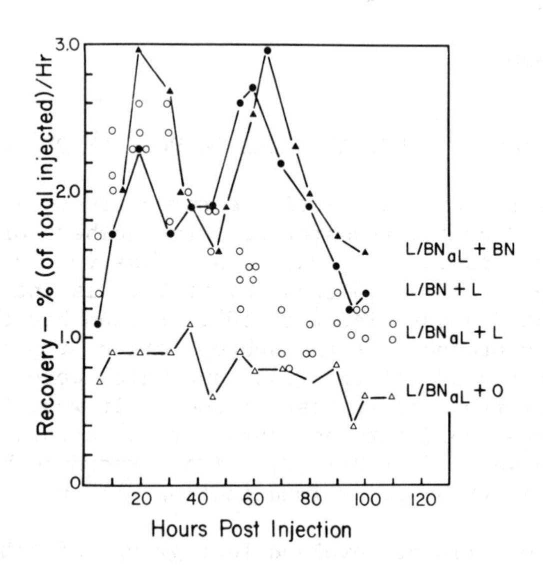

D. PROPERTIES OF THE IMMUNIZING POPULATION

Specific GVH resistance in "immunized" F_1 animals may be caused by mechanisms other than an anti-parental T cell receptor immunity. It could reflect, for example, an immune response to cell surface antigens expressed by recessive parental genes (2,9). That this is not the case has been shown by experiments employing negatively selected parental cells, specifically depleted of alloreactivity to host transplantation antigens, as the immunization population. L TDL populations negatively selected for BN alloantigens (L_{-BN}) by acute "filtration" through x-irradiated L/BN recipients (the first peak in Fig. 1) fail to generate a resistance to systemic GVH disease in L/BN hosts, but are fully effective in L/DA hosts. This experiment demonstrates the clonal distribution of the relevant immunogen and provides strong support for the possibility that specifically induced GVH resistance involves an immunity to a specific T cell receptor.

In addition, these experiments indicate the importance of T cells in the immunizing population. TDL populations negatively selected by acute filtration are comprised mainly (> 95%) of T cells since the modal blood-lymph recirculation time for B cells is comparatively longer than for T cells.

E. SPECIFICITY OF T CELL IDIOTYPES IN DIFFERENT GVH INDUCING LYMPHOCYTE POPULATIONS

A study was initiated to determine whether an anti-idiotypic T cell receptor immunity, the mechanism presumed to be operative in induced GVH resistance, could detect cross reactive idiotypes on lymphocytes of genetically different origins; i.e., whether or not GVH resistance specific for one parental cell strain would extend to at least some other strains.

Table 3 shows the results of one such experiment in which L/BN rats were immunized with L TDL, irradiated, then given GVH inducing inocula from L, BN, L/BN (control), DA, and L/DA donors. As expected, these animals were not killed by cells from L and from L/BN donors while BN cells proved to be lethal. In addition, DA cells were lethal; surprisingly, however, TDL from L/DA donors failed to kill L/BN F_1 recipients and GVH resistance to this population depends on prior immunization with L TDL. Thus, DA TDL populations, having DA anti-L and DA anti-BN receptors cause GVH disease in L/BN animals having an anti-(L anti-BN) immunity, but L/DA cells, with L anti-BN and DA anti-BN receptors, are apparently not effective.

TABLE 3

SPECIFICITY OF INDUCED GVH RESISTANCE

Pretreatment	Mortality (# dead/total) following injection with lymphocytes ($\geq$ 100m) from the following:					
	L	BN	L/BN	DA	L/DA	No cells
-	6/6	7/7	0/3	3/3	4/4	0/4
30m L-TDL	0/6	11/11	0/6	9/9	0/6	0/6

Three possible explanations for this finding seem worth considering (Fig. 2): (1) The anti-host receptors, L anti-BN and DA anti-BN, in the L/DA cell population may be present on the same cell, possibly determined by genes which are not allelically excluded. Consequently, a host immunity anti-(L anti-BN) to one receptor inactivates L/DA cells bearing both. (2) The anti-host receptors, L anti-BN, and DA anti-BN, may be cross reactive and present on different cell populations so that immunity directed to one inactivates the other. (3) L/DA T cells with the anti-host receptors, L anti-BN, and DA anti-BN, may bind soluble BN alloantigens known to be present in irradiated hosts (7). The complex may then subsequently bind soluble L anti-BN T cell receptors either from the immunizing L T cell population or from the GVH inducing L/DA T cell inocula, and a host immunity directed to the last component of the complex may inactivate both L anti-BN and DA anti-BN receptor-bearing L/DA T cells preventing GVH disease.

Further studies are required to resolve these possibilities, however, it is clear that the results can be expected to shed some light on the molecular and biological properties of T cell receptors specific for alloantigens of the major histocompatibility complex.

Fig. 2

HYPOTHESES

1. Non allelic exclusion of the T cell receptor

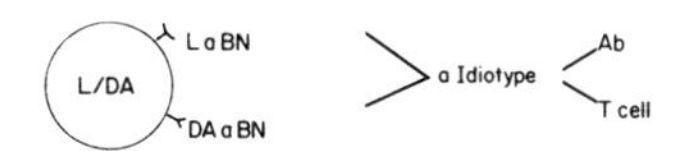

2. Cross reactive idiotypes

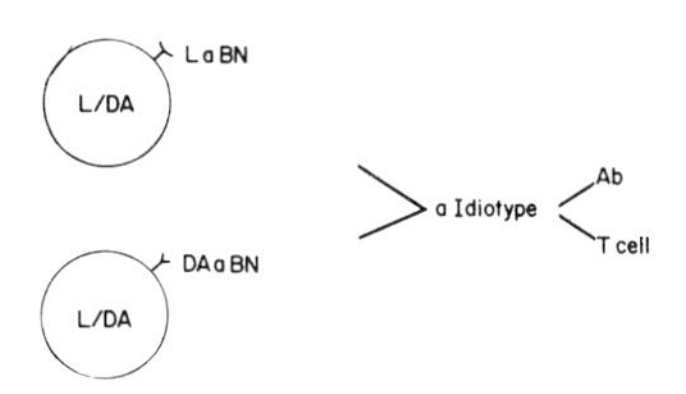

3. Immune Complexes

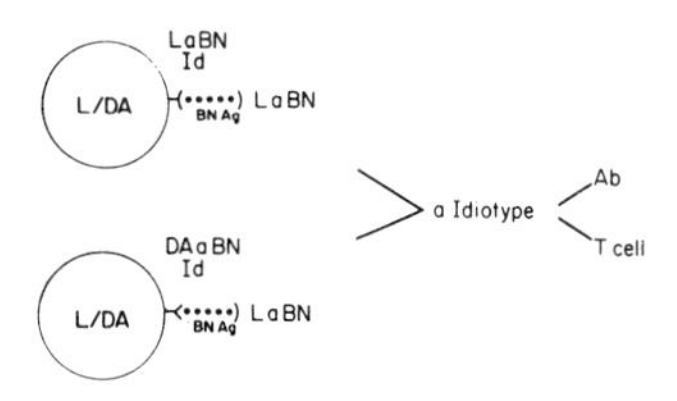

ACKNOWLEDGEMENTS

This work was supported by U.S. Health Public Service Grants AI-10191, AI-10961 and CA-15822. We wish to thank Ms.'s Karen Nowell and Dianne Wilson for help in preparation of the manuscript.

REFERENCES

1. Binz, H., and Wigzell, H. (1975) J. Exp. Med. 142, 197.
2. Cudkowicz, G., and Bennett, M. (1971) J. Exp. Med. 139, 1513.
3. Fiscus, W., Morris, B., Session, J., and Trentin, J. (1962) Ann. N.Y. Acad. of Sci. 99, 355.
4. Ford, W.L., and Atkins, R.C. (1971) Nat. New Bio. 234, 178.
5. Ford, W.L., and Hunt, S.V. (1973) In: Handbook of Exp. Imm. 2nd ed. Ch. 23. Edited by D.M. Weir, Blackwell Scientific Publications, Ltd. Oxford.
6. Gowans, J.L. (1962) Ann. N.Y. Acad. of Sci. 99, 432.
7. Hudson, L., and Sprent, J. (1976) J. Exp. Med. 143, 444.
8. Ramseier, H., and Lindenmann, J. (1969) Path. Microbiol. 34, 379.
9. Shearer, G.M., and Cudkowicz, G. (1975) Science 190, 890.
10. Wilson, D.B., Marshak, A., and Howard, J.C. (1976) J. Imm. 116, 4, 1030.

INHIBITION OF T-ANTIGEN BINDING CELLS BY IDIOTYPIC AND Ia ANTISERA

Clifford J. Bellone and Charles A. Prange

Department of Microbiology, St. Louis University School of Medicine, St. Louis, Missouri 63104

ABSTRACT. Shared idiotypy between B and T cell receptors specific for the antigen L-tyrosine-p-azophenyltrimethylammonium, tyr(TMA), was studied in an antigen binding assay. Idiotypic reagents were prepared by inoculation of rabbits with purified anti-TMA antibody raised in strain 13 guinea pigs. The idiotypic antisera (anti-Id) blocked, on the average, 80% of the antigen binding T cells (T-ABC) and 53% of the antigen binding B cells (B-ABC) from tyr(TMA) immune strain 13 lymph node cells (LNC). Nylon wool passed tyr(TMA) immune LNC were trypsin treated resulting in a 75% loss of T-ABC which returned to 92% of pre-treatment values after allowing receptor synthesis during a 16 hour culture period. Seventy-nine percent of these regenerated T-ABC could be blocked with idiotypic antisera. The antisera did not block either T- or B-ABC in L-tyrosine-p-azobenzenarsonate tyr(ABA) immune guinea pig LNC; thus, confirming the specificity of these reagents for tyr(TMA) receptors.

A variety of anti-Ig reagents, some of which block B-ABC, do not inhibit T-ABC suggesting that V regions on T cells are not linked to Ig constant regions. However, in addition to anti-Id sera, a 2 anti-13 histocompatibility serum does block T- but not B-ABC. These results suggest that the receptor on T-ABC may be V region(s) linked to an I region gene product.

INTRODUCTION

The nature of antigen recognition at T and B cell level has been intensively investigated over the past several years. While it is well documented that membrane Ig serves as the receptor on B cells (1), the nature of the T cell receptor remains open to question. Antigen binding data (2,3) and studies of humoral T cell factors (4) indicate that IgM serves as the receptor. On the other hand, reports have shown that anti-Ig reagents neither block T-ABC (5,6) nor inhibit antigen-induced T cell proliferation (7,8). Anti-sera directed to Ia determinants on lymphocytes from both guinea pigs and mice have successfully abrogated this response (9,10).

Data from several laboratories now indicate that T and B cell receptors share idiotypic determinants suggesting that

T cells employ Ig V regions in recognition. Idiotypic antisera have been able to block several T cell functions (11,12), stimulate both helper (13) and suppressor activity (14), and enumerate the number of alloreactive T lymphocytes which express a given idiotype (15).

We have recently reported a clonal expansion of B and T lymphocytes in tyr(TMA) immune guinea pigs in the apparent absence of antibody production (16). This well-defined antigen system offers a unique opportunity to analyze specific T and B cell receptors for shared V regions. We now present evidence that idiotypic antisera raised to isolated anti-TMA antibody blocks both T- and B-ABC. In addition, we show a 2 anti-13 serum blocks T- but not B-ABC. These results suggest that tyr(TMA) receptors on T-ABC may consist of V regions linked to an I region gene product(s) bearing Ia determinants.

RESULTS

Guinea pigs inoculated with 400 μg of tyr(TMA) showed an increased expansion of both Ig^+ and Ig^- lymphocytes of which 60% and 40% are B- and T-ABC respectively. Specificity of the ABC for the tyr(TMA) moiety has been shown in a previous report (16). As seen in Table I, blocking experiments with a polyvalent anti-Ig and class specific antisera effectively inhibit $TMA\text{-}BSA^{125}I$ binding to Ig^+ LNC and, furthermore, indicate that 90% of these cells are μ bearing.

TABLE I

Inhibition of $TMA\text{-}BSA^{125}I$ Binding Ig^+ and Ig^- LNC by Various Specific Antisera

Source of LNC	Antiserum	Total ABC/10^4	ABC/10^4 Lymphocytes No. Ig^{+}*	No. Ig^-
CFA	-	8	4	4
tyr(TMA)	NRS	74 ± 4.9	44 ± 3.9	29 ± 1.2
"	anti-Ig	30 ± 1.6	0 (100)	30 ± 1.5(0)
"	anti-IgM	35 ± 1.9	4 ± 0.25(90)	29 ± 1.8(0)
"	anti-IgG	70 ± 4.0	41 ± 3.6 (8)	29 ± 1.3(1.0)
"	anti-Id	27 ± 4.3	21 ± 3.7 (53)	6 ± 1.1(80)

Guinea pigs immunized 2-6 weeks previously with 400 tyr(TMA) in CFA or with CFA alone. ABC prepared as previously described (17) except LNC reacted with antisera at final dilution of 1:10 prior to and during exposure to $TMA\text{-}BSA^{125}I$. ABC/10^4 represent the mean and standard error of 8-18 experiments. No. in parentheses = $\%I = (1 - \frac{\text{Ab treated}}{\text{NRS}} \times 100)$.

*Detected by fluoresceinated rabbit anti-guinea pig Ig.

In contrast, these anti-Ig reagents did not inhibit Ig^- LNC. However, idiotypic antisera (anti-Id) raised in rabbits to isolated guinea pig anti-TMA inhibited T-ABC by 80% while reducing B-ABC by 53%.

To test for anti-Id specificity, guinea pigs were inoculated with either tyr(TMA) or tyr(ABA), a related, but non-cross reacting immunogen. As seen in Table II, the anti-Id reagent blocks both T- and B-ABC populations in tyr(TMA) immune LNC but does not inhibit ABA-BSA^{125}I binding to ABC in tyr(ABA) immune LNC.

TABLE II

Specificity of Anti-Id for Receptors of tyr(TMA) Immune LNC

Source of LNC	Antiserum	Total ABC/10^4	ABC/10^4 Lymphocytes No. Ig^+	No. Ig^-
tyr(TMA)	-	74 ± 4.9	45 ± 3.9	29 ± 1.2
"	anti-Id	27 ± 4.4	21 ± 3.7(53)	6 ± 1.1(80)
tyr(ABA)	-	43 ± 2.3	23 ± 1.2	20 ± 1.4
"	anti-Id	42 ± 3.0	22 ± 1.5(5)	21 ± 2.1(0)

Guinea pigs immunized 1-6 weeks previously with either 400 μg tyr(TMA) or tyr(ABA) in CFA. ABC prepared as described (17) except LNC reacted with anti-Ids at a final dilution of 1:10 prior to and during exposure to TMA-BSA^{125}I or ABA-BSA^{125}I. ABC/10^4 represent the mean and standard of 4-10 experiments.

To determine whether the receptors on Ig^-ABC were endogenous to this cell type or antibody absorbed passively to the surface, nylon wool passed LNC were trypsin treated, placed in culture for 16 hours, and then tested for the reappearance of receptor. Mild trypsin treatment after column passage resulted in a 75% loss of T-ABC (Table III). A 16 hour culture period results in the return of 92% of the ABC (of which 79% are inhibited by the anti-Id), indicating the idiotype bearing receptor is endogenous to T-ABC. (See following page for Table III).

To characterize further the nature of the idiotype bearing receptor on T-ABCs strain 2 anti-strain 13 lymphocyte sera was used as a blocking reagent in the ABC assay. As seen in Table IV, the antiserum inhibits T-ABC by 76% in a LNC population, but does not significantly affect the ability of B-ABC to bind TMA-BSA^{125}I. In addition, the antisera inhibited nylon wool passed ABC (75%) and peritoneal exudate lymphocyte ABC (81%), both highly enriched T cell populations.

TABLE III

Effect of Trypsinization and Resynthesis of Surface Receptor of Column Purified ABC

Source of Immune LNC	Treatment	T-ABC/10^4 Lymphocytes	% loss of T-ABC
tyr(TMA)	NRS	39 ± 2.1	--
"	anti-Id	9 ± 1.6	77
"	trypsin-45'	10 ± 0.9	75
"	trypsin-followed by 16 hr in culture	36 ± 2.0	8
"	trypsin-16 hr culture + anti-Id	9 ± 1.4	77

40 x 10^6 column purified LNC from tyr(TMA) immune animals were incubated at 37°C for 45 minutes in 1 ml MEM containing 150 μg/cc trypsin, 10 μg/cc DNAase I, and 5 mM $NgCl_2$. After trypsin treatments, cells were either immediately washed and subjected to the ABC assay (± αId) or allowed to resynthesize surface receptor by 16 hours culturing prior to ABC determination.

TABLE IV

Effect of 2 anti-13 (αIa) Sera on ABC

Treatment	ABC	ABC/10^4 Lymphocytes No. Ig^+	No. Ig^-
NGPS	105 ± 24	81 ± 23	25 ± 5
2α13	79 ± 21	73 ± 20(9)*	6 ± 2(76)
	Nylon column passage**		
--	40		
2α13	10(75)		
	Peritoneal exudate lymphocytes***		
---	78		
2α13	15(81)		

*No. in parentheses = $\%I = 1 - \frac{2\alpha 13}{NGPS} \times 100$.

**Nylon column passage as described by Julius et al. (18) resulted in a population of lymphocytes which was 94% Ig^-.

***Peritoneal exudate lymphocytes prepared as described by Rosenstreich (19) resulted in 95% Ig^- cells.

DISCUSSION

This present report indicates that V regions utilized by secreted and membrane-bound Ig serve as part of the recognition unit on T cells. Furthermore, the data suggests a physical association of the idiotype bearing receptor with an Ia bearing molecule. The evidence for shared V regions on T and B cells is based on cross reactivity between receptors as detected by idiotypic antisera. Since we have generated this sera from isolated anti-TMA antibody by affinity chromatography, it is important to know whether all of this material is Ig in nature. Analysis of the material on reducing SDS polyacrylamide gels revealed only two bands which comigrate with heavy and light chains of guinea pig Ig (data not shown) and are absorbed by anti-Ig reagents coupled to sepharose. This indicates we have made idiotypic antisera to antibody only, and not additionally to some antigen binding factor(s), possibly of T cell origin (20). This information is key to our interpretation of the results.

Since a theta-like marker is not known for T cells in guinea pigs, proof that we are in fact detecting T-ABC must lie with indirect evidence. Data from our laboratory show that guinea pig Ig^- ABC resemble mouse T-ABC in several respects (6). These include: 1) the absence of easily detectable surface Ig, 2) increased numbers of Ig^- ABC at 37° (increase inhibited in presence of sodium azide), 3) passage through nylon wool, and 4) not inhibitable by a variety of anti-Ig reagents. At the very least, we can conclude that if our Ig^- ABC is not a T cell, we are looking at a most interesting antigen specific cell type.

The nature of the T cell receptor, analogous to Ig structure, may be the product of different genes, i.e., the binding or variable region coded for by Ig V genes while the "constant" regions are coded by the I region locus located in the major histocompatibility locus of the species. Evidence for this comes from independent studies utilizing a number of different approaches. Our data, as well as previous reports, demonstrate that T and B cell receptors share idiotypy which strongly suggests utilization of Ig V regions by T cells (11,13). Other data show anti-Ig reagents do not inhibit T cell function (7,8) nor T cells from binding antigen (5,6), suggesting that V regions are not linked to known Ig constant regions. However, in most of these reports, antisera to the major histocompatibility locus of the species will block T cell function (7,8) and antigen binding (8). Recent reports show the active blocking antibody is directed to Ia determinant(s) coded by the I region in the major histocompatibility complex (9). Further work on antigen helper (21) and suppressor (22) factors indicate these molecules bear Ia determinants but as yet nothing is known concerning their idiotypic nature. Our data merge these two areas (idiotypy and Ia linked T cell product(s)), suggesting the T cell receptor may be composed of Ig V regions and Ia bearing "constant" regions. This association is based solely on blocking experiments but in view of the fluid mosiac model of the lymphocyte surface it suggests a close physical association of these determinants. These data are only indirect and rigorous proof must await biochemical and serological studies on these isolated idiotype bearing T cell receptors.

ACKNOWLEDGEMENTS

The authors wish to thank Dr. Eli Nadel and the McBride Love Foundation for support in the early steps of this project. This work was supported by N.I.H. Grant AI 13115 from the U.S.P.H.S., and aided by a grant from the National Foundation-March of Dimes.

REFERENCES

1. Warner, N. L. (1974) Adv. Immunol. 19, 67.
2. Roelants, G., Forni, L., and Pernis, B. (1973) J. Exp. Med. 137, 1060.
3. Hogg, N. M. and Greaves, M. F. (1972) Immunology 22, 967.
4. Feldman, M. (1972) J. Exp. Med. 136, 737.
5. Kennedy, L. J., Dorf, M. E., Unanue, E. R., and Benacerraf, B. (1975) J. Immunol. 114, 1670.
6. Prange, C. A., Feidler, J. J., and Bellone, C. J. Manuscript submitted.
7. Shevach, E. M., Paul, W. E., and Green, I. (1973) J. Exp. Med. 136, 1207.
8. McDevitt, H. O., Bechtol, K. B., Hämmerling, G. J., Lonai, P., and Delovitch, T. L. (1974) <u>In</u> 3rd INC-UCLA Symposium on Molecular Biology p. 597.
9. David, C. S. (1976) Trans. Rev. 30, 299.
10. Geczy, A. F., deWeck, A. L., Schwartz, B. D., and Shevach, E. M. (1975) J. Immunol. 115, 1704.
11. Binz, H. and Wigzell, H. (1975) J. Exp. Med. 142, 197.
12. Black, S. J., Hämmerling, G. J., Berek, C., Rajewsky, K., and Eichmann, K. (1976) J. Exp. Med. 143, 846.
13. Eichmann, K. and Rajewsky, K. (1975) Eur. J. Immunol. 5, 661.
14. Eichmann, K. (1975) Eur. J. Immunol. 5, 511.
15. Binz, H. and Wigzell, H. (1975) J. Exp. Med. 142, 1218.
16. Prange, C. A., Green, C., Nitecki, D. E., and Bellone, C. J. (in press) J. Immunol.

17. Davie, J. M. and Paul, W. E. (1971) J. Exp. Med. 134, 495.
18. Julius, M. H., Simpson, E. and Herzenberg, L. A. (1973) E. J. I. 3, 645.
19. Rosenstreich, D. L., Blake, J. T., and Rosenthal, A. S. (1971) J. Exp. Med. 134, 1170.
20. Binz, H. and Wigzell, H. (1975) Scand. J. Immunol. 4, 591-600.
21. Munro, A. J. and Taussig, M. J. (1975) Nature (Lond.) 256, 103.
22. Taniguchi, M., Hayakawa, K., and Tada, T. (1976) J. Immunol. 116, 542.

Workshop #18

THE T CELL RECEPTORS: Robert Cone and Charles Janeway, Yale University Medical School.

The purpose of this workshop was to bring together evidence from a variety of systems and laboratories on the molecular characteristics of receptor molecules on thymus-derived (T) lymphocytes. Surprisingly, it produced a near consensus on the structure of at least one such molecule.

Morten Simonsen introduced the workshop by describing ways in which differing protein molecules might combine on the cell surface to generate specific receptor molecules. He pointed out that the number of genes required to generate a given number of specificities decreases as more chains are involved in generating the receptor. The following attributes of T cell behavior were felt to require explanation in any theory of T cell receptors:

1) The multipotential behavior of T cells (see Workshop #1).

2) Clonal restriction of T cell responses to both MHC and non-MHC antigens.

3) The expression of allotype-linked VH gene products in functional T cells.

4) The putative production of MHC gene (I-region) products with antigen specificity.

The question of involvement of I-region (MHC) gene products in T cell receptor formation was not directly considered (see Workshop #14, FACTORS). Evidence for the involvement of I-region gene products in forming the molecules discussed here was limited to one set of experiments in which anti-I-region antibodies in guinea pigs inhibited T but not B antigen binding cells (Bellone).

A number of investigators described systems in which VH gene products, detected as idiotypic markers, were found on T cell receptors. This was true for receptors for MHC antigens (Wigzell, Binz), for carbohydrates (Eichmann), for proteins (Bona, Cazenzve), and for haptens (Rajewsky, Bellone, Augustine). Where studied, only VH and not VL markers were detected. Also, in systems using rabbit T cells Rajewsky and Bona found allotypic markers from VH (a allotypes), but did not find the b allotypic markers of Ck. Where anti-idiotypic antisera were used to purify secreted T cell products, three types of molecules were isolated: a disulfide bonded dimer of 140,000, a monomer of 70,000, and a cleavage product of 45,000 daltons. No light chains were detected by these techniques. Furthermore, it was found that T cell reactions with anti-idiotypes were tightly linked with CH but not with CL genes.

Antigen binding T cell molecules were isolated by two different techniques. Rajewsky used antigen bound to nylon nets to purify antigen binding T cell receptors for haptens, which were shown to be about 150,000 daltons in the intact form, and which were detected by haptenated phage inactivation. Maurer isolated a cell membrane molecule by binding to radiolabelled GAT which had molecular weight on reducing gels of 40-60,000 daltons. No evidence for MHC related products was found.

Isotypic characterization of the T cell receptors led to general agreement. Binz found that no anti-Ig reagents remove T cell receptor molecules, but a new rabbit serum directed at idiotype-bearing molecules from Lewis rats will bind secreted T cell receptors, and will also kill over 60% of peripheral T cells from several strains of rats and mice. Using a different approach, several investigators (Cone, Marchalonis, U. Hämmerling) could isolate an Ig-like molecule from thymocytes or T lymphoma cells. This molecule was precipitated by antisera against L chains or against Fd of conventional Ig. It differs from B cell Ig in the following ways: it is not precipitated by conventional anti-Ig reagents raised in rabbits to different isotypic markers. It is made up of two disulfide-bonded heavy chains of 70,000d which may be cleaved to 45,000d during isolation, and appears to have non-covalently bound light chains as well. It is not detectable on cell surfaces by staining or absorption of rabbit anti-kappa sera. However, antisera to murine IgM or IgG Fab'_2 fragments prepared in chickens are capable of staining both T and B cells and will bind the same T cell molecules in solution that are detected by rabbit anti-mouse L chain sera (Marchalonis, Hämmerling). These molecules can not be precipitated by anti-light chain sera (Cone) after detergent isolation if detergent concentrations greater than 0.1% are used (suggesting either a more hydrophobic molecule or detergent-sensitive antigenic determinants).

The most controversial area covered was whether or not T cell receptors contain L chains, or an L-chain-like chain. Evidence cited in favor of L chains on T cells was: 1. The isolation of L chains from T cells by surface labelling and anti-kappa chain precepitation (Cone, Marchalonis). 2. Isolation and translation *in vitro* of L chain mRNA from thymocytes and T cell tumors (Storb, Putnam). 3. The high affinity of purified antigen-binding T cell receptors (Rajewsky). Evidence against T cell L chains was: 1. Failure to detect L chains or L chain allotypes on a variety of T cell receptors isolated by anti-idiotype or by binding to antigen (Binz, Rajewsky, Bona, Eichmann). 2. Failure to prime T cells with anti-L chain idiotype reagents (Eichmann). 3. The finding that putative kappa chain mRNA hybridized at a lower CoT in T cells than in the myeloma from which the cDNA probe was derived (Milstein). This area clearly needs resolution. Possible sources of confusion are: 1. Loss of antigenic markers during isolation (Rajewsky). 2. Loss of non-covalently linked L-chains on binding to ligand or to anti-VH idiotype antibody (Marchalonis). 3. Presence of a unique T-cell L-chain isotype (Cone).

There was also disagreement about T cell avidity for antigen. Rajewsky stated that isolated T cell receptors are as avid as isolated B cell receptors from immunized mice, as measured by hapten inhibition studies of haptenated phage inactivation. By contrast, Bellone reported that T antigen binding cells are about 100 fold less avid than B antigen binding cells. Rubin pointed out that avidity may depend on the type of T cell studied, since suppressor and cytotoxic T cells (Ly23 cells) can be absorbed by antigen, while helper T cells and cells giving MLR (Ly1 cells) do not absorb to antigen. Similar data on T-rosette forming cells were communicated by Eardley. These differences could be avidity-related. Studies of function in these systems will clarify these points.

In summary there was a clear consensus that T cells do have VH encoded receptors, either single or multiple which are expressed on 70,000 dalton H chains of a new, T cell isotype. The heavy chains are disulfide bonded on the cell surface, and may be non-covalently bound to light chains. It is possible that T cells lose their L chains during maturation and activation. The involvement of Ia antigens and of Ly antigens in these receptors was not considered in detail. Hans Wigzell raised the following points in closing the session:

1. Immune T cells and non-immune T cells behave differently; this difference may or may not be related to structural differences in the T cell receptor.

2. I-region encoded factors, whether or not they are associated with antigen, have always been said to have molecular weights of 40-50,000, while the VH idiotypic molecules are 70,000. Antigen-binding and idiotypic T cell molecules are not structurally similar to I-region encoded structures. Thus, these are probably different molecules.

3. The Ly antigens should not be excluded from consideration as components of T cell receptors.

4. Functional types of T cells differ in a variety of ways; the possibility that they differ in the structure of their receptors should not be dismissed.

The chairpeople expect that the T cell receptor workshop at the next symposium will answer these and other questions and be the last of its kind.

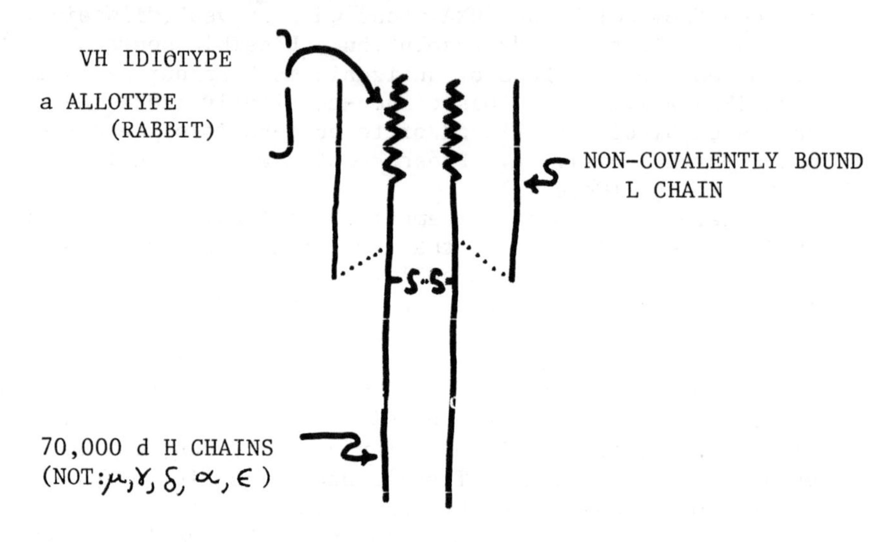

The Consensus Model for the Ig-like T Cell Receptor
(Note: Presence of L chains not resolved)

REGULATION OF IDIOTYPES EXPRESSED ON RECEPTORS OF PHOSPHORYLCHOLINE-SPECIFIC T AND B LYMPHOCYTES

Michael H. Julius, Andrei A. Augustin and Humberto Cosenza

Basel Institute for Immunology, Postfach 4005

Basel 5, Switzerland

ABSTRACT. Immunoglobulin (Ig) receptors on phosphorylcholine (PC)-specific B cells bear almost exclusively an idiotype characteristic of the Balb/c PC-binding myeloma protein TEPC 15 (T15), i.e. antibodies directed to the T15 idiotype specifically inhibit induction of a humoral anti-PC response (1). To determine whether this "idiotypic homogeneity" is also characteristic of receptors on PC-specific T cells, we developed a functional assay for PC-specific helper cells. Balb/c mice were primed with PC conjugated to an isologous myeloma protein. Splenic T cells were transferred together with bovine serum albumin (BSA)-primed splenic B cells into irradiated syngeneic recipients which were then challenged with PC-BSA. Our results show that PC-specific helper T cells collaborate with BSA-specific B cells in the induction of an anti-BSA plaque-forming cell (PFC) response. Their helper activity was completely inhibited when the irradiated recipients were injected with anti-T15 antibodies prior to cell transfer. Furthermore, we have been able to generate PC-specific helper T cells by injecting Balb/c mice with low doses of anti-T15 antibodies. When high doses of T cells are transferred, PC-helper T cells induced with anti-idiotypic antibodies are as efficient as helper cells from mice primed with PC. Thus, in parallel with their B cell counterparts, receptors on PC-specific T cells bear predominantly the T15 idiotype.

Injection of Balb/c mice with anti-T15 antibodies at birth, changes the idiotypic pattern of the anti-PC antibodies produced by adult mice from T15-positive ($T15^+$) to T15-negative ($T15^-$) (2), i.e. at the precursor cell level, the majority of PC-specific precursors express Ig receptors which bear $T15^-$ idiotypes. If neonatal suppression of the T15 idiotype also induces a parallel change in the idiotypic pattern of receptors expressed on PC-specific T cells, the helper function of PC-primed T cells from $T15^-$ mice should not be inhibited by anti-T15 antibodies. To this end, we transferred PC-specific T cells from $T15^-$ mice together with BSA-primed B cells from $T15^+$ mice into $T15^+$ irradiated recipients which were challenged with PC-BSA. The anti-BSA PFC response obtained was not inhibited by pre-treatment of the irradiated recipients with anti-T15 antibodies. Furthermore, we have failed to generate PC-helper T cells by injecting $T15^-$ Balb/c mice with low doses of anti-T15 antibodies. Thus, PC-specific T cells from neonatally suppressed mice bear $T15^-$ receptors. This parallel behaviour of PC-specific B and T cells from normal and neonatally suppressed Balb/c mice suggests that expression of idiotypes on antigen receptors of both lymphocyte lineages is governed by similar regulatory mechanism(s).

INTRODUCTION

The specificity of an immune response is due to the selective expression of immunocompetent clones which react specifically with different antigenic determinants (epitopes). Thus, precursor B cells, reactive with a given epitope, have been shown toexhibit Ig receptors bearing idiotypes identical to those present on the antibodies secreted by their progeny (1,3-6). Recently, several investigators have also demonstrated that receptors on T and B cells specific for the same epitope share similar or identical idiotypes (7-10). This observation suggests that at least some portion of the antigen-recognition structures - variable (V-) regions - are similar on both cell lineages. The implication being that T and B cells possess the same genetic repertoire of V-genes. However, whether expression of a such V-region structures is governed by the same regulatory mechanism(s) in both T and B cells, is not yet clear.

In addressing this question, we have used the response of Balb/c mice to PC as a model system. This immune response is characterized by the dominant expression of a single idiotype (3); facilitating the study of idiotypes expressed on both T and B cells under different experimental conditions. In the present report, we show that a) receptors on the majority of PC-specific T and B cells from conventional Balb/c mice bear the T15 idiotype, and b) neonatal suppression with anti-T15 antibodies induces a parallel change in the idiotypes expressed on both lymphocyte populations. These results suggest that the expression of idiotypes on receptors of PC-specific T and B cells is under similar regulatory control.

MATERIALS AND METHODS

Balb/c mice obtained from Cumberland View Farms, Clinton, Tenn., USA, were used in all experiments. The source of antigen was the heat-killed rough Pneumococcus pneumoniae, strain R36A (Pn). Pn has been shown to contain PC on its cell wall (11). Mice were immunized with 10^8 Pn cells i.v. in 0.2 ml of saline. Of the myeloma proteins used, TEPC 15 (T15) was obtained in ascites form from Litton Bionetics, Kensington, Md., USA, while MOPC 511 (M511), also in ascites form, was a kind gift of Dr. M. Potter, NCI. These are IgA (α,κ) myelomas of Balb/c origin which bind PC but exhibit individual idiotypic determinants (12). They were purified on a PC-Sepharose column as previously described (13).

Preparation of anti-idiotypic antibodies: The homologous anti-idiotypic serum reacting with T15 was raised by immunizing A/J mice according to the method described by Potter and Lieberman (12). This antiserum was used to suppress newborn Balb/c mice and to inhibit the formation of anti-PC plaques. It was absorbed on MOPC 315 [IgA (α,λ)], MOPC 104E [IgM (μ,λ)], MOPC 21 [IgG (γ_1,κ)], UPC 10 [IgG (γ_{2a},κ)], MOPC 195 [IgG (γ_{2b},κ)] and MOPC 511 sepharose columns. All these myelomas were purchased in ascites form from Litton Bionetics and they were precipitated at a final concentration of 18% Na_2SO_4 before coupling onto Sepharose.

Conjugation of sheep red cells (SRC) and proteins with PC: The p-diazonium phenylphosphorylcholine (DPPC) was synthesized according to Chesebro and Metzger (13), and it was coupled onto SRC and different proteins as described previously (14). The PC-SRC were resuspended to a 10% suspension for use in the hemolytic plaque assay (15) as modified by Cunningham and Szenberg (16).

Inhibition of plaque formation with anti-idiotypic antibodies: Inhibition of PC-specific plaques with A/J anti-T15 antibodies was used to determine whether the anti-PC antibodies secreted by single plaque-forming cells (PFC) bore the T15 idiotype. Briefly, the anti-idiotypic serum (used at a 1:500 final dilution in the chambers) was added to the cell mixture containing spleen cells, PC-SRC and guinea pig complement. The chambers were incubated at 37°C for 1 hr. All plaques were direct and could be inhibited by anti-μ antibodies.

TABLE 1

EFFECT OF NEONATAL SUPPRESSION WITH ANTI-T15 ANTIBODIES

Mice	Anti-PC PFC/10^6 cells[b]	% of anti-PC PFC[c]	
		$T15^+$	$T15^-$
Normal	345 (192-765)	96	4
Neonatally Suppressed:[a]			
8 weeks	31 (9- 57)	-	-
11 weeks	23 (11- 43)	-	-
15 weeks	187 (16-337)	11	89
26 weeks	256 (204-319)	8	92

[a] One day old Balb/c mice were injected i.p. with 0.2 ml of A/J anti-T15 serum. Along with normal littermates, after 8,11,15 or 26 weeks of life, they were challenged with Pn and their individual spleens assayed for anti-PC PFC on day 5 using PC-SRC as indicator cells.

[b] Each group represents at least six mice. Within parenthesis are given the range of the responses.

[c] The percent of $T15^-$ plaques was estimated after inhibition of $T15^+$ plaques with anti-idiotypic antibodies.

RESULTS

Effect of neonatal suppression with anti-idiotypic antibodies. The majority (95%) of PFC induced in conventional Balb/c mice upon immunization with PC, secrete antibodies bearing the T15 idiotype (3). To investigate whether suppression of $T15^+$ B cell precursors would allow the expression of $T15^-$ clones, we injected newborn Balb/c mice with anti-T15 antibodies (2). Two groups of 6 suppressed mice and their respective untreated littermates were challenged with Pn at 8 or 11 weeks of age and their individual spleens assayed for anti-PC PFC five days later. Both groups of suppressed mice responded with about 8% of the normal response. At 15 weeks after birth, 7 other suppressed mice were challenged with Pn and their responses averaged 54% of controls. Almost complete recovery from suppression was seen by the 26th week of life; their responses represented an average of 74% of the response given by six normal controls (Table 1).

When neonatally suppressed mice began to recover from unresponsiveness, the anti-PC PFC were screened for the idiotype expressed on the secreted antibodies. At 15 and 26 weeks after neonatal suppression, only 11% and 8% of the PFC secreted $T15^+$ antibodies, respectively (Table 1). In contrast, 96% of the PFC from untreated littermates produced $T15^+$ anti-PC antibodies. Thus, administration of anti-T15 antibodies at birth results in a drastic alteration of the idiotypic pattern of the anti-PC response, i.e. after regaining responsiveness, neonatally suppressed mice shift to the dominant expression of $T15^-$ anti-PC antibodies.

Idiotypic analysis of PC-precursors from neonatally suppressed mice. A possible explanation for the long lasting unresponsive state to PC in neonatally suppressed mice is that the $T15^+$ clones are deleted and the $T15^-$ clones are in such low numbers that they require a relatively long time to expand before their expression is detectable (17). To analyze whether clonal deletion could account for the absence of $T15^+$ anti-PC antibodies in neonatally suppressed mice, we assessed the proportion of $T15^+/T15^-$ precursor B cells from mice suppressed with anti-idiotypic antibodies capable of binding 100 μg or 1 μg of T15 monomer (Table 2).

Spleen cells from two mice of each group were pooled and analyzed at 9, 22 and 35 weeks after suppression, for the number of $T15^+$ and $T15^-$ PC-precursors stimulated by Pn (18,19). At 9 weeks of life, the total number of precursors detected in both groups of suppressed mice was about 10-20

TABLE 2

IDIOTYPIC ANALYSIS OF PC-PRECURSORS FROM NEONATALLY SUPPRESSED MICE

Spleen cells	IBC (μg) of anti-T15 serum[a]	Precursors/10^6 spleen cells[b]					
		9 weeks		22 weeks		35 weeks	
		Total	% $T15^-$	Total	% $T15^-$	Total	% $T15^-$
Normal	-	73	10	41	17	31	7
Neonatally suppressed	100	8	63	35	94	20	80
	1	4	25	92	6	46	13

[a] IBC = idiotype-binding capacity of the anti-T15 serum administered at birth.

[b] Precursor frequencies calculated as described previously (19). Microcultures were stimulated with Pn_4 and assayed for PC-PFC on day 4 using PC-SRC as indicator cells. Each well contained 2 x 10^4 responding cells and 3 x 10^4 irradiated (1200R) spleen cells.

fold lower than observed in normal mice. By the 22nd week of life, however, the total number of precursors in the suppressed mice was comparable to that of normal controls, i.e. the suppressed groups had recovered from the state of unresponsiveness to PC. However, idiotypic analysis of the responding PC-precursors revealed that only those mice suppressed with the higher dose of anti-T15 antibodies were chronically suppressed for precursors bearing the T15 idiotype. These results establish the existence of two concentration thresholds, defined functionally by the different effects of high and low doses of neonatally administered anti-T15 antibodies. Thus, unresponsiveness to PC can be induced by administration of extremely low doses of anti-idiotypic antibodies; however, to establish chronic suppression of $T15^+$ clones, administration of higher doses of anti-idiotypic antibodies is required. It seems unlikely, therefore, that deletion of $T15^+$ clones is the only mechanism responsible for the dominant expression of $T15^-$ idiotypes in neonatally suppressed mice.

The majority of PC-specific T cells from conventional Balb/c mice bear the T15 idiotype: To investigate whether the "idiotypic homogeneity" expressed by PC-specific precursor B cells (Table 2), is also reflected at the level of PC-specific T cells, we developed an assay for PC-helper cells. To optimize the induction of hapten specific T cells (20), we immunized Balb/c mice with isologous Ig proteins coupled either with PC (PC_5-MOPC 315) or 2,4-dinitrophenyl (DNP_9-mouse IgG) (10). Their splenic T cells were purified by the method of nylon wool filtration (21). Other Balb/c mice were primed with BSA and their spleens used as a source of B cells, after treatment with anti-Thy 1.2 serum and complement (10). The T and B cell populations were transferred alone or together into syngeneic irradiated recipients which were challenged with PC_{28}-BSA or DNP_{45}-BSA. The anti-BSA PFC response was assayed on day 8 post cell transfer, using BSA-coupled SRC (22) as indicator cells.

In previous experiments, the T cells were shown to be hapten-specific, i.e. an anti-BSA PFC response was elicited only when BSA-primed B cells and PC or DNP-primed T cells were transferred and the recipients challenged with PC-BSA or DNP-BSA, respectively (10). To assess whether the T cell receptor bears the T15 idiotype, we attempted to inhibit PC-specific helper cell function with anti-T15 antibodies. Irradiated conventional Balb/c recipients were injected with anti-T15 antibodies 16 hours prior to and at the time of

TABLE 3

INHIBITION OF PC-SPECIFIC HELPER T CELLS BY ANTI-T15 ANTIBODIES

Cells transferred ($x10^6$)[a]					
"BSA" B cells	"PC" T cells	"DNP" T cells	Antigen	Anti-T15[b] serum	Indirect anti-BSA[c] PFC/spleen
5	-	-	PC-BSA	-	800
-	10	-	"	-	700
5	10	-	"	-	9100
5	10	-	"	+	1400
5	-	-	DNP-BSA	-	900
-	-	10	"	-	300
5	-	10	"	-	8600
5	-	10	"	+	10600

[a]Cells were injected i.v. into irradiated (350R) Balb/c mice. The recipients were challenged i.v. 16 h later with 50 μg of antigen in PBS. Their spleens were assayed for anti-BSA PFC on day 8 post antigen injection.

[b]0.2 ml of a 1:2 dilution of A/J anti-T15 serum injected i.p. 16 h prior and at the time of cell transfer.

[c]Geometric mean of the responses of 4-5 animals per group. The direct PFC response in all groups was less than 5% of the total response. Indirect plaques were developed using a polyvalent rabbit anti-mouse Ig serum.

transfer of PC-primed T cells and BSA-primed B cells. The recipients were challenged with PC-BSA 16 hours after cell transfer. The data in Table 3 show that anti-T15 antibodies completely inhibited PC-helper activity, i.e. the anti-BSA response was reduced to background. As control for the specificity of inhibition, administration of anti-T15 antibodies did not inhibit DNP-specific helper activity (Table 3). Thus, comparable to their B cell counterparts, the function of PC-specific helper T cells from $T15^+$ mice can be completely inhibited with anti-idiotypic antibodies, suggesting that they bear, predominantly, the T15 idiotype.

The majority of PC-specific T cells from neonatally suppressed Balb/c mice bear $T15^-$idiotype(s): Neonatal suppression of T15-bearing clones with anti-idiotypic antibodies induces a shift in the idiotypic pattern of the anti-PC response to the dominant expression of $T15^-$ clones (Table 2). If neonatal suppression induces a parallel change on receptors of PC-specific T cells, their helper activity should not be inhibited by anti-T15 antibodies. To address this question, two sources of splenic T cells were used: conventional ($T15^+$) and neonatally suppressed ($T15^-$) Balb/c mice primed with PC-MOPC 315 (10). BSA-primed B cells and either $T15^-$ or $T15^+$ PC-primed T cells were transferred into conventional irradiated recipients which were then challenged with PC-BSA. The data in Table 4 show that when the recipients were pretreated with anti-T15 antibodies, the helper activity of PC-primed T cells from $T15^-$ donors was not inhibited, i.e. the anti-BSA response was not significantly different from control recipients not injected with anti-idiotypic antibodies. However, the helper activity of PC-primed T cells from $T15^+$ donors was again completely abolished by anti-T15 antibodies. Thus, like their B cell counterparts, PC-specific T cells from neonatally suppressed mice recognize PC with receptors bearing $T15^-$ idiotype(s).

Induction of PC-specific helper T cells with anti-T15 antibodies: It has been recently demonstrated that anti-idiotypic antibodies are able to induce priming of T and B cells bearing the relevant idiotype (8). Since receptors on PC-specific T cells from conventional Balb/c mice bear the T15 idiotype, administration of anti-T15 antibodies might induce priming of PC-specific helper T cells. To this end, $T15^+$ and $T15^-$ mice were injected i.v. with a dose of anti-T15 antibodies capable of binding 0.1 μg of T15 monomer. Eight weeks later, splenic T cells from the above two groups of mice and from untreated conventional mice were purified

TABLE 4

LACK OF INHIBITION OF T15-NEGATIVE PC-HELPER T CELLS BY ANTI-T15 ANTIBODIES

Cells transferred (x10^6)[a]					
"BSA" B cells	"PC" T cells[b]		Antigen	Anti-T15[c] serum	Indirect anti-BSA[d] PFC/spleen
	$T15^-$	$T15^+$			
5	-	-	PC-BSA	-	2050
-	10	-	"	-	200
-	-	7.5	"	-	200
5	10	-	"	-	9400
5	10	-	"	+	11100
5	-	7.5	"	-	16800
5	-	7.5	"	+	1800

[a]Cells transferred as in Table 3[a].

[b]Splenic T cells obtained from either conventional ($T15^+$) or from neonatally suppressed ($T15^-$) Balb/c mice primed with PC-MOPC 315.

[c]As in Table 3[b].

[d]Results expressed as in Table 3[c].

(21) and transferred into irradiated conventional recipients together with BSA-primed B cells. The recipients were challenged immediately with PC-BSA and the anti-BSA PFC response was assayed on day 8 post cell transfer. The results shown in Table 5 indicate that when splenic T cells from untreated conventional mice (normal T cells) or from neonatally suppressed ($T15^-$) mice primed with anti-T15 antibodies were transferred together with BSA-primed B cells, the anti-BSA PFC responses obtained were 9 and 5 fold higher than background, respectively. However, an anti-BSA response 50 fold higher than background resulted when splenic T cells from $T15^+$ mice primed with anti-T15 antibodies were transferred with BSA-primed B cells. Thus, priming with anti-idiotypic antibodies induces helper cells in $T15^+$ but not in $T15^-$ mice. These results support the previous conclusion that receptors on PC-specific T and B cells from conventional mice bear the T15 idiotype whereas T cells of the same specificity, induced in neonatally suppressed mice, lack this idiotype.

DISCUSSION

In the experiments reported here, anti-idiotypic antibodies have been used to probe for analogies between antigen recognition structures on T and B cells specific for the same epitope. To this end, we induced T cells directed to the haptens PC and DNP by immunizing Balb/c mice with isologous proteins coupled with these haptens (10). This minimizes the production of anti-hapten antibodies (20), and increases the chances of hapten recognition by T cells. We assayed for PC and DNP-helper T cells using an adoptive transfer system where hapten-specific T cells collaborate with BSA-primed B cells in the induction of an anti-BSA PFC response (Table 3). Our results indicate that helper T cells are able to discriminate between PC and DNP with a specificity equivalent to antibodies directed to these haptens (10). We were interested, therefore, in analyzing the antigen recognition structures - using idiotypes as serological markers for the V-region of antibody molecules - on PC-specific T cells in parallel with their B cell counterparts.

It is well established that the majority of PC-specific B cell precursors expressed in conventional Balb/c mice have receptors which bear the idiotype characteristic of the PC-binding myeloma protein, T15 (1, Table 2). If the V-regions of the antigen-binding structures on PC-specific helper T cells are analogous to those on PC-specific B cells, they

TABLE 5

SPECIFIC INDUCTION OF PC-HELPER T CELLS BY ANTI-T15 ANTIBODIES

Cells transferred (x10^6)[a]					
"BSA" B cells	Normal T cells	Anti-idiotype[b] "primed" T cells		Antigen	Indirect anti-BSA[c] PFC/spleen
		T15$^-$	T15$^+$		
5	-	-	-	PC-BSA	240
-	10	-	-	"	83
-	-	10	-	"	224
-	-	-	10	"	81
5	10	-	-	"	2940
5	-	10	-	"	2220
5	-	-	10	"	16100

[a]Cells were injected i.v. into irradiated (350R) conventional Balb/c mice along with 50 μg of antigen in PBS. Their spleens were assayed for anti-BSA PFC on day 8 post cell transfer.

[b]Splenic T cells obtained from conventional (T15$^+$) or neonatally suppressed (T15$^-$) Balb/c mice 8 weeks after an i.v. injection of anti-T15 serum (IBC = 0.1 μg).

[c]Results expressed as in Table 3[c].

also should reflect this "idiotypic homogeneity". This is the case, since the helper function of PC-primed T cells is completely abolished by anti-T15 antibodies (Table 3). Furthermore, PC-specific helper T cells can also be induced upon injection of $T15^+$ Balb/c mice with extremely low doses of anti-T15 antibodies (Table 5). Taken, together these results indicate that receptors on PC-specific T and B cells share the T15 idiotype, suggesting that at least part of the antigen recognition structures in both lymphocyte lineages are translation products of the same genes.

To further study analogies between receptors on PC-specific T and B cells, we investigated whether suppression of the T15 idiotype at birth altered the idiotype(s) expressed on receptors of both lymphocyte populations. Idiotypic analysis of PC-specific B cell precursors from neonatally suppressed mice demonstrated that, after recovery from unresponsiveness to PC, the majority of precursors expressed receptors bearing $T15^-$ idiotypes (Table 2). At the T cell level, PC-specific helper function induced in $T15^-$ mice was not inhibited by anti-T15 antibodies (Table 4). Moreover, in contrast to the finding in $T15^+$ mice, we failed to induce PC-helper T cells by injecting $T15^-$ Balb/c with low doses of anti-T15 antibodies (Table 5). This parallel behaviour of PC-specific receptors on T and B cells suggests that expression of antigen-binding structures on both lymphocyte lineages are subject to similar regulatory mechanisms.

These results also indicate that the $T15^+$ receptors on PC-specific helper T cells are not likely to be passively adsorbed molecules. By transferring PC-primed T cells from $T15^-$ mice together with BSA-primed B cells from $T15^+$ mice into $T15^+$ irradiated recipients, we dissociate the idiotypes present on the responding T and B cells. If the target of the anti-T15 antibodies, used to inhibit helper activity, are passively adsorbed $T15^+$ anti-PC antibodies present in the irradiated recipients (23), then the helper cell function of $T15^-$ T cells should be inhibited by anti-T15 antibodies. Since this is not the case (Table 4), anti-T15 antibodies must inhibit helper activity by interfering with the _functional_ antigen receptor on PC-specific T cells obtained from $T15^+$ mice.

In an effort to compare receptor idiotypes on functionally different populations of T cells specific for the same epitope, we have recently, in collaboration with Dr. Helmut Pohlit, developed an assay for T cells which mediate delayed-type hypersensitivity (DTH) to PC. The DTH assay used is in principle identical to that described by Miller and coworkers (24), i.e. Balb/c mice were skin painted with an ester compound of DPPC, six days later the mice were challenged in the left ear with Pn while the saline supernatant was injected into the right ear. One day later, the mice were injected with 5-[^{125}I]iodo-2'-deoxyuridine (^{125}I-UdR). The ratio of radioactivity between the left and right ear was taken as the index of DTH, 16 hours after the ^{125}I-UdR pulse. Our results indicate that the DTH reaction induced in conventional T15$^+$ mice is completely abolished when the mice are injected with anti-T15 antibodies before sensitization (induction phase) or just prior to the time they are challenged in the ear with Pn (effector phase). In contrast, injection of neonatally suppressed T15$^-$ mice with anti-T15 antibodies, at either the induction or effector phase, does not inhibit the PC-specific DTH reaction. Moreover, we have also been able to induce DTH by injecting T15$^+$ mice 8 weeks previously with low doses (IBC = 0.1 μg) of anti-T15 antibodies. Thus, functionally different populations of PC-specific T cells (helper cells and cells mediating DTH) appear to express similar idiotypes on their antigen-specific receptors. In addition, neonatal suppression of the T15 idiotype induces an analogous shift, from T15$^+$ to T15$^-$, of idiotypes expressed on both T cell populations.

Unlike B cells, receptors on T cells do not exhibit classical markers (allotypes or isotypes) located on the constant region of Ig molecules (25). Moreover, it is unclear whether idiotypes on T cell receptors are composed of Ig-like V-regions made up of the corresponding parts of heavy and/or light chains. It is possible, that although T and B cells specific for the same epitope share similar Ig-like V-region idiotypes, these structures present on T cells might be associated with molecules structurally unrelated to conventional Ig molecules (26,27).

Our experiments and those of other investigators (7-10), do not conclusively prove that the T cell receptor is an endogenous product. We are currently addressing this problem by inducing PC-specific DTH and helper T cells from T15$^-$ donors in the presence of T15$^+$ B cells, and vice versa. This should circumvent the argument that T cells acquire their receptors from the corresponding PC-specific B cells during the time they are exposed to antigen. We do not favor this view since, if the function of T cells is to be clonally restricted, it would be necessary to postulate a specific mechanism by which they acquire some B cell receptors and not others.

ACKNOWLEDGEMENTS

It is a pleasure to acknowledge the expert technical assistance of Rosemarie Lang, Heidi Haas and John Bews.

REFERENCES

1. Cosenza, H. and Kohler, H. (1972). Proc. Nat. Acad. Sci. (wash.) 69, 2701.
2. Augustin, A. and Cosenza, H. (1976). Eur. J. Immunol. 6, 497.
3. Cosenza, H. and Kohler, H. (1972). Science 176, 1027.
4. Hart, D.A., Wang, A.L., Pawlak, L.L. and Nisonoff, A. (1972). J. Exp. Med. 135, 1293.
5. Eichmann, K. (1974). Eur. J. Immunol. 4, 296.
6. Lee, W., Cosenza, H. and Kohler, H. (1974). Nature 247, 55.
7. Binz, H. and Wigzell, H. (1975). J. Exp. Med. 142, 197.
8. Eichmann, K. and Rajewsky, K. (1975). Eur. J. Immunol. 5, 661.
9. Black, S.J., Hammerling, G.J., Berek, C., Rajewsky, K. and Eichmann, K. (1976). J. Exp. Med. 143, 846.
10. Julius, M.H., Cosenza, H. and Augustin, A.A. (1977). Submitted to Nature.
11. Tomasz, A. (1967). Science 157, 694.
12. Potter, M., and Lieberman, R. (1970). J. Exp. Med. 132, 737.
13. Chesebro, B. and Metzger, H. (1972). Biochemistry 11, 766.
14. Claflin, J.L., Lieberman, R. and Davie, J.M. (1974). J. Exp. Med. 139, 58.
15. Jerne, N.K. and Nordin, A.A. (1963). Science 140, 405.
16. Cunningham, A.J. and Szenberg, A. (1968). Immunology 14, 599.
17. Kohler, H., Strayer, D.S. and Kaplan, D.R. (1974). Science 186, 643.
18. Lefkovits, I. (1972). Eur. J. Immunol. 2, 360.
19. Cosenza, H., Quintans, J. and Lefkovits, I. (1975). 5, 343.
20. Iverson, G.M. (1970). Nature 227, 273.
21. Julius, M.H., Simpson, E. and Herzenberg, L.A. (1973). Eur. J. Immunol. 3, 645.
22. Gold, E. and Fudenberg, H. (1967). J. Immunol. 99,859.

23. Lieberman, R., Potter, M., Mushinski, E.B., Humphrey, W.Jr., and Rudikoff, S. (1974). J. Exp. Med. 139, 983.
24. Miller, J.F.A.P., Vadas, M.A., Whitelaw, A. and Gamble, J. (1976). Proc. Nat. Acad. Sci. (Wash.). 72, 5095.
25. Krawinkel, U. and Rajewsky, K. (1976). Eur. J. Immunol. 6, 529.
26. Tada, T., Okumura, K. and Taniguchi, M. (1973). J. Immunol. 111, 952.
27. Munro, A.J., Taussig, M.J., Campbell, R., Williams, H., Lawson, Y. (1974). J. Exp. Med. 140, 1579.

CHANGES IN THE IDIOTYPIC PATTERN OF AN IMMUNE RESPONSE, FOLLOWING SYNGENEIC HAEMOPOIETIC RECONSTITUTION OF LETHALLY IRRADIATED MICE

A.A. Augustin, M.H. Julius and H. Cosenza

Basel Institute for Immunology, Postfach 4005,
Basel 5, Switzerland.

ABSTRACT. Upon challenge with antigen, the level of antibodies obtained in lethally irradiated bone marrow reconstituted mice is similar to that obtained in normal mice. However, the size of an immune response does not reflect which individual clones are expressed following antigenic stimulation. In order to analyze selective clonal expression, we followed the "idiotypic pattern" of an antibody response in reconstituted mice. Upon immunization of normal Balb/c mice with phosphorylcholine (PC), 95% of the anti-PC antibodies produced share an idiotypic marker with the TEPC15 (T15) myeloma protein of Balb/c origin. When irradiated Balb/c mice reconstituted with either foetal liver or bone marrow (BM) regain responsiveness to PC (at 12-15 weeks after reconstitution), the anti-PC response consists predominantly of antibodies lacking the T15 idiotype ($T15^-$). However, reconstitution of lethally irradiated mice with syngeneic spleen cells results in an idiotypic pattern similar to that of normal Balb/c mice. The same pattern is also obtained in mice reconstituted with a mixture of BM and normal spleen cells. This change of idiotypic pattern was further investigated in experiments testing: (a) possible regulatory interactions between idiotypes and anti-idiotypic antibodies and (b) differential distribution of PC-specific clones in lymphopoietic and mature compartments of the immune system.

INTRODUCTION

Assuming that idiotypes are reliable markers for B cell clones and their products, the "idiotypic pattern" of an immune response defines the clones expressed subsequent to antigenic stimulation, and reflects the degree of heterogeneity of the response.

In conventional Balb/c mice, 95% of the antibodies produced upon immunization with PC, share an idiotypic determinant(s) ($T15^+$) with the TEPC15 myeloma protein of

Balb/c origin. The predominant expression of the T15 idiotype is a constant characteristic of the idiotypic pattern of the anti-PC response in conventional Balb/c mice (1). It has been shown, however, that subsequent to neonatal administration of anti-T15 antibodies, the anti-PC response of these mice in adult life is comprised predominantly of antibodies which do not bear the T15 marker (2). To date, the only experimental protocol resulting in the expression of new idiotypes in the anti-PC and other immune responses has involved idiotypic suppression generated by anti-idiotypic antibodies.

In this report we demonstrate that reconstitution of irradiated Balb/c mice with syngeneic BM or FL cells results in a striking alteration of the idiotypic pattern of the anti-PC response, i.e., a shift to the predominant expression of T15⁻ antibodies.

METHODS

Balb/c mice obtained from Cumberland View Farms, Clinton, Tenn., USA were used in all experiments. The source of PC antigen was the heat-killed rough Pneumococcus Pneumoniae, strain R36A (Pn). The anti-T15 antiserum was raised by immunizing A/J mice according to Potter and Liebermann (3). The specificity of this antiserum has been previously reported (2). Anti-PC plaque forming cells (PFC) were assessed at day 5 after administration of antigen using PC coated sheep erythrocytes as indicator cells (PC-SRBC), prepared as previously described (2). The Cunningham-Szenberg (4) modification of the haemolitic plaque technique (5) was used. Inhibition of plaques with anti-T15 antibodies was used to determine the idiotype of the secreted anti-PC antibodies (2).

Haemopoietic cells, used for reconstitution, were obtained from adult bone marrow (BM) or foetal liver (FL) from day 16 embryos. Recipients were lethally (650R) irradiated 24 hours prior to reconstitution.

RESULTS AND DISCUSSION

When lethally irradiated Balb/c mice are reconstituted with either bone marrow or foetal liver cells, responsiveness to PC is regained 12-15 weeks later. The number of anti-PC PFC generated in these mice upon immunization is comparable to that of normal mice (1, and Table 1). However, the immune response to PC in these reconstituted animals consists predominantly of antibodies lacking the T15 idiotype ($T15^-$). In contrast, reconstitution with syngeneic spleen cells results in an idiotypic pattern similar to that of normal mice (Table 1).

TABLE 1. RECONSTITUTION OF LETHALLY IRRADIATED CONVENTIONAL Balb/c MICE

Source	No. of Cells Transfered ($x10^6$)	% of $T15^+$ anti PC-SRBC PFC *
Bone Marrow	5	6% (0-20)
Foetal liver	5	10% (0-22)
Spleen	30	87% (78-90)

*Tested 15 weeks after reconstitution - the anti-PC PFC responses were not significantly different between groups varying from 100 to 400 PFC/10^6.

In other experiments we took advantage of the fact that when Balb/c mice are neonatally suppressed with anti-T15 antibodies, they produce $T15^-$ anti-PC antibodies after immunization with PC. Transfer of spleen cells from the $T15^-$ mice into irradiated conventional $T15^+$ recipients results in the expression of $T15^-$ anti-PC antibodies (Table 2). Thus, the idiotypic pattern of the anti-PC response in recipients reconstituted with spleen cells, either $T15^+$ or $T15^-$, is identical to that of the spleen cell donor. Moreover, the anti-PC response in animals reconstituted with mixtures of spleen cells and bone marrow cells from $T15^+$ Balb/c mice follows the idiotype pattern derived from the spleen cells, i.e., $T15^+$ (Table 2). Whether the $T15^+$ antibodies in these recipients are solely derived from the reconstituting spleen cells which dominate the bone marrow cell contribution or whether the influence of the spleen cells on the bone marrow cells causes a shift in idiotype production of the latter, is not yet clear.

Conventional Balb/c mice contain high levels of naturally occurring anti-PC antibodies bearing the T15 idiotype. Since in natural circumstances, immature bone marrow cells are not exposed to the high concentration of T15 found in the circulation, perhaps their normal development is affected, resulting in the shift to the production of $T15^-$ antibodies. In order to assess the effect of circulating T15 in bone marrow reconstituted animals, we employed $T15^-$ Balb/c

TABLE 2

RECONSTITUTION OF CONVENTIONAL AND NEONATALLY T15 SUPPRESSED Balb/c MICE WITH BONE MARROW AND SPLEEN CELLS FROM "$T15^+$" AND "$T15^-$" DONORS

Recipients (irradiated with 650R)	Number of Cells Transfered ($x10^6$): Bone Marrow $T15^+$	Bone Marrow $T15^{-†}$	Spleen $T15^+$	Spleen $T15^{-†}$	% of $T15^+$ anti-PC-SRC PFC*
conventional Balb/c	4.5	-	-	-	22% (0-55)
"	-	4.5	-	-	10% (0-32)
"	4.5	-	10	-	90% (82-95)
"	4.5	-	-	10	2% (0-7)
neonatally suppressed Balb/c ($T15^-$)	5	-	-	-	6% (0-22)
"	-	-	30	-	80% (43-99)

* Tested 16 weeks after reconstitution - the range of the anti-PC responses in reconstituted animals was 150 to 250 PFC/10^6. The avidity of anti-PC antibodies produced, whether $T15^+$ or $T15^-$, were similar as assessed by inhibition of plaque formation with free hapten.

† Cells from $T15^-$, neonatally suppressed mice, at the age of 15 weeks.

as recipients. Parallel with the results shown in Table 1, transfer of BM cells from $T15^+$ mice into $T15^-$ lethally irradiated recipients again resulted in the expression of $T15^-$ anti-PC antibodies (Table 2). Thus, the presence of high levels of T15-bearing Ig in the irradiated recipients is not responsible for the shift to $T15^-$ anti-PC antibodies. In addition, the idiotypic pattern of the anti-PC antibodies, produced in $T15^-$ irradiated recipients reconstituted with spleen cells from $T15^+$ donors, was again characterized by the predominant expression of $T15^+$ antibodies (Table 2).

The T15 idiotype is considered a "germ line specificity" in Balb/c mice, since it is dominantly expressed in the immune response to PC (1) and the majority of B cell precursors reactive to PC bear the T15 idiotype (6). Since the dominance of the T15 clones in conventional Balb/c mice is considered as being favoured by the genetic background of this strain, it is intriguing that syngeneic reconstitution of irradiated mice leads to a shift from $T15^+$ to $T15^-$ antibodies in their response to PC. The persistence of low numbers of $T15^+$ antibody-forming cells, as well as the presence of $T15^+$ antibodies in the serum of the reconstituted recipients (data not shown), suggest that the T15 clone is not deleted. We consider two possible explanations of the present finding: 1) a strong suppressive mechanism against the T15 idiotype. This mechanism could emerge "de novo" in the reconstituted animal or could be already represented by potential idiotype specific "suppressor cells" preexisting in the lymphopoietic compartments of the immune system. Our experiments indicate, however, that the naturally occurring T15 idiotype does not activate or inhibit such putative suppressor cells, 2) a differential distribution of PC specific clones in the lymphopoietc and mature compartments of the immune system.

REFERENCES

1. Cosenza, H. and Köhler, H. (1972) *Science,* 140, 405.
2. Augustin, A.A. and Cosenza, H. (1976) *Eur. J. Immunol.,* 6, 497.
3. Potter, M. and Lieberman, R. (1970) *J. Exp. Med.,* 132, 737.
4. Cunningham, A.J. and Szenberg, A. (1968) *Immunology,* 11, 599.
5. Jerne, N.K. and Nordin, A.A. (1963) *Science,* 140, 405.
6. Sigal, N.H., Gearhart, P.J., Press, J.L. and Klinman, N.R. (1976) *Nature,* 259, 51.

INVESTIGATION OF THE B CELL REPERTOIRE THROUGH THE STUDY OF NON-CROSS-REACTIVE IDIOTYPES

Shyr-Te Ju and Alfred Nisonoff

Department of Biology, Rosenstiel Research Center
Brandeis University, Waltham, Massachusetts, 02154

ABSTRACT. Inoculation of rabbit antiidiotypic antibodies prevents the subsequent appearance of a cross-reactive idiotype (CRI) associated with the anti-p-azophenylarsonate (anti-Ar) antibodies of A/J mice, although such mice produce normal concentrations of anti-Ar antibodies bearing other idiotypes. We investigated the frequency of occurrence of four such "private" idiotypes, from three randomly selected mice, in 181 suppressed or nonsuppressed mice immunized with KLH-Ar. The idiotypic antibodies in suppressed mice were obtained by isoelectric focusing specifically purified anti-Ar antibodies and selecting single major peaks as ligands. The four proteins were labeled with ^{125}I and the presence of the corresponding idiotype was sought by inhibition in a radioimmunoassay. Two of the four idiotypes could not be detected in any of the 181 mice at the highest concentration tested; the concentration of idiotype was less than 1 part in 1,250 to less than 1 part in 25,000 of the anti-Ar antibody population. A third idiotype was found in three of the 181 preparations, but at a very low concentration. The fourth idiotype was present at low concentrations in 28% of the mice. The data, which are based on an assay that is far more sensitive than isoelectric focusing, support the existence of an extremely large repertoire of anti-Ar antibodies. We propose that the cross-reactive idiotype is the product of a germ-line gene(s) or a closely related gene whereas the undetectable private idiotypes that are unique to individual mice reflect a large number of somatic mutations.

INTRODUCTION

Immunization of A/J mice with keyhole limpet hemocyanin to which p-azophenylarsonate groups are conjugated (KLH-Ar) results in the production of anti-Ar antibodies, some of which share an intrastrain CRI (6); in general, 20 to 70% of the antibodies possess the idiotype. The appearance of the idiotype can be prevented by inoculation of antiidiotypic antiserum 2 to 6 weeks prior to immunization (2,3,9). A suppressed mouse does not recover from suppression once immunization has begun, although normal concentrations of anti-Ar antibodies lacking the CRI are produced. Antiidiotypic antisera can be prepared against such antibody populations and their use has resulted in the demonstration of a low frequency of occurrence of such idiotypes in the anti-Ar antibodies of other A/J mice, either suppressed or nonsuppressed (3,8). In that investigation, the radioactive ligands were purified anti-Ar antibodies from individual, suppressed mice. Such populations might contain a number of idiotypes, only some of which might be present in another mouse; the failure to obtain 50% inhibition in the radioimmunoassay might reflect a multiplicity of idiotypes in the ligand. To pursue this question more closely, we have now labeled the anti-Ar antibodies of suppressed mice with ^{125}I, subjected them to isoelectric focusing, and used the proteins in major individual peaks as ligands in the radioimmunoassays; ascitic fluids or sera of other A/J mice, containing anti-Ar antibodies, were then tested as inhibitors. The data presented here have been submitted for publication (*J. Exp. Med.*, in press).

RESULTS

Isoelectric focusing was carried out on polyacrylamide gels, using ^{125}I-labeled specifically purified anti-Ar antibodies. Fractions were separated with a Savant Auto-Gel Divider (16). To test the efficiency of the isoelectric focusing procedure, proteins in 3 of the 4 peaks used as ligand were refocused. In each case, a single sharp peak was obtained, with a pI value very close to that observed in the first isoelectric focusing procedure. Over 90% of each ligand was found to be precipitable, with a procedure using an antiglobulin reagent (6), by its rabbit antiidiotypic antiserum. To maximize the sensitivity of the radioimmunoassays, small amounts of labeled ligand (1 to 2 ng) were used. Since, as will be shown below, most anti-Ar preparations tested lacked the private idiotypes characteristic of the labeled ligands, 40 control experiments were carried out in which small amounts of the autologous, unlabeled idiotype were added to noninhibitory fluids containing anti-Ar antibodies. In each case, the expected degree of inhibition was obtained, indicating that the idiotype would have been detected if present.

The results of experiments on inhibition are shown by the data in Tables 1 to 4. It is evident (Tables 1 and 2) that

TABLE 1

DISPLACEMENT FROM ITS ANTIIDIOTYPIC ANTIBODY OF THE MAJOR PEAK (pI 6.7) OF THE ANTI-AR ANTIBODY OF THE SUPPRESSED A/J MOUSE (HIS-7)*

Inhibitor (Source of Anti-Ar Antobody)	ng Anti-Ar Antibody Required for 50% Inhibition
Autologous (HIS-7; Specifically Purified)	8
144 Nonsuppressed A/J Mice†	>10,000- >200,000
37 A/J Mice Suppressed for‡ Cross-Reactive Idiotype	>10,000- >190,000

* 1.5 ng of labeled ligand were used in the assay.
† Ascitic fluids.
‡ Sera.

two of the idiotypes studies, protein HIS-7 (pI 6.7) and HIS-13 (pI 6.9) could not be detected in any of the 181 ascitic fluids or sera tested as inhibitors. (Ascitic fluids were used in the case of nonsuppressed mice and sera in the case of suppressed, immune mice). A third idiotype (protein HIS-5, pI 6.9) was found in 3 of the 181 samples, but at very low concentrations (Table 3). The fourth idiotype (protein HIS-5, pI 6.3) was present at low concentrations in 28% of the 181 mice (Table 4).

The question arose as to whether the inhibitory capacity observed in the latter group of mice was due to the presence of idiotype or to some other factor. Tests were therefore carried out to determine whether the inhibition was caused by anti-Ar antibody. Ten randomly chosen ascitic fluids were adsorbed with Sepharose-4B to which bovine IgG-Ar was conjugated. Control adsorptions were carried out with Sepharose to which nonspecific A/J IgG was bound; this did not remove the inhibitory capacity from any of the 10 fluids tested. However, adsorption with Sepharose bearing azophenylarsonate groups substantially reduced the inhibitory capacity in each case and eliminated it almost completely from 7 of the 10 preparations. This result suggests that the inhibition was actually due to an idiotype associated with anti-Ar antibodies.

TABLE 2

DISPLACEMENT OF LABELED ANTI-AR ANTIBODIES OF A/J MOUSE HIS-13 (pI 6.9 FRACTION) FROM THEIR ANTIIDIOTYPIC ANTIBODIES*

Inhibitor (Source of Anti-Ar Antibody)	ng Anti-Ar Antibody Required for 50% Inhibition
Autologous (HIS-13; Specifically Purified)	
Unfocused	85
Focused (pI 6.9)	15
144 Nonsuppressed A/J Mice†	>10,000- >200,000
37 A/J Mice Suppressed for the Cross-Reactive Idiotype‡	>10,000- >190,000

* 2 ng of labeled ligand were used in the assay.
† Ascitic fluids.
‡ Sera.

It does not prove whether this idiotype was identical or cross-reactive with that of the labeled ligand.

Additional experiments were carried to determine whether the reactions of the 4 private idiotypes with their corresponding antiidiotypic antibodies were inhibitable by free haptens. The haptens tested were (p-azobenzenearsonic acid)-N-acetyl-L-tyrosine (PABT), p-arsanilate and the unrelated hapten, p-aminobenzoate. The latter compound was not inhibitory at the highest concentration tested, 20 mM. At the same concentration, PABT cause 74-98% inhibition of binding of each labeled ligand, whereas p-arsanilate caused 40-88% inhibition. Thus, the region of the combining site was an important idiotypic determinant in each of the anti-Ar antibodies tested.

DISCUSSION

There is a striking contrast between the frequency of occurrence of the CRI, which is found in every A/J mouse

TABLE 3

DISPLACEMENT OF LABELED ANTI-AR ANTIBODIES OF A/J MOUSE HIS-5 (pI 6.9 FRACTION) FROM THEIR ANTIIDIOTYPIC ANTIBODIES*

Inhibitor (Source of Anti-Ar Antibody)	ng Anti-Ar Antibody Required for 50% Inhibition
Autologous (HIS-5; Specifically Purified)	
Unfocused	50
Focused	15
Focused (pI 6.3)	>135
3 Nonsuppressed Mice†	4,400 to 100,000
141 Nonsuppressed†	>10,000->200,000
37 A/J Mice Suppressed for Cross-Reactive Idiotype‡	>10,000->190,000

* 1.5 ng of labeled ligand were used in the assay.
† Ascitic fluids.
‡ Sera.

immunized with KLH-Ar, and of the private idiotypes associated with the anti-Ar antibodies which arise in mice that are suppressed for the appearance of the CRI. (Suppressed and nonsuppressed mice produce approximately equal concentrations of the antibody (9)). To account for this difference, we would suggest that the CRI is the product of a germ-line gene(s) or of a gene related to a germ-line gene through a small number of somatic mutations, which take place in every A/J mouse. The private idiotypes, which are absent or present in very low frequency in other mice, may reflect a large number of somatic mutations which are unlikely to be repeated in individual mice.

The results, obtained with the anti-Ar antibodies of three randomly selected mice suppressed for CRI, provide evidence relating to the frequency of occurrence of the idiotypes arising in such mice. As ligands in the radioimmunoassay, 4

TABLE 4

DISPLACEMENT OF LABELED ANTI-AR ANTIBODIES OF A/J MOUSE HIS-5 (pI 6.3 FRACTION) FROM THEIR ANTIIDIOTYPIC ANTIBODIES

Inhibitor (Source of Anti-Ar Antibody)	ng Anti-Ar Antibody Required for 50% Inhibition
Autologous (HIS-5, Specifically Purified)	
Unfocused	37
Focused (pI 6.3)	8
Focused (pI 6.9)	>70
46 Nonsuppressed A/J Mice†	800 to 120,000
98 Nonsuppressed A/J Mice†	>10,000->200,000
5 A/J Mice Suppressed for Cross-Reactive Idiotype‡	3,000 to 150,000
32 A/J Mice Suppressed for Cross-Reactive Idiotype‡	>10,000->180,000

* 1.5 ng of labeled ligand were used in the assay.
† Ascitic fluids.
‡ Sera.

individual peaks were used from the three specifically purified anti-Ar preparations. Over 90% of each peak was reactive with its corresponding antiidiotypic antibody. The reaction of each of the four idiotypes with its antiidiotypic antibody was partially inhibited by free haptens (phenylarsonate derivatives). Two of the idiotypes could not be detected, by a very sensitive radioimmunoassay, in any of 181 A/J mice hyperimmunized with KLH-Ar. Of the mice used as a source of inhibitor, 37 had been suppressed with respect to the CRI. On the average, one molecule in 5,000 would have been detected; the lower limits vary markedly owing to the differences in anti-Ar antibody titer in the individual mice tested. The addition to noninhibitory fluids of small amounts of the autologous, unlabeled idiotype resulted in strong inhibition of binding in each of 40 such tests. A third idiotype was present in 3 of 181 mice but at extremely low concentrations. A fourth idiotype was present in 28% of the 181 mice, again at a low concentration in each case. Part or all of the inhibitory capacity in these mice was associated with anti-Ar antibodies.

Other investigators have shown that the repertoire of antibodies to an individual hapten or protein antigen in a single strain of mouse is very large (4,5,10,11). The present results confirm this, by a much more sensitive technique; i.e., idiotype analysis rather than spectrotype. The presence of 8-30 ng of idiotype would have been detected in the assay system used, whereas an average of about 50 μg was tested.

Our data would suggest that if somatic mutation or recombination is the basis of diversity, such mutations are random rather than programmed, since 2 of the idiotypes tested could not be found in any other mouse. Although we cannot formally rule out the possibility that mutations are programmed, but are periodic, the very large number of samples tested and the sensitivity of the assay would tend to argue against this possibility.

The data do not appear directly relevant to the hypothesis (1,17) that variability occurs as a result of insertion of small sections of genome encoding the hypervariable regions, except to suggest that if this mechanism is correct, the repertoire of such "episomes" must be very large.

The possibility that diversity arises through a somatic process is supported by the great diversity observed and the limited number of germ line genes suggested by mRNA-DNA (12-15), or cDNA-DNA (7) hybridization studies.

REFERENCES

1. Capra, J.D. and Kindt, T.J. (1975) Immunogenetics 1, 417.

2. Hart, D.A., Wang, A.L., Pawlak, L.L. and Nisonoff, A. (1972) J. Exp. Med. 135, 1293.

3. Hart, D.A., Pawlak, L.L. and Nisonoff, A. (1973) Eur. J. Immunol. 1, 44.

4. Köhler, G. (1976) Eur. J. Immunol. 6, 340.

5. Kreth, H.W. and Williamson, A.R. (1973) Eur. J. Immunol. 3, 141.

6. Kuettner, M.G., Wang, A.L. and Nisonoff, A. (1972) J. Exp. Med. 135, 579.

7. Leder, P., Swan, D., Honjo, T. and Seidman, J. (1977) Cold Spring Harbor Symp. Quant. Biol. (in press).

8. Nisonoff, A. and Ju, S-T. (1976) Ann. Immunol. (Inst. Pasteur) 127C, 347.

9. Pawlak, L.L., Hart, D.A. and Nisonoff, A. (1973) J. Exp. Med. 137, 1442.

10. Pink, J.R.L. and Askonas, B.A. (1974) Eur. J. Immunol. 4, 155.

11. Press, J.L. and Klinman, N.R. (1974) Eur. J. Immunol. 4, 155.

12. Rabbits, T.H., Jarvis, J.M. and Milstein, C. (1975) Cell 6, 5.

13. Schechter, I., Burstein, Y. and Spiegelman, S. (1976) Ann. Immunol. (Inst. Pasteur) 127C, 421.

14. Tonegawa, S. (1976) Proc. Nat. Acad. Sci. USA 73, 203.

15. Tonegawa, S., Steinberg, C., Dube, S. and Bernadini, A. (1974) Proc. Nat. Acad. Sci. USA 71, 4027.

16. Tung, A.S. and Nisonoff, A. (1975) J. Exp. Med. 141, 112.

17. Wu, T.T. and Kabat, E.A. (1970) J. Exp. Med. 132, 211.

REGULATION OF IgE ANTIBODY SYNTHESIS BY AUTO-ANTIIDIOTYPE IMMUNIZATION IN GUINEA PIGS

Alain L. de Weck, Andrew F. Geczy and Olga Toffler

Institute for Clinical Immunology, University of Bern, Inselspital, 3010 Bern, Switzerland

ABSTRACT. Syngeneic immunization in strain 2 and 13 guinea pigs using purified antibodies against chemically defined antigens (e.g. immunization of strain 13 animals with immuno-adsorbent-purified anti-penicilloyl-bovine gamma globulin (BPO-BGG) antibody from strain 13) produces anti-idiotypic antibodies which specifically prevent antigen-induced proliferation of immune T lymphocytes *in vitro*. In further experiments, it has now been found that passive and active immunization *in vivo* with anti-idiotypic antibodies specifically blocks IgE synthesis of guinea pigs immunized in such a way (low doses of antigen in $Al(OH)_3$) such that these animals no longer produce IgE antibodies against BPO-BGG. Various ways to achieve long lasting suppression of IgE biosynthesis by immunization with idiotypes are currently under investigation. Parallel studies have been performed in outbred animals passively or actively immunized with their own antibodies and lymphoid cells. Apart from the demonstration that IgE shares idiotypes with other Igs and with T cell receptors, these studies suggest new approaches in the management of allergic diseases.

INTRODUCTION

The existence of antibodies directed against variable portions of immunoglobulins (anti-idiotypes), in particular against antibody combining sites, and the potential role of such antibodies in the regulation of immune responses has attracted increasing attention in recent years (7,10). The induction and biological role of anti-idiotypic antibodies have been demonstrated in various experimental systems in mice, rats and guinea pigs (2,3). Anti-idiotypic antibodies have been raised by allogeneic, parent to hybrid or xenogeneic immunization using either purified immunoglobulins, myeloma proteins carrying the idiotype under investigation or lymphoid cells carrying idiotypes on their immunoglobulin receptors. However, anti-idiotypes can also be raised in an autologous or syngeneic situation, for example upon repeated immunization with antigen (8) or upon immunization with purified antibodies. In guinea pigs, we recently demonstrated that syngeneic immunization of strain 2 or 13 animals with syngeneic purified strain 2 or strain 13 antibodies against penicilloyl-bovine gamma globulin (BPO-BGG) yields anti-idiotypic antibodies capable of specifically preventing antigen-induced T lymphocyte proliferation _in vitro_ (3).

In most instances experimentally studied up to now, the passive administration or active induction of anti-idiotypic antibodies _in vivo_ has been found to have suppressive effects on the immune response (10), although in a few instances the priming or enhancing effect has been reported (2). Among the parameters of immune response studied, antibodies and antibody-forming cells of the IgG and IgM classes, T cell proliferation and graft versus host reactions (1) have been prominent. Little attention has been paid up to now to the potential of anti-idiotypes for the regulation of IgE antibody synthesis, although this antibody class plays a major role in several widespread allergic diseases. Accordingly, we have investigated the effect of passive administration and/or active induction of anti-idiotypic antibodies in guinea pigs immunized in such a way as to produce high levels of homocytotropic IgE-like (or reaginic) antibodies against a hapten-protein conjugate (BPO-BGG).

Specific inhibition of IgE response by preimmunization with idiotypes in virgin animals. Guinea pigs of strain 2 and strain 13 were immunized with syngeneic or allogeneic purified anti-BPO-BGG antibodies. These antibodies had been purified by passage through immunoadsorbent BPO-BGG columns (3). The acid-eluted purified anti-BPO-BGG antibodies were injected in complete Freunds adjuvant (CFA) in strain 2 and 13 animals. These animals were found to develop anti-idiotypic antibodies, and their serum became capable of specifically inhibiting the proliferation of immune T lymphocytes induced by BPO-BGG (3). As shown here, these animals actively preimmunized with their own idiotype are incapable of mounting an IgE-like response against BPO-BGG. As shown in _Table 1,_ strain 2 and strain 13 guinea pigs preimmunized with their own anti-BPO-BGG idiotypes show no or very low IgE-like sensitization to BPO-BGG, whereas control guinea pigs preimmunized with CFA alone develop a very strong response to BPO-BGG. The suppression affects the hapten (BPO) as well as the carrier (BGG) antigenic determinants. As specificity control, it could be shown that all idiotype preimmunized animals concomitantly produce a strong IgE response to ovalbumin (OVA). This suppression by idiotype preimmunization in virgin animals is long lasting since even 10 weeks after initiation of senistization with BPO-BGG and 38 weeks after preimmunization with idiotypes, no significant booster response to BPO-BGG could be

TABLE 1

EFFECT OF PREIMMUNIZATION WITH IDIOTYPES ON SUBSEQUENT SENSITIZATION TO ANTIGEN

Experimental: Preimmunization with idiotypes (purified anti-BPO-BGG) in CFA. 28 weeks later, beginning of immunization with BPO-BGG and OVA (1 μg each in $Al(OH)_3$, injected once weekly over 4 weeks. Boost with 1 μg BPO-BGG and OVA 9 weeks after the first injection of antigens. For further details see ref. 6.

	IgE Response	Day 0 preimmunization with BPO-BGG)	Day 21 (during immunization with BPO-BGG)	Day 63 (preboost with BPO-BGG)	Day 70 (post boost with BPO-BGG)
Strain 2					
Guinea pigs	α-BPO	0	130*	28	0
preimmunized	α-BGG	0	0	204	332
with st.2 $\bar{\alpha}$-BPO-BGG	α-OVA	0	787	875	1061
idiotypes in CFA					
Strain 13					
guinea pigs	α-BPO	0	158	61	87
preimmunized	α-BGG	0	0	0	405
with st.13 $\bar{\alpha}$-BPO-BGG	α-OVA	0	470	375	767
idiotypes in CFA					
Controls:	α-BPO		4126	1155	4480
strain 2 preimmu-	α-BGG		4768	4047	4666
nized with CFA	α-OVA		697	443	709

* reactions expressed in mm^2 PCA reactions (see ref. 6), average of 3 animals per group.

observed, in comparison to controls.

<u>Specific inhibition of ongoing antibody responses by passive administration of anti-idiotypic antibodies.</u> Anti-idiotypic antibodies from strain 2 or strain 13 guinea pigs obtained as indicated above (6) were passively injected either intravenously or in CFA in guinea pigs which had been presensitized to BPO-BGG. Two effects were noticed: (a) the antibody titers dropped immediately and recovered only after a week in the case of intravenous injection or after 2 to 4 weeks after injection in adjuvant (b) during this period of suppression, no boosting could be achieved by additional injection of antigen (BPO-BGG). An excerpt of such results, which will be published in detail elsewhere (6), is given in Table 2. In the response shown here and detected by the BPO-T4 bacteriophage test (9), the suppressed antibody response is of the IgG/IgM type.

TABLE 2

DEPRESSION OF ONGOING IMMUNE RESPONSE TO BPO-BGG BY PASSIVE ADMINISTRATION OF SYNGENEIC ANTI-IDIOTYPIC ANTIBODY

<u>Experimental:</u> Strain 2 guinea pigs immunized with 200 μg BPO-BGG in CFA, passively injected with normal guinea pig Ig or strain 2/strain 13 anti-idiotypes, obtained as described (3). Anti-BPO antibodies detected with BPO-T4 bacteriophages (9).

BPO-BGG immunized strain 2 animals		α-BPO-antibody titers
Before treatment		1:162'000 (1:72000-1:170000)
1 week after treatment with:	normal guinea pig Ig (0.5 ml iv)	>1:100'000
	$\bar{\alpha}$-strain 2 BPO-BGG idiotypes (0.5 ml iv)	1:20
	$\bar{\alpha}$-strain 13 BPO-BGG idiotypes (0.5 ml iv)	>1:100'000
2 weeks after treatment with:	normal guinea pig Ig (0.5 ml sc in CFA)	1:50'000
	$\bar{\alpha}$-strain 2 BPO-BGG (0.5 ml sc in CFA)	1:10

Development of anti-idiotypic antibodies upon repeated administration of antigen. It has been shown in some systems that repeated administration of antigen may induce the formation of anti-idiotypic antibodies (8). This was also observed in guinea pigs immunized with BPO-BGG in CFA and repeatedly boosted with 50-100 μg BPO-BGG intradermally over a period of 280 days. Among a group of seven strain 2 animals treated in this way, one became entirely negative to skin test, anti-BPO-BGG antibodies disappeared from the serum. The serum of the "suppressed" animal was shown to specifically block BPO-BGG-induced T-lymphocyte proliferation, in the same manner as anti-idiotype does. These results will be published in detail elsewhere (5).

Attempts to suppress ongoing IgE-responses with anti-idiotype immunization. For practical purposes, it would be required to suppress ongoing IgE responses in already sensitized animals. Accordingly, inbred and outbred guinea pigs were immunized with BPO-BGG in such a way as to develop long lasting and boostable IgE-like responses (low doses of BPO-BGG in $Al(OH)_3$ at regular intervals). Animals with high anti-BPO-BGG IgE levels were then treated in various ways as shown in Table 3. Since in man, suppression of the immune response by anti-idiotypes could not be done in an inbred situation, attempts to suppress the IgE response by using autologous serum, autologous antibodies or autologous lymphoid cells in outbred guinea pigs were also made. Immunized animals were followed for a period of at least one month and their IgE titers determined by PCA at repeated intervals. With the various regimens used up to now no significant suppressive effect could be demonstrated. Detailed results will be published elsewhere (6).

DISCUSSION

The experiments presented here and elsewhere (6) demonstrate clearly that syngeneic immunization with idiotypes raises antibodies which are capable of specifically suppressing antigen-induced T cell proliferation *in vitro*. In addition, preimmunization with idiotypes effect a long-lasting suppression of the immune responses against the corresponding antigen, including the formation of IgE-like antibodies. Passive administration of syngeneic anti-idiotypic antibodies has a reversible suppressive effect on IgG, IgM and IgE responses. Furthermore, repeated administration of antigen, as is performed in hyposensitization therapy, also appears (in some animals at least) to induce the formation of

TABLE 3

EFFECT OF VARIOUS IMMUNIZATIONS IN GUINEA PIGS WITH AN ONGOING IgE RESPONSE TO BPO-BGG

Experimental: Outbred and strain 2 guinea pigs immunized by repeated id. injections of BPO-BGG (1 µg of BPO-BGG in Al $(OH)_3$). When stable anti-BPO IgE response established, treatment and follow up of IgE levels for a month. For details, see (6).

Animals	Treatment	Effect on anti-BPO IgE level (PCA titer)
Outbred	Own serum (4x)	none
	Own serum (4x in CFA)	decrease
	Own serum + 1-10 µg BPO-BGG (immune complexes in vitro)	boost
	Own serum + $<$0.1 µg BPO-BGG	none
	Own leukocytes (2x)	none
Strain 2	Strain 2 $\bar{\alpha}$-BPO-BGG idiotypes (purified antibody), 2x in CFA	decrease
	dito without CFA	none
	Strain 2 lymphoblasts obtained by culture with:	
	BPO-BGG	boost
	CON A	boost
	No treatment (controls)	none
	CFA controls	decrease

suppressive anti-idiotypic antibodies.

There is therefore a compelling rationale for attempting suppression of the IgE responses with anti-idiotypic antibodies in animals with ongoing IgE responses. These attemtps are encouraged by two considerations: (a) since in guinea pigs from strain 2 and 13 anti-idiotypic antibodies could preferentially be raised in the syngeneic combination (2 idiotype into strain 2) it may well be that we are high responders to our own idiotypes; this would make sense in terms of immune response regulation; (b) immunization with a mixture of heterogeneous idiotypes appears to be effective in suppressing heterogeneous antibody responses even against a complex multideterminant antigen, since both anti-carrier and anti-hapten responses were inhibited.

Since not only purified antibodies but also lymphoid cells and lymphoblast as idiotype carriers may be used for immunization, various ways to achieve anti-idiotype immunization in a random population may be visualized. Of course, it may prove much more difficult to suppress an ongoing response rather than preventing the response by preimmunization with idiotypes. Nevertheless, the negative preliminary results reported here as far as active suppression in already immunized animals is concerned, should not deter us from further investigating this potentially new approach in hyposensitization therapy for allergic diseases. It is to be considered that the success of classical hyposensitization therapy with antigen may rest on the induction of suppressor cells and/or of anti-idiotypic antibodies.

ACKNOWLEDGEMENTS

These studies have been supported by the Swiss National Science Foundation (Grant 3.468.075).

REFERENCES

1. Binz, H.,and Wigzell, H. (1977) Progr. Allergy, 23 (in press).
2. Eichmann, K. (1975) Eur.J.Immunol. 5, 661.
3. Geczy, A.F., Geczy, C.L., and de Weck, A.L. (1976) J. exp. Med. 144, 226.
4. Geczy, A.F., and de Weck, A.L. (1977) Progr. Allergy, 22, 147.
5. Geczy, A.F.(1977) J. exp. Med. (in press).
6. Geczy, A.F.,Geczy, C.L., de Weck, A.L., and Toffler, O. (in preparation).
7. Jerne, N.K. (1974) Ann. Immunol. (Inst. Pasteur) 125 c, 373
8. Köhler, H. (1976) Transplant. Rev. 27, 24.
9. Lazary, S., Spengler, H., Schneider, C.H., and de Weck, A.L. (1972) Path. Mikrobiol. 38, 6.
10. Nisonoff, A., and Bangasser, S.A. (1976) Transplant. Rev. 27, 100.

THE SIGNIFICANCE OF MINOR CLONOTYPES IN THE DISSECTION OF B-CELL DIVERSIFICATION

Nolan H. Sigal, Michael P. Cancro, and Norman R. Klinman

Department of Pathology, School of Medicine
University of Pennsylvania
Philadelphia, Pennsylvania 19104

ABSTRACT: Minor clonotypes can be defined as those specificities which are present at too low a frequency to be identified reproducibly in all individuals of an inbred mouse strain. These specificities can be viewed either as somatic variants of germline antibodies or as encoded by germline genetic information but expressed at substantially lower frequencies than the major clonotypes. The analysis of homogeneous monoclonal antibodies generated in the splenic focus system can provide information relevant to B-cell diversification by permitting the identification, analysis, and enumeration of minor clonotypes within the B-cell pool.

Experimental systems investigated include identification of several clonotypes within the non-T15 PC-specific repertoire by idiotypic crossreactivity with murine anti-T15 serum, rabbit anti-T15 serum, or anti-idiotypic sera against McPC 603 and 167. Dissection of the B-cell repertoire specific for influenza A viral hemagglutinin by fine specificity analysis holds an even greater potential, since one can examine the response to a protein antigen bearing numerous determinants. These studies indicate that, while some clonotypes are represented by a large number of cells, the adult repertoire is enormous, perhaps consisting of greater than 10^7 specificities. The necessity to identify and enumerate minor clonotypes is discussed in terms of their implications for B-cell diversification and the potential difficulty in interpreting the results of such experiments.

INTRODUCTION

Intense interest in the origin of immunoglobulin variable region diversity has led to a widespread search for definitive methods of describing genetic markers for these regions. Perhaps the most popular tool employed in such studies has been anti-idiotypic sera. The majority of clonotypes which have been identified by these sera are readily detectable in the serum of all individuals within the relevant murine strain and are therefore acknowledged by most workers to reflect germline specificities (1, 2). Using fine specificity analysis (2) or anti-idiotypic sera (3), investigators have occasionally discovered clonotypes that appear to be unique to the individual in which they were identified. Such "minor" clonotypes, or private specificities, can be viewed either as somatic variants of germline antibody specificities or as heavy-light chain pairs that are coded for by germline genetic information but are expressed at a substantially lower frequency than are the major clonotypes. Ju et al. (4) investigated private idiotypic specificities which were generated by hyperimmunization after suppression of the predominant anti-azobenzenearsonate clonotype in A/J mice. Since some private idiotypes were not found in appreciable quantities in 181 nonimmune or immune mice, the clonotypes appeared to be truly unique to the individuals in which they were generated. The authors thus concluded that minor clonotypes must be somatically generated. Difficulty arises with this interpretation, however, both because of a low level of idiotypic crossreaction found in some individuals, and because it should not be assumed that the antibody product of clonotypes present at low frequency within the B-cell populaton will be detectable in the serum of all nonimmune and immune mice tested.

The splenic focus technique eliminates many of the ambiguities inherent in an analysis of serum antibody. First, stimulation of isolated B-cells in the presence of excess carrier-primed T cells in vitro permits the majority of specific B cells to generate detectable antibody-producing clones (5), eliminating problems created by in vivo antigen selection. Second, idiotypic crossreactions in serum antibody may indicate a diverse population of weakly crossreactive antibodies, none of which are identical to the idiotype in question. Alternatively, the relevant idiotype might be present but represent only a small proportion of the serum antibody. The analysis of homogeneous antibodies generated in the splenic focus system provides an unambiguous determination of the presence or absence of a given clonotype and can be used to determine its representation within the responding B-cell pool. In addition, even the partial reactivity of anti-idiotypic serum with a monoclonal antibody permits the identification of new clonotypes and allows the description of novel idiotypic relationships among antibody populations. This report will outline the experimental approach taken by this laboratory in the study of minor clonotypes and discuss what information may be derived from these investigations.

RATIONALE FOR THE STUDY OF MINOR CLONOTYPES

A central question which must be addressed by any theory of the generation of antibody diversity is the number of specificities within the B-cell repertoire. Presently, little agreement exists among investigators regarding this point, as estimates of the adult B-cell repertoire range from 10^5 to greater than 10^7 clonotypes (6-8). The repeat frequency of a single clonotype in a murine strain can provide an estimate of this parameter, but its representation may or may not reflect the true size of the repertoire. The T15 clonotype, which dominates the response to phosphorylcholine (PC) in BALB/c mice, is present at a frequency of approximately 1/60,000 B cells in both germfree and conventionally-reared mice (9), implying that the B-cell repertoire may be relatively small. Other investigations (10-13), including an analysis of PC-specific B cells which do not react on a 1:1 weight basis with murine anti-T15 serum (10), suggest that many specificities are present at a much lower frequency than the predominant T15 clonotype. Thus, the B-cell repertoire is composed of clones varying in size from more than 1000 cells to less than 10 cells in an individual mouse. An adequate assessment of the actual number of specificities in the adult repertoire will consequently require experiments which will: 1) assess the relative proportions of major and minor clonotypes in the B-cell population; 2) identify and enumerate the frequency of a number of minor clonotypes within the B-cell pool; and 3) establish the proportion of putative B cells which are true clonal precursor cells.

If all individuals of a mouse strain are found to share the same B-cell repertoire, including both major and minor clonotypes, this finding would imply that all specificities are encoded by germline genes. Conversely, somatic theories would be supported by the observation that while individuals of a strain share major clonotypes, which may be germline expressions, the minor clonotypes would be unique to the individual. The representation of a given clonotype within the B-cell pool may be so small, however, that the splenic focus system may not delineate all specificities within an individual's repertoire. In fact, although germline theories would predict that every H-L pair must be expressed at least once and perhaps many times within the lifetime of an individual, the repertoire may be so large that two individuals may have essentially different repertoires at any given point in time. Stated another way, the potential repertoire of an individual during his lifetime may be much greater than the number of specificities present at one point in time. If the repertoire is, in fact, so large that such a situation exists, the distinctions between germline and somatic theories become difficult to define experimentally. A detailed analysis in a number of experimental systems will be required in order to confirm or refute such heretical notions.

EXPERIMENTAL APPROACH

Experiments designed to investigate and enumerate minor clonotypes have, for the most part, centered on the analysis of specificities within the PC-specific repertoire. The anti-PC response in BALB/c mice is dominated by a single antibody specificity which is identical to the T15 myeloma protein (9), but approximately 20% of the PC-specific foci do not react with a murine anti-T15 serum (M anti-T15) (10,14). It is possible to demonstrate a heterogeneous array of PC-specific clonotypes within this subset by anti-idiotypic crossreactivity with the allogeneic anti-T15 serum and by hapten inhibition of monoclonal antibody binding to antigen (10). Using a rabbit anti-idiotypic antibody to T15 (R anti-T15), which appears to recognize a different portion of the variable region than the M anti-T15, we can identify a group of clonotypes which react with the R anti-T15 on a 1:1 weight basis but has only weak crossreactivity for the M anti-T15. This group of antibodies apparently represents several clonotypes since they display heterogeneity by degree of idiotypic crossreactivity with the allogeneic anti-T15. A maximum estimate of the frequency of this entire subset can be calculated to be one in 2×10^6 B cells, or 30-40 times fewer than the predominant T15 clonotype. In A/He and C3H mice, the R anti-T15 +, M anti-T15- subset represents approximately 1/500,000 splenic B cells, and since R anti-T15 reacts with anti-PC antibody from every individual of a variety of mouse strains (15,16), this clonotype or group of clonotypes can be considered to be encoded by highly conserved germline genes.

Another method of identifying minor clonotypes within the PC-specific repertoire is the use of anti-idiotypic antibodies against other PC-binding myeloma proteins, for example, McPC 603 and 167. In an analysis of 141 BALB/c non-T15 monoclonal antibodies, two foci were found that partially crossreacted with an allogeneic anti-McPC 603(10). While this result indicates that there were no PC-specific clones reacting on a 1:1 weight basis with both anti-McPC 603 and anti-Fab, a new clonotype subset (maybe complex) is operationally defined on the basis of this crossreaction. The frequency of this subset is approximately 1-2 in 10^7 B cells. With a rabbit anti-167 idiotypic serum, an analysis of 220 BALB/c non-T15 monoclonal antibodies demonstrated 7 clones to be reactive with this anti-idiotype. In other murine strains, some monoclonal antibodies have been found which react with anti-167 on a 1:1 weight basis, and others have been identified as crossreactive specificities.

Some of the work discussed implies that clonotypes with similar idiotypic specificities can be found in different murine strains, but that each strain expresses the clonotype with a characteristic frequency. For example, AKR mice possess a specificity that is idiotypically indistinguishable from the predominant T15 clonotype in BALB/c mice (10); this clonotype represents 1/400,000 splenic B cells in AKR mice and 1/60,000 B cells in BALB/c mice. The genetic control of the frequency with which a specificity is expressed in the B-cell pool can best be studied by examining the expression of the T15 clonotype in mice of various genetic backgrounds, since this specificity is known to be encoded by a germline gene. The Bailey C x B recombinant inbred strains and other inbred BALB/c and C57BL/6 recombinant strains provide an experimental model to test this question. Results from studies with these strains indicate that the T15 clonotype can be expressed at frequencies ranging from 1/500,000 to 1/60,000 splenic B cells. Thus, a putatively germline antibody specificity may appear as either a major or minor clonotype. The distinction between major clonotypes as defined serologically by crossreactive anti-idiotypic sera and minor, sporadically-occurring clonotypes may, therefore, be an artificial one, reflecting the genetic background of the mouse more than the actual gene complement present.

With a battery of anti-idiotypic sera, a number of clonotypes can be defined and their repeat frequencies within the B-cell population enumerated. While these experiments can establish that certain clonotypes are present at very low frequency, they only indirectly provide an estimate of the total antibody diversity that an animal can generate. In order to derive a valid estimate, one must determine the relative contribution of the large and small clones to the population as well as establish repeat frequencies for individual specificities. We can estimate the total number of specificities in the PC-specific repertoire by utilizing: 1) idiotypic crossreactivity of the monoclonal antibodies;

2) sucrose gradient isoelectric focusing; and 3) an analysis of the relative affinity of PC-specific foci for PC, glycerophosphorylcholine, and choline. Through a combination of these techniques, this system has the potential to define several thousand different anti-PC antibodies.

Delineation of clonotypes by fine specificity of antigen-antibody interactions may best be accomplished with systems which provide a wide range of closely related antigens. Analysis of the primary B-cell repertoire specific for the influenza A viral hemagglutinin (HA) in BALB/c mice utilizes this approach and provides another estimate of the total number of clonotypes responsive to a well-characterized antigen. The advantages of this system lie in the potential to examine the response to a protein antigen bearing numerous determinants. Monoclonal anti-influenza (PR8/HON1) antibodies can be characterized by their ability to crossreact with several closely related but antigenically distinct viral hemagglutinins (17). Since a small number of related HA's will provide a large number of possible crossreactivity patterns, the number of clonotypes potentially definable by this criterion is enormous. The primary PR 8 HA-specific precursor cell frequency is approximately 1/200,000 splenic B cells. Analysis of monoclonal antibodies generated by stimulation of secondary HA-specific precursor cells has identified approximately 60 clonotypes (18). Preliminary results suggest that the primary anti-HA response is equally heterogeneous so that the frequency of each of these clonotypes averages less than one per 10^7 splenic B cells.

The final approach that we have utilized in the study of minor clonotypes employs an analysis of the DNP-specific repertoire, searching for specificities which crossreact with anti-T15 idiotypic serum. As stated previously, a homogeneous antibody that shows a given idiotypic crossreactivity can be operationally defined as a distinct clonotype. While two anti-DNP monoclonal antibodies which crossreact with an anti-idiotypic serum cannot be considered identical, they would represent a subset of anti-DNP clonotypes, and the repeat frequency of this subset can be enumerated. Such a clonotype has been identified within the DNP-specific precursor cell population by idiotypic crossreactivity with anti-T15 serum. This analysis has also provided an opportunity to investigate idiotypic relationships among clonotypes. For example, as mentioned above, specificities were found within the PC-specific repertoire which reacted with the rabbit anti-T15 but not with the mouse anti-T15, but the converse, i.e., clonotypes identified as R anti-T15 negative and M anti-T15 positive were not found. However, one might predict that such a specificity may not possess PC binding properties, since the antigen-binding portion of the variable region recognized by R anti-T15 serum is not present. Therefore, it might be predicted that a family of clonotypes exists which share the same variable region framework residues with T15 myeloma protein but have combining sites sufficiently distinct to bind other antigens. These predictions have been confirmed by the identification of a R anti-T15-, M anti-T15+ clonotype within the DNP-specific repertoire at a frequency of less than $1/4 \times 10^6$ splenic B cells.

CONCLUSION

This report discusses some experimental approaches utilized in the analysis of minor clonotypes, i.e., those specificities which are present in too low a frequency to be identified reproducibly in every individual of an inbred mouse strain. The use of homogeneous antibodies generated in the splenic focus system allows one to unambiguously identify even rare specificities, study idiotypically crossreactive clonotypes, and establish repeat frequencies for these clones. Through the utilization of a number of techniques, such as anti-idiotypic identity and cross-reactivity, isoelectric focusing, and cross-binding to related antigens, and use of a variety of antigen systems, a clearer picture of the diversity within the B-cell repertoire has emerged. Studies to date indicate that, while some clonotypes are represented by a large number of B cells, the adult repertoire is enormous, perhaps consisting of greater than 10^7 specificities. Caution must be exercised in concluding that the majority of these specificities are either encoded by germline genes or somatically generated since traditional criteria for this distinction may not be valid. In addition, these studies have the potential to shed light on the mechanism of antibody diversity generation, on idiotypic relationships among clonotypes, and the extent of antibody diversity within an individual as well as within the species as a whole.

This work was supported by USPHS grants CA-15822, AI-08778 and CA-09140.

REFERENCES

1. Eichmann, K. (1975) Immunogenetics 2, 491.
2. Makela, O. and Karjalainen, K. Transplant. Rev., in press.
3. Kuettner, M.G., Wang, A., and Nisonoff, A. (1972) J. Exp. Med. 137, 22.
4. Ju, S-T., Gray, A., and Nisonoff, A. (1977) J. Exp. Med. 144, 1294.
5. Klinman, N.R., Pickard, A., Sigal, N.H., Gearhart, P.J., Metcalf, E.S., and Pierce, S.K. (1976) Ann. Immunol. (Inst. Pasteur) 127c, 489.
6. Inman, J.K. (1974) in The Immune System. E.E. Sercarz, A.R. Williamson, and C.F. Fox (eds.) Academic Press Inc., New York, p. 37.
7. Williamson, A.R. (1972) Biochem. J. 130, 325.
8. Klinman, N.R. and Press, J.L. (1975) Transplant. Rev. 24, 41.
9. Sigal, N.H., Gearhart, P.J., and Klinman, N.R. (1975) J. Immunol. 68,1354.
10. Gearhart, P.J., Sigal, N.H., and Klinman, N.R. (1977) J. Exp. Med., in press.
11. Kreth, H.W., and Williamson, A.R. (1973) Eur. J. Immunol. 3, 141.
12. Pink, J.B.L., and Askonas, B. (1974). Eur. J. Immunol. 4, 426.
13. Kohler, G. (1976) Eur. J. Immunol. 6, 340.
14. Gearhart, P.J., Sigal, N.H., and Klinman, N.R. (1975) J. Exp. Med. 141, 56.
15. Claflin, J.L. and Davie, J.M. (1974) J. Exp. Med. 140, 673.
16. Rudikoff, S., and Claflin, J.L. (1976) J. Exp. Med. 144, 1294.
17. Gerhard, W. (1976) J. Exp. Med. 144, 985.
18. Gerhard, W. (1977) Infection and Immunity, in press.

The Regulatory Network

Patricia Gearhart
Department of Biology
The Johns Hopkins University
Baltimore, MD 21218

Latham Claflin
Department of Microbiology
University of Michigan
Medical School
Ann Arbor, MI 48109

Regulatory pathways that are mediated by antibodies or lymphoid cells and are directed against antigen-binding receptors and immunoglobulins were explored. Most of the components of a proposed network, such as anti-idiotypic, anti-allotypic, and anti-isotypic regulation, have been identified in a variety of laboratory animals. In most instances, regulation of this type was an artificially induced phenomenon rather than a naturally observed one. The effect on the target immune response was usually negative, e.g. suppression, and was often, but not always, maintained by T lymphocytes.

The first portion of the session focused on those regulatory components that acted on variable region or idiotypic determinants of immunoglobulins. Weigle described the cyclical appearance of IgM and IgG anti-human gamma globulin PFC in the spleens of rabbits following a single injection of antigen. Rather than anti-idiotype-mediated regulation, Weigle proposed that this is an antigen-driven phenomenon which depends on the number of "hits" by antigen. Cells stimulated with sufficient hits develop into antibody-producing cells (first cycle). Those with insufficient antigen contact may become arrested in G_1, migrate out of the spleen, regain their receptors and then shortly return. If the half-life of antigen in the spleen is not exceeded, additional hits with antigen will initiate the second cycle of PFC production. Two examples of idiotype-specific suppression induced during the neonatal period by perinatal injection of anti-idiotypic antisera were described. Following treatment of BALB/c with A/J anti-T15 which was non-cross reactive with other PC-binding myelomas, Cosenza found a long-lived suppression of phosphorylcholine (PC)-specific PFC bearing the T15 idiotype. The suppressed state could not be transferred to normal or mildly irradiated recipients with suppressed cells. Non-T15 bearing PFC were not observed initially but recovered at about 3 months of age. By contrast Sigal, using the splenic focus assay to assess PC-specific precursors found, that although treatment of neonatal BALB/c mice with allogeneic anti-T15 completely eliminated the expression of T15-specific B cells for 4-5 months, the frequency of precursors with other identifiable idiotypes was not affected. Pierce and Klinman presented a novel system in which irradiated BALB/c recipients primed with hemocyanin (Hy) plus DNP-Hy supported only 25-30% of the primary DNP-specific B-cell response that was observed in recipients primed with Hy alone. The frequencies of secondary DNP-specific B cells and precursors to unrelated haptens, such as fluorescein, were unaffected in the DNP-Hy primed recipients. The fact that the response of CB20 donor B-cells, which differ from BALB/c mice only in the immunoglobulin heavy chain allotype locus, was unaffected by DNP-Hy priming of BALB/c recipients, suggests that a regulatory mechanism, whose target could be the variable region, is induced which specifically limits stimulation of hapten-specific primary B cells. In this context Cosenza mentioned that when the primary PC-specific PFC response of the T15 idiotype is waning, he has consistently observed a substantial rise in the frequency of anti-T15 secreting PFC. There was no clear indication at present what the biological significance of these findings was. Idiotype suppression in other systems can be mediated by T cells. Adult A/J mice given anti-idiotypic antisera to the cross-reactive idiotype (CRI) of the azophenylarsonate response develop a profoundly suppressed state which is specific and is maintained by T lymphocytes. In a new development Owen and Nisonoff showed that suppressed mice develop up to 9-10% of thy-1 positive splenic lymphoid cells which will rosette with CRI(Fab)-coated autologous erythrocytes. Other idiotypes do not rosette nor do they inhibit CRI-rosette formation. The CRI-specific T-cell receptors were of endogenous origin in that they reappeared after stripping with pronase or trypsin. The functional importance of these cells was shown by the finding that purified CRI-rosetted cells were capable of transferring suppression to normal unprimed recipients, whereas the non-rosetting fraction did not. In conclusion, the discussants agreed that a number of instances of anti-idiotypic-induced regulation have been observed, that they can affect the B-cell directly through clonal elimination or, can be mediated by T lymphocytes, but it was not certain how they contributed to a natural network of regulation.

The related area of allotype-specific suppression was introduced by Lee Herzenberg. She outlined the classic allotype-suppression of Ig-1b in (SJL x BALB)F1 mice given anti-Ig-1b at birth (L. Herzenberg, Symposium, this volume). In this system suppressor T cells (T_S) are induced which operate at the level of allotype-specific T-helper cells.

Di Pauli presented another Ig-lb allotype suppression scheme. BALB/c recipients which were given CB20 spleen cells and then were immunized with polysaccharide failed to produce antibody with the Ig-lb allotype as measured by isoelectric focusing. CB20 recipients of CB20 spleen cells did develop a normal Ig-lb response. He had not tested whether the Ig-lb B cells were affected. In the Herzenberg system the number of Ig-lb bearing cells is the same in both suppressed and non-suppressed mice. In a somewhat analogous system, Bosma suggested that the target of allotype suppressors could be the B cell. BALB/c T cells, which can specifically prevent normal Ig-lb (IgG_{2a}) allotype production by a congenic partner (CB mice), prevent the growth of and allotype production by a transplantable B-cell tumor, the plasmacytoma CBPC 101. The specificity for the Ig-lb allotype, the dependency on T cells from allotype-suppressed mice, and the lack of a requirement for helper T cells for tumor growth indicates that the B cell is the target of the suppressor T cells. Surprisingly, allotype-specific suppressor spleen cells at ratios of 20-200:1 do not kill target CBPC 101 cells (^{51}Cr-release assay). In the ensuing discussion Herzenberg mentioned that they are not certain if T_S cells are equivalent to cytotoxic T cells. The possibility was also raised that T_S pathway may be more complex than is presently evident; other cells, e.g. macrophages, Ly-1,2,3 positive, or an undefined T cell subset may be involved. It was also generally felt that negative regulation of expression of B-cell activity by suppressor T cells may be accomplished in more than one way, the target of suppression being the T_H in some instances and, the B cell in others.

A discussion of anti-isotype regulation was opened by Knight. Rabbits, heterozygous at the heavy chain variable region <u>a</u> locus (al/a2) can be readily suppressed for the al allotype. IgM and IgG molecules with the al determinant remain suppressed for up to 26 months, but IgA escapes suppression at approximately 5 months and reaches normal levels by 12 months. The reason for the emergence of IgA from suppression prior to IgG and IgM is not clear. One possibility may be that IgA precursors are compartmentalized away and protected from "suppressor cells". Mage commented that the number of B cells bearing the suppressed allotype are severely reduced compared to normal but that pre-B cells (surface Ig^-, cytoplasmic Ig^+) in the bone marrow are normal. This re-emphasizes the importance of this cell type in the B-cell maturation pathway. Leslie described isotype suppression in chickens induced by treatment with anti-μ and bursectomy. Though they have normal levels of lymphocytes, treated birds were agammaglobulinemic for all three isotypes; they are B-cell negative (surface and cytoplasmic Ig negative) but their T-cell functions are normal. Normal chicks given lymphoid cells from suppressed adults become agammaglobulinemic and suppression is eliminated by treatment of the donor cells with anti-T. Their Ig-bearing cells, however, remain at normal frequencies. Occasional isotypes escape suppression but not in an ordered fashion. In general the state of suppression of recipients of transferred cells was strikingly similar to the human immunodeficiency syndrome, common variable hypogammaglobulinemia. In reviewing a role for isotype regulation, class-specific augmentation or depression in normal responses observed by numerous investigators was noted, but the biological significance remains unclear.

In closing remarks which were acknowledged and supported by all, Prahl emphasized the importance of caution in transferring laboratory phenomenon into the realm of physiology. It is perhaps premature to examine a network or even suggest one when the rudiments of regulation and individual pathways are just now being outlined. On the other hand many components of Jerne's Network Hypothesis are there and have been defined. They have become a powerful tool to examine and dissect immune phenomenon such as suppression, differentiation, silent genes, and changes in phenotypic expression. In addition certain immunodeficiency and autoimmune diseases, e.g. common variable hypogammaglobulinemia and lupus in NZB mice, appear to be examples in class-specific suppression. In sum, the components necessary to generate a network exist and are being described; the incorporation into a biologically significant regulatory pathway, however, has not been achieved.

GENETIC CONTROL OF Ia ANTIGEN EXPRESSION

Donald C. Shreffler

Department of Genetics
Washington University School of Medicine
St. Louis, Missouri 63110

ABSTRACT. The mouse major histocompatibility complex (H-2) controls three broad classes of immunological functions, controlled by four major regions of the complex, K, I, S, D. The K and D regions determine cell-membrane structures, expressed on almost all body tissues, which appear to function as self-markers for immune surveillance mechanisms. The S region controls components of the complement system. Greatest interest currently centers on the I region, which determines cell-membrane molecules that are expressed principally on lymphocytes and macrophages. These molecules are involved in the cell-cell interactions that mediate lymphocyte activation and regulation. The serological detection and definition of an extensive set of I region gene products (the Ia antigens) has opened an important new avenue for genetic, chemical and functional definition of the molecules that mediate these mechanisms. At least four discrete Ia gene products have been defined and mapped to four discrete I subregions. The products of the I-A, I-E and I-C subregions are found on B lymphocytes and some subpopulations of T lymphocytes. I-J subregion products appear to be restricted to a T lymphocyte subpopulation that includes suppressor T lymphocytes. Ia antigenic determinants are also expressed on macrophages (although the specific molecules expressed have not yet been clearly identified) and on a number of soluble mediators produced by T cells and macrophages. The Ia antigens found on B cells are glycoproteins comprised of two polypeptides of 25,000 and 33,000 daltons. Discrete molecular products of the I-A, I-E and I-C subregions have been defined. The anti-Ia sera which detect and define these Ia antigens have been extensively tested for their capacity to inhibit immune functions, to absorb soluble mediators, or to eliminate functional subpopulations of lymphocytes. Such studies have implicated the Ia antigens in T cell helper and suppressor functions, T cell proliferative responses to antigen, macrophage interactions with T cells, and a number of other phenomena. Current evidence on map localization of these functions places genetic information for a suppression mechanism in the I-J subregion and for T-B interaction and macrophage presentation mechanisms in the I-A subregion.

INTRODUCTION

In this section, the principle considerations will center about the immune functions associated with the MHC (major histocompatibility complex). Here we will introduce and review the mouse MHC, the H-2 gene complex. Many products of the H-2 complex have been defined serologically. Major emphasis in this paper will be on a review of the immune-response-associated, serologically-defined Ia antigens and the roles of these antigens in specific lymphocyte functions.

GENETIC ORGANIZATION OF THE H-2 COMPLEX

The present state of definition of the H-2 genetic map is presented in Figure 1. The H-2 complex is located on mouse Chromosome 17, which also carries the T/t gene complex, that is important in regulation of embryological development, apparently mediated by cell-surface molecules. In the H-2-TL gene complex, ten discrete genetic loci have been defined. Each locus (except Ir-1B) controls a discrete, serologically defined protein product. Each locus has been separated from every other by one or more intra-H-2 crossovers. Each locus defines a discrete region of the H-2 complex -- a chromosomal segment defined by the crossovers that separate adjacent marker genes. A region may contain multiple genes. Recent evidence has shown that the D region has two discrete loci controlling two similar, but discrete, products (1). The interval between H-2D and Tla has at least four new loci, Qa-1, Qa-2, H-31 and H-32 (2,3). Clearly many more genes remain to be defined within this complex. The long map distance between H-2G and H-2D suggests that this is a region in which further resolution may be possible.

During the many years it has been studied, the H-2 complex has been found to control a wide variety of phenotypic traits -- not only transplantation antigens, but serum protein differences, immune responses, hormone levels and even mating preference (4,5). However, the majority of these diverse traits can be assigned to one of three general classes of functions and localized to one of four major genetic regions of the complex. The functions and the serologically detected products associated with each region are summarized in Figure 2. Clearly, the majority of H-2 gene products are cell memgrane structures. All of the well-defined functions of the H-2 gene products are immune functions. We will concentrate on the I region functions, but it should be strongly emphasized that the S region controls components of the complement system (6) and the K and D regions are very directly involved in mechanisms of cell-mediated cytotoxicity (Shearer, these Proceedings).

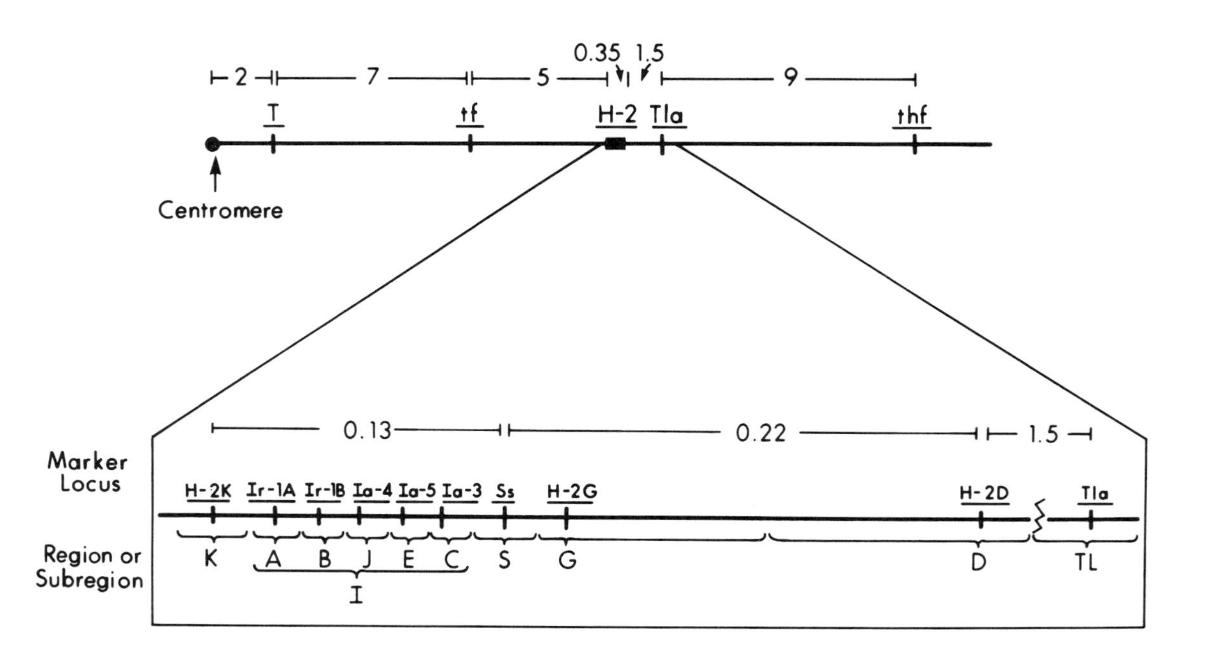

Fig. 1. The current genetic map of mouse chromosome 17 and the H-2-TL gene complex.

Region	K	I	S	G	D	TL
Marker Locus	H-2K	Ir-1	Ss	H-2G	H-2D	Tla
Products	Cell Membrane Molecules	Cell Membrane Molecules; Mediators	Serum Proteins	Erythrocyte Membrane Molecules	Cell Membrane Molecules	Cell Membrane Molecules
Histocompatibility Role	Cytotoxic Targets	MLR-GVHR Stimulation	None Defined	None Defined	Cytotoxic Targets	Transplantation Antigens
Function	Marker for Cytotoxicity vs. Deviant Cells	Immune Response; Cell-Cell Interactions	Complement Components	Unknown	Marker for Cytotoxicity vs. Deviant Cells	Unknown

Fig. 2. A summary of the major regions of the H-2 complex and their products and functions.

THE I REGION AND THE Ia ANTIGENS

A great diversity of traits and functions have been mapped to the I region. These can be classified into six major categories: 1) susceptibility to oncogenic viruses (the Rgv-1 gene); 2) differential antibody or cellular responses to specific thymus-dependent antigens (the Ir-1 genes); 3) the antigens that stimulate the MLR (Lad genes); 4) factors mediating cell-cell interactions among T cells, B cells and macrophages (Ci genes); 5) histocompatibility antigens (H-2I); 6) lymphocyte alloantigens (Ia).

These traits undoubtedly reflect the actions of

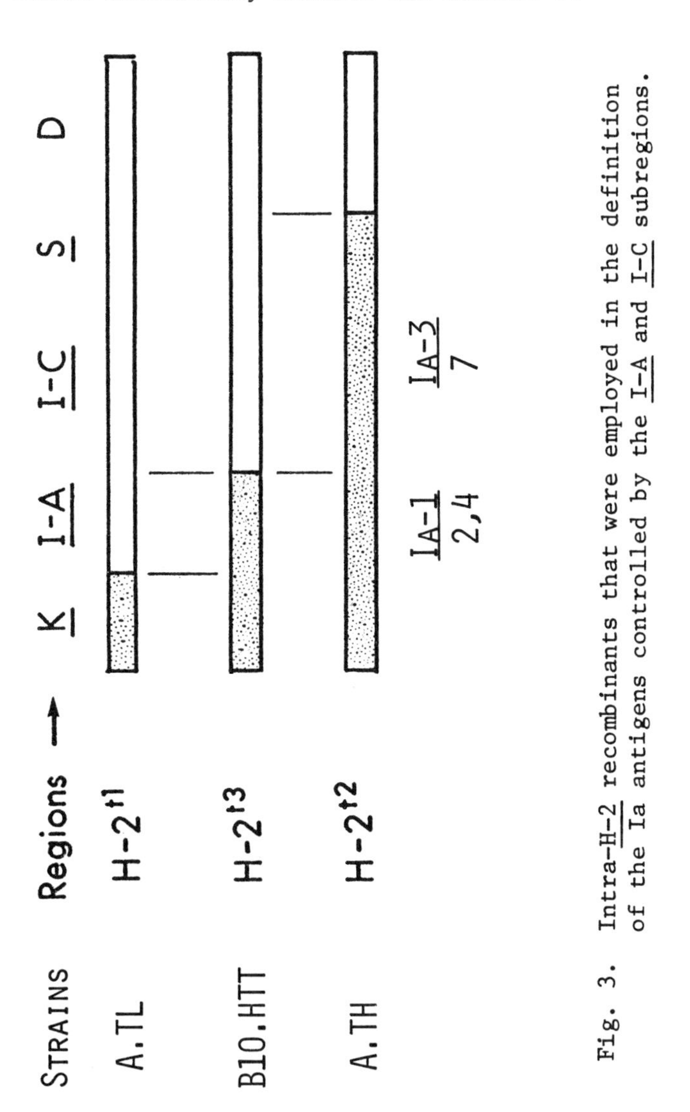

Fig. 3. Intra-H-2 recombinants that were employed in the definition of the Ia antigens controlled by the I-A and I-C subregions.

multiple genes, but in some cases the product of a single gene could be involved in several traits. The most fundamental question about I region function concerns the relationships among the viral susceptibility, immune response, cell interaction and MLR-stimulating traits. Because they offer an approach to direct detection of the products of I region genes, anti-Ia sera are of great potential value in "sorting out" the various traits, functions, products and genes.

The detection of Ia antigens first became possible with the development of some key congenic resistant inbred strains carrying recombinant H-2 haplotypes that were identical in the K and D regions, but different in all or part of the I region. An illustration of the key strains for work in our laboratory (4) is shown in Figure 3. Reciprocal immunizations among these strains yielded the earliest anti-Ia sera (7), and eventually defined two discrete Ia loci, Ia-1 and Ia-3, defining subregions I-A and I-C, and characterized by specificities Ia.2, 4 and 7 (4,8), detected in the Cr^{51}-release and dye exclusion microcytotoxic tests.

Work in our laboratory (4,9) and in those of other workers has defined multiple Ia specificities. Analyses of intra-H-2 recombinants have permitted the mapping of these specificities to four discrete loci in four discrete subregions (Figure 4). In addition to Ia-1 and Ia-3, recent work has defined two new Ia loci, Ia-4 and Ia-5, mapping in subregions I-J and I-E respectively (10,11,12). The tissue expression of the Ia antigens may be summarized as follows (4,9,12): 1) Positive cells are B lymphocytes, some T lymphocytes (especially cortisone-resistant or Con A-stimulated thymocytes, suppressor cells and some helper cells), macrophages, sperm and epidermal cells; 2) negative tissues are erythrocytes, platelets, liver, kidney and brain; 3) the predominant antigens seen by cytotoxicity and immunoprecipitation (products of I-A, I-E, I-C are found on B lymphocytes; some T lymphocytes also express these antigens; 4) at least one Ia molecule (I-J) is restricted to T (suppressor) cells; 5) no allelic or genic exclusion of the I-A and I-C B lymphocyte antigens has been found.

One of the very important adjuncts to serological analysis by cytotoxicity has been the immunoprecipitation analyses of Cullen et al. (13). These studies have shown that: 1) Each gene product is a glycoprotein molecule of 58,000 daltons, with two polypeptides of 33,000 and 25,000 daltons; 2) no association with H-2K, H-2D, immunoglobulin or β_2-microglobulin molecules has been demonstrable; 3) the Ia antigenic difference is in the polypeptide; there is no effect of glycosidases; peptide map differences have been demonstrated; 4) three discrete molecules have been found; these are controlled by the I-A, I-E, I-C subregions; 5) multiple specifi-

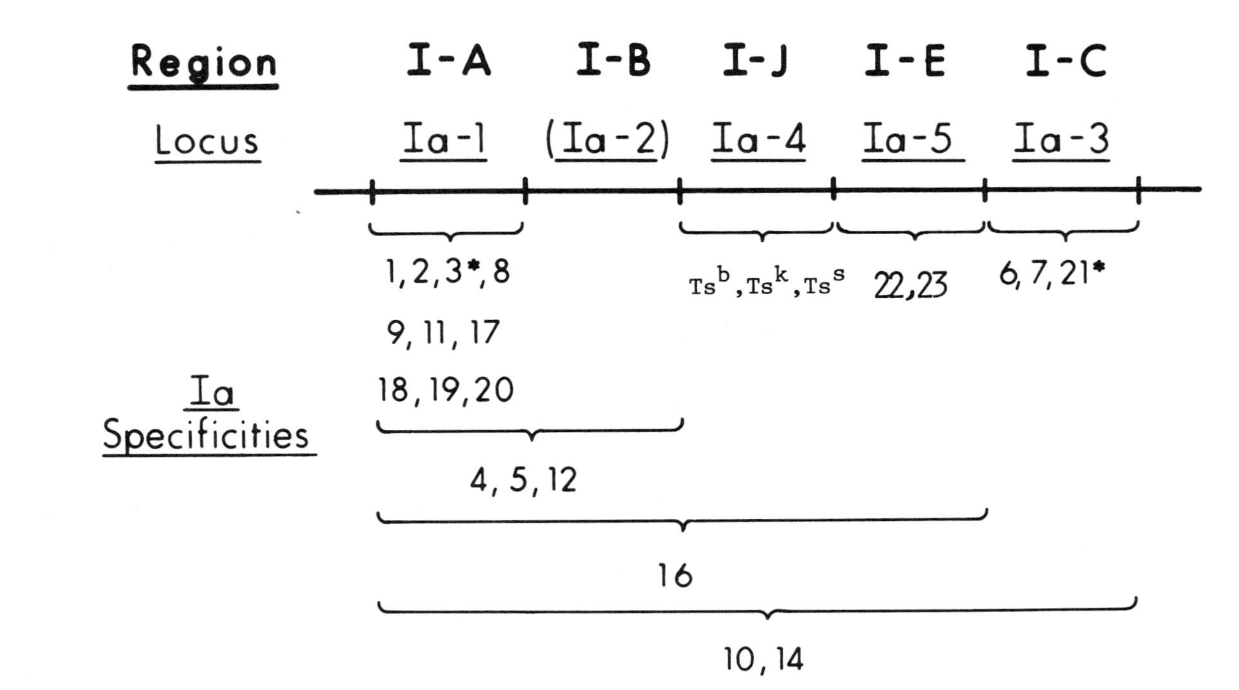

Fig. 4. Localization of Ia specificities to I subregions. Those marked by an asterisk were localized by immunoprecipitation studies.

cities may occur on the same molecule; 6) all cytotoxic specificities have been detectable by immunoprecipitation; 7) allelic I-A specificities occur on separate molecules, i.e. there are no hybrid molecules in heterozygotes.

The roles of the Ia antigens in histocompatibility and immune functions have been investigated by experiments which have evaluated the capacity of anti-Ia sera to: 1) inhibit assays of immune function; 2) to absorb soluble effector factors; 3) to eliminate (in the presence of complement) functional subpopulations of lymphocytes. Results of such studies are summarized in Table 1. These show very clearly that at least some Ia antigens play an important role in immune mechanisms. Many of these relationships will be discussed more fully in subsequent papers in this volume.

TABLE 1

EFFECTS OF ANTI-Ia SERA ON IMMUNE FUNCTIONS

1) Inhibition of:
   - *In vitro* primary and secondary PFC responses
   - *In vitro* antigen-induced T cell proliferation.
   - B and T cell acceptors for T cell effector factors.
   - B and T cell Fc receptors.
   - MLR-stimulating antigens.
   - LPS stimulation.
2) Absorption of:
   - T cell helper or enhancing factors.
   - T cell suppressive factors.
   - Macrophage T helper cell-inducing factor.
3) Elimination (with complement) of:
   - Suppressor T cells.
   - Helper T cells.
   - IgG-producing B cells.

Finally, in Figure 5, I have attempted to organize and summarize a variety of genetic, serological, biochemical and functional data, in order to relate specific traits to specific genetic loci. Four discrete Ia loci and products are defined, as discussed above. MLR stimulation is mediated primarily by the I-A subregion and to a lesser extent by I-C. Recent work in our laboratory (14) has shown moderate MLR stimulation by purified T cells in the B10.A(3R)-B10.A(5R) and B10.S(9R)-B10.HTT combinations which differ only in the I-J subregion. Both in vivo immune response (15) and in vivo immune suppression (16) appear to result in at least some cases from interactions among two discrete genes within the I region. It will be noted that the I-B subregion is defined

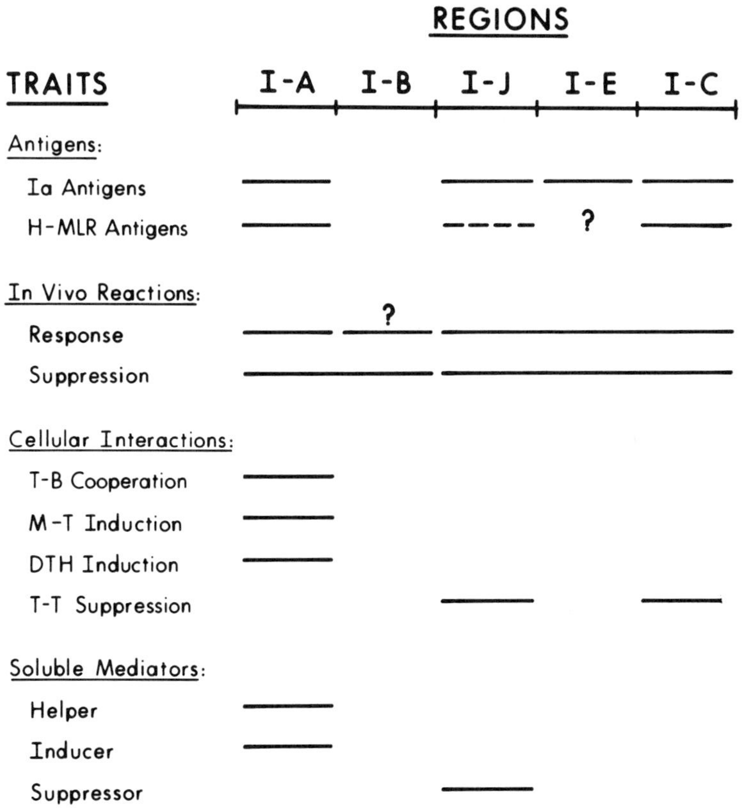

REGIONS

TRAITS I-A I-B I-J I-E I-C

Antigens:

Ia Antigens

H-MLR Antigens ?

In Vivo Reactions:

Response ?

Suppression

Cellular Interactions:

T-B Cooperation

M-T Induction

DTH Induction

T-T Suppression

Soluble Mediators:

Helper

Inducer

Suppressor

Fig. 5. Summary of current evidence localizing certain I region traits to specific subregions.

only by in vivo immune responses. This could be an artifact of interaction between genes in the I-A and I-J, I-E, I-C segments, as previously discussed (12).

Cellular interactions in T cell-B cell cooperation, macrophage-T cell interaction and delayed hypersensitivity reactions all map to the I-A subregion although not necessarily controlled by the same genes (17,18,19). T-T suppressive interactions map to I-J (10,11) or to I-C in the case of suppression of MLR-responsive cells (20). Soluble T helper (10, 17) and macrophage T-inducer factors (18) map to I-A, while a soluble suppressor factor maps to I-J (10). Since the antibodies reactive with the soluble I-A region enhancing factor described by Tada are absorbed only by T cells, this factor must be controlled by a gene that is discrete from Ia-1 (which controls B cell antigens), but also located in I-A. The relationship of the macrophage factor to these two products remains to be determined.

These results indicate that genes of the I region determine a number of discrete immunoregulatory functions. A number of the gene products involved are detectable by anti-Ia sera. These combined genetic, serological, biochemical and functional approaches have resulted in the "sorting out" of several discrete functions and should facilitate the further resolution of the I region role in immune mechanisms.

ACKNOWLEDGEMENTS

Supported by U.S.P.H.S. Research Grant AI 12734. I am grateful to Mrs. Carol Jones for her excellent assistance in preparation of the manuscript.

REFERENCES

1. Hansen, T.H. and Cullen, S.E. (1977) J. Immunol. 118, 1403.

2. Stanton, T.H. and Boyse, E.A. (1976) Immunogenetics 3, 525.

3. Flaherty, L. (1976) Immunogenetics 3, 533.

4. Shreffler, D.C. and David, C.S. (1975) Adv. Immunol. 20, 125.

5. Yamazaki, K., Boyse, E.A., Thaler, M.V., Mathieson, B.J., Abbot, J., Boyse, J., Zayas, Z.A. and Thomas, L. (1976) J. Exp. Med. 144, 1324.

6. Shreffler, D.C. (1976) Transpl. Revs. 32, 140.

7. David, C.S., Shreffler, D.C. and Frelinger, J.A. (1973) Proc. Nat. Acad. Sci. 70, 2509.

8. Meo, T., David, C.S., Nabholz, M., Miggiano, V. and Shreffler, D.C. (1973) Transpl. Proc. 5, 1507.

9. David, C.S. (1976) Transpl. Revs. 30, 299.

10. Tada, T. (1977) These Proceedings.

11. Murphy, D.B., Herzenberg, L.A., Okumura, K., Herzenberg, L.A. and McDevitt, H.O. (1976) J. Exp. Med. 144, 699.

12. Shreffler, D.C., David, C.S., Cullen, S.E., Frelinger, J.A. and Niederhuber, J.E. (1977) Cold Spring Harbor Symposium on Quantitative Biology 41 (in press).

13. Cullen, S.E., Freed, J. H. and Nathenson, S.G. (1976) Transpl. Revs. 30, 236.

14. Okuda, K., David, C.S. and Shreffler, D.C. (submitted for publication).

15. Dorf, M.E., Stimpfling, J.H. and Benacerraf, B. (1975) J. Exp. Med. 141, 1459.

16. Debre, P., Kapp, J.A., Dorf, M.E. and Benacerraf, B. (1975) J. Exp. Med. 142, 1447.

17. Munro, A.J. and Taussig, M.J. (1975) Nature 256, 103.

18. Erb, P. and Feldman, M. (1976) J. Exp. Med. 142, 460.

19. Vadas, M.A., Miller, J.F.A.P., Whitelaw, A.M. and Gamble, J.R. (1977) Immunogenetics 4, 137.

20. Rich, S.S. and Rich, R.R. (1976) J. Exp. Med. 143, 672.

STRUCTURAL STUDIES ON MURINE THYMOCYTE Ia ANTIGENS

Benjamin D. Schwartz*, Anne M. Kask, Susan O. Sharrow, Chella S. David, and Ronald H. Schwartz

Laboratory of Immunology, National Institute of Allergy and Infectious Diseases and the Immunology Branch, National Cancer Institute, Bethesda, Maryland 20014

Department of Genetics, Washington University School of Medicine, St. Louis, Missouri 63110

ABSTRACT. Because of the contradictory interpretation of results from previous chemical studies, we undertook a study to resolve the question of whether Ia antigens exist on thymocytes. Contaminating cells within the thymus cell population were removed using the fluorescence activated cell sorter, and the resulting thymocytes were tested for the presence of Ia antigens using chemical techniques. Both Ig^- thymus cells and thymocytes selected with a rabbit anti-mouse brain antiserum were shown to synthesize Ia antigens. These antigens were on molecules composed of two chains of 33,000 and 25,000 daltons respectively, similar to the molecules bearing Ia antigens derived from spleen. It was estimated that thymocytes have 1-2% as much Ia as B cells.

INTRODUCTION

The discovery of Ir genes within the I-region controlling T dependent immune responses (1) and the discovery of Ia genes within the same region coding for cell surface Ia antigens (2) raised the possibility that Ia antigens were on the surface of T lymphocytes and were closely associated with the mechanisms by which immune responses were regulated. A number of functional studies have supported this possibility (3,4,5), while serological and chemical studies have given contradictory results (6-13). Fathman, et. al. utilized the fluorescence activated sorter (FACS) to increase the sensitivity of detection of fluorescence, and demonstrated that as many as 50% of thymus cells reacted with anti-Ia antisera (14). Murphy et.al. have found a new set of Ia antigens which are apparently only on suppressor T cells (15). Delovitch and McDevitt (12) and Goding et. al. (13) used chemical techniques, and both groups demonstrated small quantities of Ia antigens in their thymus cell populations. The former group felt that these antigens were most likely derived from contaminating B cells within the thymus, while the latter group felt that the Ia antigens

*Present address: Depts. of Medicine and Microbiology
Washington University School of Medicine
St. Louis, Missouri 63110

were derived from thymocytes since they could demonstrate T cell but not B cell immunoglobulin in their preparation.

We have attempted to resolve the question of the existence of Ia antigens on T cells using the FACS to purify thymus T cells free of contaminating cells, and then using chemical methods to test the sorted cells for the presence of Ia antigens.

METHODS

Thymuses were removed from C3H/HeN mice (5-8 wks old) and single cell suspensions made. The cells were reacted with either the fluorescein-conjugated $F(ab)_2$ fragments of a goat anti-mouse gamma globulin (GAMGG) (the kind gift of Dr. Irwin Scher) or with the fluorescein-conjugated $F(ab)_2$ fragments derived from a rabbit anti-mouse brain (RAMB) (the gift of Dr. Thomas Chused) that had been absorbed with mouse red blood cells, acetone liver powder, bone marrow cells, and nude spleen cells, and then passed through the FACS. 5×10^7 cells were collected over a period of seven hours. The sorted cells were labeled with ^3H-leucine for 16 hours. The labeled antigens were solubilized from the cells using 0.5% of the nonionic detergent NP-40, freed from particulate matter by ultracentrifugation, partially purified by lentil lectin affinity chromatography, isolated by indirect immunoprecipitation using Cowan I strain Staph. aureus and analyzed by sodium dodecyl sulfate polyacrylamide gel electrophoresis (SDS-PAGE).

RESULTS

Figure 1A shows the FACS analysis profile when C3H thymus cells were reacted with fluorescein-conjugated $F(ab)_2$ fragments of GAMGG to stain any Ig^+ cells. Increasing intensity of fluorescence (gain = 8) is plotted on the x-axis, and the number of cells at each intensity is plotted on the y-axis. 98% of thymus cells showed virtually no fluorescence with these $F(ab)_2$ fragments (channels 0-100), 1% of thymus cells demonstrated minimal fluorescence (channels 100-200), and 0.3-0.5% of cells demonstrated a fluorescence intensity similar to that seen with B lymphocytes reacted with this reagent (channels 200-1000, B cell data not shown). We concluded that our thymus cells were contaminated with approximately 0.5% B cells. To ensure that all Ig^+ cells would be eliminated from the population subsequently studied, we discarded the 4% most intensely staining cells, and radiolabeled the remaining 96% of cells (channels 0-61). 0.2% of this population were classified as macrophages by latex ingestion.

Fig. 1B shows the FACS analysis profile following reaction of other C3H thymus cells with the fluorescein conjugated $F(ab)_2$ fragments derived from the absorbed RAMB serum. The axes and gain are the same as in Fig. 1A. The vast majority of cells stain with this reagent. However, only the 93% brightest

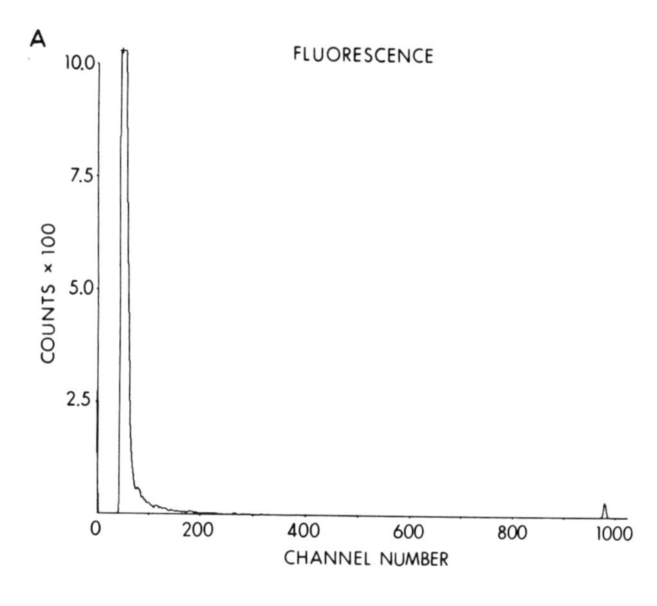

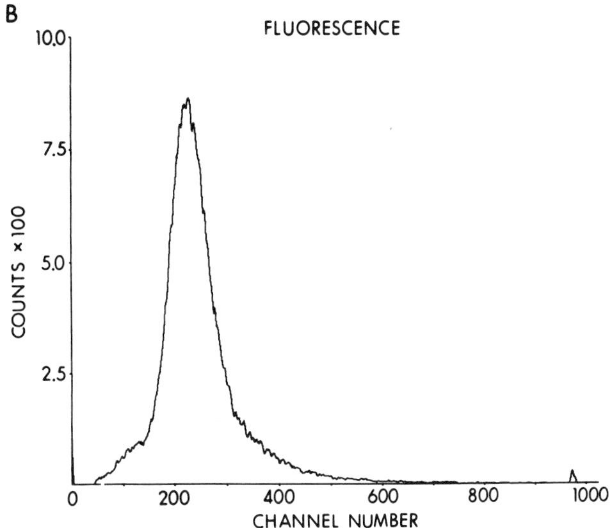

Fig. 1. FACS analysis profiles of thymocytes stained with fluorescein conjugated $F(ab)_2$ fragments of GAMGG (A), or with fluorescein conjugated $F(ab)_2$ fragments of RAMB (B). Fluorescent intensity at a gain of 8 is plotted on the x-axis, and the number of cells at each intensity is plotted on the y-axis.

cells (channels 155-1000) were selected as positive. On analysis, this sorted population had <0.05% Ig^+ cells as assessed by staining with rhodamine conjugated rabbit anti-mouse gamma globulin, and 0.2% macrophages. This compared to 0.5% Ig^+ cells and 0.5% macrophages in the unsorted population.

Figure 2 shows the SDS-PAGE patterns of the Ia antigens derived from thymocytes after removal of Ig^+ cells (Fig. 2A), and from RAMB positively selected thymocytes (Fig. 2B). In both cases, A.TH anti-A.TL (anti-Ia^k) antiserum has reacted with Ia antigens which migrate as two components of 33,000 and 25,000 daltons respectively, a pattern identical to that obtained with Ia antigens derived from spleen cells.

The Ia antigens of unsorted thymocytes and spleen cells, and the H-2K antigens of RAMB sorted thymocytes, unsorted thymocytes, and spleen cells were also analyzed as above (data not shown). The Ia/H-2K ratio for unsorted thymocytes was 0.55, for RAMB sorted thymocytes 0.63 (values identical within the limits of our experimental error), and for spleen cells 6.33. Since we had calculated these Ia/H-2K ratios, it was possible to estimate the thymocyte/spleen cell Ia ratio by determining the thymocyte/spleen cell H-2K ratio. The last ratio was obtained by reacting thymocytes or spleen cells with A.TL anti-A.AL (anti-H-2K) antiserum, washing the cells, adding ^{125}I-labeled Protein A, washing the cells again, and then assessing the amount of ^{125}I-labeled Protein A bound to each cell type. We thus determined that the thymocyte/spleen cell H-2K ratio was approximately 0.23. Then by a simple calculation,

$$\frac{\text{Thymocyte Ia}}{\text{Spleen cell Ia}} = \frac{\text{Thymocyte (Ia/H-2K)}}{\text{Spleen cell (Ia/H-2K)}} \times \frac{\text{Thymocyte H-2K}}{\text{Spleen cell H-2K}}$$

$$= \frac{0.63}{6.33} \times 0.23 = 0.023$$

Thus, thymocytes possess approximately 2% the amount of Ia as spleen cells. If the majority of spleen cell Ia is derived from B cells, and if B cells constitute 50% of the spleen cell population, then thymocytes may have as little as 1% as much Ia as B cells.

DISCUSSION

The results presented here strongly indicate that thymocytes synthesize Ia antigens and bear them on their cell surface. This interpretation of our results was made possible only by eliminating contaminating B cells from our thymocyte population by use of the FACS prior to radiolabeling the Ia antigens. That this purification step is necessary is suggested by the contradictory interpretations of results in the previous two studies (12,13), and by our data which suggest that as little as 0.5-1% B cells in the thymus can contribute enough Ia to obfuscate the meaning of the results (see below).

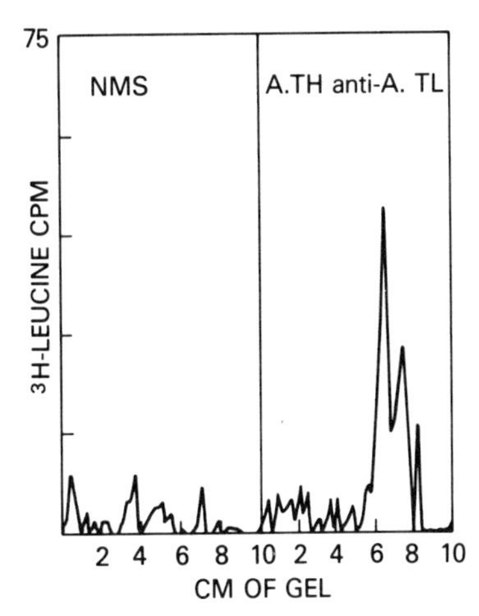

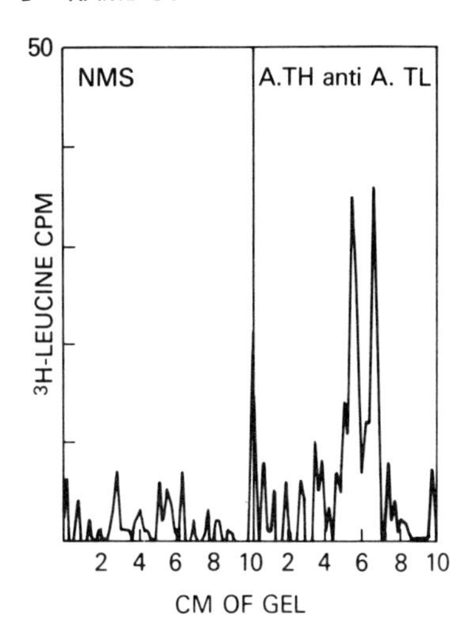

Fig. 2. SDS-PAGE patterns of Ia antigens isolated from thymocytes devoid of Ig^+ cells (A), and from thymocytes positively selected with RAMB (B), and analyzed under reducing conditions the Ia antigens in each case migrate as two components of 33,000 daltons and 25,000 daltons respectively.

Once it has been determined that the sorted populations of thymocytes do synthesize Ia, it is of interest to compare the RAMB sorted and unsorted populations. Since the unsorted population contained 0.5% B cells, if we use the estimated value of 100/1 for the B cell/thymocyte Ia ratio, then the B cells in the unsorted population could be contributing as much as 33% of the Ia antigens chemically detected in the unsorted population. This percentage has to be considered a maximal value, since the 100/1 ratio is an estimated maximum. However, even using this value, thymic T cells must be contributing 67% of the total Ia detected. The RAMB sorted population contained at most 0.05% B cells, and the thymic T cells are therefore responsible for 97% of the Ia antigens detected.

Although both our sorted populations contained 0.2% macrophages, a previous study (16) suggests that the methods employed could not detect the Ia antigens contributed by 3×10^4 macrophages, the maximum absolute number present in the population. Thus, we concluded that thymocytes actually synthesized the Ia antigens. Together with the data of Fathman,et. al. (14) using the cell sorter, our data suggest that approximately 50% of thymocytes have Ia antigens on their surface in low density.

ACKNOWLEDGEMENTS

We wish to thank Laverne Keys for expert secretarial assistance.

REFERENCES

1. McDevitt, H.O. and Sela, M. (1969) Science 163, 1207.
2. David, C.S., Shreffler, D.C. and Frelinger, J.A. (1973) Proc. Natl. Acad. Sci. U.S.A. 70, 2509.
3. Schwartz, R.H., David, C.S., Sachs, D.H., and Paul, W.E. (1976) J. Immunol. 117, 531.
4. Munro, A.J. and Taussig, M.J. (1975) Nature 256, 103.
5. Takemori, T. and Tada, T. (1975) J. Exp. Med. 142, 1241.
6. Sachs, D.H. and Cone, J.L. (1973) J. Exp. Med. 138, 1289.
7. Dickler, H.B. and Sachs, D.H. (1974) J. Exp. Med. 140,779.
8. Hammerling, G.J., Deak, B.D., Mauve, G., Hammerling, V., and McDevitt, H.O. (1974) Immunogenetics 1, 68.
9. Frelinger, J.R., Neiderhuber, J.E., David, C.S., and Shreffler, D.C. (1974) J. Exp. Med. 140, 1273.
10. Gotze, D., Reisfeld, R.A., and Klein, J. (1973) J. Exp. Med. 138, 1003.
11. David, C.S., Meo, T., McCormick, J., and Shreffler, D.C. (1976) J. Exp. Med. 143, 218.
12. Delovitch, T.L. and McDevitt, H.O. (1975) Immunogenetics 2, 39.
13. Goding, J.W., White, E., and Marchalonis, J.J. (1975) Nature 257, 230.

14. Fathman, C.G., Cone, J.L., Sharrow, S.O., Tyrer, H. and Sachs, D.H. (1975) J. Immunol. 115, 584.
15. Murphy, D.B., Herzenberg, L.A., Okumura, K., Herzenberg, L.A., and McDevitt, H.O. (1976) J. Exp. Med. 144, 699.
16. Schwartz, R.H., Dickler, H.B., Sachs, D.H. and Schwartz, B.D. (1976) Scand. J. Immunol. 5, 731.

GENETIC ANALYSIS OF Ia DETERMINANTS EXPRESSED ON CON A REACTIVE CELLS

Gerald B. Ahmann, David H. Sachs and Richard J. Hodes

Immunology Branch, National Cancer Institute,
National Institutes of Health, Bethesda, Maryland 20014

ABSTRACT. Pretreatment of mouse lymphoid cells with anti-Ia sera and complement abrogated the proliferative response to Con A and LPS but not to PHA. Treatment with (B10.AxA) anti-B10 antibodies and complement abrogated the ability of BALB/c ($H\text{-}2^d$) cells to respond to Con A. Absorption of this reagent with B10 or B10.A(5R) cells, but not B10.A(4R) cells, removed this activity. B10.A(4R) anti-B10.A(2R) antibodies and complement inhibited the proliferative response of B10.BR ($H\text{-}2^k$) cells to Con A. These data demonstrate that at least two different I subregions code for antigens expressed on Con A reactive cells.

INTRODUCTION

During the last 5 years, I-region-associated (Ia) antigens have been associated with many immunologically-related phenomena, including mixed lymphocyte reactivity (MLR), graft-versus-host reactivity, and T-B cell interactions (for review, see Ref. 1, 2). Early studies established that Ia determinants are expressed predominantly on B cells (3, 4). However, several lines of evidence now indicate that some or all Ia antigens are also expressed on at least a subset of T cells (5, 6).

Niederhuber and Frelinger and coworkers have analyzed the effects of anti-Ia sera on the response to mitogens (7-10). Anti-Ia sera with or without complement were found to inhibit LPS-induced proliferation of mouse B cells (7). However, for the T cell specific mitogens, PHA and Con A, anti-Ia sera alone did not affect proliferative responses, while pretreatment with anti-Ia and complement markedly suppressed the response to Con A but not to PHA (8). Only a small proportion of the T cells were found to be directly responsive to Con A, and these "promoter cells" apparently recruited Ia-negative cells to respond to Con A (9). Using appropriate intra-H-2 recombinant strains, data were obtained indicating that Ia antigens on Con A promoter cells were determined only by the I-J region (10).

We have also examined the effect on mitogen response of treatment of mouse lymphocytes with anti-Ia reagents and have confirmed the finding that Con A but not PHA responsiveness can be abrogated by pretreatment with anti-Ia serum and complement. However, genetic studies reported here suggest that Ia determinants expressed on Con A reactive cells are coded for by at least two different subregions of the major histocompatibility complex (MHC).

METHODS

Animals. The strains of mice used and their respective H-2 haplotypes are listed in Table 1.

TABLE 1

DISTRIBUTION OF H-2 REGIONS OF STRAINS USED

Strain	Haplotype Designation	Regions & Subregions							
			I						
		K	A	B	J	E	C	S	D
A, B10.A	a	k	k	k	k	k	d	d	d
C57BL/10	b	b	b	b	b	b	b	b	b
BALB/c	d	d	d	d	d	d	d	d	d
B10.A(2R)	h2	k	k	k	k	k	d	d	b
B10.A(4R)	h4	k	k	b	b	b	b	b	b
B10.A(5R)	i5	b	b	b	k	k	d	d	d
B10.BR	k	k	k	k	k	k	k	k	k
A.TL	t1	s	k	k	k	k	k	k	d
A.TH	t2	s	s	s	s	s	s	s	d

Antisera. Antisera production and immune ascites preparations were carried out as previously described (3, 6). Immune ascites reagents were used for all of the studies reported here. The reagents employed and their cytotoxic titers (as measured by two-stage dye exclusion) on whole spleen cells were as follows: (B10.AxA) anti-B10 (1:64 on $H\text{-}2^d$); A.TH anti-A.TL (1:256 on $H\text{-}2^d$ and 1:512 on $H\text{-}2^k$); and B10.A(4R) anti-B10.A(2R) (1:16 on $H\text{-}2^k$). Quantitative absorptions were performed as previously described (11).

Cell Suspension and Culture Conditions. Single cell suspensions were prepared from mouse spleen, lymph node, and nylon nonadherent spleen cells in MEM (Eagle's) containing antibiotics and fetal calf serum (FCS) according to published procedures (12, 13). Nylon nonadherent spleen cells were always ≤8% Ig-positive. Mitogen responses were assayed by culturing 4 x 10^5 viable cells per flat-bottomed microtiter well in 0.2 ml of serum-free MEM containing 2-mercaptoethanol. Appropriate mitogens were added and cultures done in triplicate. Cells were cultured for 72 hrs and pulsed with [^{3}H] thymidine 20 hrs before culture termination. Cultures were then harvested with a multiple automated sample harvester and counted in a liquid scintillation counter (14).

Mitogens. P-PHA was obtained from Difco. Con A (3X crystalized) was purchased from Miles. After filtration, the concentration of Con A was determined by ultraviolet absorption using $E^{mg/ml}_{280} = 1.3$. LPS was a generous gift of Dr. J. L. Ryan.

Antiserum Treatment of Cells. Cells were suspended at 5 x 10^6/ml in MEM-FCS containing the anti-Ia reagent at a dilution of 1:10 (except for A.TH anti-A.TL which was used at 1:30) and incubated for 30 min at 37°C. They were then washed and a second incubation with rabbit complement was performed for 30 min. This was followed by a wash and subsequent resuspension at 2 x 10^6 viable cells/ml in serum-free MEM prior to culturing.

RESULTS

Initial experiments were done to determine the effect on the mitogenic response of BALB/c ($H\text{-}2^d$) lymph node cells of treatment with (B10.AxA) anti-B10 or A.TH anti-A.TL antibodies and complement. (B10.AxA) anti-B10 contains known cytotoxic antibody against the determinant Ia.8, an I-A subregion antigen shared by $H\text{-}2^b$ and $H\text{-}2^d$, whereas A.TH anti-A.TL contains antibodies against Ia.7, an I-C subregion antigen of $H\text{-}2^k$ and $H\text{-}2^d$. It must be emphasized that although these are the only presently defined cross-reactive determinants recognized by these sera which should be present on BALB/c ($H\text{-}2^d$) cells, both reagents could potentially contain antibodies against cross-reactive determinants encoded by other BALB/c I-region genes. The results of treatment of BALB/c lymph node cells with these reagents are shown in Figure 1. It can be seen that treatment with either A.TH anti-A.TL or (B10.AxA) anti-B10 significantly reduced the response of BALB/c lymph node cells to Con A at all mitogen doses tested. In contrast, cells resistant to these two reagents and complement had a PHA response similar to cells treated with normal ascites and complement (data not shown). Treatment of the cells with immune ascites alone (without complement) had no effect on the Con A response, and the response of an inappropriate target [e.g., B10.BR ($H\text{-}2^k$) with (B10.AxA) anti-B10 and complement] was likewise unaffected.

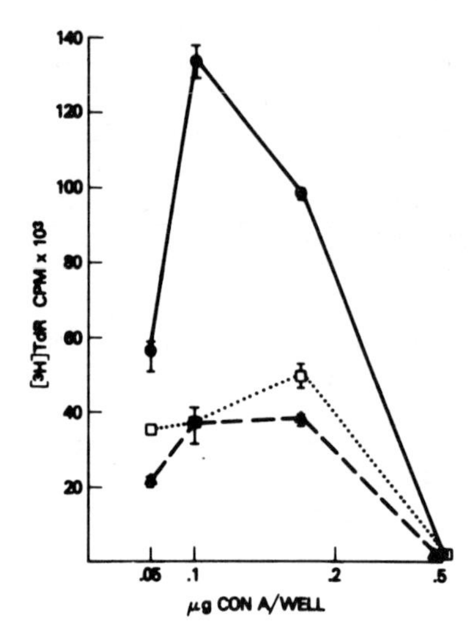

Fig. 1. Con A stimulated proliferative response of BALB/c (H-2^d) spleen cells. Cells were treated with: normal ascites (●———●); A.TH anti-A.TL ascites (□······□); and (B10.AxA) anti-B10 ascites (▲- - - - -▲). Each point represents the mean counts per minute (CPM) of three cultures ± SEM.

Absorption studies were then undertaken to map the genes responsible for the target antigen detected by (B10.AxA) anti-B10 on Con A reactive BALB/c cells. The immune ascites was absorbed with B10, B10.A(4R), or B10.A(5R) spleen cells, and the absorbed reagents were then tested for complement-dependent cytotoxicity to BALB/c spleen cells. Initial studies demonstrated that three absorptions with 4 x 10^8 B10 spleen cells were required to remove all of the cytotoxicity for BALB/c spleen cells. Subsequent studies using B10.A(4R) and B10.A(5R) spleen cells at this concentration revealed that B10.A(5R) cells were equivalent to B10 cells in their ability to absorb cytotoxic activity, while B10.A(4R) absorption did not significantly affect cytotoxicity. These results are consistent with the suggestion that (B10.AxA) anti-B10 has cytotoxicity on H-2^d targets only against Ia.8 which has been mapped to the I-A subregion. The cytotoxic activity of anti-Ia reagents [including (B10.AxA) anti-B10 employed here] has been demonstrated to be predominantly directed to B cells, with no detectable killing of T cells. Therefore, the effect of anti-Ia and complement treatment on Con A reactivity (a T cell dependent response) was determined using the same absorbed antisera. The results for responding BALB/c lymph node cells are depicted in Figure 2 and were essentially identical to those of the cytotoxicity studies: B10 or B10.A(5R) cells, but not B10.A(4R) cells, removed the ability of the (B10.AxA) anti-B10 to abrogate the Con A response. In this experiment, and in a series of five experiments, no consistent effect on the PHA response was observed. These

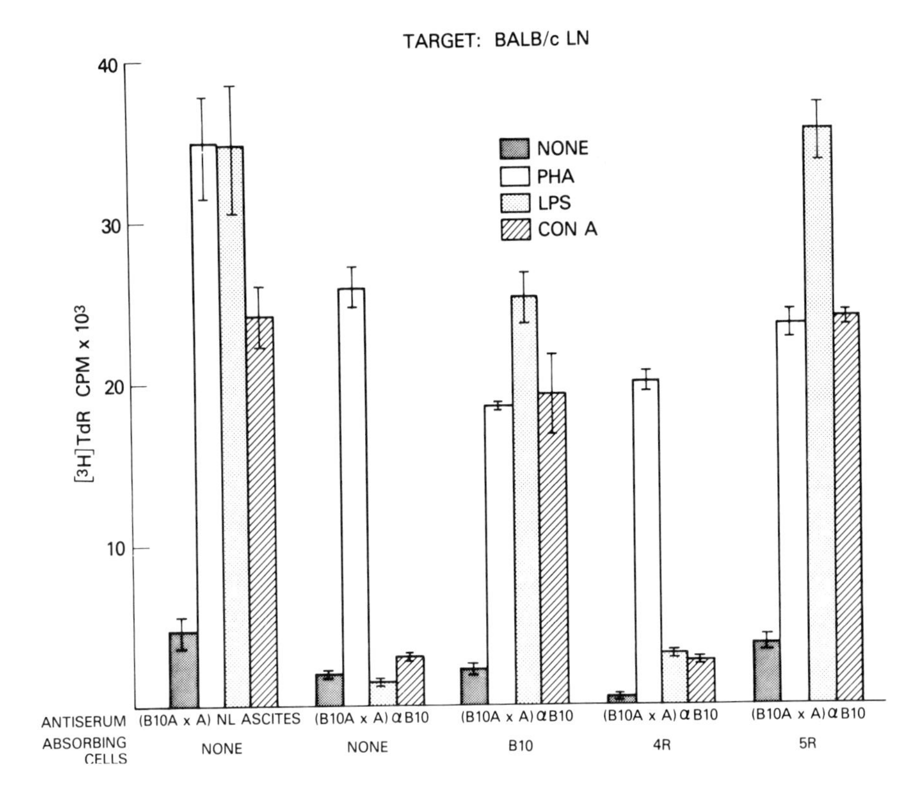

Fig. 2. Con A, PHA, and LPS proliferative responses of BALB/c (H-2d) lymph node (LN) cells after treatment with (B10.AxA) akti-B10 and complement. The (B10.AxA) anti-B10 reagent was used unabsorbed or after absorption with B10, B10.A(4R), or B10.A(5R) spleen cells. 4×10^5 viable cells were cultured for 72 hrs with .5 μg P-PHA, .1 μg Con A, 10 μg LPS or no mitogen. Each bar represents the mean response of three cultures expressed as CPM ± SEM.

absorption results, demonstrated schematically in Figure 3, strongly suggest that cells essential to Con A reactivity express Ia determinants encoded by genes in the I-A subregion and not solely by the I-J subregion.*

To test the possibility that subregions other than I-A could code for Ia antigens expressed on Con A reactive cells,

*No recombinant is readily available to permit exclusion of the K region. However, since there is no H-2-like reactivity detectable in the reagent employed here (by serologic and chemical precipitation criteria), we shall presume that the responsible gene is in the I-A subregion.

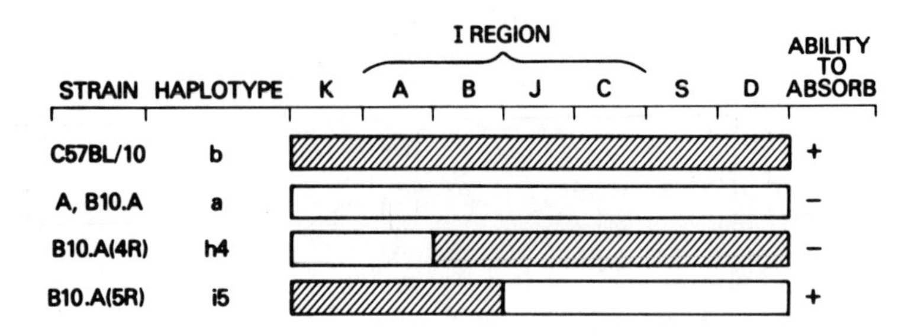

Fig. 3. Schematic representation of the ability of selected B10.A recombinant strains to absorb the activity of (B10.AxA) anti-B10 for Con A reactive cells. Shading of haplotypes indicates the strain of origin of each subregion.

the effect of a B10.A(4R) anti-B10.A(2R) reagent and complement on the Con A reactivity of B10.BR ($\underline{H-2}^k$) cells was evaluated. This reagent could contain antibodies against products of genes in or between I-B and S. The results are presented in Figure 4 and demonstrate that treatment with B10.A(4R) anti-B10.A(2R) and complement does reduce the Con A reactivity of B10.BR nylon nonadherent spleen cells. This suggests that cells necessary for Con A reactivity also express antigens coded in the MHC in or between I-B and S.

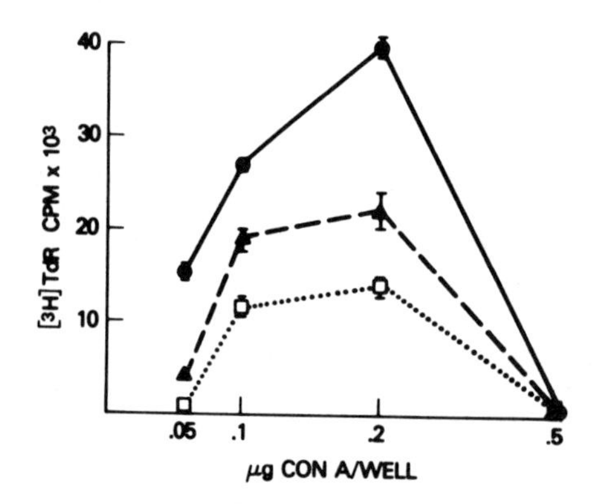

Fig. 4. Con A-stimulated proliferative response of B10.BR (H-2^k) nylon nonadherent spleen cells. Cells were treated with: normal ascites (●——●); A.TH anti-A.TL ascites (□·····□); or B10.A(4R) anti-B10.A(2R) ascites (▲- - - -▲) and complement.

Cumulatively, these data suggest that at least two different I subregions code for antigens expressed on Con A reactive cells. One of these subregions appears to be I-A and the other to the right of I-A (i.e., in the I-B through S interval).

DISCUSSION

Our data confirm the findings of Niederhuber and Frelinger et al. (7-10) that pretreatment of cells with anti-Ia sera and complement abrogates their ability to respond to Con A (and LPS) in culture but has no effect on the PHA response. The abrogation of Con A reactivity was confirmed over a dose-response curve and was most dramatic at suboptimal and optimal doses. These results were reproducible with either lymph node cells or nylon nonadherent spleen cells (a T cell-enriched population) as the responding population. It should be noted that although Con A reactivity is regarded as a T cell property, non-T cells have been shown to be essential to Con A reactivity (15, 16). Preliminary experiments from our laboratory have in fact suggested that a non-T cell may be involved in anti-Ia abrogation of Con A reactivity. Further work is in progress to clarify this point.

These data also provide information concerning the subregion genetic mapping of the genes responsible for anti-Ia effects on Con A reactivity. The ability of B10.A(5R), but not B10.A(4R), spleen cells to absorb the activity of the (B10.AxA) anti-B10 reagent on BALB/c target cells maps the relevant Ia determinant to the I-A subregion. The positive effect of the B10.A(4R) anti-B10.A(2R) reagent on the Con A reactivity of B10.BR cells suggests that an additional subregion or regions outside the I-A subregion (between I-B and S) is also involved in Con A reactivity. These findings are in contrast to the conclusions of Frelinger et al. (10) who have reported Con A abrogating activity to be directed exclusively to antigens encoded by the I-J subregion. Many of our studies were performed on nylon nonadherent, T cell-enriched populations analogous to those employed by Frelinger et al. with culture conditions apparently similar. The explanation for this descrepancy is therefore unclear. It does seem possible, however, that the use of a single dose of Con A as presumably used in the data of Frelinger et al. may have led to failure in detecting an effect from anti-Ia treatment. In our hands, for example, slightly supraoptimal doses of Con A at times obliterated the difference between responses of control and of anti-Ia treated populations, although differences were apparent over the remainder of the dose-response curve. In addition, different reagents and different strains of mice were used by Frelinger and coworkers from those employed in the experiments reported, and this might explain the discrepancies.

Work is presently in progress to further delineate which particular subregion between I-B and S is expressed on Con A reactive cells.

REFERENCES

1. Shreffler, D.C. and David, C.S. (1975) *Adv. Immunol.* 20, 125.
2. Sachs, D.H. (1976) *Contemp. Top. Mol. Immunol.* 5, 1.
3. Sachs, D.H. and Cone, J.L. (1973) *J. Exp. Med.* 138, 1289.
4. Hämmerling, G.J., Deak, B.D., Mauve, G. Hämmerling, U. and McDevitt, H.O. (1974) *Immunogenetics* 1, 68.
5. Frelinger, J.A., Niederhuber, J.E., David, C.S. and Shreffler, D.C. (1974) *J. Exp. Med.* 140, 1273.
6. Fathman, G.C., Cone, J.L., Sharrow, S.O., Tyrer, H. and Sachs, D.H. (1975) *J. Immunol.* 115, 584.
7. Niederhuber, J.E., Frelinger, J.A., Dugan, E., Coutinho, A. and Shreffler, D.C. (1975) *J. Immunol.* 115, 1672.
8. Niederhuber, J.E., Frelinger, J.A., Dine, M.S., Shoffner, P., Dugan, E., and Shreffler, D.C. (1976) *J. Exp. Med.* 143, 372.
9. Niederhuber, J.E. and Frelinger, J.A. (1976) *Transplant. Rev.* 30, 101.
10. Frelinger, J.A., Niederhuber, J.E. and Shreffler, D.C. (1976) *J. Exp. Med.* 144, 1141.
11. Sachs, D.H., Winn, H.J. and Russell, P.S. (1971) *J. Immunol.* 107, 481.
12. Hodes, R.J. and Terry, W.D. (1974) *J. Immunol.* 113, 39.
13. Hodes, R.J., Handwerger, B.S. and Terry, W.D. (1974) *J. Exp. Med.* 140, 1646.
14. Nadler, L.M. and Hodes, R.J. (1977) *J. Immunol.*, in press.
15. Rosentreich, D.L., Farrar, J.J. and Dougherty, S. (1976) *J. Immunol.* 116, 131.
16. Mills, G., Monticone, V. and Paetkau, V. (1976) *J. Immunol.* 117, 1325.

WORKSHOP 2 - MAJOR HISTOCOMPATIBILITY COMPLEX

P. Jones, B. Schwartz, and C. Terhorst - Conveners

The structure of HLA molecules was discussed by Cox Terhorst. The larger subunit of most HLA specificities (except HLA-A2 and HLA-WA28) has an acid-labile peptide bond between two intrachain disulfide bridges. Additional cleavage with CNBr and NTCB reveals that the distance between the two half-cystines of each bridge is of the order of 60 amino acids. The amino acid sequence around one of the half-cystines (CyS_3) shows homology with immunoglobulins. These data suggest that HLA molecules have a basic structure which is like two immunoglobulin domains. The asparagine-linked glycan is located within a stretch of 100 amino acids N-terminal to this disulfide bridge. As shown for H-2 the carbohydrate moiety does not contribute to the antigenic determinants of the molecule.

Ted Hansen presented genetic, serological, and sequential precipitation evidence for the existence of a third H-2-like molecule, termed D'. The molecule is reactive with an antiserum which detects public H-2 specificities, but not with antisera directed against H-2 private specificities. The D' molecule is encoded by a gene which maps to the right of H-2G and to the left of Qa-2. The molecule has molecular weight of 45,000 daltons, and may or may not be associated with β_2 microglobulin. A spontaneous mutation which resulted in loss of expression of the D' molecule was found in a BALB/c mouse.

Mike Cecka presented sequence data of murine Ia molecules. He showed that the sequences of the α and β chains of the $I\text{-}A^k$ Ia molecule are not similar to each other nor to the α chain of the $I\text{-}EC^k$ Ia molecule. However, there was a striking homology between the $I\text{-}EC^k$ α chain and human p34 (5 of a possible 6 comparisons showed identical residues).

Ben Schwartz described partial N-terminal amino acid sequence data for the 25,000 dalton chains of both the strain 13 guinea pig Ia.3,5 molecule, in which the 33,000 dalton and 25,000 dalton chains are non-covalently associated, and the strain 2 guinea pig Ia.4,5 molecule, in which the two chains are disulfide bonded. The internal homology of the two 25,000 dalton chains was striking since identical amino acids were found at 4 of a possible 10 comparisons. There was also the indication of homology with human p29 (identifiable at 3 of 8 possible comparisons) and the β chain of the mouse $I\text{-}A^k$ Ia molecules (identifiable at 2 of 4 possible comparisons).

Cox Terhorst stated that the N-terminal sequences of papain-solubilized human p30 and p23 were the same as the sequences of detergent-solubilized p34 and p29, respectively, indicating that the N-terminal ends of both chains are exposed. He also stated that peptide maps of ^{125}I-labeled p34 and p29 suggested that p34 has two more peptides than p29, raising the possibility that p29 is a degradation product of p34. However, Freed and Nathenson using peptide maps of internally labeled murine Ia antigens have shown substantial differences between the α and β chains of Ia. They concluded that the β chain was not a degradative product of the α chain.

Jim Prahl discussed studies of the structure of human complement components. Both C'3 and C'5 are composed of two polypeptide chains, while C'4 has three. Preliminary amino acid sequence studies have been done on the α (93,000 dalton), β (75,000 dalton) and γ (35,000 dalton) subunits of C'4. No homology was found to C'3 or C'5. The γ chain shows no sequence homology to other molecules coded for by the HLA complex, despite its size similarity to human Ia-like molecules. Preliminary biosynthetic studies of guinea pig C'4 suggest that all three subunits may be synthesized as a single gene product.

Charles Wild presented studies of mouse serum proteins bearing Ss and Slp determinants. Anti-Ss precipitated molecules indistinguishable from human C'4 in size and subunit structure; the human and mouse molecules also cross-react antigenically. When tested on preparations from DBA/2 mice, which are Ss^{hi} and Slp^{+}, antisera against Ss and Slp clear for each other, suggesting that these determinants are on the same molecule. Mouse serum was reported to contain an enzymatic activity which degrades Ss from 200,000 daltons to a 145,000 dalton molecule; this activity may be similar to human C4b inactivator.

Don Shreffler confirmed that both anti-Ss and anti-Slp precipitate molecules containing three subunits. However, in sequential precipitation studies anti-Slp reacted with only 20-80% of the molecules recognized by anti-Ss.

Pat Jones presented two-dimensional polyacrylamide gel analyses of mouse Ia molecules immunoprecipitated from NP-40 extracts of ^{35}S-methionine-labeled spleen lymphocytes. Combined isoelectrofocusing and SDS electrophoresis of Ia produce patterns of spots, detectable by autoradiography, which are haplotype-specific and independent of the genetic background. Molecules coded for by the I-A subregion from b, k, and d haplotypes fall into two groups: very acidic molecules with molecular weights of 33,000-36,000 daltons and very basic molecules of less than 30,000 daltons. Molecules precipitated by antisera directed against the E and C subregions of k haplotype were shown to be determined by a locus (loci) in the E subregion because they produce patterns identical to those generated using the same antisera from a haplotype mice. Since the a haplotype has E^k and C^d, these B cell molecules must be coded for by loci in the E subregion.

Lee Hood led a speculative discussion of three models for the evolution and maintenance of complex allotypes such as H-2D and H-2K products. The first model is the classical allelic model in which mutations are accumulated and fixed. The second model is one of gene duplication and then deletion. The third model, discussed in most detail, is a regulatory model where a cis-dominant regulatory mechanism allows expression of only one gene of a multigene family. Precedents for this model cited from outside the H-2 system include phage insertion into bacterial DNA, the "jumping genes" of maize, and the complex allotypes of immunoglobulins. Evidence for the model from within the H-2 system includes the fact that viral infection stimulates expression of "wrong" H-2 specificities, that the H-2K locus has a high mutation rate, and that the mutant H-$2D^{bd}$ molecule differs from the wild type H-$2D^b$ molecule by several tryptic peptides. This model postulates that within an individual inbred strain, multiple genes for "wrong" H-2 specificities exist, and are expressed at a low frequency within that strain. The model therefore allows for immediate selection of "wrong" specificities if environmental changes favor such selection. Each inbred strain need only contain a subset of all possible H-2 genes; thus, all mice together would contain all possible H-2 genes. The hope was expressed that experiments at the DNA level would soon become practicable to test this model.

Workshop Summary - Workshop Number 7

Expression of Ia subregion antigens on immune-related subpopulations

Chairmen Donal Murphy and David H. Sachs

Ever since the discovery of Ia antigens several years ago, it has been intriguing to speculate that this series of cell surface antigens might be involved in cellular interactions in the immune response and thereby explain *Ir* gene function. However, early studies indicated that these antigens were predominantly expressed on B cells, and that every B cell appeared to have a full complement of all the Ia antigens of its haplotype. It was hard to formulate an hypothesis by which such uniformly expressed antigens could give rise to the apparently high level of specificity indicated by studies of *Ir* genes.

However, as methods of detection have become more sensitive, it has become clear that Ia antigens are also expressed on subpopulations of T cells and probably on other immune related subpopulations as well. In addition, it appears that different subregion antigens may be differentially expressed on different subpopulations of cells, with at least quantitative if not qualitative differences in distribution. It was therefore the purpose of this workshop to focus on recent developments regarding the differential expression of Ia antigens determined by genes in each of the *I* subregions on different subpopulations of cells.

The first topic discussed was the expression of Ia antigens on macrophages. Dr. Neiderhuber presented evidence for the expression of Ia antigens determined by several *I* subregions (at least *I-A*, *I-J*, *I-E*, and/or *I-C*) on macrophage subpopulations. His system involved *in vitro* reconstitution experiments, and most of the subregions were detected only by cytotoxicity, i.e., killing of the macrophage population with the appropriate anti-Ia antibody and complement. However, blocking with anti-Ia reagents in the absence of complement was obtained in several instances if the anti-Ia antibodies had the potential to react with *I-J* subregion determinants. Several other participants in the workshop also mentioned confirmatory data indicating that anti-Ia antibodies could inhibit antigen presentation by subpopulations of macrophages in their systems.

Dr. R. Schwartz presented blocking studies in an *in vitro* T cell proliferation assay using anti-Ia antisera against defined subregion antigens. In cases of immune responses which appear to be controlled by multiple I-region genes (alpha and beta) it appeared that antibodies against more than one subregion were capable of blocking in this assay. This led to the interesting speculation that the Ir gene control in such systems depends on cell surface products determined by the same subregions to which the different Ir gene functions have been mapped. The actual cell on which the anti-Ia antibody exerts its effect in this system has not been unequivocally determined. However, it appears likely that at least part of the effect may also be at the level of the macrophage.

A fairly extensive discussion ensued on the possibility that many of the effects of anti-Ia antisera in functional assays which have been ascribed to T cells may in fact have reflected effects of the antisera on macrophages. The consensus was that while some of the effects may indeed be at the macrophage rather than the T cell level, it would be premature to conclude that all effects are at the macrophage level. In particular, effects of anti-I-J antisera on suppressor T cells appeared to be independent of macrophages. Further, even in certain macrophage-dependent assays, the addition of fresh macrophages after anti-Ia treatment frequently is not sufficient to reverse the observed effects.

Two of the papers presented led to a discussion of the present difficulties in the definition of the I-E versus I-C subregion, in particular the evidence for a recombination event between these two subregions in the $H-2^a$ haplotype. Dr. David presented evidence for a new Ia specificity (Ia.23) which appears to define an $I-E^d$ allele. If this result is confirmed by chemical studies, it will support the notion of two distinct I region genes to the right of I-J. However the origin of these two genes in the $H-2^a$ haplotype will still remain in question. Dr. S. Rich presented evidence, in a system involving suppressor factors, that effective suppression requires homology in the I-C subregion between the factor producing cells and the target cells for suppression. In this system $H-2^a$ appears to possess the $I-C^d$ allele. It remains possible, however, that these functional assays are measuring a cell surface product other than the serologically detected I-C region Ia antigens. If so, these results could indicate a recombination in the $H-2^a$ haplotype within the presently defined I-C subregion such that such genes coding for serologic Ia specificities fall to the left of the point of recombination and the genes involved in suppressor factor specificities fall to the right. However, at present, there is likewise no clear evidence that the I-C subregion of $H-2^a$ does not derive from the $\underline{H-2}^d$ haplotype.

Dr. J. Frelinger presented data on the effect of anti--Ia antisera on LPS mitogen induced responses, but found no evidence for predominant effects of antibodies to any one subregion. Dr. R. Hodes presented data on the effects of anti-Ia antisera and complement on Con A mitogen induced responses. Contrary to previously published reports, these data indicated that antibodies directed to the products of more than one I-region gene were capable of blocking Con A induced responses. These data therefore indicate that such blocking is not an exclusive property of anti-I-J antibodies.

The last two papers in the workshop concerned the detection of Ia antigens on somatic lymphoid hybrid cells (Dr. B. Osborne) and lymphoid tumor cells (Dr. B. Chesebro). In the case of somatic hybrids between T cells and AKR thymoma cells, evidence was presented for the expression of restricted Ia antigens. Preliminary absorption studies indicated that I-E or I-C antigens were probably being detected. On the other hand, the B cell tumors appeared to express all of the Ia antigens of their haplotype of origin, in that absorption with these tumors appeared to remove all Ia activity from the antisera tested. These approaches are exciting in that they may permit further dissection of the fine structure of the I-region through differential absorption of antibodies to differentially expressed Ia genes. However, such results must be interpreted with extreme caution. When a tumor cell reacts with an anti-Ia antiserum but removes only a portion of the known Ia reactivity, it is important to show by independent criteria that the antibodies removed were indeed directed to Ia antigens and not to contaminating viral antigens. Absorption analyses and chemical isolation studies are among the correlative criteria suggested.

This summary is intended only as an overview of what was presented in this workshop and is not intended to be used as a primary reference. Of necessity, we have not attempted to provide formal references to any of the information presented, and only the names of speakers have been indicated despite the fact that most of the presentations represented collaborative efforts. We apologize in advance for these unavoidable deficiencies.

SOMATIC CELL HYBRIDS WITH T CELL CHARACTERISTICS*

Richard A. Goldsby[†], Barbara A. Osborne[§], Donal B. Murphy[§], Elizabeth Simpson[Ψ], Jim Schröder[§] and Leonard A. Herzenberg[§]

ABSTRACT. Hybrid populations have been obtained by polyethylene glycol assisted fusion of BW5147, a hypoxanthine-aminopterin-thymidine-sensitive mouse lymphoma with immunized spleen cells and subsequent selection in hypoxanthine-aminopterin-thymidine medium. Most of the hybrids express Thy-1, a characteristic T cell surface antigen. This contrasts with the complete absence of Thy-1 positive hybrids obtained from fusions of spleen cells with myeloma cells. A search for the presence of Ly-1,2 and I-region markers revealed that some of the hybrid clones isolated also express these differentiation antigens. Karyotypic analysis of a number of hybrid clones revealed a tendency to retain three or more of certain chromosomes including those specifying Ly-1, Thy-1 and the H-2 complex.

These studies demonstrated that T-lymphoma X spleen cell hybrids provide a means of producing cloned populations of cells selectively expressing characteristic T cell surface markers. The possibility exists that such hybrid populations may carry out specific T cell functions.

INTRODUCTION

The hybridization of differentiated cells with established cell lines is a useful strategy for the production of clonal cell lines expressing differentiated properties and possessing unlimited growth potential. The fusion of lymphoid cells with appropriate established cell lines to yield clonally pure, histiotypically differentiated, cell populations could be particularly useful in dissecting and understanding the many networks of specific cellular interactions that produce and control the organism's repertoire of immune

*This work supported, in part, by grants from the National Institutes of Health (CA-04681 and AI-07757) and from the National Foundation.

†NASA, Ames Research Center, Laboratory of Planetary Biology, MS 239-10, Mountain View, CA, 94035

§Depts. of Genetics and Medicine, Stanford University School of Medicine, Stanford, CA, 94305

ΨClinical Research Center, Transplantation Biology Section, Harrow, Middlesex, HA1 3UJ, England responses. In this regard a number of studies employing somatic cell hybridization have made contributions to our knowledge of antibody synthesis. Of considerable importance is the demonstration by Köhler and Milstein (1) that the fusion of in vitro cultivated mouse myeloma cells with spleen cells from immunized mice gives rise to hybrid cell lines which secrete monoclonal antibody to the immunizing antigen. Interestingly, even though 30-40% of the spleen cells used as partners for hybridizations with myeloid cells were T cells, none of the hybrids recovered had the surface antigen Thy-1, a characterisitc marker for mouse T cells. In contrast to the results obtained by Köhler and Milstein in myeloma fusion studies, we find that fusion of spleen cells with a mouse T lymphoma cell line (BW5147) gives rise to hybrid cell lines which express T cell markers and do not secrete immunoglobulins.

METHODS

a) Hybridization. Hybrid populations were obtained by fusing spleen cells from (BALB/c x SJL)F_1 mice with BW5147 (kindly provided by Dr. Robert Hyman, Salk Institute) which were hypoxanthine guanine phosphoribosyl transferase (HGPRtase) negative, sensitive to hypoxanthine-aminopterin-thymidine (HAT) (2) and resistant to 1 x 10^{-3} M ouabain. Spleen cells always came from either recently dinitrophenylated keyhole limpet hemocyanin (DNP-KLH) boosted mice or from mixed lymphocyte cultures in which cells from CBA or AKR donors were stimulated by allogenic donors.

For fusion a mixture of 1 x 10^8 spleen or MLC derived cells and 1 x 10^7 BW5147 cells in Dulbecco's modified Eagle's medium was removed and fusion initiated by the gradual addition of 2 ml of 50% (w/w) polyethylene glycol (PEG) 1000 or PEG 1540 in serum-free DME at pH 7.8. After 2 min at 37°C, the suspension was diluted to 50 ml by the stepwise addition of 37°C serum-free DME (over the course of 10 min), pelleted by centrifugation, resuspended and cultured in 0.2 or 2.0 ml at 1-2 x 10^6 cells/ml in selective medium (DME + 10% FCS, 100 μM hypoxanthine, 10 μM aminopterin, 30 μM thymidine and 3.3 x 10^{-4} M ouabain). After 14 days, selective medium was removed and replaced with DME + 10% FCS containing 100 μM hypoxanthine and 30 μM thymidine. Hybrid populations appeared in 10-18 days and were harvested 21 days after seeding. Clonal isolates were obtained from hybrid populations by limiting dilution in 0.25 ml wells of microtiter dishes.

b) Serology. Surface markers were detected by use of appropriate antisera and a two-stage dye exclusion microcytotoxicity test (3).

c) Karyotyping. Chromosome preparations were made by conventional methods (4) and a modification of the trypsin Giemsa (5) banding technique was used to identify individual chromosomes.

RESULTS

a) Surface Markers. The expression of Thy-1 alleles on the cell lines obtained by the application of the hybridization protocol described demonstrates directly that these populations arose as a result of fusion of normal spleen cells and the BW5147 lymphoma (see Table 1). Of the 27 clones tested from the (BALB/c x SJL)F_1 x BW5147 hybridization, 18 expressed both the Thy-1.1 allele carried by BW5147 and the Thy-1.2 allele carried by the (BALB/c x SJL)F_1 cells. Similarly, of 12 populations tested from the CBA x BW5147 hybridization, 11 carried the BW5147 Thy-1.1 allele and the Thy-1.2 allele from CBA.

Table 1

QUANTITATIVE STUDIES OF Thy-1 DISTRIBUTION IN HYBRID POPULATIONS

Hybridization	No. of Hybrid Clones Expressing Thy-1 Alleles			
	1.1 + 1.2	1.1	1.2	$-^{ve}$
Lymphoma x (SxB)F_1 (Thy-1.1) (Thy-1.2)	18	7	0	2
Lymphoma x CBA MLC (Thy-1.1) (Thy-1.2)	11	1	0	0
Lymphoma x AKR MLC (Thy-1.1) (Thy-1.1)	0	13	0	0
Myeloma* x spleen (Thy-1 Neg) (Thy-1.2)	0	0	0	>10

*Köhler & Milstein (8)

In contrast, hybridization of BW5147 with a Thy-1.1 carrying strain, AKR, gave rise only to populations carrying Thy-1.1 Thus, expression of Thy-1.2 in the first two sets of hybrids represents expression of genetic information derived from the spleen cell parent rather than (a) expression of a gene carried cryptically by BW5147, or (b) non-specificity of the anti Thy-1.2 typing. Specificity of typing was further confirmed by quantitatively absorbing the anti Thy-1.2 with hybrid cells and then testing the absorbed antiserum on thymocytes to demonstrate removal of anti Thy-1.2 activity.

Examination of clones from the hybridization of BW5147 with spleen cells from DNP-KLH primed and boosted (BALB/c x SJL)F_1 mice revealed that some express I-region antigens. Population 12 is killed by an A.TL anti-A.TH serum in the dye exclusion microcytotoxicity assay (Fig. 1). Antibody reactive with the hybrid is specific, since normal spleen cells from strain B10.S(7R) completely absorb the cytotoxic activity, while cells from strains B10.BR, B10.A, and B10.HTT fail to absorb. The locus controlling the antigenic determinant expressed in the hybrid maps between the I-J and D regions. We assume that this locus maps in the I region (in either I-E or I-C), since no cytotoxic activity has yet been reported utilizing reagents against the S and G regions. Experiments are in progress to confirm this map location.

* * * * * * * *

	H-2 Haplotype									
		I								
Strain	K	A	B	J	E	C	S	G	D	TL
A.TL	s	k	k	k	k	k	k	k	d	c
Anti A.TH	s	s	s	s	s	s	s	s	d	a

Absorbed with:			Kills Population 12
B10.BR	________________	?	Yes
B10.S(7R)	----------------	---	No
B10.A	________________	---	Yes
B10.HTT	------__________	?	Yes

____ denotes reactivity remaining

---- denotes reactivity absorbed

Figure 1. Preliminary mapping of I-region determinants expressed on population 12

* * * * * * * *

Finally, in addition to examination for the presence of Thy-1 and I-region antigens, several of the hybrids were screened for the expression of the lymphoid tissue alloantigens Ly1,2,3. These differentiation antigens have been found only on thymus-derived lymphocytes. Even though the tumor parent, BW5147, does not express Ly determinants, preliminary screening has found several hybrid clones which display these T cell histiotypic antigens.

b) Karyology. Table 2 presents a detailed comparison of the karyotypes of BW5147 and 15 hybrid clones. The BW5147 parent has a modal chromosome number of 43, most of which are morphologically identical to those found in the normal mouse complement. In contrast to the relative karyological constancy of the tumor parent, considerable variability is seen among different hybrid clones. In spite of the considerable karyotypic variation found among the 15 clones examined, similarities appeared which argue against a completely random pattern of chromosome loss. Inspection of Table 2 shows that certain chromosomes, Nos. 1, 9, 10, 17 and 19, occurred in more than a third of the hybrid clones. This is of particular interest since three of these, 9, 17 and 19, code for cell surface markers present on lymphocytes.

DISCUSSION

From the experiments reported here it can be concluded that the hybridization of spleen cells with the appropriate established cell line can yield continuous populations with characteristic T cell surface antigens. The expression of Thy-1.1 and Thy-1.2 in those hybridizations where allelic differences existed between the parents demonstrates that the genetic information which determines such histiotypic characteristics as Thy-1 can be expressed by both parental genomes. Detection of s haplotype I-region markers and preliminary results from tests for the presence of Ly antigens are consistent with the suggestion that the hybrids bearing these markers are derived from the fusion of normal spleen cells and BW5147.

In retrospect, it is difficult to identify which subpopulations of splenic lymphocytes fused with the lymphoma to yield hybrid populations. However, the frequency of Thy-1-bearing hybrids obtained from the different hybridizations summarized in Table 1 suggests that they are probably derived from fusions between T cells and BW5147. Note that significantly more Thy-1.1 and Thy-1.2 positive hybrids were obtained from the hybridization of cells from a 5-day MLC which contains nearly all T cells, than from the hybridization of spleen cells which contains only 30-40% T cells.

As expected the examination of a number of hybrid clones revealed the diversity of karyotypes usually seen among hybrid clones of the same parental lines. However, the frequency with which the chromosomes, coding for Thy-1 (No. 9), Ly-1 (No. 19) and the H-2 complex (No. 17) are retained in three or more copies would not be expected if chromosome loss was purely random. The association between polysomy and the presence of

TABLE 2

KARYOTYPE OF HYBRID CLONES AND BW5147

Chromosome No.	Clones Having 1 or 2 of Indicated Chromosome	Clones Having ≥ 3 of Indicated Chromosome
1	4	11
2	13	2
3	14	1
4	13	2
5	13	2
6	11	4
7	14	1
8	12	3
9	5	10
10	9	6
11	13	2
12	13	2
13	12	3
14	11	4
15	11	4
16	11	4
17	7	8
18	15	0
19	9	6
X	15	0

information for lymphocyte markers is clear for these three chromosomes and one is tempted to wonder about the possibility that some other chromosomes, such as 1 or 10, may be retained in larger numbers because they bear information which aids lymphocyte viability and growth in these media.

It is clear that somatic cell hybrids obtained by the fusion of splenic lymphocytes and BW5147 provide a means of producing, in large numbers, if desired, clonal populations of cells expressing _I_-region antigens and the T cell surface markers Thy-1 and Ly. Such clonal populations may be expected to provide a convenient source of these antigens for biochemical studies. The isolation of clonal populations expressing Ly and _I_-region markers may be particularly useful for studies of T cell functions. In populations of normal lymphocytes the selective expression of Ly and Ia determinants define subpopulations responsible for various T lymphocyte functions (6,7). The expression of markers characteristic of functional T cell subsets by some of the hybrid populations reported here encourages us to examine them further for the presence of T cell functions. Studies are now in progress to determine if these and other hybrid populations we have obtained display such functions as help, suppression or antigen binding. It is apparent that the expression of any of these functions by clonal populations of continuous cell lines will greatly facilitate the analysis of these T cell mediated functions.

ACKNOWLEDGMENTS

We gratefully acknowledge the capable technical assistance of Miss Joann Williams and the encouragement of Dr. Adrian Mandel. We are also grateful to Ms. Jean Anderson for patient preparation of this manuscript.

REFERENCES

1. Köhler, G. and Milstein, C. (1975) Nature 256, 495.
2. Littlefield, J. W. (1963) Proc. Natl. Acad. Sci. (USA) 50, 568.
3. Murphy, D. B. and Shreffler, D. C. (1975) J. Exp. Med. 141, 374.
4. Hungerford, D. A. (1965) Stain. Technol. 40, 333.
5. Seabright, M. (1971) Lancet 11, 971.
6. Cantor, H. and Boyse, E. A. (1976) J. Exp. Med. 141, 1376.
7. Okumura, K., Herzenberg, L. A., Murphy, D., McDevitt, H., and Herzenberg, L. A. (1976) J. Exp. Med. 144, 1.
8. Köhler, G., Pearson, T. and Milstein, C. (1977) Somatic Cell Genetics, in press

T AND B CELL HYBRIDS

César Milstein and Len Herzenberg, Convenors

Hybrid cell lines or hybridomas (those cell lines grown as tumours) made by fusion of spleen cells from immunised mice and a myeloma often produce both antibody, with specificity for the immunising antigen, and the myeloma protein. This finding of Köhler and Milstein (1) provides an important new means of producing apparently limitless amounts of monoclonal antibodies with preselected specificities.

This workshop began with one of us (C.M.) giving a bit of the history and the current status of work in his laboratory, with myeloma hybrids, which started with a fusion between rat and a mouse myeloma (2). This was done using standard somatic cell hybridisation procedures (Sendai virus as fusing agent and the Littlefield HAT selection system). In the hybrids obtained, all the myeloma polypeptide chains of both parental cells were made but the V-C attachments did not change. Later when a myeloma with an antibody specificity was being sought from which variants could be selected, the idea arose: "Why not manufacture our myeloma to order?" The first experiments worked and except for improvements in technical details, still serve as the prototype for making myeloma antibody producing cell hybrids or B cell hybridomas.

The current technique is to use a myeloma cell line deficient in hypoxanthine-guanine phosphoribosyl transferase (HGPRTase) selected by growth in azaguanine to fuse with immune spleen cells. Hybrids are grown out in HAT (hypoxanthine, aminopterin, thymidine) medium. This medium kills the HGPRTase negative parent. The spleen cells do not grow *in vitro* after several days and are eliminated on continued transfer. The hybrids are cloned in soft agar or by limiting dilution in microcultures.

About 60% of the clones secrete immunoglobulins with the myeloma heavy and light chains [called G (gamma) and K (kappa) respectively, when the MOPC 21 myeloma parent is used] and with a new heavy and light chain, H and L, contributed by the spleen cell parent. These hybrids, designated H L G K, secrete the mixture of polymeric molecules obtainable by chain associations of H, L, G and K. Conveniently, on subcloning and resubcloning, a usable frequency of loss variants are found which make only three, two, one or no chains. Thus variants making only H and L can be selected with only moderate effort. These produce antibody molecules only of the kind made by the original parent spleen cell.

Participants in the workshop told of making hybridomas making monoclonal antibodies to a variety of antigens including sheep and goat erythrocytes, lysozymes, lactic dehydrogenase, Streptococcal A carbohydrate, Staphilococcal DNAase, DNP-KLH, NIP and an antigen of the rat major histocompatibility complex. Clearly the technique is widely applicable to the making of extremely useful serologic reagents of monoclonal specificity.

Several people told of their extensive unsuccessful attempts to make hybrids producing anti-H-2 activity. Many hundreds of hybrids tested were reported. It was not clear whether lack of success is due to the low frequency of anti-H-2 (and Ia) producing cells even in hyperimmunised mice.

Suggestions for improving the recovery efficiency of antibody-producing hybrids included varying immunisation protocols, improving tissue culture technique by such means as prescreening of sera for supporting clonal growth, etc. The recovered hybrid efficiency several laboratories obtained was from 10^{-4} to 10^{-6} per spleen cell with an input ratio of 10:1 spleen cells to myeloma parent. One laboratory has used LPS blasts as normal parent cells with apparently highly efficient hybrid recovery, thus opening the possibility of preselecting the B cells to be fused. Some laboratories are using polyethylene glycol (PEG) (M.W. 1,500-6,000; 33-50% w/v) as fusing agent while others continue with Sendai virus. The former is more convenient and is at least as effective. With myeloma-myeloma hybrids a very efficient recovery was reported by including a few per cent of whole blood in the cultures.

T cell hybrids: None of the myeloma hybrids express Thy-1 antigens. Presumably, therefore, none of these hybrids express other T cell specific gene products or carry out T cell functions. The production of T cell hybrids which do produce T cell specific products was reported and some characterisation of these hybrids described. These T hybrids could provide useful sources for biochemical and physiological studies of specific T cell subpopulations.

Hybrids were made using PEG and essentially the same fusing and selecting system (HAT) as described for producing myeloma hybrids. However, a T lymphoma line (BW 5147 - derived by R. Hyman) was one parent and immune T cells of various origins were the other parents in different crosses.

Other experiments were reported in which a fibroblast line or other T lymphomas were fused with cells from an MLC. In this case gel electrophoresis was used to follow the T cell character of the hybrids.

In the hybrids made with BW 5147, greater than 90% of the recovered hybrids expressed Thy-1 antigen of at least one parent. Thus T cell function lost by myeloma fusion is retained when the phenotypically similar T lymphoma is used.

When unfractionated Thy-1.2 spleen cells were the lymphocyte parent some 60% of hybrids were Thy-1.2 bearing. When the lymphocyte parent was first stimulated for four or five days in an MLC greater than 90% were Thy-1.2 positive. Only Thy-1.1 was found on hybrids when both parents were Thy-1.1.

Karyotype and DNA content analysis revealed that over several months the hybrids often lost many chromosomes they had soon after fusion. They seemed to stabilise with a few more chromosomes than the lymphoma parent, whose marker (metacentric) chromosomes are often kept in the hybrid.

T cell specific Ly and Ia antigens are selectively expressed on different clonal hybrid cell lines. A Ly-1 antigen not found in either parent appears to be expressed in some T cell hybrid lines. Could this indicate that the Ly-1 alleles described so far are regulatory genes?

The selective expression of different T cell Ia antigens on hybrid cells was reported. This opens up the possibility of isolation and structural analysis of these substances. It also makes it possible to map the genes controlling them without having to make new recombinant congenic mouse strains. A word of caution was raised about possible false results due to viral antigens on some hybrids. It was emphasised that all alloantigenic analysis should be done or confirmed by absorptions to eliminate this possibility.

The enormous potential for characterisation of the antigens coded by genes of the MHC of mice and other species, including Man, provided by myeloma and T cell hybrid availability was pointed out. The myeloma hybrids provide the exquisitely specific antibodies and the T cell hybrids the expanded populations of each differentiated phenotype (subset of T cells).

Participants included: B. Osborne, T. Mossman, D. Secher, E. Cohen, K. Rajewsky, N. Klinman, L. Sherman, F. Bach, B. Mathieson, A. Williamson, C. Milstein, L. Herzenberg.

REFERENCES

1. Köhler, G. and Milstein, C. (1975) Nature 256, 495.
2. Cotton, R.G.H. and Milstein, C. (1973) Nature 244, 42.

EXPRESSION OF ALA-1 ON T AND B EFFECTOR CELLS

Ann J. Feeney*
Department of Biology, University of California, San Diego
La Jolla, California 92093

ABSTRACT. The expression of Ala-1 (activated lymphocyte antigen-1), a cell surface alloantigen originally detected on mitogen stimulated T and B lymphocytes, was studied on a number of functional subsets of T and B cells. Ala-1 was shown to be expressed on _in vivo_ primed IgM and IgG PFC, _in vivo_ primed helper T cells, _in vivo_ and _in vitro_ primed CTL, and proliferating cells in a MLC. In contrast, Ala-1 was not found on precursors of IgM PFC, helper T cells, or cytotoxic T cells. These results indicate that Ala-1 is expressed on T and B effector cells but is absent from their non-activated precursors.

INTRODUCTION. To study interactions among different subsets of lymphocytes, stringent criteria are necessary to delineate and separate the various T and B cell subpopulations. Antisera produced against lymphocyte surface antigens have therefore become increasingly important in defining subpopulations and are commonly used in positive and negative selection procedures to purify T or B cell subpopulations. Differential expression of Ly antigens has provided a clear discrimination between helper T cells and amplifier T cells ($Ly1^{+}2^{-}$) on one hand, and cytotoxic T cells and suppressor T cells ($Ly1^{-}2^{+}$) on the other (1,2). These antigens are present on T cells both before and after antigenic stimulation.

In contrast to the Ly antigens, we have described a murine alloantigen, Ala-1, which is present in both the T and B cell lineages, and whose expression is restricted to cells which have been activated by mitogens or antigens (3,4). Ala-1 is expressed in >90% of Concanavalin A (Con A)-stimulated lymph node cells and >90% of lipopolysaccharide-stimulated splenic B cells (3). In contrast, only 0-15% of normal lymph node and spleen cells express Ala-1, and it is never detected on thymocytes. Thymocytes cultured with Con A do not express Ala-1, although they morphologically resemble the Con A-stimulated peripherial lymphocytes. Hence, Ala-1 does not appear to be a blast cell antigen, but rather a differentiation antigen characteristic of effector lymphocytes.

We have previously shown that Ala-1 is expressed on some antigen-stimulated effector populations and is absent from their precursors (4). The present study extends the number of effector populations found to express Ala-1 and the number of precursor populations which are $Ala\text{-}1^{-}$.

METHODS. _Antiserum_: Ala-1.2 antiserum is produced by monthly

* Supported by ACS Grant J321.

immunization of C3H/An mice with increasing numbers of phytohemagglutinin-stimulated spleen and lymph node cells from C58 mice. The C3H/An anti C58 sera is exhaustively absorbed with several non-activated lymphocyte populations to remove any contaminating alloantibodies or autoantibodies before use (3).

General strategy: The procedure used in the following experiments was to investigate whether removal of Ala-1^+ cells by treatment with Ala-1 antiserum and complement (C, selected non-toxic rabbit serum) would affect the relevant function. The residual activity of these cells was compared to that of other aliquots of the cell population which were treated with C alone, anti Thy-1 and C (if a T cell function was being investigated) or were untreated. In all cases, the percent reduction of function after anti Ala-1 treatment was made in comparison to a sample treated with C alone. The untreated sample served as a control to show that the C was not toxic. To control for the specificity of the Ala-1.2 antiserum, experiments were also performed on mice bearing the Ala-1.1 allele; and in all cases there was no decrease in functional activity in these mice after treatment with anti Ala-1.2 and C.

Procedures for obtaining primed populations will be outlined below. Further details of materials and methods have been published (4).

RESULTS

(1) IgM and IgG plaque-forming cells (PFC) are Ala-1^+. C57BL/6(B6) mice were immunized intravenously with 10^8 sheep red blood cells (SRBC). Five or six days later their spleens were removed, and the splenocytes were treated with anti Ala-1.2 and C. The cells were then assayed on SRBC in the Jerne hemolytic plaque assay (5), as modified by Mishell and Dutton (6), to enumerate IgM PFC surviving after mass cytolysis. The number of PFC was reduced on the average by 92% after treatment with anti Ala-1 and C (4)(Fig. 1).

IgG PFC were also shown to be Ala-1^+. Primed spleen population from mice hyperimmunized with SRBC were treated with anti Ala-1.2 and C. The number of IgG PFC were reduced by an average of 74%, as compared to primed spleen cells treated with C alone (4) (Fig. 1).

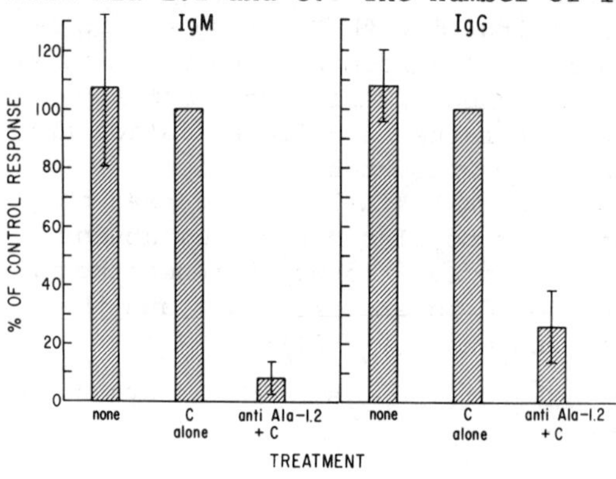

Fig.1. Effect of anti Ala-1 pretreatment on IgM and IgG PFC. The left panel shows the arithmetic mean of 6 experiments; the right panel, 10 experiments. Vertical bars indicate standard error. [Taken from J. Immunol. (4)]

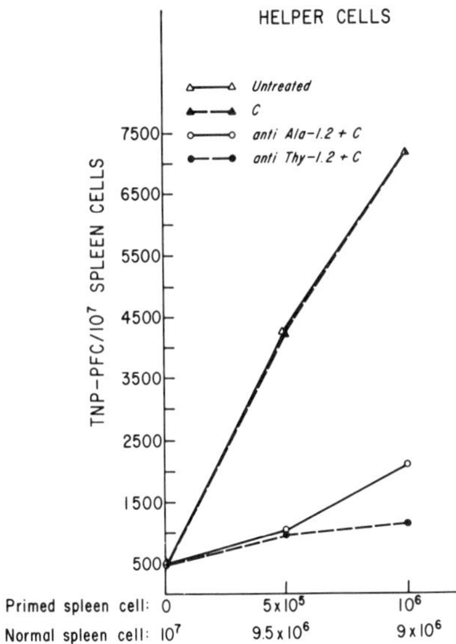

Fig.2. Effect of anti Ala-1 pretreatment on helper T cell activity. Primed B6 spleen cells were treated with anti Ala-1.2 and C (O), anti Thy-1.2 and C (●), C alone (▲), or untreated (Δ), and were then co-cultured in varying proportions with normal spleen cells for 4 days with TNP-SRBC added to all cultures.

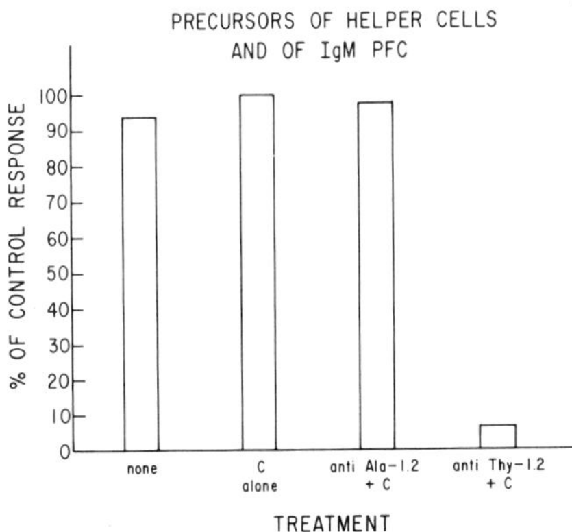

Fig.3. Effect of anti Ala-1 pretreatment on precursors of helper T cells and of IgM PFC.

By the criterion of trypan blue uptake, only a low proportion of cells were killed with anti Ala-1 and C, usually only 0-15% greater than the proportion obtained with C alone.

(2) Primed helper T cells are Ala-1$^+$. Spleen cells from B6 mice primed with 10^6 SRBC were used as a source of helper T cells. The titration analysis was performed according to Hoffmann and Kappler (7). After treatment with various antisera and C, the primed cells were titrated with normal spleen cells, keeping the total cell number constant, and were cultured with SRBC conjugated with trinitrophenyl (TNP) in Mishell-Dutton cultures (8,9). After four days, the cells were assayed for TNP-PFC on TNP-conjugated horse erythrocytes. In seven such experiments, pretreatment of the carrier-primed spleen cells with anti Ala-1.2 and C reduced helper activity by an average of 79% (range 59-94%) although, as in the case of the PFC, low numbers of cells were lysed (4). Anti Thy-1.2 pretreatment resulted in almost complete elimination of helper cell activity. A typical experiment is shown in Fig. 2.

(3) Precursors of helper cells and of IgM plaque-forming cells are Ala-1$^-$. Normal spleen cells of B6 mice were treated with anti Ala-1.2 and C, and were then cultured with SRBC in Mishell-Dutton cultures. Four days later, the cells were assayed to determine the number of PFC generated against SRBC. A reduction in the number of PFC

would indicate that either the precursors of helper cells or the precursors of IgM PFC were depleted. Treatment with anti Ala-1.2 and C did not reduce the number of PFC as compared with the sample treated with C alone (Fig.3). Anti Thy-1.2 pretreatment gave the expected reduction of plaque-forming cells, since the precursors of helper cells are Thy-1$^+$. These results are consistent with the absence of Ala-1 from unstimulated lymphocyte populations (4).

(4) Cytotoxic T lymphocytes (CTL) are Ala-1$^+$. CTL were obtained from the peritoneal cavity of B6 mice immunized with P815 tumor cells. After purification by passage through a nylon fiber column (10), the CTL were added to ^{51}Cr-labelled P815 cells and co-cultured for 2 hrs. The ratio of untreated CTL to target cells used (50:1) resulted in approximately 80% lysis. Mass cytolysis of the CTL population with anti Ala-1.2, or anti Thy-1.2, both of which lysed over 80% of the population, eliminated >95% of the CTL activity (4).

CTL generated *in vitro* in a mixed lymphocyte culture (MLC), were also shown to be Ala-1$^+$. B6 spleen cells were cultured with allogeneic DBA/2, C3H, or BALB/c mitomycin C-treated spleen cells at a 4:1 ratio. After 5 days of culture, the cells were treated with anti Ala-1 and C, and then titrated into a constant number of ^{51}Cr-labeled target cells (DBA/2, C3H, BALB/c, or B6 spleen cells cultured for 2 days with Con A) and were incubated for 4 hrs. In seven such experiments, anti Ala-1 pretreatment resulted in an average of 32-fold inhibition (range: 6-fold to complete inhibition) of lymphocytotoxicity as compared to samples treated with C alone (Fig.4).

(5) Cells proliferating in a MLC are Ala-1$^+$. It has been shown that two populations of cells are involved in the generation of CTL in the MLC: the Ly2$^+$ precursors of the CTL, and the Ly1$^+$ amplifier T cells. The amplifier cells cooperate with Ly2$^+$ precursor cells to augment the generation of CTL,

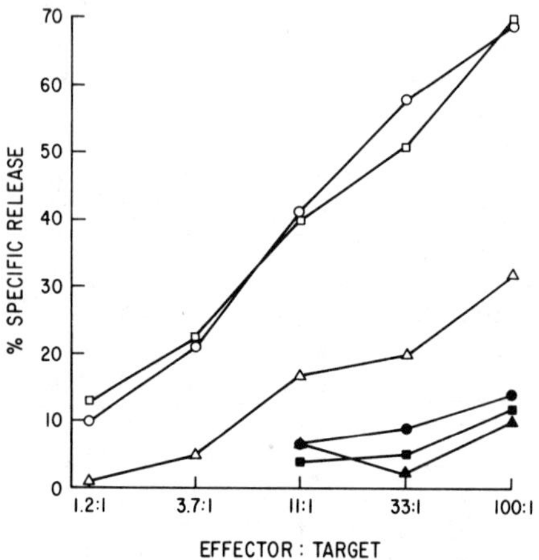

Fig.4. Effect of anti Ala-1 pretreatment on cytotoxic T cells generated *in vitro*. B6 cells were cultured with DBA/2 mitomycin C-treated cells for 5 days, and then aliquots were treated with anti Ala-1.2 and C (Δ), C alone (□), or were left untreated (O). The cells were assayed for lymphocytotoxicity on DBA/2 targets (open symbols) and on B6 targets (closed symbols).

and are responsible for a large proportion of the proliferation observed in a MLC (11,12). In order to determine if the cells proliferating in a MLC are Ala-1$^+$, B6 and (B6 x DBA/2)F_1 spleen cells were co-cultured at a 4:1 ratio in a MLC. Cultures of B6 spleen cells alone were also set up. After 4 days, aliquots of the MLC were treated with anti Ala-1.2 and C, C alone, or were left untreated. In one experiment, the cells were also treated with anti T cell serum [rabbit anti mouse brain (13)] and C. Usually, the cultures of B6 cells alone were left untreated, however, in two cases, an aliquot was treated with anti Ala-1 and C. After the mass cytolysis, the cells were cultured in fresh medium for an additional 18-24 hrs, during which time ^{125}IUdR was present. The cultures were then assayed for incorporation of ^{125}IUdR with a multiple automatic sample harvester. The results are shown in Table 1. Treatment of the cells with anti Ala-1 resulted in an average of 89% reduction of total ^{125}IUdR incorporation, as compared to the sample treated with C alone. Anti Thy-1 pretreatment eliminated 95% of the incorporation. The ^{125}IUdR incorporation seen in the cultures of B6 cells alone was also eliminated by anti Ala-1 and C. This incorporation seen in control B6 cultures may be due, at least partially, to antigenic or mitogenic stimulation by components of the fetal calf serum (6,14). Thus, reduction of this incorporation by treatment with anti Ala-1 and C is not unexpected.

TABLE 1

EFFECT OF ANTI ALA-1 PRETREATMENT ON MLC ACTIVITY

Cells cultured	Treatment	Experiment number 1		2		3	
		CPM	%*	CPM	%*	CPM	%*
B6 + BDF$_1$	Untreated	26,247		5,646		13,624	
"	C	24,010		5,835		11,247	
"	Anti Ala-1 + C	1,131	95	210	96	2,637	77
"	Anti T cell serum + C	1,319	95				
B6	Untreated	12,155		3,584		7,559	
"	Anti Ala-1 + C			144	96		

* % reduction.

B6 and (B6 x DBA/2)F_1 spleen cells were co-cultured for 4 days. Aliquots were then treated as described in the table, and were cultured for an additional 18-24 hours in the presence of ^{125}IUdR.

(6) Precursors of CTL are Ala-1$^-$. Aliquots of 10 x 10^6 normal B6 spleen cells were treated with anti Ala-1.2 and C, C alone, or were left untreated. The cells were then cultured with 4 x 10^6 DBA/2 or C3H mitomycin C-treated spleen cells for 5 days, and tested for cytotoxic activity as described above (Section 4). In four experiments, cytotoxic activity of

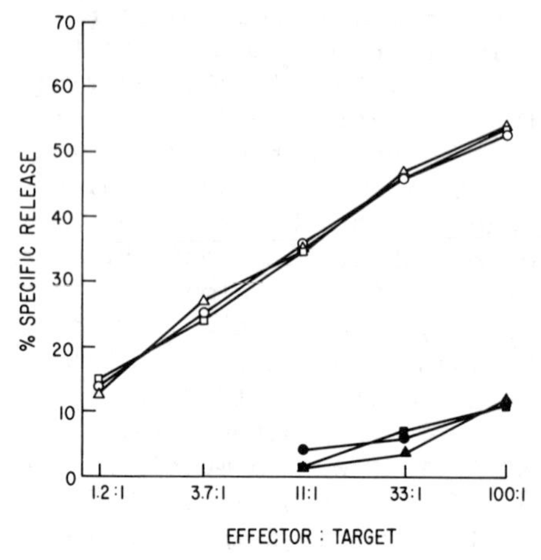

Fig.5. Effect of anti Ala-1 pretreatment on the precursors of cytotoxic T cells. B6 cells were treated with anti Ala-1.2 and C (Δ), C alone (□), or were untreated (O). The samples were then cultured with DBA/2 mitomycin C-treated cells for 5 days, and assayed for lymphocytotoxicity on DBA/2 targets (open symbols) and B6 targets (closed symbols).

cultures treated on day 0 with anti Ala-1 and C was equal to the activity of control cultures (Fig.5). These results show conclusively that Ala-1 is not expressed on the precursors of CTL.

Cantor and Boyse have shown that elimination of Lyl$^+$ cells prior to a MLC results in diminished cytotoxic activity compared to control MLC, although the Lyl$^+$ cells themselves are not the precursors of the CTL (11). The fact that anti Ala-1 pretreated cultures and control cultures generated equal cytotoxicity suggests that Ala-1 is not expressed on the precursors of the amplifier cells. In order to directly test whether Ala-1 is present on the precursors of cells proliferating in a MLC, B6 spleen cells will be treated with anti Ala-1 and C before co-culturing them with allogeneic cells in a MLC, and then determining whether there is a decrease in the incorporation of ^{125}IUdR.

DISCUSSION

T and B lymphocytes differentiate in their respective lineages to a stage where they are poised to express, upon antigenic or mitogenic stimulation, new genetic programs characteristic of effector cells. Although the program will necessarily be different for each specific functional subset of lymphocytes, one common feature of the effector cell stage of lymphocyte differentiation appears to be the expression of the Ala-1 antigen. We have found that Ala-1 is expressed on IgM and IgG PFC, helper T cells, CTL, and cells proliferating in a MLC, even though each of these subsets has a distinct immunological function in their activated state. In contrast, lymphocytes in a preactivated state are Ala-1$^-$, as was shown by the absence of Ala-1 from precursors of IgM PFC, helper T cells and CTL. Hence, the expression of Ala-1 can be used to distinguish effector T and B lymphocytes from their non-activated precursors.

ACKNOWLEDGEMENTS

I would like to express my sincere thanks to Dr. Ulrich Hämmerling, in whose laboratory this research was begun, for his help and guidance during the course of the work. I would also like to thank Peter Lichty for excellent technical assistance; Sarita Uhr, who performed some of the experiments on MLC-generated cytotoxic cells; and Dr. Richard Dutton, in whose laboratory this work is presently being done. This work was supported by Grants CA 16889, CA 17085, and CA 08748 from the National Cancer Institute; AI 08795 from the United States Public Health Service; and IM-1 from the American Cancer Society. The secretarial work of Kathy Wong is gratefully appreciated.

REFERENCES

1. Kisielow, P., Hirst, J.A., Shiku, H., Beverley, P.C.L., Hoffmann, M.K., Boyse, E.A. and Oettgen, H.F. (1975) *Nature* 253, 219.
2. Cantor, H. and Boyse, E.A. (1975) *J. Exp. Med.* 141, 1376.
3. Feeney, A.J. and Hämmerling, U. (1976) *Immunogenetics* 3, 369.
4. Feeney, A.J., and Hämmerling, U. (1977) *J. Immunol.* in press.
5. Jerne, N.K., and Nordin, A.A. (1963) *Science (Wash.D.C.)* 140, 405.
6. Mishell, R.I. and Dutton, R.W. (1967) *J. Exp. Med.* 126, 423.
7. Hoffmann, M. and Kappler, J.W. (1973) *J. Exp. Med.* *137, 721.*
8. Rittenberg, M.B. and Pratt, K.L. (1969) *Proc. Soc. Exp. Biol. Med.* 132, 575.
9. Kettman, J. and Dutton, R.W. (1970) *J. Immunol.* 140, 1558.
10. Julius, M.H., Simpson, E. and Herzenberg, L.A. (1973) *Eur. J. Immunol.* 3, 645.
11. Cantor, H. and Boyse, E.A. (1975) *J. Exp. Med.* 104, 1390.
12. Bach, F.H., Segall, M. Zier, K.S., Sondel, P.M., Alter, B.J. and Bach, M.L. (1973) *Science* 180, 403.
13. Linthicum, D.S., Sell, S., Wagner, M. and Trefts, P. (1974) *Nature* 252, 173.
14. Shiigi, S.M. and Mishell, R.I. (1975) *J. Immunol.* 115, 741.

DEMONSTRATION AND PARTIAL CHARACTERIZATION OF MURINE B AND T CELL SURFACE IMMUNOGLOBULINS USING AVIAN ANTIBODIES

J. J. Marchalonis,* G. W. Warr,* C. Bucana,* L. Hoyer,* A. Szenberg,† and N. L. Warner‡

*Basic Research Program, Frederick Cancer Research Center, Frederick, Maryland 21701; †Walter and Eliza Hall Institute of Medical Research, Parkville, Victoria 3052, Australia; ‡Section of Immunology, Department of Pathology, University of New Mexico, Albuquerque, New Mexico 87131

ABSTRACT. Chicken antibodies produced against the $(Fab)_2$ fragment of normal mouse IgG_2 and purified by binding to and elution from IgG-Sepharose 4B give strong indirect fluorescence with murine T cells and cultured T lymphoma cells. Erythrocytes and nonlymphoid tumor cells of macrophage and other myeloid types do not bind this reagent, even though they bear avid F_c receptors. Chicken antibodies, absorbed with myeloma-derived κ chains coupled to Sepharose, did not bind to either B or T cells. Surface Ig-like material of cells of the T lymphoma WEHI-22 was demonstrated by immunoelectronmicroscopy using ferritin-labelled rabbit antibody to chicken Ig as the developing reagent. Surface Ig of radioiodinated B cells isolated using chicken anti-mouse Fab consisted only of 7S IgM and IgD-like molecules. ^{125}I-labelled Ig isolated from the culture fluid of WEHI-22 cells contained light polypeptide chains and a heavy chain (nominal mass approximately 60,000 daltons), which migrated in the region between the γ and μ chain standards. The major intact WEHI-22 component had an apparent mass of 140,000-150,000 daltons and was distinct from both intact 7S IgM and the IgD-like molecule. These data show that T cells express a surface Ig containing determinants that at least cross-react with B cell-derived κ chains.

INTRODUCTION

Although immunoglobulin-like material has been isolated from the membranes of mammalian T cells (1-7), it has been generally difficult, if not impossible, to visualize this material on the cell surface (8). In contrast, virtually all lymphocytes of lower species (9-11), including putative T-type helper cells (12), express endogenously synthesized Igs readily detectable with rabbit antibodies raised against circulating Ig. Moreover, surface material bearing idiotypic determinants shared with circulating antibodies has been demonstrated on subpopulations of mammalian T cells (13). These observations, taken together, suggest that antibodies raised in phylogenetically divergent species might react with V-region determinants and/or other determinants shared by Igs of B and T cells and, therefore, allow direct visualization of Ig-like material on T cells. Consistent with this hypothesis, chicken antibodies to the $(Fab)_2$ fragment of murine serum Ig have been shown by autoradiographic and immunofluorescent methods to bind to both B and T cells (14), and the binding was established to be due largely to interaction with κ chains. Chicken antibodies cross-reactive with murine μ chains (15) or the Fab-fragment of human IgG (16) were previously shown to detect Ig on murine and human T cells, respectively.

In the present communication, we report the visualization in the electron microscope of Ig-like material on murine T cells using chicken antibodies and provide partial characterization of Ig of B cells and that of the monoclonal T lymphoma WEHI-22 isolated using avian antibodies.

METHODS AND MATERIALS

Cells. Suspensions of thymus and spleen lymphocytes of normal or nu/nu CBA mice were prepared as described elsewhere (17). Monoclonal, continuously cultured tumor cell lines were obtained from Dr. A. W. Harris.

Mouse IgG and $(Fab)_2$. Mouse IgG was prepared from serum as previously described (1,14). $(Fab)_2$ was prepared by digestion with pepsin (18); intact IgG and F_c fragments were removed by binding to protein A-Sepharose (19).

Immunization and preparation of antibody. White Leghorn Australorps hybrid chickens were immunized by i.m. injections of murine $(Fab)_2$ in complete Freund's adjuvant (14). Immune serum was absorbed on a column of mouse IgG coupled to Sepharose and specific antibodies were eluted with glycine-HCl buffer, pH 2.5. Control preparations consisted of chicken IgY of newly batched chickens, normal adult IgY passed through a mouse IgG-Sepharose column, and purified chicken antibodies to the $(Fab)_2$ fragments of human or rabbit immunoglobulins.

Immunofluorescence analysis. This was performed using a "sandwich" system with fluorescein or rhodamine-labelled rabbit antibodies to chicken Ig as the developing reagent (14).

Immunoelectronmicroscopy. Immunoelectronmicroscopy (IEM) was done at the transmission electron microscopy (TEM) and scanning electron microscopy (SEM) levels. The samples were processed as follows: 1 x 10^5 WEHI-22 cells were incubated with 10 μg chicken anti-mouse $(Fab)_2$ globulin for 30 minutes at 4°C, centrifuged through an albumin gradient, washed 2 times with serum-free Eagle's minimal essential medium (EMEM) (Grand Island Biochemicals Co., New York), and then incubated with 1:4 dilution of ferritin-conjugated rabbit anti-chicken gamma globulin (Cappel Laboratories, Inc., Cockranville, Pa.) for 30 minutes at 4°C. The cells were washed 3 times with serum-free EMEM and fixed with 2.5% glutaraldehyde in cacodylate buffer. The cells were allowed to settle on polylysine-coated coverslips for 45 minutes at 4°C, washed with buffer, and processed according to the O-T-O technique described by Kelley _et al._ (20). The samples were dehydrated with a graded series of ethanol and were processed for either TEM or SEM. Samples for TEM were flat embedded in Spurr's low viscosity media and polymerized in a 70°C oven. The coverslip was removed from the polymerized block by hydrofluoric acid. Thin sections were cut with a diamond knife in an LKB ultrotome, stained with Reynold's lead citrate and examined in a Hitachi HU-12A electron microscope. Samples for SEM were substituted with Freon 113 and critical point dried through Freon 13 and coated with platinum palladium alloy and examined in a Hitachi field emission scanning electron microscope operating at 25 kV.

Radioiodination and isolation of immunoglobulin. Spleen lymphocytes of CBA-nu/nu mice were surface radio-iodinated with (^{125}I)iodide in a lactoperoxidase-catalyzed reaction (21), and labelled membrane proteins were solubilized in 0.1% Triton X100 (22) in phosphate buffered saline, pH 7.3. Proteins in an aliquot of culture fluid in which WEHI-22.1 T lymphoma cells had been grown were labelled with (^{125}I)iodide by means of lactoperoxidase (23). Ig was isolated from these samples using a "sandwich" coprecipitation system in which cell lysate or labelled culture fluid was incubated either with chicken antibody to mouse Fab (specific system) or chicken IgY, which had been passed through mouse-IgG-Sepharose to remove antibodies to mouse Ig (control system). Precipitates were developed by adding rabbit antiserum to chicken Ig. Precipitates were washed, solubilized and subjected to polyacrylamide gel electrophoresis (PAGE) in the presence of sodium dodecyl sulfate (SDS) as described elsewhere (17).

TABLE 1

BINDING OF CHICKEN ANTI-MOUSE FAB TO MOUSE CELLS

Cell Type	Comments	IMMUNOFLUORESCENCE WITH CHICKEN ANTI-MOUSE FAB			
		Untreated	Absorbed With κ Chain Sepharose	Absorbed With (λα) Sepharose	Absorbed With Unconjugated Sepharose
Nu/nu Splenocytes	CBA	+++	ND	ND	ND
Thymocytes	CBA	++	-	++	++
WEHI-265, Myeloid Leukemia	F_c-Receptor	-	ND	ND	ND
P815, Mastocytoma	F_c-Receptor	-	ND	ND	ND
Erythrocytes		-	ND	ND	ND
WEHI-279, B Lymphoma	κ,μ	+++	-	+++	+++
2 PK-3, B Lymphoma	κ,γ	+++	-	+++	+++
EL 4, Thymoma	Ig - (Mammalian Reagents	-	ND	ND	ND
WEHI-7.1, T Lymphoma	Ig - (Mammalian)	++	-	++	++
S49.1, T Lymphoma	Ig - (Mammalian)	++	-	++	++

RESULTS

Table 1 presents a selection of the murine cell types that were tested for binding of chicken anti-mouse Fab by immunofluorescence. This reagent binds only to lymphocytes; nonlymphoid cells such as the mastocytoma P815 and the myeloid leukemia WEHI-265 were negative, even though both possess avid F_c-receptors. One lymphoid cell line studied, the thymoma EL 4, was also negative using chicken anti-mouse Fab. Binding was eliminated by absorption of the chicken antibody with κ-chains (from the myeloma HPC-114), bound to Sepharose 4B, but not by absorption with λα-Sepharose (MOPC 315) or unconjugated Sepharose. In other experiments (not shown), binding was eliminated by absorption with murine $(Fab)_2$ fragments coupled to Sepharose.

Figure 1 illustrates binding of chicken anti-mouse Fab to cells of the WEHI-22 T lymphoma line as visualized by IEM. Figures 1A and 1C represent control cells using SEM or TEM respectively. Figures 1B and 1D demonstrate localization of ferritin on WEHI-22 cells incubated with chicken anti-mouse Fab followed by ferritin-labelled rabbit antibodies to chicken Ig. The cells express C-type virus-like particles associated with membrane (indicated by arrow). The surface Ig of WEHI-22 is continuously distributed over the cell membrane and is not associated with the viral-like particles.

PAGE patterns of radioiodinated intact and reduced B cell surface Ig are shown in Figure 2. The patterns are identical to those obtained using rabbit antibodies to mouse Ig (24,25) in that 7S IgM and IgD-like molecules are present (upper pattern). Upon reduction, light chains and μ and δ-like heavy chains are resolved (lower figure). Molecules migrating in the position of H-2 antigens (R_F 0.5-0.6) do not occur at a detectable level.

PAGE patterns of radioiodinated intact and reduced WEHI-22 Ig isolated from culture fluid are depicted in Figure 3. The intact molecule migrates on 5% gel slightly ahead of the position of the IgG standard (M.W. 150,000 daltons) and is clearly distinct in mobility from 7S IgM (see Figure 2). Specific counts are also present at the end of the gel. Upon reduction, heavy chains with a mobility between those of μ and γ, and light chains migrating at the slow end of the standard light chain envelope are resolved. Both B and T lymphoma patterns contain small amounts of aggregated material (R_F approximately 0.1). The T lymphoma pattern shows, in addition, trace amounts of material migrating in the heavy chain region;

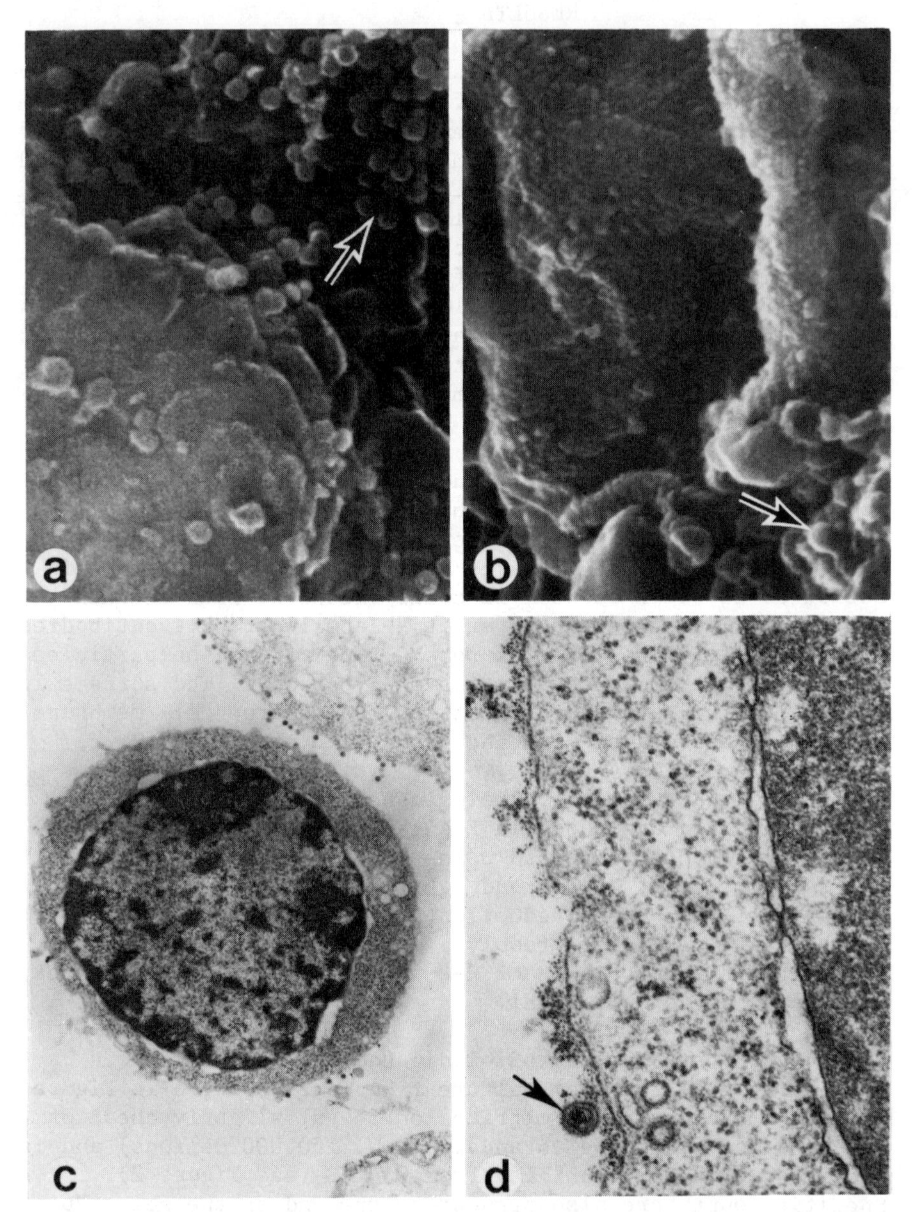

Fig. 1 - Immunoelectronmicroscopy of WEHI-22 T lymphoma cells incubated with Ig of normal neonatal chicken serum (a and c) or chicken anti-mouse Fab (b and c) followed by incubation with ferritin-labelled rabbit antibody to chicken Ig. a. SEM (control) x 30,000; b. SEM (experimental) x 30,000; c. TEM (control) x 8,000; d. TEM (experimental) x 50,000. Arrows indicate virus-like particles.

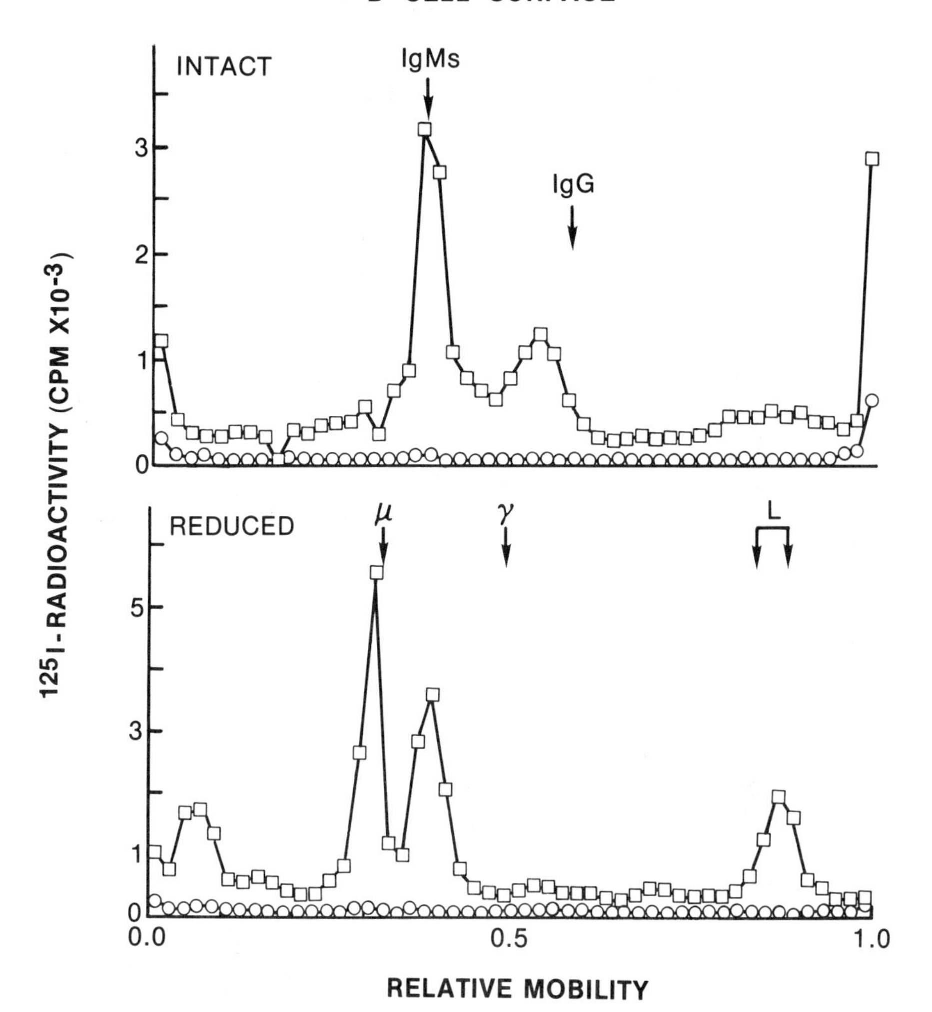

Fig. 2 - PAGE analysis of ^{125}I-labelled surface Ig of nu/nu spleen cells. Upper figure, unreduced Ig resolved on 5% gel. IgM indicates position of 7S IgM; IgG indicates position of mouse IgG (M.W. 150,000). Lower pattern, reduced Ig resolved on 10% gel. μ, γ, and L indicate positions of standard mu, gamma and light chains, respectively. (□), specifically precipitated Ig (chicken anti-mouse Fab + rabbit anti-chicken Ig). (○), counts associated with control precipitate (normal chicken Ig absorbed with mouse IgG-Sepharose + rabbit anti-chicken Ig).

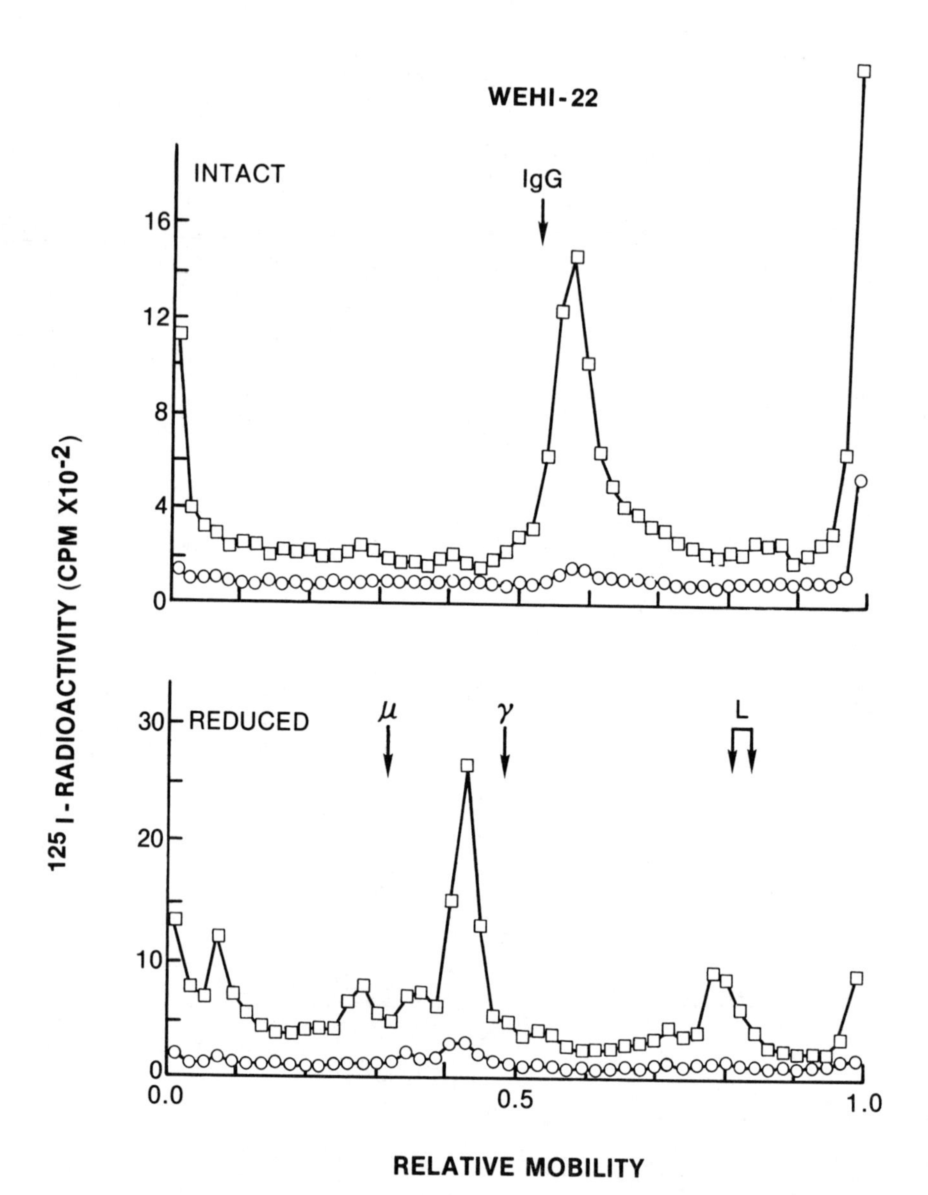

Fig. 3 - PAGE analysis of ^{125}I-labelled Ig released into culture fluid by WEHI-22 T lymphoma cells. Upper pattern, intact sample resolved on 5% gel. Lower pattern, reduced sample resolved on 10% gel. (□), specifically precipitated material; (○), control precipitate. Standards and systems have same meanings as in Figure 2.

one component resembled μ chain, the other was slightly faster. No detectable components were seen in the H-2 region. In another experiment to determine the specificity of the chicken anti-mouse Fab, the antibody was covalently bound to Sepharose 4B by the CNBr-method for use as a solid-phase immunoabsorbent. Isolation of ^{125}I-labelled WEHI-22 Ig by this immunoabsorbent was blocked by incubation with purified mouse IgG but not by incubation with bovine serum albumin or fetal bovine serum (data not shown).

DISCUSSION

These data show that chicken antibodies raised against the $(Fab)_2$ fragment of murine IgG and bearing specificity for κ-chains enable the immunofluorescent and immunoelectron-microscopic demonstration of surface Ig on thymus cells and monoclonal T lymphoma cells. Although it is possible that these antisera might react with cell surface antigens such as histocompatibility antigens, this is unlikely because non-lymphoid H-2 bearing cells do not react with the antibodies. Interestingly, virus-like particles budding from the surface of the T lymphoma WEHI-22 were uniformly negative for Ig-like material.

Surface molecules isolated from B cells using chicken anti-mouse Fab consist only of Igs and these are the 7S IgM and IgD-like Igs routinely precipitated by mammalian antisera. Ig-like material released into culture supernatant by WEHI-22 cells differed from B cell surface IgM and IgD in PAGE mobility of the intact molecule as well as its heavy chains. The data obtained using chicken anti-mouse Fab are consistent with recent results, which indicate that T lymphoma Ig exists as a dimer of heavy chains of apparent mass 60,000-70,000 daltons covalently linked via disulfide bonds that exists in non-covalent association with κ-like light chains (26). The possibility has not been excluded that the relatively rapid mobility of the heavy chain results from limited proteolysis by enzymes present in the culture fluid. Similar data have been reported for Ig-like material isolated from normal T cells (2,22). We suggest that T cell Ig represents a new isotype of murine Igs, which occurs predominantly in association with the T cell surface. Combining our data with those of studies using anti-idiotypic reagents (27) and certain antisera reactive with κ-chains and Fd regions (26), we would predict that this T cell molecule shares V regions with serum and B cell surface Igs, but differs from these in the constant regions of its heavy chains and possibly its light chains. Chicken antibodies might react with determinants on κ-chains

(V region framework markers?) distinct from antigenic determinants detected by rabbit or sheep antisera to κ chains. We found that the latter reagents did not bind to thymus cells or to any of the T lymphoma cells tested. Detailed amino acid sequence analysis is required to ascertain conclusively the relationship between T cell Ig-like proteins and serum Igs.

ACKNOWLEDGEMENTS

We thank Ms. Pat Smith and Gabriel Marton for expert technical assistance. Work performed at the Walter and Eliza Hall Institute was supported in part by grants CA-20085 to JJM and AM-11234 to N.L.W. from the U.S.P.H.S. and AHA75-877 to JJM from the American Heart Association. Work at the Frederick Cancer Research Center was supported by the National Cancer Institute under Contract No. NO1-CO-25423 with Litton Bionetics, Inc.

REFERENCES

1. Marchalonis, J.J., Cone, R.E. and Atwell, J.L. (1972) *J. Exp. Med.* 135, 956.
2. Moroz, C. and Lahat, N. (1974) *Cell. Immunol.* 31, 397.
3. Boylston, A.W. and Mowbray, J.F. (1974) *Immunology* 27, 855.
4. Rieber, E.P. and Riethmuller, G. (1974) *Z. Immun. Forsch* 147, 276.
5. Haustein, D. and Goding, J.W. (1975) *Biochem. Biophys. Res. Commun.* 65, 483.
6. Hammerling, U., Pickel, H.G., Mack, C. and Masters, D. (1976) *Immunochemistry* 13, 533.
7. Smith, W.I., Ladoulis, C.T., Misra, D.N., Gill, T.J., III and Bazin, H. (1975) *Biochim. Biophys. Acta* 382, 506.
8. Warner, N.L. (1974) *Adv. Immunol.* 19, 67.
9. DuPasquier, L., Weiss, N. and Loor, F. (1972) *Eur. J. Immunol.* 2, 366.
10. Emmrich, F., Richter, R.F. and Ambrosius, H. (1975) *Eur. J. Immunol.* 5, 76.
11. Warr, G.W., DeLuca, D. and Marchalonis, J.J. (1976) *Proc. Nat. Acad. Sci., U.S.A.*, 73, 2476.
12. Ruben, L.N., Warr, G.W., Decker, J.M. and Marchalonis, J.J. (1977) *Cell. Immunol.*, in press.
13. Binz, H., Bachi, T., Wigzell, H., Ramseier, H. and Lindenmann, J. (1975) *Proc. Nat. Acad. Sci U.S.A.*, 72, 3210.
14. Szenberg, A., Marchalonis, J.J. and Warner, N.L. (1977) *Proc. Nat. Acad. Sci. U.S.A.*, in press.
15. Hammerling, U., Mack, C., and Pickel, H.G. (1976) *Immunochemistry* 13, 525.

16. Jones, V.E., Graves, H.F. and Orlans, E. (1976) *Immunology* 30, 281.
17. Cone, R.E. and Marchalonis, J.J. (1974) *Biochem. J.* 140, 345.
18. Edelman, G.M. and Marchalonis, J.J. (1967) In *Methods in Immunology and Immunochemistry* (C.A. Williams and M.W. Chase, eds.) Academic Press, New York. Vol. 1, 405.
19. Goding, J.W. (1976) *J. Immunol. Meth.* 13, 215.
20. Kelley, R.O. Dekker, R.A.F. and Bluemink, J.G. (1973) *J. Ultrastruct. Res.* 45, 254.
21. Marchalonis, J.J., Cone, R.E. and Santer, V. (1971) *Biochem. J.* 124, 921.
22. Cone, R.E. and Brown, W.C. (1976) *Immunochemistry* 13, 571.
23. Marchalonis, J.J. (1969) *Biochem. J.* 113, 299.
24. Vitetta, E.S. and Uhr, J.W. (1975) *Science* 189, 964.
25. Marchalonis, J.J. (1976) *Contemp. Top. Mol. Immunol.* 5, 125.
26. Moseley, J.M., Marchalonis, J.J., Harris, A.W. and Pye, J. (1977) *J. Immunogenetics*, in press.
27. Binz, H. and Wigzell, H. (1976) *Scand. J. Immunol.* 5, 559.

PARTIAL CHARACTERIZATION OF MEMBRANE IMMUNOGLOBULINS ON RAINBOW TROUT LYMPHOCYTES

Karen Yamaga, Howard M. Etlinger and Ralph T. Kubo

National Jewish Hospital and Research Center
Denver, Colorado 80206

ABSTRACT. The reported presence of membrane immunoglobulins (Ig) on the thymic lymphocytes of lower vertebrates and the current controversy concerning the nature of the T cell antigen receptor led us to the characterization of membrane components which were reactive with rabbit anti-Ig reagents on thymic and splenic lymphocytes from rainbow trout. By the indirect immunofluorescent staining technique, greater than 90% of both splenocytes and thymocytes bound anti-Ig sera with an apparent equivalent intensity. Quantitative estimates of the amounts of Ig present on these cells using a radioimmunoinhibition assay, however, revealed that thymocytes possessed at least 10-fold lower amounts of Ig as compared with splenocytes. Characterization of the membrane Ig determinants by the lactoperoxidase iodination method and by SDS-polyacrylamide gel analysis showed the presence of 3 major components on splenocytes with approximate molecular weights of 100,000, 65,000 and 25,000 daltons. In contrast, thymocytes contained several high molecular weight components (70 to 150,000 daltons) and a minor 40,000 dalton peak but lacked radioactive peaks in the 25,000 dalton region. The lack of any radioactivity in the light chain region suggested that the reactivity between anti-trout Ig and thymocytes was not due to Ig but to other antibodies present in the serum. This suggestion was further supported by the finding of an antigenic cross-reactivity between trout serum Ig and keyhole limpet hemocyanin (KLH). This cross-reactivity appeared to be due primarily to L-fucose specificities. Membrane Ig from trout splenocytes did not cross-react with KLH whereas the 100,000 dalton component from splenocytes did. Furthermore, both anti-trout Ig and anti-KLH appeared to react with similar membrane components on trout thymocytes. Our data suggest that although thymus cells react with anti-trout Ig reagents, the molecular structures of thymic Ig reactive determinants are distinct from the membrane Ig found on splenic lymphocytes and are most probably carbohydrate in nature.

INTRODUCTION

In a number of mammalian species, B lymphocytes, in contrast to T lymphocytes, possess readily detectable immunoglobulins on the membrane which have been demonstrated by a variety of techniques (1). Several laboratories (2-5) have reported that in lower vertebrates both thymocytes and splenocytes are reactive with anti-Ig reagents by immunofluorescent staining techniques. Furthermore, there have been recent reports on the characterization of Ig on fish thymocytes (6-8). We have undertaken a study of the reactivity of an antiserum to trout Ig with trout spleen and thymus cells. We present here data which suggest that the reactivity of anti-trout Ig with trout thymocytes appears to be due to cross-reactive specificities, probably carbohydrate in nature, rather than to Ig specific antigenic determinants.

MATERIALS AND METHODS

The membrane proteins of trout splenocytes and thymocytes were radioiodinated by the lactoperoxidase procedure (9), reacted with rabbit antisera and the immune complexes adsorbed onto Protein A-bearing Staphylococci (10). The complexes were solubilized with sodium dodecyl sulfate (SDS) and analyzed on SDS-polyacrylamide gels in reduced or unreduced form (11). Rabbit anti-trout Ig sera were prepared by multiple injections of purified trout Ig in incomplete Freund's adjuvant. Trout Ig was isolated from whole trout serum either by gel filtration and starch block fractionation or by specific purification of trout anti-dinitrophenyl antibodies on an immunoadsorbent. Rabbit anti-Ig sera gave a single precipitin line against whole trout serum by immunoelectrophoresis.

For antigen binding determinations, a radioimmunoassay was employed. To determine the inhibitory effects of proteins, monosaccharides or viable cells, 200 µl of the inhibitory material and 200 µl of the antiserum dilution that bound 50% of the labeled antigen were mixed overnight at 4°C (12). When cells were tested, both the antiserum and the cells were diluted in L-15 medium (Microbiological Associates) containing 10% fetal calf serum, 100 µg/ml cyclohexamide to inhibit protein synthesis and 0.03 M sodium azide. After overnight incubation, the cells were centrifuged and checked for viability which was greater than 90% for thymocytes and greater than 85% for splenocytes. The supernatant fluid was then tested for binding to the labeled antigen and the percent inhibition calculated.

RESULTS AND DISCUSSION

Indirect immunofluorescent staining of trout splenocytes and thymocytes by rabbit antisera raised against trout Ig revealed that at least 90% of the cells were reactive. By this technique, there appeared to be no difference between the splenocytes and thymocytes in staining intensity. However, quantitative determinations of the amount of Ig present on the membranes of splenocytes and thymocytes using a radioimmunoinhibition assay indicated that whereas splenocytes had 7 to 25 ng of Ig per 10^6 cells, thymocytes consistently showed at least 10-fold lower amounts of Ig. In order to investigate why the immunofluorescent and radioimmune assays appeared to give discrepant results, further characterization of the anti-trout Ig reactivity on splenocytes and thymocytes was performed.

By the lactoperoxidase surface labeling technique, splenocytes were shown to possess 3 major components that were reactive with anti-trout Ig sera with molecular weights of approximately 100,000, 65,000 and 25,000 daltons as determined by SDS-polyacrylamide gel electrophoresis analysis. In contrast, thymocytes contained multiple high molecular weight components (about 70,000 to 150,000 daltons) and a minor 40,000 dalton component but lacked a radioactive peak in the 25,000 dalton region. Thus it appeared that the typical polypeptide chain structures readily detected on splenocytes were not demonstrable on thymocytes.

In our previous studies with the lactoperoxidase technique on murine and human lymphocytes, anti-keyhole limpet hemocyanin (KLH) antiserum was often employed as one of the controls for nonspecific binding. To our surprise, when this antiserum was used with trout lymphocytes, extensive reactivity with the trout cells was observed. Subsequently it was established by radioimmunoassay that anti-KLH showed antigenic cross-reactivity with serum trout Ig and that conversely, antisera raised against trout Ig cross-reacted with KLH. Other antisera raised against murine, human, guinea pig and chicken immunoglobulins did not show significant binding to KLH. To establish that anti-KLH reacted with trout Ig and not to a possible contaminant, an immune precipitate consisting of anti-KLH and labeled trout Ig was reduced and electrophoresed on SDS-polyacrylamide gel electrophoresis and yielded typical H and L chains of serum Ig.

Fairless *et al* (13) reported the presence of relatively high amounts of methyl pentoses in KLH preparations. In order to determine if carbohydrates might be the basis of the cross-reactivity between serum trout Ig and KLH, the binding of either anti-KLH to trout Ig or anti-trout Ig to KLH was inhibited with a variety of pentoses and hexoses. In both cross-reacting systems, L-fucose completely inhibited the cross-reactivity whereas D-fucose and L-rhamnose showed only partial inhibition at much higher concentrations. For example, 50% inhibition of the binding of KLH to anti-trout Ig was obtained with 7×10^{-9} moles of L-fucose whereas 1.5×10^{-4} moles of D-fucose was necessary to obtain comparable inhibition. All other sugars tested at 2×10^{-4} moles were not inhibitable. Comparison of the degree of inhibition and the slopes of the inhibition curves obtained with fucose, serum trout Ig and KLH indicated that L-fucose was the predominant, if not sole, specificity responsible for the cross-reactivity.

In order to estimate the amount of KLH-trout Ig cross-reactive specificities present on splenocytes and thymocytes, the cross-reacting radioimmunoassay was inhibited with trout cells. As shown in Fig.1, 8.5×10^3 thymocytes inhibited 50% of the reactivity of KLH to anti-trout Ig whereas 4×10^4 splenocytes were required to effect the same inhibition (Fig. 1B). Similarly, 5-fold less thymocytes than splenocytes were needed to inhibit the binding of trout Ig to anti-KLH (Fig. 1C). Thus, thymocytes appeared to contain more cross-reacting specificities than splenocytes. These data contrast with the homologous system mentioned above in which splenocytes possessed more Ig molecules than thymocytes (Fig.1A). Furthermore, trout erythrocytes contained low amounts of cross-reacting specificities. Human tonsil cells, murine splenocytes and chicken erythrocytes did not show any inhibition with 2×10^7 cells.

The molecular nature of the membrane determinants recognized by either anti-trout Ig or anti-KLH reagents were determined by the lactoperoxidase technique. When anti-trout Ig or anti-KLH immunoprecipitates were analyzed in unreduced form by SDS-polyacrylamide gel electrophoresis (5% gels),only a single radioactive peak with an approximate molecular weight of 200,000 daltons was observed (Peaks 1,Fig.2A and 2B). The anti-trout Ig reagent was more reactive than anti-KLH for splenocyte membrane determinants. Materials were eluted from each gel, reduced and re-electrophoresed on 7.5% SDS-polyacrylamide gels. As shown in Fig. 2C, the radioactivity of the anti-trout 200,000 dalton material was resolved into 3 peaks with approximate molecular weights corresponding to 100,000, 65,000 and 25,000 daltons, respectively. In contrast, the anti-KLH 200,000 dalton material yielded under the same analysis, a pattern showing a single polypeptide chain of about 100,000 daltons (Fig.2D). Thus, anti-KLH does not appear to react with trout membrane Ig although it does react with serum Ig. A likely explanation for the inability of anti-KLH to react with membrane trout Ig would be the lack of

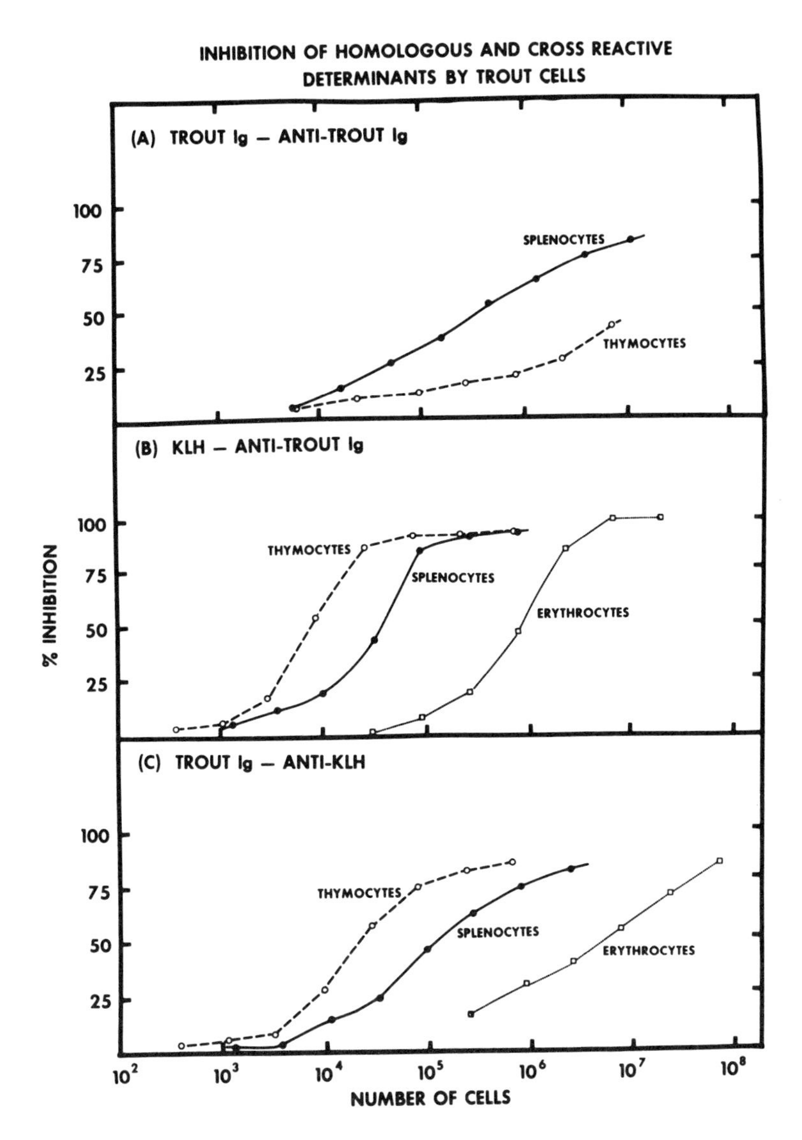

Fig. 1. Inhibition of homologous (A) and cross-reactive (B,C) determinants by trout splenocytes (closed circles), thymocytes (open circles) and erythrocytes (open squares).

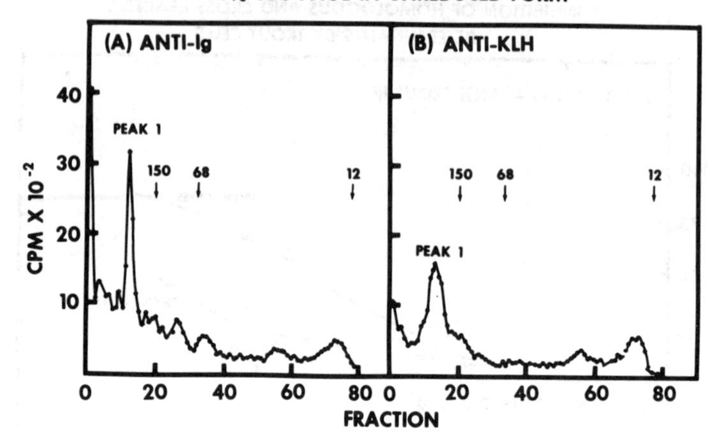

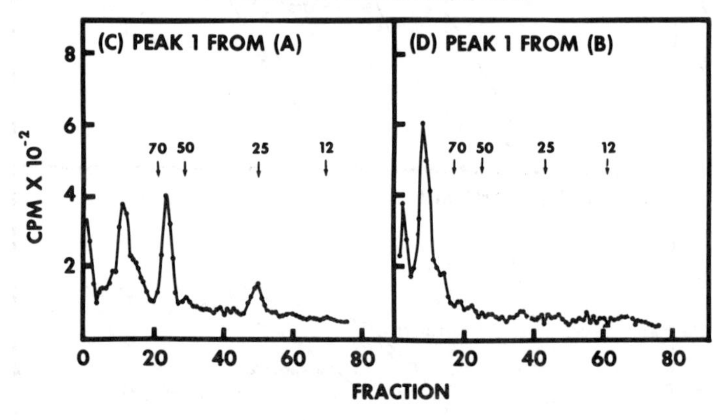

Fig. 2. Iodinated membrane proteins from trout splenocytes were precipitated with rabbit anti-trout Ig (A) or rabbit anti-keyhole limpet hemocyanin (B). Arrows indicate the leading edges of dansylated human IgG (150,000 daltons), BSA (68,000 daltons) and cytochrome C (12,000 daltons). Radioactive materials from Peak 1 of (A) and Peak 1 of (B) were eluted, reduced and re-electrophoresed on 7.5% SDS-PAGE. Peak 1 from anti-Ig gels is shown in (C) and Peak 1 from anti-KLH gels is given in (D). Arrows indicate the leading edges of dansylated μ (70,000 daltons), γ (50,000 daltons) and L chain (25,000 daltons) and of cytochrome C (12,000 daltons).

L-fucose on membrane Ig. In support of this possibility, it has been shown that murine membrane Ig from B cells does not contain fucose whereas secreted Ig does (14). That anti-KLH and anti-Ig react against two distinct membrane components was confirmed by first removing the complexes with anti-KLH and subsequently analyzing the determinants that bound to anti-Ig. Radioactivity was found only in the 65,000 and 25,000 dalton regions. Thus, it seems coincidental that in the unreduced form both cross-reacting and Ig specific components electrophorese as 200,000 dalton molecules.

The analysis of the reactivity of the anti-trout Ig and anti-KLH reagents on trout thymocytes by the lactoperoxidase method revealed that immune precipitates of both reagents did not show significant differences by SDS-polyacrylamide gel electrophoresis analysis in reduced or unreduced forms. As mentioned above, no radioactivity in the 25,000 dalton region was seen and multiple high molecular weight components were evident. Thus, it would appear that the basis for the reactivity of anti-trout Ig with trout thymocytes was due to specificities which were also reactive with anti-KLH antiserum. It is suggested that the reactivity of both anti-trout Ig and anti-KLH on trout thymocytes was due to carbohydrate specificities including fucose. It is not yet clear, however, if L-fucose is the sole carbohydrate specificity responsible for trout lymphocyte membrane cross-reactivities. Studies are currently in progress to further characterize these specificities.

ACKNOWLEDGEMENTS

We thank Howard M. Grey for his suggestions throughout the course of this study and for his critical review of the manuscript. We also gratefully acknowledge the secretarial assistance of Charlene Griffiths. These studies were supported by NIH Research Fellowship AI-05334, NIH Grant AI-12136 and American Heart Grant 74856.

REFERENCES

1. Warner, N.L. (1974) Adv. Immunol. 19,67.
2. Du Pasquier, L., Weiss, N. and Loor, F. (1972) Eur. J. Immunol. 2,366.
3. Emmvich, F., Richter, R.F. and Ambrosius, H. (1975) Eur. J. Immunol. 5,76.
4. Ellis, A.E. and Parkhouse, R.M.E. (1975) Eur. J. Immunol. 5,726.
5. Charlemagne, J. and Tournefier, A. (1975) Adv. Exp. Biol. Med. 64,251.

6. Warr, G.W., DeLuca, D. and Marchalonis, J.J. (1976) Proc. Natl. Acad. Sci. USA. 73,2476.
7. Fiebig, H. and Ambrosius, H. (1976) In "Phylogeny of Thymus and Bone Marrow-Bursa Cells", (Wright, R.K. and Cooper, E.L., eds.), p 195, Elsevier North Holland Bio Medical Press, Amsterdam, The Netherlands.
8. Cuchens, M., McLean, E. and Clem, L.W. (1976) In "Phylogeny of Thymus and Bone Marrow-Bursa Cells", (Wright, R.K. and Cooper, E.L., eds.), p 205, Elsevier North Holland Biomedical Press, Amsterdam, The Netherlands.
9. Grey, H.M., Kubo, R.T. and Cerottini, J.-C. (1972) J. Exp. Med. 136,1323.
10. Kessler, S.W. (1975) J. Immunol. 115,1617.
11. Maizel, J.V.,Jr. (1971) In "Methods in Virology", (Maramorsch, K. and Kaprowski, H., eds.), p 179, Academic Press, New York.
12. Rabellino, E., Colon, S., Grey, H.M. and Unanue, E.R. (1971) J. Exp. Med. 133,156.
13. Fairless, B., Hornish, D. and Bartel, A.H. (1967) Immunochemistry 4,116.
14. Melchers, F. and Andersson, J. (1973) Transpl.Rev. 14,76.

STRUCTURE AND FUNCTION OF MOUSE CELL SURFACE IMMUNOGLOBULIN

R.M.E. Parkhouse, Erika R. Abney[1] and A. Bourgois[2]

National Institute for Medical Research
Mill Hill, London NW7 1AA, England

ABSTRACT. Mouse surface Ig was digested by trypsin and the products analyzed by immunochemical techniques. The products of tryptic hydrolysis were similar to those previously observed when human IgD and IgM were digested. The fact that the Fab fragment of both human and mouse IgD is similarly hydrolysed suggests a structural homology.

A major emphasis is placed on the finding that surface IgD is extremely susceptible to proteolysis whereas surface IgM is very resistant. We have assumed that these different characteristics relate to respective biological roles and consequently suggested a hypothetical model. The crux of the model is that IgD, in whole or in part, is lost from the B-cell as consequence of immunogenic presentation. In contrast, the protease resistant IgM, and later other isotypes, remains on the cell surface. The released IgD functions to stimulate a specific regulatory anti-idiotype response which acts by recognition of protease resistant cell surface Ig, still remaining on the relevant cell.

INTRODUCTION

A striking difference between the class distribution of Ig in serum and on the surface of B-lymphocytes has recently been revealed. Unlike the situation in serum, IgG and IgA are poorly represented on the surface of cells. Instead, we find monomeric IgM and an Ig which has been described as the murine counterpart of IgD (1,2,3,4). These findings have led to general speculation that the precursors of IgG and IgA-producing cells may have IgD receptors, a question that is referred to in more detail elsewhere in this volume (Abney et al; Vitetta et al). Whatever the answer to this

[1] Present address: Unidad de Biología Experimental, Facultad de Medicina, Universidad Nacional Autónoma de México, C.U. Apartado Postal 70343, México 20, D.F.

[2] Present address: Centre d'Immunologie de Marseille-Luminy, 13288 Marseille Cédex 2, France.

question, however, any suggested role for IgD must take into account the fact that it is frequently, perhaps always, found in association with another Ig isotype on the surface of B-lymphocytes. On an individual B-cell, however, there is only one heavy chain variable region expressed, and this is therefore shared by the two (and even three) constant regions synthesised. This consideration, together with our studies on the differential susceptibility of surface IgM and IgD to trypsin proteolysis (reported below) have prompted the formulation of a suggested role for IgD (5), namely the elicitation of a regulatory anti-idiotype response. Thus, IgD is designed to function primarily off the B cell, whereas IgM (and possibly IgG and IgA), which remains on the cell, is the specific target of the anti-idiotype response. Whether IgM and, or, IgD act as specific and direct signalling devices for immunogenic stimulation of the B-cell is not considered.

MATERIALS AND METHODS

Cell suspensions were prepared and labeled externally with ^{125}I. Radioactive cell-surface Ig was precipitated from Nonidet P-40 lysates with antibody and characterized by SDS gel electrophoresis using an internal marker of ^{131}I-labeled proteins.

For the digestion experiments, trypsin was added (final conc., 40 μg/ml) to lysates of radiolabeled cells for various times and temperatures. At the end of the incubation period, the sample was adjusted to contain a fivefold excess of soybean trypsin inhibitor, 1 mM phenylmethylsulfonyl fluoride, and 10 mM iodoacetamide. The proteolytic cleavage products of surface immunoglobulin were isolated using goat anti-rabbit IgG and sequential additions of rabbit antibodies to mouse K chain, μ chain and δ chain. Precipitates were then submitted to SDS gel analysis with or without a prior reduction step. In some experiments, and in order to reduce the complexity of gel patterns, radioactive cells lysates were passed through Sepharose-linked anti-μ or anti-δ prior to treatment with trypsin.

These procedures have already been described in detail elsewhere (1,4,5).

RESULTS AND DISCUSSION

The initial stimulus to study the effect of trypsin on surface Ig was the work of Jefferis (6). He showed that although human IgD is hydrolysed very rapidly to Fc and Fab, further incubation with trypsin results in degradation of Fab rather than Fc. An unexpected fragmentation product of the

Fab results, that is a V_H and C_L complex, disulphide linked. The object of the work was therefore to look for a similar degradation pattern of the mouse surface IgD-like molecule.

As has been recently demonstrated by others (7) we found the mouse IgD equivalent to be markedly susceptible to trypsin (5). An 80% conversion to Fab and Fc occurred in 10 min at 0°C. However, when digestion was continued for several hours at room temperature a fragment of 15,000 M.Wt. was released from Fab. This was assumed to be an extended V_H, as reported for human IgD (6). The similarity in susceptibility to trypsin of human IgD and its presumed mouse counterpart is therefore evidence for a structural homology.

The surface IgM was, on the other hand, resistant to the hydrolytic action of trypsin at room temperature. Even 5 hr incubation failed to result in appreciable digestion and several hours (3-5 hr) at 37°C were necessary to give complete fragmentation. Under these conditions the surface IgM was degraded to F(ab')2μ, Fabμ and Fcμ (M.Wts. of 125,000, 50,000 and 50,000 respectively).

Our results therefore emphasize the difference in susceptibility of surface IgM and IgD to proteolysis. We will assume that this is a clue to the biological roles of these molecules. Like Vitetta and Uhr (8), we could postulate proteolysis of IgD on the lymphocyte surface after interaction with antigen.

Vitetta and Uhr, however, proposed that the proteolysis of IgD is a prerequisite for the exposure on the Fcδ fragment of a site necessary for cellular triggering. An alternative hypothesis is that the Fabδ-Ag complex liberated plays the main role and serves to elicit a specific regulatory response, either by T- or B-cells, directed at the idiotype of the released Ig fragment. The cell surface IgM, which is resistant to proteolysis, is presumed to remain on the cell and be the target for the anti-idiotype response. Circumstantial evidence to support the proposal is the fact that B-cells stimulated by LPS lose surface IgD faster than the other isotypes (9,10). Thus, IgD is lost from B-cells soon after stimulation. Indeed little IgD is found on LPS-stimulated cells after one day of culture, although this is not a feature of unstimulated B-cells in culture.

Furthermore, it now appears that mouse IgD is even more susceptible to proteolysis than previously suspected. A careful study has shown that mouse δ chain comigrates with mouse μ-chain on SDS-polyacrylamide gel electrophoresis (11). Thus, the material that has previously been called mouse δ chain, and which migrates faster than mouse μ-chain is a degradation product, lacking a fragment at the C-terminal end of the molecule. It is, therefore, conceivable that scission at this extremely susceptible point is the biologically significant event. If so, then the fragment of IgD released upon immunogenic challenge would be divalent rather than monovalent.

ACKNOWLEDGEMENTS

E.R. Abney is a recipient of a fellowship award from CONACyT, Mexico and A. Bourgois is a Fellow of the European Molecular Biology Organization. We thank I.R. Hunter and Gillian Clark for excellent assistance in the work. R.M.E. Parkhouse thanks the Director of the Centro de Investigación y de Estudios Avanzados del Instituto Politécnico Nacional for hospitable facilities during the preparation of this article, in particular the secretarial assistance of Ana Alba Gómez.

REFERENCES

1. Abney, E.R. and Parkhouse, R.M.E. (1974) Nature, 252, 600.
2. Melcher, U., Vitetta, E.S., McWilliams, M., Lamm, M.E. Philips-Quagliata, J.M., and Uhr, J.W. (1974). J. Exp. Med. 140, 1427.
3. Vitetta, E.S., Melcher, U., McWilliams, M., Lamm, M.E., Philips-Quagliata, J.M. and Uhr, J.W. (1975). J. Exp. Med. 141, 206.
4. Abney, E.R., Hunter, I.R. and Parkhouse, R.M.E. (1976) Nature 259, 404.
5. Bourgois, A., Abney, E.R. and Parkhouse, R.M.E. (1977) Europ. J. Immunol. In the press.
6. Jefferis, R. (1975) Immunol. Comm. 4, 477.
7. Vitetta, E.S. and Uhr, J.W. (1976) J. Immunol. 117, 1579.
8. Vitetta, E.S. and Uhr, J.W. (1975) Science 189, 964.
9. Cooper, M.D., Kearney, J.F., Lawton, A.R., Abney, E.R., Parkhouse, R.M.E., Preud'homme, J.L., and Seligman, M. (1976) Ann. Immunol. (Inst. Pasteur) 127C, 573.
10. Bourgois, A. and Askonas, B.A. (1977) Eur. J. Immunol. In the press.
11. Sitia, R., Corte, S., Ferrarini, M. and Bargellesi, A. (1977). Manuscript submitted for publication.

A MODEL FOR THE DEVELOPMENT OF IMMUNOGLOBULIN ISOTYPE DIVERSITY

Erika R. Abney, M.D. Cooper, J. F. Kearney, A.R. Lawton and R.M.E. Parkhouse*

Unidad de Biología Experimental, Facultad de Medicina, UNAM. Apdo. Postal 70343, México 20, D.F.; Departments of Pediatrics and Microbiology, University of Alabama in Birmingham, Ala., U.S.A.; Departamento de Bioquímica, Centro de Investigación y de Estudios Avanzados del I.P.N., Apdo. Postal 14-740, México 14, D.F.

Using differential fluorescence and specific antisera to heavy chain isotypes we have made a systematic study of the distribution of immunoglobulin classes on lymphocytes from various tissues of mice taken at various ages. In new born mice IgD is absent and the majority of cells bearing sIg express only IgM. At about five days of age sIgD appears on a small number of sIgM positive lymphocytes and thereafter the number of sIgD cells increases, reaching adult values at 4-6 weeks. B lymphocytes with surface IgD are absent in mice treated from birth with anti-mouse μ chain. Administration of anti-δ, however has failed to eliminate production of IgG or IgA.

With the first appearance of sIgD positive cells, even greater numbers of cells bearing sIgG and sIgA can also be detected. All of these cell types invariably also express IgM; they rarely have IgD. Later in development IgG and IgA can occur singly on cells or in combination with IgD, and, or, IgM; i.e.,there appear to be cells with three isotypes simultaneously present. Cells expressing only IgA or IgG were usually larger than cells bearing IgA or IgG in combination with IgM or IgD.

An essentially similar set of observations were made with lymphocytes from nude, germ free mice. Finally, surface IgD, but not surface IgM, is rapidly lost from LPS-stimulated lymphocytes, and murine plasma cells containing IgD are never encountered. Taken together these observations suggest a model for the development of Ig-diversity (Fig. 1). Thus sIgM-bearing cells give rise to cells committed to synthesis of IgM, IgG and IgA. All of these cells express their commitment to their respective Ig classes by synthesis of the

* On leave of absence from the National Institute for Medical Research, London, England.

appropriate sIg, but continue to express sIgM. Around this time, sIgD as an additional antigen receptor begins to be expressed on the surface of all subclasses of B-lymphocytes. Upon antigen stimulation sIgD is lost, sIgM may or may not be lost, depending on the number of cell divisions, and cytoplasmic Ig will reflect the primary Ig commitment of the activated B cell.

The similarity in results obtained with normal and germ free, nude mice indicates that generation of isotype diversity on lymphocyte surfaces can proceed in the absence of T-cell influence. Also suggested, but not proven, is that isotype diversity can arise in the absence of antigenic stimulation, although there is no doubt that clonal expansion and differentiation to high rate Ig-secreting cells is antigen-driven.

The model presented raises a number of questions. First, it is worth noting that the current confusion regarding the class of Ig expressed on B-memory cells can be resolved if these cells bear multiple Ig classes. Further, it seems likely that all lymphocyte types represented in the scheme can be stimulated by antigen under appropriate conditions. Therefore, the choice of experimental conditions, in particular the method used to produce memory cells for testing, will influence the observations recorded. Finally, the genetic basis for simultaneous expression of 2, and even 3, heavy chain genes by one lymphocyte remains to be defined, but we would favour an explanation based on simultaneous integration of a selected V gene with all constant genes.

ACKNOWLEDGMENTS

This work was supported by Grant Number CA 16673, awarded by the National Cancer Institute, DHEW; Grant Number 1-354, awarded by The National Foundation, March of Dimes; Grant Number AI 10502, awarded by NIAID, USPHS.

REFERENCES

Due to space limitations and the ready availability of relevant material in this volume, only the following recent discussion paper, intended to serve as an introduction, is quoted.

Cooper, M.D., Kearney, J.F., Lawton, A.R., Abney, E.R., Parkhouse, R.M.E., Preud'homme, J.L. and Seligman, M. (1976) Ann. Immunol. (Inst. Pasteur) 127C, 537-581.

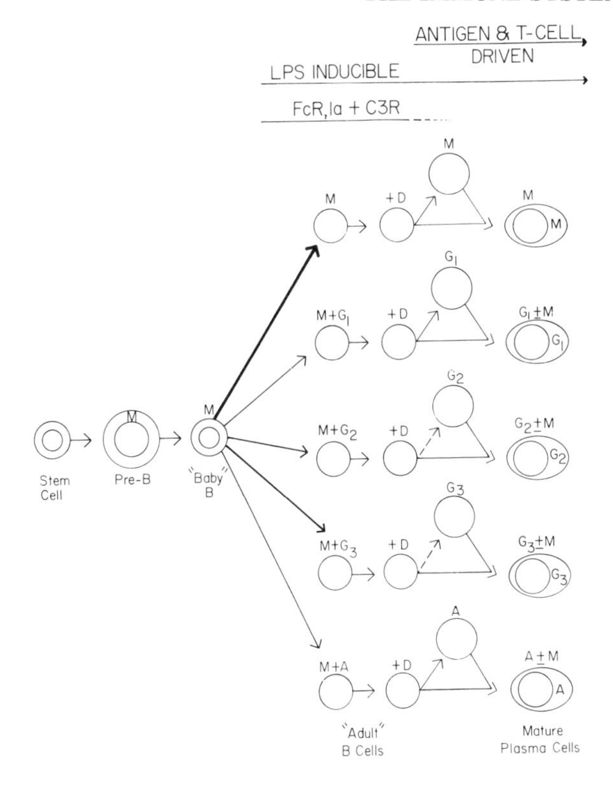

Rapidly dividing "pre-B" cells lacking functional antibody receptors are the precursors of immature sIgM+ cells. The easily tolerizable "baby" B cell is pivotal in the switch to expression of other Ig classes. During development, small percentages of $sIgM^+$ cells begin to express one of the IgG subclasses or sIgA. "Doubles" may become "triples" when they later begin to express sIgD. With stimulation by LPS, and presumably by antigen, sIgM and sIgD are lost with proliferation by B cells that express a single IgG subclass or IgA. Athymic "nude" mice raised under germfree conditions have the same frequencies of the various isotype combinations on their B cells as do conventional mice. Some IgG and IgA plasmablasts and mature plasma cells continue to express sIgM but lose sIgD.

MULTIPLE IMMUNOGLOBULIN HEAVY CHAIN EXPRESSION BY LPS STIMULATED MURINE B LYMPHOCYTES

J. F. Kearney, A. R. Lawton and M. D. Cooper

Departments of Microbiology and Pediatrics
and the
Comprehensive Cancer Center
University of Alabama in Birmingham
Birmingham, Alabama 35294

ABSTRACT. IgG and IgA secreting plasma cells which differentiate in LPS stimulated cultures of murine lymphoid cells are derived from precursors which express IgM. Using this *in vitro* model we have analyzed some morphologic and biochemical events involved in switching of C_H genes during B cell differentiation. Induction of surface membrane IgG (sIgG) on sIgM positive cells occurs rapidly and does not require DNA synthesis. During early stages of culture, some cells simultaneously synthesize cytoplasmic IgM and IgG (cIg), while restriction is the rule at days 6-7. Inhibitors of cell proliferation, including suboptimal concentrations of calf serum in the medium, substantially increase the numbers of cytoplasmic doubles in culture. These results suggest that induction of expression of $C\gamma$ genes in sIgM positive cells requires transcription of new mRNA, but not DNA synthesis. Restriction to synthesis of IgG alone is dependent upon several cycles of proliferation.

INTRODUCTION

During differentiation of B cells, alterations in gene expression result eventually in the appearance of plasma cells which synthesize and secrete different classes of immunoglobulin. From many different observations it is apparent that the same V_H and V_L genes can be expressed with different C_H genes, so that each clone, defined by variable region markers, contains members synthesizing immunoglobulins of the same antigenic specificity but of different heavy chain classes (1-3).

Recognition of the phenomenon of switching has given rise to two interrelated questions. From the biological standpoint, is this process regulated by exposure to antigens, with or without factors supplied by accessory cells (4-6), or is it mediated by intrinsic signals supplied in specialized microenvironments (7, 8). Perhaps a more fundamental question concerns the relationship of switching to intergration of V and C genes. Is switching a manifestation of sequential insertion of a single V gene into a series of linked C_H genes or does it involve integration of multiple copies of V_H genes followed by regulation of expression of V-C gene combinations?

In this paper we present observations arising from studies of LPS-stimulated differentiation of murine B lymphocytes to plasma cells secreting immunoglobulins of each major class which bear upon these questions.

Previous evidence that an IgM bearing cell is the LPS sensitive precursor for IgG synthesizing cells. Inclusion of purified goat antibodies to mouse μ chains in the LPS stimulated cultures of both newborn and adult lymphoid tissues completely inhibited the production of plasma cells of all classes (9). Kinetic studies, in which IgM and IgG or IgA on the surface and within the cytoplasm of cells differentiating in response to LPS were detected using two color immunofluorescence, showed that in the newborn essentially all cells expressing surface or cytoplasmic IgG or IgA (cIgG, cIgA) initially expressed surface IgM (sIgM) which was subsequently lost during the period of rapid proliferation and development of cells restricted to synthesis of only cIgG or cIgA (Fig. 1). Parallel studies with adult B lymphocytes have revealed a similar progression from cells expressing sIgM and sIgG to cells synthesizing only cIgG (10, 11).

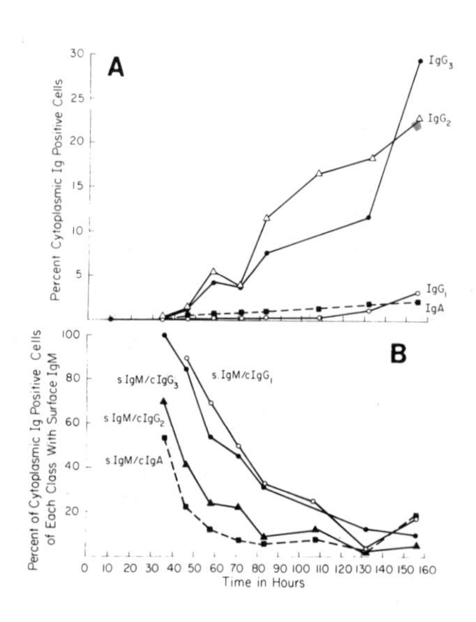

Fig. 1. Presence of surface IgM on cells containing cytoplasmic IgG_1, IgG_2, IgG_3, and IgA in LPS stimulated spleen cultures of 5-6 day CBA/J mice. A. Total frequency of cytoplasmic Ig positive cells. B. Percent of $cIgG_1$, -0-; $cIgG_3$, -●-; $cIgG_2$, -▲-; and cIgA, --■--; which also stain for sIgM.

Role of proliferation in the development of plasma cells restricted to IgG or IgA synthesis. In LPS stimulated cultures of adult spleens there are cells which synthesize both cIgM and cIgG early (2-4 days) but by 7 to 10 days virtually none are present (11). A chance observation in early experiments was that LPS-stimulated cultures established with certain batches of fetal calf sera proliferated poorly but contained many cIgG positive cells that also stained for cIgM. This suggested the possibility that proliferation might be involved in the loss of IgM synthesizing potential by IgG and IgA precursors. Cell cultures were therefore established in media containing 2.5 to 30% of a single batch of FCS and stimulated with LPS. Proliferation of total cells and of cIgG positive cells rose with increasing FCS concentrations. At low concentrations of fetal calf

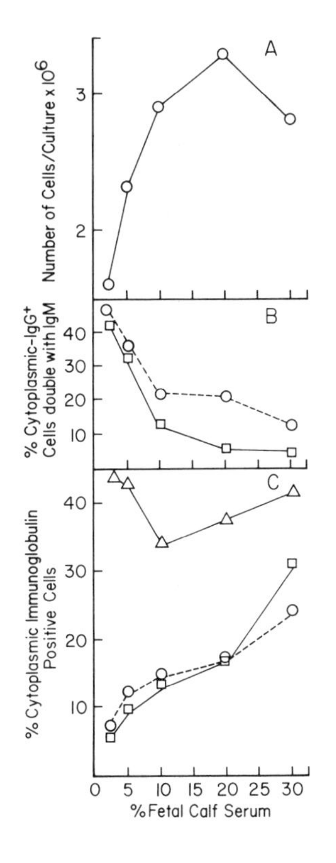

Fig. 2. Effects of various concentrations of fetal calf serum on LPS dependent proliferation and differentiation of cytoplasmic immunoglobulin containing cells.

A. Cell numbers/culture determined by Coulter Counter on day 5.

B. Percent of $cIgG_3$, --0--; or $cIgG_2$, —□— positive cells which also stain for cIgM.

C. Total percent of cIgM, —△— ; $cIgG_3$, --0--; and $cIgG_2$, —□— positive cells.

serum large proportions of $cIgG_2$- and $cIgG_3$- containing cells also stained for cIgM, while doubles were rare at higher concentrations (Fig. 2). In experiments, to be reported in detail elsewhere, the effects of more specific inhibitors of cell proliferation on the LPS-dependent development of isotype restricted IgG plasma cells from IgM bearing precursors were studied. Thymidine, at concentrations >40 μg/ml, and sodium butyrate (1mM) markedly inhibited proliferation of LPS stimulated mouse B lymphocytes but permitted differentiation of cells containing cytoplasmic immunoglobulins. In the presence of increasing concentrations of thymidine, proliferation of total cells and of cIgG positive cells declined in parallel, while the proportion of cIgG positive cells also stained for cIgM rose from 0.5% (control cultures) to 63% (200 μg/ml thymidine). This suggested that restriction to exclusive IgG synthesis by plasma cells was dependent on proliferation of the IgM-producing precursors. Blockade of LPS stimulated newborn cell cultures with 100 μg/ml of thymidine up to 33 hrs after the start of culture resulted in the differentiation of few cells containing $cIgG_1$, $cIgG_2$, $cIgG_3$ and cIgA, of which all were brightly stained for cIgM.

In another experiment, 1mM sodium butyrate was used as an inhibitor of DNA synthesis (12). This inhibitor was added to cultures at 24 hr intervals following initiation of LPS-stimulated cultures. Replicate cultures were counted at the time of addition of the inhibitor, and again at 7 days when numbers of cIgM, cIgG, and doubles were determined. The dose of butyrate used severely limited proliferation after its addition to cultures, so that cells were in a sense frozen after varying numbers of cycles. Cultures blocked at time zero contained few cIgG positive cells, and most were doubles. Addition of the inhibitor up to day 5 resulted in an increase of cIgG positive cells by 1000-fold, and a reduction in doubles to 10%. Nevertheless, the absolute number of doubles per culture increased by a factor of 10. In the absence of inhibitor, less than 0.1% of cIgG cells were doubles. These results suggest that LPS is able to induce a degree of stable differentiation with minimal proliferation, as indicated by persistence of cells containing both cIgM and cIgG for seven days in the presence of the mitotic inhibitor. They further suggest that most or all IgG precursors responding to LPS go through a phase during which both IgM and IgG are synthesized at a cytoplasmic level. Finally, they reemphasize the point that restriction is accomplished by proliferation.

It is of interest that Con A, which induces T cell dependent suppression of B cell differentiation in response to antigen (13, 14) or LPS (15), has effects similar to those of inhibitors of DNA synthesis. Of the few cIgG positive cells developing in cultures containing Con A and LPS, most also contained cIgM.

An issue raised by these results is whether the cells containing cIgM and cIgG are derived from B lymphocytes bearing both of these isotypes, or from cells bearing sIgM, or sIgM and sIgD. Pernis recently reported induction of sIgG on sIgM positive cells following brief (6 hr) exposure to LPS (11). We have performed similar studies using newborn cells, in which sIgM is virtually the only isotype expressed on B lymphocytes. One day old CBA/J liver and spleen cells were cultured alone in suspension for 6-7 days then cleared of non-viable cells over Ficoll-Hypaque gradients. This procedure gave a population of $\sim$ 80-90% $sIgM^+$ B lymphocytes many of which were presumably "young" or immature; they appeared morphologically to be medium or large lymphocytes with very bright staining for surface IgM. These cells had a moderately high intrinsic ^3H-thymidine incorporation. The other population studied was obtained from the thoracic duct of adult (SPF) nude mice; >90% of these cells were typical small $sIgM^+$ B lymphocytes which also expressed various other non-μ heavy chains ($\gamma 1$, 0.4; $\gamma 2$, 16.4%; $\gamma 3$, 4.7% and α, 2.7%). These adult B cells did not incorporate significant amounts of ^3H-thymidine during a 12 hr labeling pulse.

These highly enriched populations of $sIgM^+$ B cells were exposed to LPS (50 μg/ml) for 12 hrs in the presence or absence of inhibitors of DNA, RNA and protein synthesis. Cells were then harvested and stained for surface immunoglobulin with various combinations of purified goat antiodies specific for mouse μ, γ_1, γ_2, γ_3 and α chains. Cell smears were examined with a combination of phase contrast and fluorescence microscopy to give the proportion of $sIgM^+$ cells which also stained for sIgG or sIgA in the presence or absence of metabolic inhibitors.

In experiments with young B lymphocytes, within 12 hrs LPS induced a 2-5 fold increase in the proportions of $sIgM^+$ cells which expressed another immunoglobulin class. These increases occurred in the absence of DNA synthesis (blocked by cytosine arabinoside) and were due to new protein synthesis since they were blocked by cycloheximide. Actinomycin D also inhibited expression of non-μ sIg. None of the inhibitors reduced the numbers of cells which stained for sIgM or the other classes from pre-culture values. There were no significant changes in numbers of cells or viability during the course of these experiments. Limited studies on other combinations, (e.g., staining cells with anti-IgG_1 and anti-IgG_2 etc.) revealed very few doubles, suggesting (i) that the apparent increase of $sIgM^+$ cells expressing other classes was not due to nonspecific binding of goat antibodies (or there would have been more doubles of this kind) and (ii) that during this 12 hr induction period new expression of only one other Ig class usually occurred per sIgM positive cell.

Other interesting observations were obtained by autoradiographic analysis of cultures which had been continuously labeled with ^{3}H-thymidine during LPS induction. Of cells expressing sIgG or sIgA in combination with sIgM$^+$, only a very small proportion incorporated ^{3}H-thymidine during this period, substantiating the evidence obtained by induction in the presence of cytosine arabinoside. The very rare single sIgG$^+$ or sIgA$^+$ cells detected were usually labeled with ^{3}H-thymidine as were many single sIgM$^+$ cells. These findings support our other observations that DNA synthesis and/or cell division is necessary for the development of Ig class restriction but not for the induction process.

In contrast to the results with newborn B cells, we were unable to detect an increase in the proportion of sIgM positive cells bearing sIgG or sIgA in short term LPS-stimulate cultures of nude TDL. Since this population already contained substantial numbers of "doubles", a small increase might have been missed.

DISCUSSION

How do these observations bear on the genetic mechanisms for control of immunoglobulin heavy chain expression by the members of the B cell line? The induction of new immunoglobulin classes on a population of young IgM-bearing B cells occurs rapidly, requires protein synthesis, and does not require DNA synthesis. The fact that actinomycin D blocks induction might suggest that synthesis of mRNA is required, but the multiplicity of effects of this compound makes this interpretation speculative. As a rule only one IgG subclass or IgA appeared on individual IgM-bearing cells, suggesting that possibility that the induced Ig was already programmed for expression. The failure of sIgM bearing nude TDL to be induced may favor pre-committment, since they indicate that not all sIgM positive cells have the capacity to express a different class.

Proliferation was necessary to obtain full development of restriction of IgG or IgA synthesis by plasma cells in LPS-stimulated cultures from both newborn and adult animal; inhibition of proliferation resulted in accumulation of cells which synthesized both IgM and IgG. There are two interpretations of these results which bear on the relationship of switching to V-C gene integration. Assuming that the majority of B cells express a single V_H product, and that V-C integration occurs at the level of transcription rather than translation, doubles can result in only two ways: simultaneous transcription of two different V-C genes, or transcription of only a single V-C gene at one point in time with the second immunoglobulin resulting from translation of a long-lived mRNA. In either case, repeated proliferative events might result in restriction. In the first instance, this would presumably be accomplished by repression of the V-Cμ gene, and in the second, by dilution and functional inactivation of the mRNA for IgM. We believe the weight of current evidence, particularly the existence of stable clones of malignant B cells in which two immunoglobulin classes are produced by the same cell (16-17) favors the former interpretation. However, the long lived mRNA hypothesis has not, in our opinion, been excluded for normal cells.

Simultaneous transcription of two V-C genes by single immunocytes would place important constraints on some of the mechanisms proposed for V-C intergation. In particular, it would require the existence of several copies of the particular V_H gene to be expressed by a given clone for models based on episomal insertion. Models of the "cut and splice" variety, which have had the attraction of relating switching to sequential integration of a single V gene into a different C_H gene, would be excluded. If simultaneous transcription of two (or more) V-C gene complexes is indeed the explanation for the existence of multiple classes of immunoglobulin on the surface or within the cytoplasm of B cells, our other results suggest two further interpretations. First, integration of V and C genes must occur early in B cell differentiation, since induction occurs with young cells. Secondly, switching appears to be highly regulated, since induction clearly does not result in expression of every class of Ig in individual cells.

ACKNOWLEDGEMENTS

The studies described herein were supported by NIH grants AI 11502, CA 16673 and CA 13148. A. R. Lawton is a recipient of a Research Career Development Award AI 70780. We wish to thank Ms. Lanier Ager for excellent technical assistance and Mrs. Nancy Perry and Mrs. Mary Huckabee for help in the preparation of this manuscript.

REFERENCES

1. Potter, M. (1972) *Physiol. Rev.* 52, 631.
2. Weigert, M., Cesari, M., Yonkovich, S. J., and Cohn, M. (1970) *Nature* 228, 1045.
3. Press, J. L., and Klinman, N. R. (1973) *J. Exp. Med.* 138, 300.

4. Pierce, C. W., Asofsky, R., and Solliday, S. M. (1972) *Fed. Proc.* 32, 41.
5. Warner, N. L. (1972) *In: Contemporary Topics in Immunobiology, Vol. 1.* M. D. Cooper, and M. G. Hanna, Jr., editors. (Plenum Press, New York) p. 87.
6. Vitetta, E. S., and Uhr, J. W. (1975) *Science* 189, 964.
7. Kincade, P. W., Lawton, A. R., Bockman, D. E., and Cooper, M. D. (1970) *Proc. Natl. Acad. Sci. USA* 67, 1918.
8. Cooper, M. D., Lawton, A. R., and Kincade, P. W. (1972) *Clin. Exp. Immunol.* 11, 143.
9. Kearney, J. F., Cooper, M. D., and Lawton, A. R. (1976) *J. Immunol.* 116, 1664.
10. Kearney, J. F., Cooper, M. D., and Lawton, A. R. (1976) *J. Immunol.* 117, 1567.
11. Pernis, B., Forni, L., and Luzzati, A. L. *In: Cold Spring Harbour Symposium "Origins of Lymphocyte Diversity"*, in press.
12. Kyner, D., Zabos, P., Christman, J., and George, Acs (1976) *J. Exp. Med.* 144, 1674.
13. Rich, R. R., and Pierce, C. W. (1973) *J. Exp. Med.* 132, 205.
14. Dutton, R. W. (1975) *Ann. N. Y. Acad. Sci.* 249, 23.
15. Piguet, P. F., Dewey, H. K., and Vassalli, P. (1976) *J. Immunol.* 117, 1817.
16. Pernis, B., Brouet, J. C., and Seligmann, M. (1974) *Eur. J. Immunol.* 4, 776.
17. Fu, S. M., Winchester, R. J., Feizi, T., Walzer, P. D., and Kunkel, H. G. (1974) *Proc. Natl. Acad. Sci. USA* 71, 4487.

REGENERATION OF SURFACE Ig AS A MEASURE OF IMMUNOLOGICAL MATURATION OF CBA/N MICE

I. M. Zitron, I. Scher and W. E. Paul

Laboratory of Immunology, NIAID, NIH;
Naval Medical Research Institute,
Bethesda, Md. 20014

ABSTRACT. The CBA/N mouse strain carries an X-linked defect, manifested in its B cells, which gives rise to an inability to respond to certain thymus-independent antigens. On the basis of analysis of amount and class of membrane Ig, expression of Lyb5 and Mls stimulatory determinants, and pattern of immunologic responsiveness, the defects in these mice are believed to arise from an arrest in the maturation of B cells or of a B cell sub-line. Recently, it has been reported that Ig-bearing cells of young mice of normal strains fail to re-express membrane Ig after overnight incubation with anti-Ig. This has been suggested as a mechanism by which immunologic tolerance is achieved in neonatal mice and might account for the failure of CBA/N mice to make certain immune responses. We have confirmed the finding that splenic B cells of neonatal, non-defective animals fail to re-express sIg after overnight incubation with anti-Ig reagents. Taking advantage of the X-linked nature of the CBA/N defect, we have compared adult (CBA/NxDBA/2)F_1 male [defective] mice with their female [non-defective] littermates, with respect to this function. Both the male and female animals are capable of the re-expression of membrane Ig. This has been shown to hold true when initial incubation involved either a polyvalent anti-Ig or an anti-μ chain reagent. Thus the maturational arrest in the defective animals, if it occurs as a single discrete defect, must occur at a stage such that their B cell population is more mature than that of the neonatal mouse. The lack of certain responses in CBA/N mice is, then, not simply ascribable to their being at a stage of maturation such that Ig receptors are irreversibly modulated due to encounter with a polyvalent ligand.

INTRODUCTION

CBA/N mice are of considerable immunological interest, since they carry an X-chromosome linked defect manifested in the lymphocytes of the B lineage. The expression of this defect has been characterized (1-6) and a striking similarity has been observed between the B cells of the defective adult mice and those of neonatal animals of non-defective strains. Among the similarities noted have been a lack of responsiveness to a number of thymus-independent (TI) antigens (7), a failure to express minor lymphocyte stimulating determinants (8) a failure to express Lyb5 (9) and a characteristic population distribution of the amount of surface immunoglobulin (sIg) per cell, which is associated with a low ratio of surface δ to μ chains (1,6).

An independent functional parameter of B cell maturity, viz. the re-expression of sIg after exposure overnight to an anti-Ig reagent, has recently been reported (10,11). Under the conditions described, B cells from adult animals are capable of re-expressing sIg, while those from foetal or neonatal animals fail to do so. Another characteristic of immature B cells is their susceptibility to tolerance induction by certain ligands (12,13). While no precise mechanism has been demonstrated for this, it is tempting to associate the lack of sIg re-expression, after prolonged incubation with anti-Ig, with the increased susceptibility to tolerance induction. Since immune-defective CBA/N mice and F1 males derived from CBA/N mothers demonstrate immature B cell characteristics, we have performed the experiments described in order to test the hypothesis that the defective mice are non-responsive due to tolerization of their B cells by a mechanism analogous to sIg receptor loss subsequent to cross-linking by anti-Ig or antigen and consequent endocytosis of the complex.

MATERIALS AND METHODS

The antisera used for _in vitro_ capping and endocytosis of sIg were a polyvalent rabbit anti-mouse Ig (RaMIg) and a goat anti-mouse μ chain reagent. Prior to use, all reagents were absorbed with normal mouse spleen cells to remove any cytotoxic activity, and then heat-inactivated. The staining reagents were a fluoresceinated-(FL)-$F(ab')_2$ fragment of a polyvalent anti-Ig and FL-goat anti-mouse μ. All intact anti-Ig reagents were deaggregated by ultracentrifugation immediately prior to use.

The experimental design was to expose spleen cells for 24 hours, at a concentration of 10^7 cells/ml, to either one of the anti-Ig reagents or a control serum (all at 1/100 final concentration). After the capping and endocytosis had taken place, the cells were harvested, washed three times and then replated (at a concentration of $5x10^6$ viable cells/ml) in medium containing no anti-Ig. The medium used throughout was RPMI 1640 supplemented with 10% foetal calf serum, penicillin, streptomycin and glutamine. After a further overnight incubation, the cells were harvested, washed and the dead cells removed by centriguation on Ficoll-Hypaque. Cell yields, in terms of both total and viable cells were measured. The cells were stained for sIg (on ice, in the presence of sodium azide) and analyzed either visually for the frequency of sIg-positive cells or by the Fluorescence Activated Cell Sorter (FACS) for the distribution of sIg densities.

RESULTS

Confirmation of the difference between adult and neonatal mice in the ability of their spleen cells to re-express sIg. Spleen cells from adult (8 weeks) or neonatal (4 days) non-defective mice were exposed for the first 24 hours *in vitro* to either normal rabbit serum or rabbit anti-mouse Ig and analyzed after an additional 24 hours in culture. The data (shown in Table 1) confirm previous reports (10,11) that B cells from immature animals fail to re-express sIg under these conditions, in contrast to their adult counterparts.

TABLE 1

Expression of sIg at 48 hours by splenic lymphocytes exposed to polyvalent rabbit anti-mouse Ig during the first 24 hours in culture

Spleen Cell Source	Treatment *in vitro* 0-24 hrs.	Ig-positive Cells/ no. cells examined	% Ig-positive Cells ± rel. s.d.
Female F_1 (8 weeks)	none	101/219	46.1 ± 10.0
"	n.r.s.	102/228	44.7 ± 9.9
"	RaMIg	114/287	39.7 ± 9.4
CBA/CaHN (4 days)	n.r.s.	90/293	30.7 ± 10.5
"	RaMIg	0/73	0

Ability of spleen cells from female and male (CBA/N x DBA/2)F_1 mice to re-express sIg after capping and endocytosis.

In the experiment shown in Table 2, rabbit anti-Ig was used for capping and endocytosis and the FL-polyvalent anti-Ig reagent for analysis.

TABLE 2

Expression of sIg at 48 hours by splenic lymphocytes exposed to polyvalent anti-Ig during the first 24 hours

(CBA/NxDBA/2)F_1 Spleen Cell Donor	Treatment in vitro 0-24 hrs.	Ig-positive Cells at 48 hours/ no. Cells Examined	% Ig-positive cells (± rel. s.d.)
Female (8 wks)	none	103/200	51.5 ± 9.9
"	n.r.s.	105/249	42.2 ± 9.8
"	RaMIg	105/265	39.6 ± 9.8
Male (8 wks)	none	103/341	30.2 ± 9.9
"	n.r.s.	104/307	33.9 ± 9.8
"	RaMIg	108/437	24.7 ± 9.6

The data show quite clearly that, under these conditions, both male and female F1 mice re-express sIg. The reported lower frequency of sIg-positive cells in the male, as compared to the female, is maintained through the culture procedure; final yields of viable cells of each population (not shown) were comparable, so it is unlikely that the results can be explained by the selective death of either sIg positive or negative cells.

The cells which had re-expressed sIg appeared, upon visual examination, to be generally "duller" staining than those which had been exposed to control sera. Accordingly, we have performed a series of experiments using the FACS to analyze cell populations in terms of distribution of sIg densities.

In the experiment shown here (Table 3), the reagent used for capping and endocytosis was the goat anti-mouse μ and the cells were analyzed with the anti-μ probe.

A number of points may be made from the data in Table 3:

1. The splenic B cells from both female and male F1 mice are capable of re-expressing sIgM after capping and endocytosis by anti-μ antibody.
2. With neither male nor female cells was there complete re-expression of sIgM after capping and endocytosis, in terms of the frequency of sIgM-positive cells (channels 80-1000), when compared to cells exposed to normal goat serum during the first 24 hours. It is possible that this reflects delayed re-expression and that, given time, the total

TABLE 3

FACS ANALYSIS OF THE FREQUENCY OF sIgM-POSITIVE CELLS AND DISTRIBUTION OF sIgM DENSITIES IN SPLEEN CELLS FROM MALE AND FEMALE F_1 MICE AFTER CAPPING AND ENDOCYTOSIS WITH ANTI-μ

(CBA/NxDBA/2) F_1 Cell Donor	Capping/endocytosis Reagent	% Positive Cells (Channels 80-1000)	Distribution of sIgM Densities		
			Channels 80-200	Channels 200-400	Channels 400-1000
female - fresh spleen	-	58.0	21.9	21.4	14.7
female	normal goat serum	57.7	29.5	18.2	10.0
female	goat anti-mouse μ	39.5	25.7	9.4	4.4
male - fresh spleen	-	46.2	10.4	16.4	19.4
male	normal goat serum	33.1	12.6	9.4	11.1
male	goat anti-mouse μ	26.8	13.8	8.8	4.2

frequency of Ig-positive cells in the population which had undergone the endocytosis-re-expression cycle would reach that of the normal serum controls; we have no data on this. An alternative explanation is that the spleens of even adult mice contain a set of immature B cells, of the type found as the exclusive population in neonatal spleen, which fail to re-express sIgM after capping and endocytosis. Interestingly, if the latter is the case, the data would suggest that cells judged to be immature by this criterion are, in fact, somewhat less frequent in the spleens of immune-defective F1 males than in their female littermates.

3. Normal goat serum-treated F1 male cells demonstrated a relatively higher frequency of sIgM-positive cells with high densities of sIgM (channels 400-1000) when compared to F1 female cells, as previously reported (6). The distribution of sIgM on both male and female cells which had undergone capping and endocytosis of sIgM followed by re-expression, showed similar differences. Finally, as we had anticipated from visual examination of the cell populations, there was a reduction in the frequency of cells with high densities of sIgM.

DISCUSSION

These experiments demonstrate that the male F1 mouse, while expressing an X-linked defect which causes its B cells to resemble in many respects those of the neonatal animal, does not have in common with the neonate the inability to re-express sIg after capping and endocytosis. If the defect in the CBA/N is due to a maturational arrest in a single lineage of developing B lymphocytes, then this arrest must be placed beyond the time at which non-defective B cells mature sufficiently to acquire this ability (about 7 day of age). There are alternative schema for B cell differentiation, involving multiple pathways; in such models, the maturational arrest in the immuno-defective male F1 would involve one pathway.

The FACS analysis showed that both male and female B cells exhibited a diminished frequency of high density sIgM-positive cells following capping and endocytosis with anti-μ and subsequent re-expression. The possible explanations for the downward shift in staining intensity may be related to the time allowed for re-expression, but may also reflect a maturation process occurring *in vitro* which results in a decrease in sIg density.

Finally, since there is no difference in the ability to re-express sIg between the defective and normal B cells, the lack of antigen-responsiveness by F1 male cells is unlikely to be due simply to tolerance induction by sIg receptor loss. Further, any abnormalities in tolerance induction in the immune-defective F1 males are unlikely to be due to a mechanism involving the irreversible receptor loss characteristic of neonatal cells.

ACKNOWLEDGMENT

We should like to acknowledge the excellent technical assistance of Ms. Susan Sharrow.

REFERENCES

1. Finkelman, F.D., Smith, A., Scher, I. and Paul, W.E. (1975) J. Exp. Med. 142: 1316.
2. Amsbaugh, D.F., Hansen, C., Prescott, B., Stashak, P., Barthold, D. and Baker, P. (1972) J. Exp. Med. 136: 931.
3. Scher, I., Frantz, M.M. and Steinberg, A.D. (1973) J. Immunol. 110: 1396.
4. Scher, I., Ahmed, A., Strong, D., Steinberg, A.D. and Paul, W.E. (1975) J. Exp. Med. 141: 788.
5. Scher, I, Steinberg, A.D., Berning, A. and Paul, W.E. (1975) J. Exp. Med. 142: 637.
6. Scher, I., Sharrow, S. and Paul, W.E. (1976) J. Exp. Med. 144, 507.
7. Mosier, D.E., Scher, I. and Paul, W.E. (1976) J. Exp. Med. 117: 1363.
8. Ahmed, A. and Scher, I. (1976) J. Immunol. 117: 1922.
9. Ahmed, A., Scher, I., Sharrow, S., Smith, A., Paul, W.E., Sachs, D.H. and Sell, K.W. (1977) J. Exp. Med. 145:101.
10. Sidman, C.L. and Unanue, E.R. (1975) Nature 257: 149.
11. Raff, M.C., Owen, J.J.T., Cooper, M.D., Lawton, A.R. III, Megson, M. and Gathings, W.E. (1975) J. Exp. Med. 142: 1052.
12. Metcalf, E.S. and Klinman, N.R. (1975) J. Exp. Med. 143: 1327.
13. Nossal, G.J.V. and Pike, B.L. (1975) J. Exp. Med. 141: 904.

PURIFICATION AND CHARACTERIZATION OF ANTIGEN BINDING CELLS FOR SHEEP RED BLOOD CELLS

James J. Kenny and Robert F. Ashman

Department of Microbiology and Immunology
UCLA School of Medicine, Los Angeles, California 90024

ABSTRACT. A two step centrifugation procedure has been established to isolate sufficient numbers of antigen binding cells (ABC) to permit measurement of biochemical parameters of activation. The first step involves partially separating sheep erythrocyte rosettes from unrosetted lymphocytes by their difference in buoyant density on Ficoll-Hypaque. Subsequent passage through a linear 5-10% Ficoll gradient produces further purification of rosettes by sedimentation velocity. Approximately 4.5 x 10^6 ABC can be obtained at 50-100% purity from 10^9 immune spleen cells (5 days post-immunization) and 1 x 10^5 ABC at 20-40% purity from 10^9 non-immune spleen cells. The purified ABC from 5 day immune animals are 80-90% B cells, and 0.1% plaque forming cells (PFC), while 1% of them have intracellular immunoglobulin (Ig).

Compared to unfractionated and rosette-depleted cell populations, purified ABC from 5 day immune animals have a 3-5 times higher rate of ^{14}C-choline incorporation into phospholipids and ^{14}C-acetate incorporation into neutral lipids.

INTRODUCTION

Most reported methods for producing populations enriched for specific ABC involve adherence of ABC to solid surfaces which have been coated with hapten protein conjugates (1-3) or directly derivatized with hapten (4,5). Drawbacks of these procedures include cell loss and damage during enzymatic digestion of the substrate or mechanical removal of bound cells. This publication describes a method for isolating highly purified ABC in quantity without the need for special equipment, which in principle should be applicable to any antigen which can be firmly coupled to a red blood cell.

METHODS AND RESULTS

Purification and characterization of ABC. CBA/J or CBA/St mice 8-12 weeks of age were immunized intraperitoneally with sheep erythrocytes (SRBC) and cell suspensions were prepared 5 days later as previously described (6), except that mouse red blood cells were lysed with Gey's solution (7). Dead cells, granulocytes, and mature macrophages were removed on a glass wool column. The resulting suspension (>90% viable by trypan blue) was adjusted to 2.5 to 3.0 x 10^7 cells/ml and rosetted with SRBC at 50 xg. Six ml of lymphocytes (containing 1-2% rosettes) were then layered onto 4 ml of Ficoll-Hypaque and spun at 1800 xg for 20 minutes as described by Parish et al (8). In 15 experiments, the pellet contained 57-100% of the rosette forming cells (RFC) but also 10-20% of the original lymphocyte population including all the dead cells and most of the free SRBC. The interface contained the remaining 60-90% of the lymphocytes with 1-23% of the rosettes. Up to 80% of the PFC present in a 5 day immune spleen can be lost through glass wool filteration and Ficoll-Hypaque centrifugation, with the remainder equally divided between the pellet and the interface.

After the pellet for the Ficoll-Hypaque gradient has been washed and rerosetted, 2-6 x 10^7 cells in 2-3 ml were layered onto each 40 ml (5-10%) linear Ficoll gradient. Fig. 1 shows the distribution of cells and rosettes in this gradient following centrifugation at 50 xg for 10 minutes.

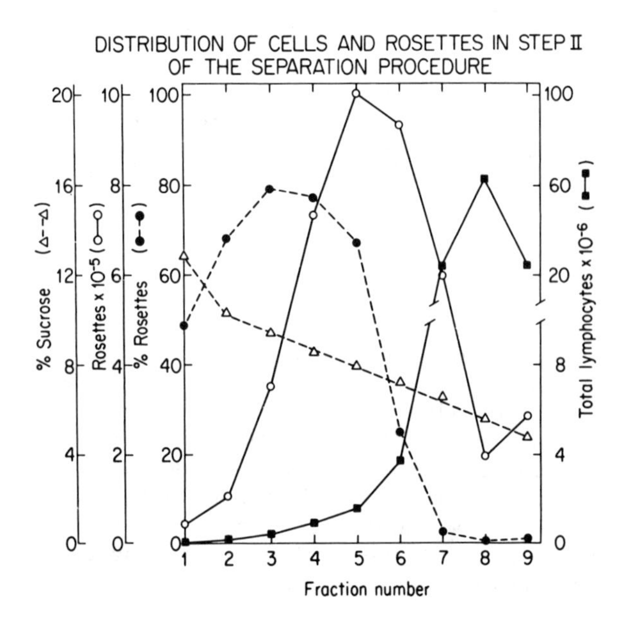

Five ml fractions were collected and the total number and purity of rosettes in each fraction determined. The largest number of rosettes were recovered in fractions 5 and 6. Because fraction 6 was usually less than 25% ABC and contained a large number of dead cells, fractions 1-5 were collected as the ABC enriched population. This population contains 50% of the rosettes and 10% of the cells put onto the linear gradient representing 30% of the initial ABC population but 1% of the original spleen cells. Forty to 50% of the PFC put onto linear gradients are also obtained in the enriched pool; however, these PFC represent only 0.1% of the purified ABC. (Approximately 1% of the purified ABC have intracellular Ig as judged by fluorescence using anti-IgM and anti-J chain antisera.) The purified ABC are 80-90% B lymphocytes and 10-20% T lymphocytes by indirect immunofluorescence with rabbit anti-T and anti-IgM. When the same two step centrifugation procedure was applied to splenic RFC from normal unimmunized animals, 10^9 spleen cells containing 0.05% RFC yeilded about 10^5 ABC at 20-40% purity.

Lipid synthesis in purified ABC. Purified 5 day immune ABC, rosette depleted cells and unfractionated spleen cells were pulse labeled with ^{14}C-choline or ^{14}C-acetate for 4 hours in vitro. The lipids were then extracted and run on TLC plates according to Pratt et al (9). Table 1 shows that the purified ABC have a 3-5 times higher rate of incorporation of choline into membrane phospholipids and acetate into neutral lipids than unfractionated or rosette depleted spleen cells.

TABLE 1: LIPID SYNTHESIS IN PURIFIED ANTIGEN BINDING CELLS

Cellular Fraction	Phospholipids (^{14}C-choline - CPM)		Neutral Lipids (^{14}C-acetate - CPM)	
	Exp. A	Exp. B	Exp. A	Exp. B*
Unfractionated	1148	1601	287	208
Rosette Depleted	1553	1697	336	214
Rosette Enriched	5012	6515	1068	1150

* Digitonin precipitation of cholesterol from 2 x 10^6 cells.

Since ABC are about 85% B cells whereas spleen is about 40% B cells, 5 day immune spleen cells were separated into T and B cells via nylon wool columns (10) and EA monolayers (11) to see whether differences in lipid turnover between B and T cells provided a partial explanation for this result. No difference in choline or acetate incorporation was seen among the T cell (5% Ig^+) and B cell (70% Ig^+) fractions and the unfractionated population.

DISCUSSION

Other researchers have enriched for ABC using rosette formation followed by either velocity sedimentation (12,13) or density fractionation (14), but have produced neither large numbers of cells nor highly purified populations of ABC. However, they have shown that the ABC which they obtain can perform both B (PFC precursor) and T (helper cell) functions (13, 14). The method described above should yield sufficient numbers of ABC to study some of the biochemical events in ABC differentiation. For example, Resch and Ferber (15) have shown that within a few hours of mitogenic activation, lymphocytes increase synthesis and turnover of membrane lipids about 3-fold en route to cell division and differentiation into functional T and B cells. The purified ABC from 5 day immune animals also show a 3-5 fold greater rate of ^{14}C-choline and ^{14}C-acetate incorporation into lipid than non-antigen-binding cells (Table 1). However, this may reflect recent cell division rather than differentiation towards a mature PFC. Experiments are in progress to resolve this point.

ACKNOWLEDGEMENTS

The expert technical assistance of Heinrich Kolbel was greatly appreciated. We would like to thank Dr. I. Weissman for his generous gift of anti-T cell serum and Dr. M. Koshland and Dr. V. Ewan for their anti-J chain antiserum. This research was supported by NIH grant CA12800 and James J. Kenny was a recipient of a predoctoral National Institutional Research Service Award CA9120-02.

REFERENCES

1. Haas, W. et al (1974) *Eur. J. Immunol.* 4, 565.
2. Choi, T.K. et al (1974) *J. Exp. Med.* 139, 761.
3. Maoz, A. et al (1976) *Nature* 260, 324.
4. Edelman, G.M. et al (1971) *Proc. Nat. Acad. Sci. USA* 68, 2151.
5. Kiefer, H. (1975) *Eur. J. Immunol.* 5, 624.
6. Ashman, R.F. (1973) *J. Immunol.* 111, 212.
7. Dresser, D.W. and Greaves, M.F. (1973) In: *Handbook of Experimental Immunology,* Ed. D.M. Weir, Vol. 2, Chapt. 27.
8. Parish, C.R. et al (1974) *Eur. J. Immunol.* 4, 808.
9. Pratt, H.P.M. et al (1977) Submitted for publication.
10. Julius, M.H. et al (1973) *Eur. J. Immunol.* 3, 645.
11. Kedar, E. et al (1964) *J. Immunol.* 112, 1231.
12. Elliott, B.E. et al (1975) *J. Exp. Med.* 141, 5.
13. Kontiainen, S. and Andersson, L.C. (1975) *J. Exp. Med.* 142, 1035.
14. Bach, J.F. (1970) *Nature* 227, 1251.
15. Resch, K. and Ferber, E. (1975) In: *Immune Recognition,* Ed. A. Rosenthal.

WORKSHOP 3: IMMUNOGLOBULIN ISOTYPES ON B LYMPHOCYTES

Conveners: R. Asofsky and E. Vitetta

The workshop was divided into 4 distinct, but somewhat overlapping sections: (1) identification and characterization of immunoglobulin isotypes (classes and subclasses) on the surface of lymphocytes; (2) functional properties of lymphocytes bearing different isotypes or sets of isotypes; (3) changes in B lymphocytes following inductive signals; and (4) presentations of models of B cell differentiation.

Studies of lymphocytes from mice and from humans indicate that the major population of B lymphocytes bears both IgM and IgD at the same time. Minor populations express IgM alone or IgD alone. A very small population expresses IgG, sometimes alone, sometimes with IgM or IgD. There was no agreement as to whether or not 3 isotypes (IgM, IgD, and IgG) could be found at the same time on a single cell. Some of the data presented suggests the existence of such "triples". Data was presented on IgD from an Abelson virus-induced lymphoma. The δ chains from this tumor migrated identically in SDS-polyacrylamide gels with μ chains, suggesting that the smaller δ chains from normal lymphocytes have been partly digested by some unknown enzyme. Although there was general agreement that both IgG and IgA were expressed on some B cell membranes, there was dispute as to whether the expression of such molecules appeared before or after IgD in development. Evidence was presented that the expression of IgM preceded expression of IgD in ontogeny. Earlier appearance of IgM has been observed on developing B cells of humans and of mice and rats.

Studies on whether functional correlations could be made between subsets of B lymphocytes bearing different kinds of isotypes were of the following kind: (1) treatment with anti-immunoglobulin (anti-Ig) during stimulation to see if functions were blocked or increased; (2) removal of subsets by treatment with anti-Ig and complement (C'); (3) sorting of subsets with fluorescein-labeled anti-Ig; (4) treatment of cells with proteolytic enzymes which differentially remove different isotypes from the membrane; (5) correlations of function with ontogenetic appearance of different subsets, and (6) studies of homing of cells labeled with fluorescent antibody. Human lymphocytes stimulated with pokeweed mitogen in the presence of anti-IgM showed suppression of ultimate IgG synthesis; IgM synthesis was variably affected. Under identical conditions, cells from the same individuals stimulated with LPS showed marked suppression of IgM synthesis. In a study using adoptive transfer, mouse spleen cells treated with anti-μ and C' in the presence of azide were depleted of primary but not secondary IgM responses; treatment with anti-γ removed secondary IgG but not IgM responses; treatment with both anti-γ and anti-μ removed secondary IgM and IgG responses. Almost identical work, but without azide added to antibody and C' showed depletion of only IgG_1 responses.

There were a variety of studies using cells in adoptive transfer that had been selected on the fluorescence activated cell sorter (FACS). Several experiments can be summarized as follows: Cells selected for the presence of IgD gave all of a primary response to DNP, but only 50% of a secondary response (both IgM and IgG). Cells which bore no IgD gave 50% of a secondary DNP response, and this response was of the highest affinity. It appeared that, although cells with surface IgG in combination with other isotypes were responsible for long term immunologic memory, the cells bearing IgG only gave rise to cells which synthesized antibody of the highest affinity. Treatment of B cells with papain, an enzyme that selectively removes IgD under appropriate conditions, made these cells very susceptible to induction of tolerance by T-dependent antigens in vitro. It is therefore unlikely that IgD transmits tolerogenic signals.

During ontogeny, cells from 2 week old mice (lacking IgD) could be stimulated to make anti-TNP antibody by TNP-LPS and TNP-brucella, but not by TNP-ficoll or TNP-dextran. In addition, these cells could not respond to SSS-III polysaccharide or phosphorylcholine-brucella. The latter four antigens stimulated cells of older animals, at times when they expressed IgD. An allotypic anti-IgD suppressed TNP-ficoll responses but not TNP-LPS responses. Cells from the intestine which bore surface IgA preferentially "homed" to lamina propria and mesenteric lymph node, and differentiated into IgA forming cells. Incubation of BALB/c mice with ascaris extract increased the numbers of IgA bearing cells, but reduced the proportion of cells which would make the TEPC 15 idiotype.

Two kinds of inductive stimulus were used: (1) polyclonal, such as LPS; and (2) specific antigens. Neonatal mouse spleen cells stimulated in vitro with LPS acquire surface IgG and IgA, followed by expression of IgD. All of these events are blocked by anti-μ and by treatment with a large dose of thymidine. Cells from fetal liver show a similar sequence of acquisition of immunoglobulins, but this was not blocked by thymidine or hydroxyurea, but rather by actinomycin-D and cyclohexamide. It appears that fetal liver cells differentiate directly while neonatal spleen cells must divide to express this series of isotypes. An α 1 $\rightarrow$6 dextran gives a polyclonal response in neonatal mouse spleen cells from which the dextran antibody response is absent. The latter appears late (2-3 months) in maturation. Anti-IgM blocks synthesis of all classes of antibody to sheep cells in mouse spleen cell cultures, and acts from 0-48 hours. Anti-IgG blocks IgG responses only, and acts from 48-72 hours, suggesting an antigenically driven switch from IgM to IgG. A single clone of anti-DNP antibody producers is blocked from differentiating by anti-μ and not by anti-γ, anti-α, or anti-∂, suggesting that only IgM is expressed in this clone. A study of MOPC-315, a myeloma synthesizing an IgA anti-nitrophenyl specificity, indicated that this tumor's rate of differentiation in millipore chambers was inducible with DNP-KLH. Work on production of J chain by rabbit spleen cells indicates that synthesis of this polypeptide increases more rapidly than μ chains following immunization.

Four major conclusions were made at this workshop: (1) IgM is always the first immunoglobulin produced in development; (2) IgM or IgD bearing cells can give rise to plasma cells producing any immunoglobulin class; IgG bearing cells give rise only to IgG producers; (3) there is an ordered development of immunoglobulin isotypes and other markers on B cell membranes; and (4) as far as present knowledge is concerned, the development of both human and mouse B cells appears to be very similar.

The evening ended with several notes of caution. First, since most of the work makes use of monospecific anti-immunoglobulin antisera, specificity must be carefully demonstrated in the particular experimental system under investigation. Second, cells with low density of immunoglobulin may escape detection by any of the assays now in use. Third, even when density is sufficiently high, orientation (presentation) of surface immunoglobulins and the extent to which they are masked by other neighboring surface molecules may limit detection. Finally, the mere presence of an immunoglobulin does not prove that it is synthesized by the cell on which it is found, because many classes of immunoglobulins are cytophilic for cells via specific receptors.

Workshop Summary: B Cell Differentiation

Although considerable research has been performed on functional heterogeneity of T cell populations, it is only relatively recently that attention has become focused upon possible heterogeneity within B cell populations. This workshop accordingly centered on several aspects of the problem of B cell differentiation, and included studies of the CBA/N mouse, which shows abnormal B cell differentiation; on cell surface markers (including I region gene products) that may be used to characterize different B cell populations; and on possible functional heterogeneity within B cells in terms of susceptibility to tolerance induction or to immune activation.

The session was introduced by Nossal, who proposed that B cell heterogeneity could be considered as a temporal sequence of differentiation with at least 10 distinct stages. These included the initial hemopoietic stem cell: a progenitor B cell that has been committed to the B cell pathway but does not express immunoglobulin genes: the cell defined by Melchers to synthesize immunoglobulin but bears little detectable membrane immunoglobulin; the cell defined by Osmond and Nossal which is particularly susceptible to tolerance induction; the typical B cell with high density membrane immunoglobulin of IgM type; a variety of cell stages activated by various environmental stimuli, that respond differentially to T dependant and T independant antigens; memory B cells; and varying stages leading to the mature plasma cell.

Studies on B cell differentiation have been particularly aided by the use of the CBA/N mouse, which was originally proposed to show maturational arrest in B cell differentiation. On the basis of studies reported by various investigators including Schur, Kessler, Huber, Kincade, and Mosier, the immune responsiveness of the CBA/N mouse might be characterized as follows: Normal responses to sheep erythrocytes, TNP-Brucella, TNP-LPS, T cell mediated responses to mitogens and to MHC associated antigens. In contrast, defective responses have been to certain B cell mitogens, and to several T independant antigens including SIII, TNP-Lys Ficol, TNP dextran and phosphoryl choline Brucella. The mice also show an abnormal ratio of δ/μ bearing cells at the cell surface immunoglobulin level. It cannot however be concluded yet whether these cells differ in the proportion of the cell surface immunoglobulin classes or in actual amount of membrane immunoglobulin of the different classes.

The discussion on these mice particularly centered around the cellular level and nature of the differentiation abnormality. The original concept suggested that a block in differentiation occurred at some point in a single temporal sequence of B cell differentiation, which would accordingly suggest that the mice had a relatively high proportion of cells at immature stages in the B cell pathway. Discussion then centered on three general observations that however were not in accord with this concept. Mosier stressed that CBA/N mice are in fact quite resistant to tolerance induction with some antigens, which contrasts to the increased suseptibility of tolerance induction in neonatal mice which do show high proportions of immature cells. He also stressed that CBA/N mice respond to Brucella but not to Ficoll and to certain haptens as noted above. These observations which all deal with T independant antigens, therefore do not suggest the existence of different stages in differentiation but rather alternative pathways of differentiation. This was further stressed by Kincade who showed that CBA/N mice totally lacked the ability to produce B cell colonies in vitro. This technique was originally thought to represent the clonal growth of B cell precursors in general, but now appears to be selectively detecting a certain subpopulation, namely one that is lacking from CBA/N mice. Accordingly at this stage, the data appears more compatible with the view that one particular arm of B cell differentiation is missing from the CBA/N mice, rather than there being a single temporal B cell sequence, which is arrested at a certain point in these mice.

The next major area of discussion concerned the use of these mice for the production of antisera that are specific for B cell surface markers. It is apparant that a variety of different antibodies can be produced by immunizing CBA/N mice with cells from various sources. At least two different antisera have been specifically named, namely Lyb-3 and Lyb-5, and indications were presented by Kessler of at least two other antibody activities that could be produced by the defective mice. The first antiserum described by Huber termed Lyb-3, is clearly distinct from other Lyb sera in that no genetic polymorphism was detected by this antiserum. In this sense it is more analogous to an isotypic antiserum rather than an alloantiserum. In normal mice, at least 90% of IgM cells were positive for this marker whereas only 40% of IgM/IgD bearing cells are positive for the marker. Thus this antiserum appears to distinguish between subpopulations of B cells in normal mice. Schur described the Lyb-5 serum, which does detect a genetic polymorphism, and 75% of immunoglobulin positive and CRL positive cells expressed this marker, with again heterogeneity being observed in that some B cells were negative. Kessler described another antiserum prepared in CBA/N mice against CBA/J cells, which by cell surface iodination and gel analysis, detected a component that was also present on B cells, as shown by enriched expression on Ig positive C3 receptor positive cells, and also showed genetic polymorphism. Correlative studies with various strains showed that this determinant was not that of Lyb-3 or Lyb-5, and might preliminarily be termed Lyb-6. Mosier further confirmed heterogeneity of cells bearing Lyb-5 markers, in that in normal mice anti-Lyb5 serum totally abolishes the TNP-Ficoll responses but leaves anti-TNP Brucella responses intact.

It is thus becoming increasingly evident that there are a wide number of B cell expressed cell surface markers that are encoded by polymorphic genes, and are expressed on different subpopulations of B cells. The use of CBA/N mice as recipients permits the detection of many of these surface markers, and it will clearly be required for the various groups producing different antisera to attempt to correlate their various observations in order that a clear definition of the different specificities and molecular association of these determinants be obtained.

A variety of other B cell surface markers have also been described recently in the literature and were briefly discussed here. The Ly-4 and Ly-7 markers described by McKenzie were reported by Warner to show differential expression on murine B and T cell tumors, in that the Ly-4 marker is present on several B cell tumors of early differentiation stages, and the Ly7 marker is born by all B cell tumors studied, and with only one clear exception, is not present on T cell tumors.

The use of heteroantisera for detecting B cell subpopulations was introduced by North, who described a heteroantiserum prepared against Abelson virus induced tumors, which reacted with only approximately 50% of B cells, further indicating that heteroantisera may also be a potential method for distinguishing B cell subpopulations.

Further studies that will combine these various serological approaches of distinguishing B cell subpopulations, in combination with various functional tests, may thus eventually lead to a clearer definition of the range of B cell heterogeneity and the functional subpopulations.

N.L. Warner
J. Press

CELL SEPARATION AND CHARACTERIZATION

Workshop Conveners: Len Herzenberg and Leon Wofsy

The challenge of lymphocyte diversity has fostered the development of cell sorting as a very significant research tool in immunology and potentially in other areas of cell biology. But technical advances do not match the pace at which subpopulations of lymphocytes and other cell types that make up the immune network are being defined. No one all-purpose technique of cell purification has emerged or is likely to be perfected. Instead, as this workshop demonstrated, the application of cell sorting to more and more problems has made it necessary to employ a variety of diverse methods. The workshop devoted itself mainly to considering experiences of those present in tackling problems where cell fractionation is a useful approach, evaluating the effectiveness of particular techniques both for selection of cell subsets and for study of their function.

The success of one laboratory in isolating suppressor T cells specific for keyhole limpet hemocyanin (KLH) is a striking example of the application of multiple separation techniques: 1) KLH-primed mouse spleen cells were first depleted of B cells by passage through an anti-Ig column; 2) specific purification was achieved with a KLH-Sephadex G-200 affinity column (and verified by demonstrating highly enriched suppressor activity in adoptive transfers); 3) the isolated cells, representing $< 0.5\%$ of the starting population of 10^9 spleen cells, were analyzed and further purified on the Fluorescence Activated Cell Sorter (FACS) by selecting for lymphocytes bearing surface antigen(s) coded in the *I*-J subregion.

A range of hapten-modified "affinity" materials continue to be used with varying success in isolating antigen binding B and T cells. Because of difficulties encountered in eluting cells from affinity columns, a significant advance appears to be the use of hapten-modified gelatin surfaces, although in one case the two methods are being used in sequence. Cells selected with hapten-gelatin can be brought to highest purity (for antigen binding as well as for function) by further procedures employing rosetting techniques and FACS. The use of hapten-modified nylon for cell purification has yielded a remarkable and valuable byproduct, namely, the apparent isolation of T cell antigen binding receptor molecules.

Results presented at the workshop showed the use of rosetting, followed by serial application of buoyant density and sedimentation velocity gradients, to isolate red blood cell-antigen binding lymphocytes from immune and normal mice. Cells were recovered in apparently high purity and sufficient numbers to permit some biochemical as well as functional studies. Target-specific cytotoxic lymphocytes have also been highly enriched by gradient centrifugation as effector-target conjugates and characterized in a single-cell cell-mediated cytotoxicity assay.

While much attention is given to purification of antigen binding cells, at least as great challenge attaches to the need to secure particular functional categories of B and T lymphocytes (and accessory cells) in high purity and high yield. This, too, involves approaches that take advantage of characteristic cell surface markers, as well as less specific methods that focus on differences in the physical properties of functionally distinguishable cell subsets.

One report showed the essential role of several different procedures to obtain highly purified human T cells. These cells were used to obtain antisera that, after absorption with autologous tumors, made possible the identification of subsets of human T cells. These subsets seem analogous, with respect to surface markers and function, to some of the murine T cell compartments.

Several reports demonstrated the fractionation of both B and T lymphocytes into a number of distinctive functional subpopulations. Using primarily velocity sedimentation methodology, one laboratory separated two apparently different types of primary AFC-progenitor B cells, a less mature set revealed in adoptive transfer assays and a more mature set detected in cell culture assays. Another laboratory used a ficoll velocity sedimentation application to enrich for suppressor, helper, and cytotoxic T cells in populations stimulated by Con-A, MHC antigens, or conventional antigens. Another report described an FACS separation of B cells into fractions with different densities of surface immunoglobulin, which could be further distinguished on the basis of complement receptor and non-H2 linked minor lymphocyte stimulating antigens.

The FACS is clearly the most versatile and generally effective means now in use for specific cell purification and analysis, although other techniques seem required when it is necessary to process very large numbers of cells in reasonable time periods. Technical advances have extended the multiparameter basis on which cells are sorted by the FACS, and it is now feasible to separate fluorescein-labeled from rhodamine-labeled cells. It was stressed that the obvious limitation in experiments with the FACS often is the failure to verify stringently the serological specificity of reagents used to attach fluorescent cell markers.

A new development in specific cell fractionation is the use of hapten-sandwich labeling techniques in conjunction with various established rosetting and affinity separation methods. Antibodies (including mouse alloantibodies) against any cell surface antigen can be conjugated extensively with a hapten without significant loss of antibody activity. Lymphocytes specifically labeled with such hapten-modified antibodies spontaneously form stable rosettes with anti hapten antibody-coupled red cells, permitting very rapid and effective separation by gradient centrifugation. Anti-hapten antibody-coupled gelatin surfaces, including beads, may be especially useful for isolating highly purified populations of hapten-sandwich labeled cells.

Amoung the participants in discussion at the workshop were: K. Okumura, D. Scott, J. Kenny, E. Grimm, K. Shortman, S. Sharrow, T. Hartman, C. Scott, B. Rubin, R. Ashman, B. Rotman, H. Tse, S. Schlossman, L. Herzenberg, and L. Wofsy.

REGULATION OF THE ANTIBODY RESPONSE BY T CELL PRODUCTS DETERMINED BY DIFFERENT *I* SUBREGIONS

Tomio Tada

Laboratories for Immunology, School of Medicine,
Chiba University, Chiba, Japan

ABSTRACT. Two distinct antigen-specific T cell factors, one suppressive (TsF) and the other enhancing (TaF) the antibody response were identified as the products of *I* region of *H-2* complex. The gene which codes for the TsF was mapped in *I-J* subregion, and that for the TaF in *I-A* subregion. Both molecules are selectively expressed on T cells but not on B cells, and are antigenically distinct from classical B cell Ia antigens. The TsF is a product of the Ly $2^+,3^+$ T cell, and the TaF is produced by the Ly 1^+ T cell. The acceptor sites for these T cell factors were found to be present on Ly $1^+,2^+,3^+$ nylon adherent T cells but not on B cells, and they are also encoded by genes in *I-J* and *I-A* subregions.

The mechanism whereby the TsF suppresses the antibody response was analysed. The available evidence suggests that the TsF acts on Ly $1^+,2^+,3^+$ nylon adherent T cells to induce the new suppressor T cells from the latter. The newly 'induced' suppressor T cells have Ly $2^+,3^+$ phenotype, and in the presence of corresponding antigen they exert a nonspecific suppressor effect on the antibody response against unrelated antigen. It is suggested that there exists an amplification loop in the suppressor system in which a small amount of TsF initially present can turn on a potent nonspecific suppressor mechanism.

INTRODUCTION

Studies on the genetic control of the immune response have been achieved mainly through two different approaches which are now closely inter-related; 1) the genetic analysis of the ability of animals to mount an immune response to given antigen, and 2) the demonstration and characterization of products of *I* region genes which mediate the regulation of cell interactions. The first approach has led to the discovery of specific Ir and Is genes, and gene complementations as reviewed by Dr. Benacerraf (in this volume). I shall mainly discuss the second approach which is concerned with cell surface materials and 'factors' bearing *I* region gene products. The major points which I would like to deal with are; 1) the genetics of these antigen-specific T cell factors in relation to the well defined B cell Ia antigens, 2) the expression of these molecules on different subsets of T cells, 3) the inter-relationship between these T cell factors, and 4) the mechanism whereby these T cell factors regulate the immune response.

A FAMILY OF ANTIGEN-SPECIFIC T CELL FACTORS AS IMMUNOREGULATORY MOLECULES

Recent reports from various laboratories indicate that there exists a family of immunoregulatory molecules derived from antigen-stimulated T cells, which share certain common characteristic features. They are antigen-specific in terms of both the binding affinity for antigen and the dependence of their function on the presence of relevant antigen. Despite this antigen-specificity, they have no known immunoglobulin determinants, and their molecular weight being around 50,000 daltons. The most remarkable feature of these molecules, which has recently been determined, is that they are the products of genes mapped in *I* region of major histocompatibility complex (MHC). Furthermore, it has been shown that the receptor (acceptor) sites for these T cell factors on target cells are also encoded by *I* region genes. Thus it appears that *I* region products including these factors play an integral role in the maintenance of the regulatory network consisting of different populations of lymphoid cells for performing the well organized immune response to antigen.

Table 1 summarizes some properties of three known family members of antigen-specific T cell factors. It is apparent that there are some differences between these factors with respect to the genetics, site of action and mode of their functions. The helper (cooperative) factor (ThF) described by Taussig, Munro and Mozes (1-3) is produced by Ly 1^+ helper T cell, and is coded for by *I-A* subregion gene. It acts directly on B cells across strain and species barriers (4). The suppressor factor (TsF) to keyhole limpet hemocyanin (KLH) is produced by Ly $2^+,3^+$ T cell and acts on another subset of T cell (Ly $1^+,2^+,3^+$ T cell, see below) with a strict *H-2* histocompatibility requirement between the TsF and responding cells, and is coded for by *I-J* subregion gene. Kapp *et al.* (5) showed that a similar factor obtained from non-responder strains to GAT can act on target cells of other non-responder strains of different haplotype, and thus under certain circumstances the *H-2* restriction appears to be not absolute. The amplifying (enhancing) T cell factor (TaF) is a product of the Ly 1^+ T cell, but is not a helper T cell-replacing factor as is Taussig's cooperative factor. It is an *I-A* subregion gene product and can interfare with the response of *I-A* subregion compatible strains in the presence of helper T cell. Thus there

TABLE 1

A FAMILY OF ANTIGEN-SPECIFIC T CELL FACTORS AS IMMUNOREGULATORY MOLECULES

Class	Subclass	Genetics	Comment
Antigen-specific T cell factor	Helper (ThF)	*I-A*	Produced by Ly 1 T cell, Target: B cells, No *H-2* restriction (Taussig, Munro, Mozes)
	Suppressor (TsF)	*I-J*	Produced by Ly 2,3 T cell, Target: Ly 1,2,3 T cell, Highly *H-2* restricted (Tada, Kapp)
	Amplifier (TaF)	*I-A*	Produced by Ly 1 T cell, Target: Ly 1,2,3 T cell, Highly *H-2* restricted (Taniguchi, Tada)

exist sufficiently overt differences between these T cell factors even through they apparently belong to the same family of biologically active molecules. In addition, similar types of molecules both suppressive and helpful have been demonstrated in other systems (6,7) which also have some unique features differing from those sited above.

In view of the multiplicity and complexity in the mode of action of these T cell factors, questions now to be asked are 1) what is the inter-relationship between these T cell factors ? and 2) how can we reconcile the apparent discrepancies observed in different experimental systems ? I shall try to answer these questions by referring to a novel T cell type having Ly $1^+,2^+,3^+$ alloantigens, which perhaps serves an intermediary role between different compartment of the immunoregulatory system, each of which utilizes different T cell factors as messengers.

SUBREGION ASSIGNMENT OF GENES CODING FOR THE SUPPRESSIVE AND AMPLIFYING T CELL FACTORS IN *H-2* COMPLEX

Our previous studies demonstrated that two distinct antigen-specific T cell factors can be extracted from thymocytes and splenic T cells of mice that were immunized with a relatively high dose of carrier antigen (8). One of these factors, designated as the antigen-specific suppressive T cell factor (TsF), can suppress both primary and secondary antibody responses of the syngeneic or *H-2* compatible recipient against a hapten coupled to the same carrier by which suppressor T cells were generated (9). The same antigen-specific suppressive activity is also demonstrable in an *in vitro* secondary antibody response, if the extract was added to the primed spleen cells at the start of cultivation (10).

The other factor, which is designated as the antigen-specific amplifying factor (TaF), was found present in the same carrier-primed T cell extract (8). The activity was demonstrable if the extract was added to the cultured spleen cells one to three days after the onset of cultivation. The function of TaF is also genetically restricted in that the factor obtained from one strain of mice can suppress the response of only syngeneic or *H-2* histocompatible spleen cells. Figure 1 shows representative results of the secondary IgG antibody response of DNP-KLH-primed C3H spleen cells on which both suppressive and enhancing effects of the thymocyte extracts from KLH-primed mice were manifested depending on the time when the extracts were added to the culture.

Evidence that these suppressive and amplifying effects of the T cell extracts are, in fact, associated with two discrete molecules came from an observation that a strain of mice (A/J),

FIG. 1.

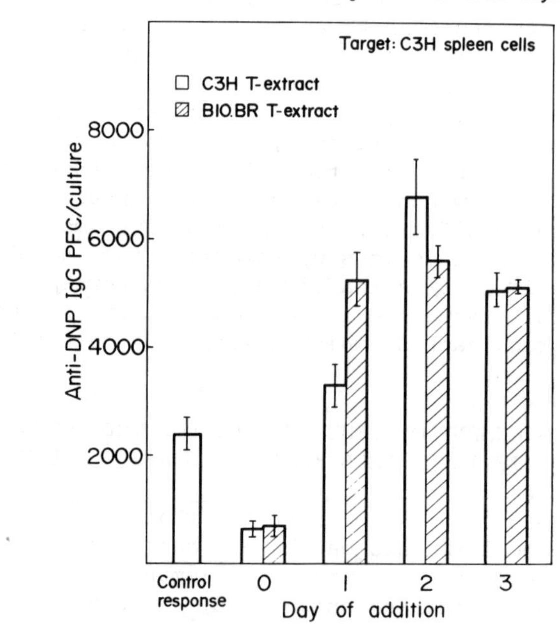

which does not produce the TsF to KLH, does produce the TaF (11). Furthermore, some anti-Ia antisera which successfully remove the TsF do not affect the TaF activity, and vise versa, and thus it is suggested that these factors are coded for by genes mapped in different *I* subregions (8).

In collaboration with Dr. Chella S. David, we have attempted to map the genes which code for the TsF and TaF among *I* subregions. We prepared immunoadsorbent columns composed of various anti-Ia antisera directed at restricted *I* subregions. The thymocytes or spleen cell extract of KLH-primed C3H mice was passed through one of these immunoadsorbents, and the residual suppressive or enhancing activity in the absorbed material was assessed by adding the effluents to the cultured syngeneic spleen cells on day 0 (suppression) or on day 2 (enhancement).

Figure 2 merely summarizes the results of experiment in which we explored the subregion specificity of alloantisera which remove the suppressive activity. It was first shown that antisera reactive with *I-A* or *I-C* subregions of $H\text{-}2^k$ were incapable of absorbing the TsF, while those having specificity for the chromosomal segment putatively present in the interval between *I-B* and *I-C* subregions could invariably remove the suppressive activity. This notion coincided with the discovery of Ia-4 locus by Murphy *et al.* (12), which marks a cell surface antigen on the allotype suppressor T cell. By the occurrence of two pairs of recombinants, i.e., B10.A(3R) and B10.A(5R), and B10.HTT and B10.S(9R), we were able to map the gene which codes for the TsF in a newly defined *I* subregion (*I-J*) intercalated between *I-B* and *I-C* subregions (13).

FIG. 2.

Absorption of $H\text{-}2^k$ suppressive T cell factor with anti-*I* region antisera

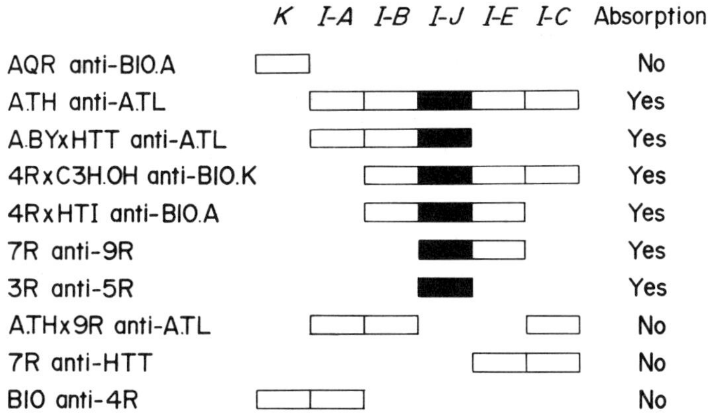

K I-A I-B I-J I-E I-C Absorption

AQR anti-B10.A No
A.TH anti-A.TL Yes
A.BYxHTT anti-A.TL Yes
4RxC3H.OH anti-B10.K Yes
4RxHTI anti-B10.A Yes
7R anti-9R Yes
3R anti-5R Yes
A.THx9R anti-A.TL No
7R anti-HTT No
B10 anti-4R No

The mapping of the gene which codes for the TaF was likewise attempted by adding the absorbed material to the cultured spleen cells on day 2. The simplified results are shown in figure 3. It is clear that antisera reactive with the products of *I-A* subregion are capable of absorbing the amplifying activity, whereas those lacking the specificity for *I-A* subregion are not. Hence, it is apparent that the TsF and TaF, both having the same specificity for KLH and almost identical physicochemical properties, are encoded by genes in *I-J* and *I-A* subregions, respectively.

FIG. 3.

Absorption of *H-2^k^* enhancing T cell factor with anti-*I* region antisera

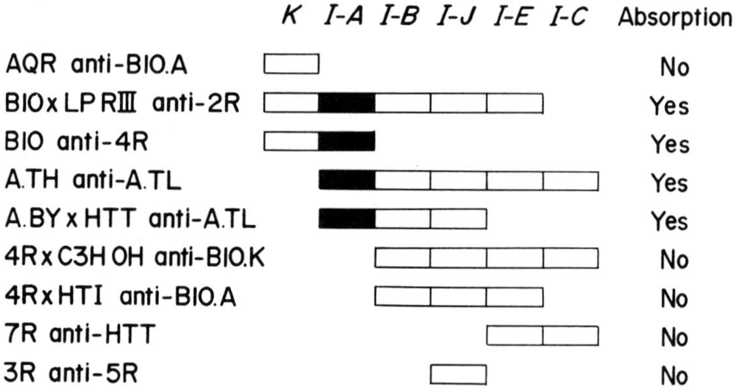

	K	I-A	I-B	I-J	I-E	I-C	Absorption
AQR anti-B10.A							No
B10 x LP RIII anti-2R							Yes
B10 anti-4R							Yes
A.TH anti-A.TL							Yes
A.BY x HTT anti-A.TL							Yes
4R x C3H OH anti-B10.K							No
4R x HTI anti-B10.A							No
7R anti-HTT							No
3R anti-5R							No

A question arises as to whether these Ia-bearing molecules are, in fact, antigenically identical to classical Ia antigens detectable mostly on B cells. More specifically, if these T cell factors are classified as members of a class of immunoregulatory molecules, it is conceivable that they are selectively expressed on functionally different subsets of T cells but not on B cells. It has been shown that the *I-J* subregion product is only detectable on suppressor T cells in the allotype suppression (12). It has also been demonstrated by Taussig *et al.* (14) that the antigen-specific helper factor could not be absorbed by antisera crossreactive with B cell Ia antigens of different haplotype. The same was true for the TaF in our own experimental system (unpublished). If the above postulate holds true, there are two distinct types of molecules determined by *I* region genes; one is the B cell Ia antigen and the other the immunoregulatory molecule selectively expressed on T cells.

In order to clarify the above problem, we absorbed the antiserum A.TH anti-A.TL, which was capable of removing both TsF and TaF, either with B or T cells, and the residual absorbing capacity for both suppressive and enhancing activities was assessed. The serum was extensively absorbed with splenic B cells of C3H and B10.BR, which had been prepared by treating the spleen cells with anti-brain antiserum (anti-BAT) and C, until no cytotoxic activity for splenic B cells was detectable. On the other hand, an aliquot of the same antiserum was absorbed only once with 10^8/ml of splenic T cells which were separated by passing the spleen cells through a nylon wool column. The B cell cytotoxicity of the A.TH anti-A.TL was not affected by the single absorption with T cells. Gamma globulin fractions of these absorbed antisera were coupled to cyanogen bromide-activated Sepharose 4B, and the crude extract of thymocytes from KLH-primed C3H mice was absorbed by passing through the column of these immunoadsorbents. The residual suppressive and enhancing activities were assayed by adding the absorbed material to the culture of syngeneic primed spleen cells stimulated with an appropriate amount of antigen.

Table 2 summarizes the results of above experiments. The crude KLH-primed thymocyte extract suppressed or enhanced the *in vitro* secondary antibody response depending on the time when it was added to the culture. Both effects were abrogated by absorption of the extract with the immunoadsorbent of original unabsorbed A.TH anti-A.TL. This absorbing capacity of antiserum was found well retained after absorption with B cells even though the cytotoxic anti-Ia activity had almost completely disappeared. On the contrary, whereas the absorption of antiserum with splenic T cells did not affect the B cell cytotoxicity, the antiserum was found to have lost the ability to remove both enhancing and suppressive T cell factors from the KLH-primed thymocyte extract.

The conclusions derived from above experiments are: (1) The absorbing capacity of anti-Ia antiserum for both enhancing and suppressive T cell factors is removed by absorption with T cells but not with B cells, and thus both factors should derive from T cells: (2) Both factors carry determinants coded for by genes in *I* region, but nevertheless these determinants are antigenically distinct from Ia antigens previously determined by B cell cytotoxicity: (3) Since the enhancing T cell factor is a product of *I-A* subregion gene and is distinct from B cell Ia antigen, *I-A* subregion may have at least two loci, one coding for B cell Ia antigens and the other for the TaF (and probably for the cooperative T cell factor). The results indicate that both *I-J* and *I-A* subregions have loci which are specialized in determining the new class of biologically active molecules on subsets of T cells, which are different from serologically detectable B cell Ia antigens.

TABLE 2

T BUT NOT B CELLS ABSORB THE REACTIVITY OF ANTI-Ia ANTISERUM TO THE SUPPRESSIVE AND ENHANCING T CELL FACTORS

T-extract absorbed[a] with	Experiments	
	Suppression (C3H)	Enhancement (B10.BR)
Control (No T-extract)	2493 ± 355[d]	672 ± 63[d]
Unabsorbed	118 ± 41	1512 ± 234
A.TH anti-A.TL (anti-I^k)	2577 ± 163	607 ± 10
A.TH anti-A.TL absorbed with B[b]	2422 ± 338	672 ± 137
A.TH anti-A.TL absorbed with T[c]	293 ± 124	1473 ± 182

a. KLH-primed T-extract was absorbed with immunoadsorbents composed of anti-Ia antisera.
b. C3H and B10.BR spleen cells treated with anti-brain antiserum and C.
c. Nylon wool column purified T cell.
d. Number of anti-DNP IgG PFC/culture.

DIFFERENT SUBSETS OF T CELLS PRODUCE AND ACCEPT TsF AND TaF

Since it has been determined that different *I* subregion genes code for TsF and TaF, it is probable that these are expressed on functionally different subsets of T cells. In collaboration with Dr. H. Cantor, the Ly phenotype of the cells which produce TsF and TaF was determined.

The spleen cells of C57BL/6J mice primed with KLH were treated with anti-Ly 1 or anti-Ly 2,3 antiserum together with rabbit complement, and live cells were collected by floating the cells on fetal calf serum. The cells were then disrupted by sonication to obtain the extract. Both the suppressive and enhancing activities of the extract from anti-Ly-treated spleen cells were assayed by the protocol stated above. Representative results are shown in Table 3. The extract of the cells which had been treated with anti-Ly 1 + C exhibited a strong suppressive activity, whereas after treatment with anti-Ly 2,3 the suppressive activity was completely lost. On

TABLE 3

LY PHENOTYPE OF T CELLS WHICH PRODUCE TsF AND TaF

Spleen cells[a] treated with	Experiments	
	Suppression (given on day 0)	Enhancement (given on day 2)
Control (No T-extract)	1520 ± 108	531 ± 54
C alone	813 ± 73	1017 ± 128
Anti-Ly 1 + C	797 ± 62	485 ± 139
Anti-Ly 2,3 + C	1558 ± 305	1163 ± 124

a. KLH-primed C57BL/6J spleen cells as the source of T cell factors.

the other hand, the extracted material from anti-Ly 1-treated spleen cells could not amplify the antibody response, whereas that after treatment with anti-Ly 2,3 showed a comparable enhancing activity to that of untreated spleen cells (Table 3). These results indicate that the TsF derives from Ly $2^+,3^+$ T cells, while the TaF is a product of Ly 1^+ T cells.

One of the remarkable features of both TsF and TaF is that these factors do not directly act on B cells. Especially for the TaF, it has been shown that the TaF cannot replace the helper T cell function in the response of T cell-depleted spleen cells to DNP-KLH, the fact which clearly distinguishes it from the antigen-specific helper factor (ThF) of Taussig and Munro (1). In fact, in order for the function of TaF, the presence of primed T cells in the responding cell population was found definitely necessary (see below).

One further characteristic feature of the TsF and TaF is that there exists a very strict histocompatibility requirement for the effective suppression and enhancement in cross-strain experiments (11), which is not observed in the ThF effect (4). For the effective suppression, the identity of the haplotype in *I-J* subregion between the donor of the TsF and responding spleen cells is both necessary and sufficient (13). The same type of genetic restriction is also noted in the TaF effect, in which the identity of *I-A* subregion is strictly required. These findings suggest that unlike the acceptor for ThF which is expressed on B cells, the acceptor sites for the TsF and

TaF are expressed only on certain T cells, and that such acceptor sites are also encoded by genes in *I-J* and *I-A* subregions.

Our recent studies have proven the above postulate to be the case: Both the TsF and TaF activities were removed by incubating the extract with normal spleen cells of the same haplotype for 1 hr at 4°C. This absorbing capacity of spleen cells was completely abrogated by the treatment of cells with anti-Thy 1.2 and C.

An interesting fact is that the cells capable of absorbing the TsF are adherent to nylon wool column (13). We separated splenic T cells by passage through a nylon wool column and tested for their absorbing capacity for TsF. To our surprise, the T cell fraction, which passed through the nylon wool column, could not absorb the suppressive activity, whereas the cells adhered to the column removed the activity effectively. Again, if the adherent cells were treated with anti-Thy 1.2 and C, this absorbing capacity for both TsF and TaF was completely abolished.

In order to learn the regulatory significance of the nylon adherent T cell which has the acceptor site for the TsF, we separated KLH-primed spleen cells into two populations which are adherent (A) and non-adherent (P) to the nylon wool. The population P (denoting nylon column passed cells) consisted of more than 90% T cells, and the population A contained usually around 10% of T cells which were killed by anti-Thy 1.2 and C. These cells were admixed with syngeneic DNP-primed splenic B cells (DNP-KLH-primed spleen cells treated with anti-BAT and C), and the mixture was cultured for 5 days in the presence of 0.1 μg/ml of DNP-KLH.

As was expected, nylon-purified KLH-primed T cells (P) exerted a strong helper effect giving rise to a good secondary anti-DNP-IgG antibody response (Table 4). However, the addition of KLH-specific TsF to the mixture of B and nylon-passed T cells (P) in the absence of adherent (A) cells displayed no significant suppression of antibody response. On the other hand, the mixture of B and adherent (A) cells did not produce the secondary antibody response, indicating that most specific helper T cells were present in the nylon passed cell fraction.

If the same B cells were mixed with nylon-passed (P) helper T cells and a small number (2.5×10^6) of nylon adherent (A) cells, a good secondary antibody response is obtained (Table 4). The addition of the KLH-specific TsF to this mixture now induces a significant suppression of antibody response. Since the proportion of T cells in the adherent cell population is usually about 10% (probably in the order of 10^5), the results indicate that the presence of a small number of nylon-adherent T cells is both necessary and sufficient for

TABLE 4

REQUIREMENT OF THE NYLON ADHERENT (A) T CELLS FOR THE SUPPRESSION OF ANTIBODY RESPONSE BY THE T CELL FACTOR

Cells	KLH-primed T cell factor	Anti-DNP IgG PFC/culture
B	——	11 ± 4
B + P	-	2231 ± 245
B + P	+	2345 ± 350
B + A	-	231 ± 23
B + A	+	261 ± 70
A + P + A	-	2176 ± 354
B + P + A	+	687 ± 154

B: DNP-KLH-primed spleen cells treated with anti-brain antiserum and C (5×10^6), P: KLH-primed nylon wool column-passed T cells (5×10^6), A: KLH-primed nylon wool column adherent cells (2.5×10^6).

the induction of the effective suppression. Other studies indicated that adherent cells from unprimed mice could not mediate the above effect, and thus the suppression was achieved via the intermediary of a special T cell type which had to be primed with the antigen. It was further found that the acceptor site on this intermediary T cell for the TsF was blocked by alloantisera having specificity directed to *I-J* subregion. Similarly, the treatment of the same adherent T cells with anti-*I-A* subregion products specifically blocked the acceptance of TaF leaving the acceptor site for TsF unaffected. These results indicate that suppression and enhancement of antibody response are, in fact, achieved via the intermediary of the third T cell type, which is adherent to nylon wool and expresses Ia antigen coded in *I-J* or *I-A* subregion.

The Ly phenotype of the intermediary cell was likewise determined by treating the adherent cells with anti-Ly 1 or anti-Ly 2,3 and C (Table 5). The residual adherent cells were added to the culture of DNP-primed B cells and nylon-purified KLH-primed T cells with or without the KLH-specific T cell factor. It was found that treatment of adherent T cell ei-

TABLE 5

LY PHENOTYPE OF THE ADHERENT ACCEPTOR CELL

Adherent T cell[a] treated with	KLH-primed T cell factor	Anti-DNP IgG PFC/culture
——	-	4129 ± 334
——	+	814 ± 352
Anti-Ly 1	+	3448 ± 556
Anti-Ly 2,3	+	4283 ± 738
(Anti-Ly 1)+(Anti-Ly 2,3)[b]	+	4180 ± 328

a. A toral of 2.5×10^6 adherent cells before treatment.

b. Mixture of cells which were treated with anti-Ly 1 and anti-Ly 2,3.

ther with anti-Ly 1 or anti-Ly 2,3 results in the loss of suppressor function in the presence of TsF. The mixture of anti-Ly 1-treated and anti-Ly 2,3-treated cells could not reconstitute the effect of intermediary cell type. The results indicate that the nylon-adherent T cells which accept the TsF are Ly $1^+,2^+,3^+$ T cells. The direct proof for the presence of such cells is presented by Okumura *et al*. (in this volume).

THE MECHANISM OF SUPPRESSION

Then, how does the adherent T cell after acceptance of the TsF exert the suppressive effect ? A series of experiments was carried out to solve this problem. In the first experiment, the KLH-primed adherent T cell was cultured with KLH-specific TsF and antigen (KLH) for 48 hr, washed thoroughly, and then added to the culture of the mixture of DNP-primed B cells and KLH-primed nylon passed T cells (P) in the absence of TsF. As shown in Table 6, even though the TsF was not present during the subsequent culture period, the TsF-pretreated adherent cells (as indicated by 'C alone' in the Table 6) exerted a significant suppressive effect on the anti-DNP antibody response mounted by B and P cells. The simplest explanation is that the adherent T cell after accepting the TsF was turned to become a new suppressor T cell during the first culture period with the TsF and antigen. This 'induced' suppressor T cell now appears to suppress the antibody response

TABLE 6

LY PHENOTYPE OF THE 'INDUCED' SUPPRESSOR T CELL

Cultured adherent[a] cells treated with	Anti-DNP IgG PFC/culture	Suppression
Control	1577 ± 163	——
C alone	166 ± 43	+
Anti-Ly 1	195 ± 55	+
Anti-Ly 2,3	1314 ± 260	-

a. Total of 2.5 x 10^6 adherent cells were treated either one of alloantisera and C, and then added to the culture of DNP-primed B cells (5 x 10^6) plus KLH -primed nylon purified T cells (5 x 10^6) in the presence of 0.1 μg/ml of DNP-KLH.

in the absence of initial TsF.

The Ly phenotype of the 'induced' suppressor T cell was studied by treating the adherent cells with anti-Ly antisera after the first 48 hr culture with TsF and antigen. A striking difference was found in the sensitivity to anti-Ly antisera of the adherent cells before and after 48 hr cultivation with TsF. As I described in the previous section, the acceptor T cell was sensitive to either anti-Ly 1 or anti-Ly 2,3. By contrast, the 'induced' suppressor T cell which was obtained after cultivation with TsF and antigen was killed only by anti-Ly 2,3 but not with anti-Ly 1 (Table 6). Although it is premature to draw a conclusion, it seems that the Ly $1^+,2^+,3^+$, acceptor T cell became Ly $1^-,2^+,3^+$ T cells during the 48 hr cutivation period with the TsF and antigen. This 'induced' suppressor T cell by itself actually inhibits the response of DNP-primed B cell and KLH-primed helper T cell. Thus it appears that the TsF is not the suppressor itself but the inducer of the actual suppressor T cell.

The next question we asked is whether the effect of the 'induced' suppressor T cell is antigen-specific or not. We took the DNP-primed B cell and egg albumin (EA) primed nylon purified (P) helper T cell as responding cells. To this mixture, we added KLH-primed nylon adherent T cell which is known to accept only the KLH-specific TsF. The cell mixture was cultured with DNP-EA and free KLH, which gave rise to a good secondary anti-DNP antibody response. With this secondary antibody response, we asked whether or not the KLH-specific TsF can suppress the antibody response which is mounted essentially by DNP-primed B cell and EA-primed helper T cell. The results of experiments are summarized in Table 7. In the presence of KLH-primed adherent T cell, KLH-specific TsF could suppress the secondary response to DNP-EA, even though the TsF is not specific for EA (Table 7). In the absence of free KLH, however, the KLH-specific adherent cells could not induce the suppression even though KLH-specific or EA-specific TsF was present (data not shown). Therefore, even though the final effector phase of suppression is nonspecific, antigen is needed to stimulate the 'induced' suppressor T cell. The same conclusion was derived from a reverse experiment in which the response to DNP-KLH was suppressed by EA-primed adherent T cell together with EA-specific TsF. The results imply that the TsF acts on the nylon adherent T cell to induce a new suppressor T cell, which in turn exerts nonspecific suppressor effect on the antibody response to unrelated antigen, provided the relevant antigen for the adherent T cell coexists. The results indicate that a small number of the initial antigen-specific suppressor T cell would turn on a very potent amplification loop recruiting the new suppressor T cell from nylon adherent Ly $1^+,2^+,3^+$ population, which finally leads to a

TABLE 7

REQUIREMENT OF THE IDENTITY OF SPECIFICITY BETWEEN ADHERENT T CELLS AND TsF

T cells		T cell factor specific for	Anti-DNP IgG PFC/culture
P primed with	A primed with		
a) EA	KLH	——	1259 ± 136
EA	KLH	KLH	310 ± 120
EA	KLH	EA	1203 ± 88
b) KLH	EA	——	1848 ± 320
KLH	EA	EA	260 ± 31
KLH	EA	KLH	1623 ± 187

Stimulating antigens:a) DNP-EA + KLH, b) DNP-KLH + EA.

strong nonspecific suppression of the immune response against unrelated antigen. Our more recent studies indicated that the 'induced' suppressor T cell acts across the *H-2* barrier to suppress responses of B and P cells derived from other strains (unpublished).

Figure 4 illustrates our hypothetical schema of the immunoregulatory system in which two compartments are intentionally separated. In the regulatory compartment, antigen-specific T cell factors produced by Ly 2,3 (TsF) and Ly 1 (TaF) T cells would act on Ly 1,2,3 T cells perhaps via complementary interaction between factors and acceptor sites both of which are encoded by genes in the same *I* subregions. At least in the suppressor system, it was suggested that the Ly 1,2,3 T cells after accepting one of the factors would become actual suppressor T cells being associated with the phenotypic changes in Ly antigens. In the presence of initial antigen by which these acceptor cells were primed, they will exert antigen-nonspecific effects on the second compartment in which classical T-B cooperation takes place. This second regulatory step may be mediated by various nonspecific T cell factors already reported by others, and has no *H-2* restriction. In addition, the presence of the antigen-specific cooperative factor (ThF) operative probably in the second compartment has been identified by Taussig and Munro (1), which directly acts on B cells without *H-2* restriction.

If the above postulates are taken collectively, one can explain some of the conflicting results obtained from various experimental systems in which multiple T cell factors are per-

FIG. 4.

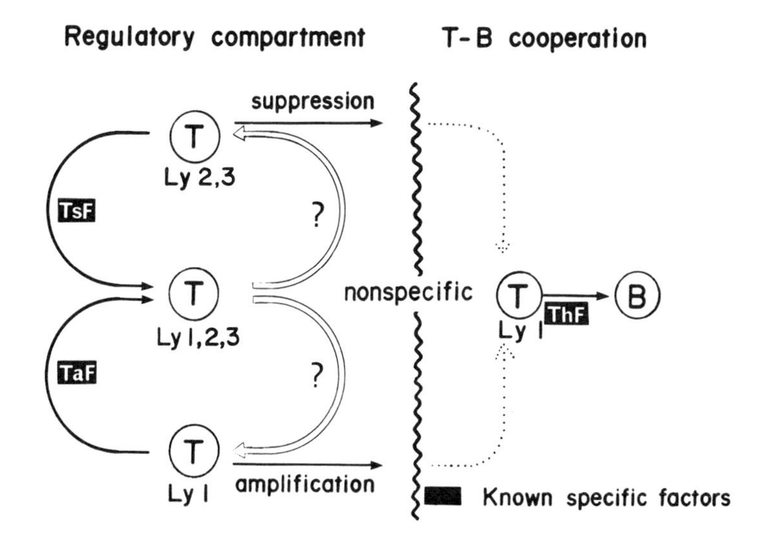

forming different roles in different compartments. Some of them may be strictly antigen-specific and the other *H-2* restricted, although they would belong to a single class of biologically active molecules. The future chemical and functional analyses of these T cell factors would give the final answer how the immunoregulatory system maintains its integrity by these genetic materials.

ACKNOWLEDGEMENTS

The works presented here have been performed in collaboration with many friends in different countries as is in the immunoregulatory system. Notably, Drs. C. S. David, B. Benacerraf, H. Cantor and D. H. Sachs provided us with many of the antisera with invaluable advise. Drs. L. A. Herzenberg, H. O. McDevitt and D. C. Shreffler gave us a lot of important criticism. I have been well informed by other friends who kindly sent me preprints and unpublished data. I would like to acknowledge my sincerest thanks for their warmest support and friendship, without which I have never been able to follow the progress of immunology. I would like to acknowledge with a proud the dedicating and tireless collaborations by my colleagues in the Laboratories for Immunology, Chiba University, Drs. M. Taniguchi, K. Okumura, T. Takemori, T. Tokuhisa and K. Hayakawa who participated in the works cited herein. The technical and secretarial assistances by Mr. H. Takahashi and Ms. Yoko Yamaguchi are also gratefully acknowledged. This work was supported by grants from the Ministry of Education, Culture and Science, and from the Ministry of Health.

REFERENCES

1. Taussig, M.J., and Munro, A.J. (1974) *Nature (Lond.)* 251, 63.
2. Munro, A.J., Taussig, M.J., Campbell, R., Williams, H., and Lawson, Y. (1974) *J. Exp. Med.* 140, 1579.
3. Taussig, M.J., Mozes, E., and Isac, R. (1974) *J. Exp. Med.* 140, 301.
4. Taussig, M.J., Munro, A.J., Campbell, R., David, C.S., and Staines, N.A. (1975) *J. Exp. Med.* 142, 694.
5. Kapp, J.A., Pierce, C.W., DeLa Croix, F., and Benacerraf, B. (1976) *J. Immunol.* 116, 305.
6. Zembala, M., and Asherson, G.L. (1974) *Eur. J. Immunol.* 4, 799.
7. Erb, P., and Feldmann, M. (1975) *Eur. J. Immunol.* 5, 759.
8. Tada, T., Taniguchi, M., and David, C.S. (1977) *In: XLI Cold Spring Harbor Symposium on Quantitative Biology.* in

press.
9. Takemori, T., and Tada, T. (1975) *J. Exp. Med.* 142, 1241.
10. Taniguchi, M., Hayakawa, K., and Tada, T. (1976) *J. Immunol.* 116, 542.
11. Taniguchi, M., Tada, T., and Tokuhisa, T. (1976) *J. Exp. Med.* 144, 20.
12. Murphy, D.B., Herzenberg, L.A., Okumura, K., Herzenberg, L.A., and McDevitt, H.O. (1976) *J. Exp. Med.* 144, 699.
13. Tada, T., Taniguchi, M., and David, C.S. (1976) *J. Exp. Med.* 144, 713.

THE I REGION GENES IN GENETIC REGULATION

Baruj Benacerraf, Carl Waltenbaugh, Jacques Thèze,

Judith Kapp*, and Martin Dorf

Department of Pathology, Harvard Medical School

Boston, Massachusetts 02115

ABSTRACT. The I region of the major histocompatibility complex of mammals is concerned with the regulation of immune responses to thymus-dependent antigens. For many antigens, the control of immune responses results from the interaction of Ir genes in the I-A (β genes) and in the I-C subregions (α genes). These genes complement in both the cis and trans position. Moreover, evidence of coupled complementation has been obtained, indicating that asymmetric complementation patterns occur between different α and β alleles. Evidence has also been obtained in several systems and different laboratories that the I region gene products play a critical role in the specificity of antigen presentation by macrophages to specific T cells. These systems will be discussed and their relevance to Ir gene defects evaluated. In addition, many, but not all, genetic non-responder strains develop specific suppressor T cells following injection with the relevant antigen. The genetic control of specific T cell suppression has also been ascribed to the I region of the murine H-2 complex and the same phenomena of coupled complementation has been observed for specific immune suppression genes as for Ir genes. A specific suppressor factor can be obtained from thymocytes or peripheral T cells of specifically suppressed mice. Such a factor has been obtained from DBA/1 or A.SW strains for the terpolymer GAT and from BALB/c and B10.BR mice for the copolymer GT. These factors suppress specific immune responses of selected allogeneic as well as syngeneic strains. The most informative example of suppression across H-2 concerns the behavior of BALB/c GT factor in A/J mice. This factor suppresses the response of A/J mice to GT-MBSA

* Present address: Department of Pathology
The Jewish Hospital of St. Louis
St. Louis, Missouri 63110

although this and other strains bearing the H-2^a haplotype are not suppressed by GT alone, indicating two steps in the generation of GT suppressor T cells. Moreover, evidence has been obtained in both the GAT and GT system that specific suppressor extracts are able to stimulate the generation of specific suppressor T cells. The immunochemical properties of specific suppressor factors are being investigated. The GAT and GT factors can be specifically retained by an antigen-Sepharose column, and by anti-I immunoadsorbent but not by an anti-mouse Ig column. The factor can be eluted from the antigen column. The B10.Br (H-2^k) GT suppressor factor is specifically removed by an anti-I-J^k immunoabsorbent. The data to be reported illustrate, therefore, that the I region regulates immune responses specifically by two correlated processes: 1) the stimulation of specific helper or suppressor T cells by controlling the manner in which antigen is presented to these cells, and 2) the production of antigen specific T cell factors endowed with helper or suppressor effects on the response of other immunocompetent cells.

INTRODUCTION

1. The H-linked Immune Response Genes.

The I region of the MHC is critically important in the specific regulation of immune responses. Our understanding of the gene products of this region and of the functions they perform in the immune system has progressed at an accelerated pace as more investigations are concerned with these problems and the appropriate genetic strains are available, thanks to the fundamental contributions of transplantation biologists. It appears appropriate, both historically and in order to analyse how our perceptions have evolved of this very complex field, to present a brief account of the facts and issues concerning I region genes.

The I region was initially identified and named as the genetic region where immune response (Ir) genes are coded which either individually or through complementation control specific immune responses (1,2). These Ir genes have been discovered in all mammalian species investigated and behave identically in all species (3). What is their function? They control the capacity to develop cellular immune responses to specific antigens as well as the ability to mount sustained humoral responses to these antigens (4). These antibody responses were shown to depend upon the ability of Ir genes to control the development of specific helper T cells against the antigens concerned.

It was indeed established that: a) Ir genes control the response to the carrier moiety of hapten-conjugates and not to the haptens (5), and b) the specific defect in a non-responder may be overcome by immunization with the antigen coupled to an immunogenic carrier for which appropriate Ir gene function exist (6). Therefore, the prediction could be made which proved to be correct, that H-linked Ir genes would be found to control specifically the response to thymus dependent antigens, and conversely that thymus independent antigens are not under H-linked Ir gene control (2,4).

2. The H-linked Immune Suppressor Genes.

The H-linked Ir genes appear, therefore, to control in an antigen specific manner, the development of cellular immunity and helper T cell activity to T dependent antigens. It became rapidly apparent, however, as a consequence of the pioneering studies of Gershon and Kondo (7) and of Cantor and Boyse (8) that the regulatory activity of T cells on immune responses could also be manifested by antigen specific suppression, and that the specific helper and suppressor T cells belong to distinct T cell classes bearing different Ly phenotypes. The discovery of the negative regulatory activity of suppressor T cells raised the issue whether these specific suppressor T cell responses were also controlled by genes in the I region of the MHC in a manner similar to the control by other I region genes of specific helper T cell function. This proved to be the case. A new experimental approach was developed to document the genetic control of specific suppression by I region genes. Use was made of the ability of non-responder animals for the copolymers Glu^{60}, Ala^{30}, Tyr^{10} (GAT) (9) or Glu^{50}, Tyr^{50} (GT) (10) to make antibody responses to these antigens coupled to an immunogenic carrier such as methylated bovine serum albumin (MBSA). Previous injection of GAT or GT suppressed specifically the GAT-MBSA or the GT-MBSA responses respectively in the mouse strains bearing the appropriate H-2 haplotype (11). This suppression was shown to be caused by specific Ly 2,3 suppressor T cells and to be controlled by genes in the I region of the mouse H-2 complex (12,13).

Table 1

MAPPING OF COMPLEMENTING *Is* GENES RESPONSIBLE FOR SUPPRESSION

Strain	*H-2* Haplotype	*H-2* Region Formulae I	I-A	I-B	I-J	I-C	S	D	% Suppression of GT-MBSA response
B10	*b*	*b*	*b*	*b*	*b*	*b*	*b*	*b*	—
B10.D2	*d*	*d*	*d*	*d*	*d*	*d*	*d*	*d*	64
B10.BR	*k*	*k*	*k*	*k*	*k*	*k*	*k*	*k*	82
B10.S	*s*	*s*	*s*	*s*	*s*	*s*	*s*	*s*	87
B10.A	*a*	*k*	*k*	*k*	*k* |	*d*	*d*	*d*	—
5R	*i5*	*b*	*b*	*b* |	*k*	*d*	*d*	*d*	81
D2.GD	*g2*	*d*	*d* |	*?*	*b*	*b*	*b*	*b*	—
HTG	*g*	*d*	*d*	*d*	*d*	*d*	*d* |	*b*	80
7R	*t4*	*s*	*s*	*s*	*s*	*s*	*s* |	*d*	88
9R	*t4*	*s*	*s*	*?* |	*k*	*d*	*d*	*d*	—
A.TL	*t1*	*s* |	*k*	*k*	*k*	*k*	*k*	*d*	71
B10.HTT	*t3*	*s*	*s*	*s*	*s* |	*k*	*k*	*d*	—
			*Is-2** (I-A, I-B)			*Is-1* (I-C, S)			

Vertical bars refer to positions of crossing-over.

* Immune suppressor (*Is*) genes are provisionally termed *Is-2* (mapping in the *I-A* or *I-B*) and *Is-1* (mapping in the *I-C* or *S* subregion).

3) <u>Other Immunological Phenomena Controlled by I Region Genes.</u>

Other important immunological phenomena are controlled by I region genes besides specific immune responsiveness and suppression. These are:

a) The lymphocyte activating determinants which stimulate *in vitro* MLR and graft-versus-host reactions. There are a large population of T cells precommitted to these I region controlled specificities (14).

b) The interaction of helper T cells or their products with B cell clones for optimal responses to T dependent antigens (15) (controlled by the I-A subregion).

c) The expression on cells of the immune system of a new class of molecules, the Ia molecules. These are the specificities which stimulate the MLR and GVH (16,17).

d) The presentation of T dependent antigens by macrophages to T cells for adequate helper function (18) or delayed type hypersensitivity (DTH) (19) and the contribution of the Ia molecule on the macrophage together with the immunizing antigen to the specific selection of the T cell clones activated (20).

e) These Ia molecules are coded by different sub-regions of the I region, I-A, I-J, I-C, I-E, as defined by well-documented recombinant events. These Ia molecules coded by I-A and I-C subregions are expressed on many but not all B cells (16,17) and also on a significant fraction of macrophages (21). It is controversial whether helper T cells bear these gene products, but if they do they are present in much reduced amount. The Ia molecules coded by the I-J region are present on suppressor T cells (22). No Ia molecule is detected on killer T cells constituting a clear difference between these two Ly 2,3 cells.

f) The control of antigen specific factors from T cells. These factors have either helper properties and I-A determinants (23) or suppressive properties and I-J determinants (24).

To summarize the issues at this stage: Several, as yet, unrelated sets of immunological phenomena are controlled by I region genes such as: The development of specific helper or suppressor function to T dependent antigens, and the production of specific helper or suppressor factors and on the other hand the presence of Ia molecules predominantly on B cells, but also in lesser amount on macrophages and still lesser amount on T cells with helper or suppressor properties. An attempt will be made in the following sections to correlate these functions with the cellular and molecular assignment of I region controlled molecules.

4) The Cells In Which H-linked Ir Genes are Expressed.

The assignment of Ir gene function to cell types has depended upon the demonstration of the presence of the nonresponder defect in one or another of the interacting cells of the immune system. Macrophage, T and B cells.

Initially, based upon the restriction of Ir genes to the control of the response to T dependent antigens and to cellular immunity, the Ir genes were believed to be expressed exclusively on T cells (2). Furthermore, their antigen specific functions suggested that they code for the elusive T cell receptor. However, this early interpretation was an unwarranted oversimplification, since apparent T cell defects may also result from faulty antigen presentation by macrophages (25) and/or from the incapacity of B cell clones to be responsive to the regulatory activity of T cells (26,27,28). Examples of macrophage and B cell Ir gene controlled defects have now been well documented in the guinea pig (for macrophage) (25) and in the mouse (macrophage and B cell) in several systems.

The original observation of Rosenthal and Shevach (18) that the I region gene product on responder guinea pig macrophages was essential for the stimulation of the proliferative response to antigen of primed responder T cells was the first indication of a macrophage Ir gene defect in guinea pigs. The studies in mice of Schwartz and ourselves with the GLØ copolymer (28,29), which require the activity of complementing β+ and α+ genes in I-A and I-C (30), established the requirement of both β+ and α+ GLØ genes in macrophages as well as in B cells for responsiveness to this antigen. As to the T cells no data has been obtained which establishes or rules out the necessary expression of H-linked Ir genes in these cells at this time.

It is becoming increasingly evident, therefore, that specific Ir gene function is expressed in more than one cell of the immune system, and that an Ir gene defect can reflect an incapacity to respond at one of several levels or even an abnormally high suppressor T cell response (9,31) as will be discussed later. We may, therefore, formulate a general scheme to explain the function of Ir genes and their contribution to specificity. This model accounts for the identification of Ia molecules on macrophages and B cells and their possible role as Ir gene products.

The studies of Zinkernagel and Doherty (32), of Shearer (33) and of our laboratory on the specificity of cytolytic T cells (CTL) (34) has provided an indication how a) T cells are committed and restricted to react with antigen on cell membranes in relation to MHC products and how b) the same T cell reacts both with an antigen and an MHC molecule. In the case of CTL, the MHC molecule for which T cell receptors exist are the K and/or D gene products or their variants. (The I region products can also serve as weaker targets) (35,36). But, the same phenomena are indeed observed when we consider the manner in which helper (18) and DTH (19) T cells respond to T dependent antigens on macrophages in relation to the Ia molecules which these macrophages bear. The recent studies of Paul and associates (20) have convincingly demonstrated that immunization of $(2 \times 13)F_1$ guinea pigs with ovalbumin stimulates two distinct families of clones which are capable of responding to either the antigen on strain 2 macrophages or to the antigen on strain 13 macrophages. Only $(2 \times 13)F_1$ macrophages bearing ovalbumin can stimulate both clones. An Ir gene defect in such a system could result from the inability of macrophage Ia molecules to interact effectively or meaningfully with antigen to stimulate appropriate class of T cells, as initially proposed by Shevach *et al.* (37). A considerable restrictive specificity can be contributed by the Ia molecules and, therefore, by the Ir gene in this type of model. With respect to the Ia molecules on B cells and the demonstration alluded to earlier of Ir gene expression on B cells, the model postulates that the interaction of B cells with the appropriate helper T cell depends (particularly in primary responses) upon the ability of the appropriate T cell to recognize and select B cell clones on the basis of the antigen which the B cell has bound and of the capacity of the T cells to interact with the appropriate Ia-antigens complex on the B cell. Numerous instances have been presented of hapten-specific B cell clones being selected on the basis of the carrier molecules used for primary immunization. This is particularly the case for molecules under Ir gene control as shown by Schlossman (38) and recently in our laboratory by Kipps (39) for the response of DNP-GLØ. This hypothesis offers an interpretation for the long recognized and poorly explained observations (40) that Ir gene function affects the specificity of the antibodies produced, a finding well substantiated by the recent elegant studies of Berzofsky *et al.* (41). This ability of helper T cells to regulate and, thereby, select B cell clone must be considered to be one aspect of a fundamental mechanism whereby T cells also select for class, allotype and possibly idiotype in a specific manner (42).

5) Gene Complementation for Ir and Is Genes. Coupled Complementation.

The restrictive specificity of Ir gene function whether exercised at the level of macrophages, T or B cells as discussed earlier is well documented and implies the ability of an Ir gene product to interact with individual antigens or groups of antigens specifically in a manner which imparts unique specificity to the reaction product. It is not surprising, therefore, that gene complementation of Ir genes and Is genes (43) has been observed in many laboratories (see Table 2) and more particularly that such complementation between responder alleles at two distinct loci is restricted so that all β+ genes do not necessarily complement all α+ genes, a phenomena which we have termed coupled complementation and which has been reported in detail elsewhere (43). For the purpose of this discussion, we would like to stress that coupled complementation indicates a) that all responder Ir genes for a given antigen in different H-2 haplotypes are not all identical, and b) that the mechanism of interaction between Ir gene products may be productive of considerable heterogeneity and, therefore, of the required molecular framework for the specificity and restriction observed in Ir gene function.

6) Specific T Cell Factors Bearing I Region Specificities.

One of the most exciting findings of the past few years in the area of regulation of the immune response by T cells is the identification by several laboratories of antigen specific factors in the mouse believed to be produced by T cells and shown to bear antigenic specificities coded by I subregions. These factors display helper

TABLE 2

EXPERIMENTAL SYSTEMS INVOLVING DUAL H-LINKED *Ir* GENE CONTROL

Species	Antigen	Experimental Evidence		Type of Complementation	
		F_1 hybrids	Recombinants	α—β	β—β
Mouse	GLØ	+	+	+	+
	GLT^{15}		+	+	
	GLT^{5}		+	+	
	GLleu	+	+	+	
	H-2.2	+	+	+	
	LDH_B	+	+	+	
	(T,G)-A--L	+			+
	ℓ-OVA	+			
	HUL		+	+	
	HEL		+	+	
Rat	(T,G)-A--L	+			

activities *in vivo* (44,45) on antibody responses or suppressor activity *in vivo* and *in vitro* on antibody responses (46,47,48) and *in vivo* on contact sensitivity (49,50). These factors have many common properties and several distinctive ones.

a) They are produced as a consequence of antigen stimulation.

b) They are obtained from T cell populations and not from B cells or macrophages. Their synthesis by T cells, however, has not been formally demonstrated.

c) They are antigen specific and are selectively adsorbed by antigen-Sepharose columns.

d) They lack immunoglobulin constant region determinants of either L or H chain.

e) Their molecular weight by gel filtration is estimated to be between 40,000 and 50,000 daltons.

f) They bear distinct I region controlled specificities. The helper factor is removed by anti-I-A immunoadsorbent (23); the suppressor factors in three distinct systems bear determinants coded for by the I-J subregion (24,51,52).

7) Properties of GAT and GT Specific Suppressor Factor.

The first specific suppressor factor (SF) was described by Tada (46). He obtained active suppressor supernatants from the thymuses and spleens of mice injected with tolerogenic doses of conventional immunogens by subjecting the cell suspensions to ultrasonication and centrifugation. This SF was tested on the response to hapten-carrier conjugates and was found: a) to be carrier specific in its suppressor activity, b) specifically absorbed by an antigen column, c) to lack Ig determinant and d) to possess determinants of the I-J subregion of the H-2 complex (24). This region was, indeed, initially defined by its ability to code for specificities present on the surface of suppressor T cells (22) and on SF (24). The Tada SF was strain specific and did not suppress the response of mice which differed at I-J from the mice in which the factor was prepared.

Our own laboratory took advantage of the fact that: a) The predominant response of genetic nonresponder mice to GAT immunization is the production of GAT-specific suppressor cells detected in the thymus and spleen from 3 days to 21 days following injection with 10 μg GAT in alum (Maalox) and capable of suppressing GAT-MBSA antibody responses. b) Similarly the predominant response of "selected" GT nonresponder strains with the GT suppressor genotype is the selective development, following injection with 10 or 100 μg of GT, of suppressor T cells capable of suppressing the IgM and IgG responses of these strains to GT-MBSA (11). Using the methods described by Tada (46), soluble extracts were prepared from the thymuses and spleens of DBA/1 mice injected with GAT (47) or BALB/c and B10.BR mice injected with GT (53). Highly specific GAT-SF was obtained from DBA/1 mice and GT-SF from BALB/c and B10.BR thymus and spleen cells.

In our hands, the properties of the GAT or GT suppressor factors are identical, as far as they have been investigated, except for their distinct specificity for the respective copolymers. We shall list briefly the biological properties of these factors and their initial molecular characterization:

A. BIOLOGICAL PROPERTIES

a) The DBA/1 GAT SF can be obtained from thymuses from which the thymic lymph nodes have been removed and also from spleen of GAT primed mice. The spleen cells were further fractionated into adherent cells, and nonadherent cells. Then the B and T cells were separated with anti-mouse Fab columns. The suppressor activity was only observed in extracts prepared from T cells (47).

b) The suppressive activity was highly specific for the anti-GAT response.

c) The suppression was effective *in vivo* when 0.5 ml of the extract up to 1/8 dilution was administered intravenously on the day of immunization with GAT-MBSA (The extract was prepared from 6×10^8 thymus and spleen cells/ml). When tested on *in vitro* IgG responses to GAT-MBSA, the extract was generally suppressive at a dilution below 1/1000 (47).

The same range of activity was observed with the BALB/c GT-SF by *in vivo* or *in vitro* assays (53). In addition, if the extract was administered *in vivo* to BALB/c mice one week prior to immunization with GT-MBSA, it was suppressive at a significantly lower dilution than when administered on the day of immunization with GT-MBSA.

B. IMMUNOCHEMICAL PROPERTIES.

a) GAT and GT suppressor factors are selectively removed by GAT-Sepharose and GT-Sepharose respectively although some cross-reactive adsorbtion was observed between the two immunoadsorbents. Neither factors are removed by BSA-Sepharose (48).

b) A study was made of the elution of the DBA/1 GAT-SF from GAT-Sepharose with a KCl gradient. The peak suppressive activity as tested *in vitro* was found to elute at a 0.6 M KCl concentration, which is precisely the concentration of KCl which causes the elution of anti-GAT antibodies made by DBA/1 mice immunized with GAT-MBSA. Thus, both GAT-SF and GAT-antibodies appear to have similar affinity for the antigen. The GAT-SF was substantially purified by elution with KCl (48).

c) The GAT-SF from DBA/1 mice was not adsorbed by a rabbit anti-mouse Ig immunoadsorbent specific for the K light chain and for the γ, γ_2, M and α heavy chains (48).

d) The DBA/1 GAT-SF was absorbed by anti-H-2^q immunoadsorbent and more precisely by an anti-I^q immunoadsorbent (48).

e) The DBA/1 GAT-SF purified by elution with KCl was very suppressive *in vitro* of the GAT-MBSA response. Such a purified factor was selectively adsorbed by an anti-H-2^q immunoadsorbent, indicating that the same molecular complex has specificity for GAT and I region determinants (48).

f) The DBA/1 GAT suppressor activity in the crude extract is retained by an anti-GAT immunoadsorbent indicating that the factor in the crude extract is combined with antigen. However, the purified suppressor factor eluted by KCl from GAT-Sepharose is not adsorbed by an anti-GAT immunoadsorbent showing that the purified factor is free from antigen (48).

g) Because of the lack of appropriate alloantisera specific for the determinants of the I subregions of the H-2^q haplotype, we investigated the ability of B10.BR (H-2^k) GT-SF to suppress the IgG PFC response of BALB/c spleen cells to GT-MBSA in culture. The factor was suppressive at very high dilutions. The suppressive activity was specifically adsorbed by an immunoadsorbent prepared with the gamma globulin fraction of an anti-I-J^k antiserum (3R anti 5R) and not by an immunoadsorbent specific for H-2K^k + I-A^k determinants (51). Thus, the GT-SF bears the same antigenic specificities controlled by the I-J subregion detected by Tada and associates in their suppressor factor (24). This same anti-I-J^k immunoadsorbent was shown by Pierres and Greene, in our laboratory (52), to remove the suppressor activity for contact sensitivity to picryl chloride of a supernatant prepared from the lymph node cells of CBA (H-2K) mice injected previously with trinitrophenylsulfonate to suppress their sensitivity to picryl chloride. Thus, H-2K specific suppressor factors active either on antibody responses to GT-MBSA or on delayed sensitivity to picryl chloride bear specificities controlled by the I-J^k region. Mozes and associates (23) have verified that the same antiserum used to prepare these active I-J^k specific immunoadsorbents did not absorb a helper factor specific for (T,G)-Pro--L. This helper factor was removed by an anti-K^k + I-A^k immunoabsorbent which was inactive against either the GT or picryl suppressor factors.

We can conclude, therefore, that the suppressor factors active on both specific humoral and cellular responses are antigen specific and bear determinants of the I-J but not the I-A subregions. In addition, the factor in the crude extract is found bound to antigen and the possibility must be considered that the active suppressor material is a complex of specific factor with a very small amount of native antigen or processed antigen. This possibility is strongly supported by the data in the next section showing that the specific suppressor factor stimulates the development of suppressor cells.

8) Effect of Suppressor Extract in Syngeneic and Allogeneic Strains.

The attempts to suppress GAT-MBSA with H-2^q DBA/1 GAT-SF in allogeneic strains revealed that this factor is active not only in DBA/1 mice but also on A.SW and (DBA/1 x SJL)F_1 nonresponder mice (55). This result contrasts with the I region restrictions reported by Tada and associates for the activity of KLH-SF on secondary *in vitro* responses. We have, therefore, investigated the suppressive activity of H-2^d BALB/c GT-SF for BALB/c mice and for selected allogeneic strains, A/J (H-2^a) and B10.BR (H-2^k). BALB/c GT-SF suppressed very effectively the GT-MBSA IgG primary responses of these allogeneic strains. No indication of I region restrictions were observed (53). We must emphasize several important points: 1) GT-SF was administered on the day of immunization with GT-MBSA at a relatively high concentration (equivalent to 15 x 10^7 cells). (However, in other experiments mentioned earlier, B10.BR GT-SF was assayed on the response of BALB/c spleen cells to GT-MBSA in culture and found to be suppressive at very high dilutions (51).) 2) The BALB/c GT-SF is not active on all allogeneic strains, since in other experiments, it was not able to suppress the response of SJL mice to GT-MBSA (53). These genetic restrictions, however, are not frequent and are not explained solely on the basis of H-2 genotype.

9) Specific Suppressor T Cell Factor Stimulates the Development of Suppressor T Cells.

A most intriguing observation described in section 8 is the ability of BALB/c GT-SF to suppress the responses to A/J mice to GT-MBSA (53), considering that preimmunization of this strain with GT, either intravenously or intraperitoneally in a very wide dose range, does not succeed in suppressing its response to GT-MBSA (Table 3).

TABLE 3

EFFECT OF GT OR BALB/c (H-2^d) GT-SF ON THE GT-MBSA RESPONSE OF A/J (H-2^a) MICE

	GT-Specific IgG PFC per Spleen[a]	P Value
Control	8,700 ± 1,300 (12)[b]	
100 μg GT	8,200 ± 1,000 (12)	N.S.
Maalox Extract	7,700 ± 1,100 (16)	N.S.
BALB/c GT-SF (15 x 10^7 cells equivalent)	700 ± 250 (16)	< .001

a. 10 μg GT as GT-MBSA in Maalox-Pertussis.

b. Number of mice per group.

Furthermore, A/J mice treated with BALB/c suppressor extract are able to transfer suppression of GT-MBSA responses to normal A/J recipients,(Table 4) indicating that BALB/c GT-SF stimulates the production of GT-specific suppressor cells in this strain (54).

This raises the issue whether the genetic defect in A/J mice resides in their inability to produce GT-SF. An attempt was, therefore, made to prepare to GT-suppressor extract from A/J mice using the technique which had been successful for the production of GT-SF in BALB/c mice or of GAT-SF in DBA/1 or A.SW mice. The A/J GT extract was tested on both syngeneic A/J mice and allogeneic BALB/c mice.

TABLE 4

EFFECT OF BALB/c GT-SF ON THE STIMULATION OF SUPPRESSOR CELLS IN A/J MICE

	GT-Specific IgG PFC/Spleen[a]	P Value
A/J Spleen cells 20 x 10^6		
None	5,500 ± 700	
From mice treated with BALB/c Maalox extract	6,700 ± 900	N.S.
From mice treated with BALB/c GT-SF	400 ± 100	< .001

a. Twelve mice per group were immunized with 10 μg GT as GT-MBSA in Maalox-pertussis.

The A/J GT extract was unable to suppress the GT-MBSA responses of either of these two strains (53). Since BALB/c mice are readily suppressed either by GT-preimmunization or by BALB/c GT-SF, these results indicate that A/J mice do not produce GT-SF although they respond to BALB/c GT-SF and produce GT-specific suppressor cells. These data can also be considered to provide an essential control to exclude the possibility that small amounts of unprocessed GT in the BALB/c suppressor extract might be responsible for the suppression of the response observed in A/J mice and by extension also in BALB/c mice.

The finding that BALB/c GT-SF suppresses the response of "GT-nonsuppressor" A/J mice led to the conclusion that the factor stimultes the production of suppressor cells (53). In most of our experiments, the extracts were administered on the day of immunization with GT-MBSA based on our earlier assumptions that the GT-SF was indeed the product of the suppressor T cells and possibly their effector material. We have re-examined this question and investigated whether BALB/c GT-suppressor extract can suppress GT-MBSA responses when administered one week or even three weeks before GT-MBSA challenge. Highly effective suppression of GT-MBSA responses was achieved in animals injected with the suppressor extracts one or more weeks before challenge (54).

The conclusion that the BALB/c GT-suppressor extract blocks the response to GT-MBSA by stimulating the production of GT suppressor T cells requires also the demonstration of the adoptive transfer of the suppression to normal recipients with spleen cells with BALB/c mice injected one week earlier with BALB/c suppressor extracts. Twenty x 10^6 spleen cells from mice injected with BALB/c GT-SF one week previously were found to suppress the response of normal BALB/c recipients to GT-MBSA. Furthermore, the suppressor cells were found to be sensitive to treatment with anti-θ anti-serum and complement and, therefore, to be T cells (54).

In our opinion, the most remarkable observation presented in these experiments is the stimulation of specific suppressor cells by BALB/c GT-suppressor extract. This conclusion is based upon the transfer of GT-specific suppression with syngeneic spleen cells from mice treated with BALB/c GT-suppressor extract, in both BALB/c and A/J mice. The use of nonresponder strains to study specific suppressor T cells and the properties of specific suppressor factor has many advantages, *i.e.*, the absence of specific antibody synthesis and of specific helper activity, the ease of induction of specific suppression in certain strains with small doses of antigen administered by any route even in adjuvants. There is, however, a serious problem that has to be controlled for: the ease with which low doses of antigen induce suppression in nonresponder strains. This is particularly critical in experiments which demonstrate the stimulation of the production of suppressor cells by lymphoid extracts from GT-primed mice. The burden of proof requires that we establish the inability of the traces of antigen in the active extracts to stimulate suppressor cells by themselves, in the absence of suppressor factor. Several lines of evidence can be presented to this effect: 1) The amount of antigen present in the extract is insufficient to stimulate suppressor cells at the low dilution where the extract is effective in syngeneic BALB/c recipients (54); 2) A/J mice are specifically suppressed by BALB/c GT-SF but not by preimmunization with GT, and the suppression can be adoptively transferred to normal A/J mice by their spleen cells (53); 3) Cyclophosphamide treated BALB/c mice are not suppressed by GT preimmunization but are suppressed by BALB/c GT-suppressor extract (31); 4) Strong evidence has, therefore, been presented that a population of T cells sensitive to cyclophosphamide in BALB/c mice produces a GT-suppressor factor and that this factor is able to stimulate the response of a distinct population of suppressor T cells. The biological significance of this mechanism for the potentiation and maintenance of specific suppression is evident. The possible role of antigen or antigen fragments complexed to antigen-specific I region products with helper or suppressor activity in the regulation of immune responses is raised by our experiments and constitutes a critical problem for further investigation together with the study of the mode of action of these factors and of their relationships.

REFERENCES

1. McDevitt, H.O., Deak, B.D., Shreffler, D.C., Klein, J., Stimpfling, J.H., and Snell, G.D. (1972) *J. Exp. Med.* 135, 1259.
2. Benacerraf, B., and McDevitt, H.O. (1972) *Science* 175, 273.
3. Benacerraf, B., and Dorf, M.E. (1974) *Progress in Immunol.* 2, 181.
4. Benacerraf, B., and Katz, D.H. (1975) *Advances in Cancer Research* 21, 121.
5. Levine, B.B., Ojeda, A., and Benacerraf, B. (1963) *Nature* 200, 544.
6. Green, I., Paul, W.F., and Benacerraf, B. (1966) *J. Exp. Med.* 123, 859.
7. Gershon, R.K., and Kondo, K. (1971) *Immunology* 21, 903.
8. Cantor, H., and Boyse, E.A. (1975) *J. Exp. Med.* 41, 1376.
9. Kapp, J.A., Pierce, C.W., and Benacerraf, B. (1973) *J. Exp. Med.* 138, 1121.
10. Debré,P., Kapp, J.A., and Benacerraf, B. (1975) *J. Exp. Med.* 142, 1436.
11. Benacerraf, B., Kapp, J.A., Debré, P., Pierce, C.W., and de la Croix, F. (1975) Transplantation Rev. 26, 21.
12. Debré, P., Kapp, J.A., Dorf, M.E., and Benacerraf, B. (1975) J. Exp. Med. 142, 1447.
13. Benacerraf, B., Dorf, M.E. (1977) Proc. of Cold Spring Harbor Symposia on Quantitative Biology, in press.
14. Klein, J. (1975) Biology of the Mouse Histocompatibility Complex. Springer-Verlag, New York, New York.

15. Katz, D.H., Graves, M., Dorf, M.E., Dimuzio, H., and Benacerraf, B. (1975) *J. Exp. Med.* 141, 263.
16. David, C.S., Shreffler, D.C., and Frelinger, J.A. (1973) *Proc. Nat. Acad. Sciences U.S.A.* 70, 2509.
17. Hauptfeld, V., Klein, D., and Klein, J. (1973) *Science* 181, 167.
18. Rosenthal, A.S., Lipsky, P.E., and Shevach, E.M. (1975) *Fed. Proc.* 34, 1743.
19. Miller, J.F.A.P., and Vadas, M.A., Whitelaw, A., and Gamble, J. (1975) *Proc. Nat. Acad. Sci. U.S.A.* 72, 5095.
20. Paul, W.E., Shevach, E.M., Pickeral, S., Thomas, D.W., and Rosenthal, A.S. (1977) *J. Exp. Med.* 145, 618.
21. Unanue, E.R., Dorf, M.E., David, C.S., and Benacerraf, B. (1974) *Proc. Nat. Acad. Sci. U.S.A.* 71, 5014.
22. Murphy, D.B., Herzenberg, L.A., Okumura, K., Herzenberg, L.A., and McDevitt, H.O. (1976) *J. Exp. Med.* 144, 699.
23. Isac, R., Mozes, E., and Dorf, M.E. (1977) *Immunogenetics*, in press.
24. Tada, T., Taniguchi, M., and David, C.S. (1976) *J. Exp. Med.* 144, 713.
25. Shevach, E.M., and Rosenthal, A.S. (1973) *J. Exp. Med.* 138, 1213.
26. Katz, D.H., Hamaoka, T., and Benacerraf, B. (1973) *J. Exp. Med.* 137, 1405.
27. Benacerraf, B., Kapp, J.A., Pierce, C.W., and Katz, D.H. (1974) *J. Exp. Med.* 140, 185.
28. Katz, D.H., Dorf, M.E., and Benacerraf, B. (1976) *J. Exp. Med.* 143, 906.
29. Schwartz, R.H., Dorf, M.E., Benacerraf, B., and Paul, W.E. (1976) *J. Exp. Med.* 143, 897.
30. Dorf, M.E., and Benacerraf, B. (1975) *Proc. Nat. Acad. Sci. U.S.A.* 72, 3671.
31. Debré, P., Waltenbaugh, C., Dorf, M.E., and Benacerraf, B. (1976) J. Exp. Med. 144, 277.
32. Zinkernagel, R.M., Doherty, P.C. (1975) *J. Exp. Med* 141, 1427.
33. Shearer, G.M., Rehn, G.R., and Garbarino, G.A. (1975) *J. Exp. Med.* 141, 1348.
34. Burakoff, S.J., Germain, R.N., and Benacerraf, B. (1976) *J. Exp. Med.* 144, 1621.
35. Chiang, C.L., and Hauptfeld, V. (1977) *J. Exp. Med.* 145, 450.
36. Billings, P., Burakoff, S., Dorf, M.E., and Benacerraf, B. (1977) *J. Exp. Med.*, in press.

37. Rosenthal, A.S., and Shevach, E.M. (1976) In: The Role of Products of the Histocompatibility Gene Complex in Immune Responses., D.H. Katz and B. Benacerraf, eds., Academic Press, New York, New York.
38. Civin, C., Levin, H., Williamson, A., and Schlossman, S. (1976) *J. Immunol.* 116, 1400.
39. Kipps, J.T., Dorf, M.E., and Benacerraf, B. Unpublished data.
40. Bluestein, H.G., Green, I., Maurer, P.H., and Benacerraf, B. (1972) *J. Exp. Med.* 135, 98.
41. Berzofsky, J.A., Schechter, A.N., Shearer, G.M., and Sachs, D.H. (1977) *J. Exp. Med.* 145, 123.
42. Paul, W.E., and Benacerraf, B. (1977) *Science*, in press.
43. Dorf, M.E., Stimpfling, J.H., Cheung, N.K. and Benacerraf, B. In: Proceedings of the Third Ir Gene Workshop. H.O. McDevitt, ed., Academic Press, New York, New York (1977).
44. Taussig, M.J., Munro, A.J., Campbell, R., David, C.S., and Staines, N. (1975) *J. Exp. Med.* 142, 694.
45. Mozes, E. (1976) In: The Role of Products of the Histocompatibility Complex in Immune Responses. D.H. Katz and B. Benacerraf, eds., Academic Press, New York, New York.
46. Tada, T., and Taniguchi, M. (1975) *Transplantation Rev.* 26, 106.
47. Kapp, J.A., Pierce, C.W., and Benacerraf, B. (1977) *J. Exp. Med.*, in press.
48. Thèze, J., Kapp, J.A., and Benacerraf, B. (1977) *J. Exp. Med.*, in press.
49. Zembala, M., Asherson, G.L., Mayhew, B., and Krejci, J. (1975) *Nature* 253, 72.
50. Greene, M.I., Pierres, A., and Benacerraf, B. (1977) *Arthritis and Metabolism*, in press.
51. Thèze, J., Waltenbaugh, C., and Benacerraf. Unpublished data.
52. Pierres, A., Greene, M.I., and Benacerraf, B. Unpublished data.
53. Waltenbaugh, C., Debré, P., and Benacerraf, B. (1977) *J. Immunol.*, in press.
54. Waltenbaugh, C., and Benacerraf, B. (1977) In: Proceedings of the Third Ir Gene Workshop. H.O. McDevitt ed., Academic Press, New York, New York.
55. Kapp, J.A., Pierce, C.W., de la Croix, F., and Benacerraf, B. (1976) *J. Immunol.* 116:305.

A COMPARISON OF I REGION ASSOCIATED FACTORS INVOLVED IN ANTIBODY PRODUCTION

Marc Feldmann, Marilyn Baltz, Peter Erb, Sarah Howie, Sirkka Kontiainen and Jim Woody

ICRF Tumour Immunology Unit, Department of Zoology, University College London, and Institute of Microbiology, University of Basel

ABSTRACT. The properties of three mediators involved in cell interactions which are under study in our laboratory were compared. These are macrophage genetically related factor, abbreviated GRF, which is involved in the induction of T helper cells; T helper factor, involved in B cell induction; and T suppressor factor, which inhibits the helper pathway and hence the antibody response. GRF is a complex of Ia antigen and immunogen, with a MW of 55,000 which is pseudo-antigen-specific, due to the antigen it carries with it. Antigen specific T helper factor, obtained from supernatants of *in vitro* activated T helper cells induces antibody responses in T cell depleted lymphoid populations *in vitro* or *in vivo*, provided macrophages are present. It binds to immunoadsorbents of antigen and some anti-IgM antisera. Suppressor factor, SF, obtained from supernatants of *in vitro* activated T suppressor cells, is antigen specific and inhibits via action on T cells, probably helper cells.

INTRODUCTION

There is overwhelming evidence now that the immune system is organised as a network. The details of this network are not clear, nor is the role of idiotype anti-idiotype interactions within this network. We have been concerned with the cell interactions involved in antibody production and have identified multiple stages, each with complex cell interactions, in the induction of helper cells, of suppressor cells, and in the final triggering of B cells. Each of these stages of cell interaction involves two or three cells which must communicate information. By the use of double chamber tissue culture flasks and by collecting cell free supernatants it has been shown that many aspects of cell to cell communication is by mediators, and not by membrane contact.

The properties of these cell free mediators of cell interaction have been studied by many workers (1-10). Several of these have been reported to bind to immunoadsorbents of anti-Ia antisera - factors from macrophages (GRF), T helper cells and T suppressor cells have all been reported to contain Ia antigen, as characterised by the use of anti-Ia antisera. In this communication, we compare the properties of these three I region controlled factors, as studied in our laboratory, and relate these to findings from other laboratories.

METHODS

These have all been, or will soon be, published elsewhere (1,2,8).

RESULTS AND DISCUSSION

A. Comparison of GRF, HF and SF

The key properties are compared in Table 1. It is known that these factors are functionally distinct, i.e. cannot be used interchangeably, and thus there is a multiplicity of mediators or factors. Because these are all associated with the I region, there is a tendency to consider these factors as a family. This generalisation is probably justified, but there is still far too little biochemical information to know how close is the family resemblence. The I-A origin of GRF, and T helper factor, the latter determined by Munro and Taussig (3), in contrast to the I-J origin of SF, determined by Okumura, Tada, etc. (6), suggests the generalisation that 'help' and 'suppression' are controlled by different subregions, I-A and I-J respectively, and that the I region is functionally subdivided. This generalisation may still be premature, as antigens not under I-A control may have their GRF or HF controlled by other I subregions.

The targets of the factors have been studied by functional tests, as well as by absorption. The differences in the cellular targets has emphasised that these are three different factors. Regrettably, the definition of the cellular targets is solely of the functional target, and it is not possible to exclude the possibility that other cells have receptors for these factors - either less, or of lower affinity.

The question of genetic restrictions in cell cooperation is currently a vexed one. There are reports that T-B, T-M and suppressor (T-T) interactions are under H-2 restriction (11,1,6). However, there are exceptions in each of these categories. Potency of factor preparations regrettably need not reflect the number of active molecules present. It probably is usually a reflection of the sensitivity of the assay used.

TABLE 1

COMPARISON OF I ASSOCIATED FACTORS

PROPERTIES	FACTOR		
	GRF	HF	SF
SOURCE	Macrophage	Ly-1$^+$ cell	Ly-2$^+$3$^+$ T cell
NATURE	Ia-Ag	Ia-Ig-Ag combining site	Ia(?)-Ag combining site
I subregion	I-A	I-A (Munro & Taussig)	I-J (Tada)
Ag specificity	pseudo	+	+
MW	55,000	?	?
EFFECT	Induces helper cells	Induces B cells IgM IgG	Suppresses T cells
TARGET	Ly-1$^+$2$^+$3$^+$ T cell	Macrophage	T
Receptor	T only, I-A control	M (?B also)	T only (?I-J, Tada)
Genetic restriction	+	-	-
POTENCY	$\sim 10^{-1}$-10^{-2}	$\sim 10^{-4}$	$\sim 10^{-2}$-10^{-3}
STABILITY	+	+	+

Data from refs. 1, 2, 8, and unpublished.

B. Macrophage GRF

This material replaces the function of macrophages in helper cell induction *in vitro* (1). As yet it is not known whether it acts *in vivo*, or whether it is always controlled by the I-A subregion. The question of the antigen specificity of the Ia component is unresolved, although preliminary results suggest that the Ia component of GRF_{KLH} will combine with other antigenic fragments (Erb, unpublished data). This would imply that it is not antigen specific, and further defines the Ia-Ag bond as non-covalent. However, there is some evidence that macrophages are the site of specific Ir genes (12,13) which may suggest that GRF, as a reflection of macrophage Ir genes, may have a certain degree of antigen specificity. The relative contributions of Ia and antigen to the total MW of 55,000 is not yet known, nor whether the Ia is two chains or one. The relationship to B cell Ia, with its α and β chains is also not yet known.

Studies performed with GRF so far have emphasised its genetic restrictions, which mapped on the I-A subregion. The reasons for this restriction have not been fully explored. It appears to be a restriction of the *primary* response as antigen has not been deliberately introduced, and it seems unlikely that all our mice have been pre-primed by environmental antigen to three different antigens such as KLH or synthetic polypeptides such as (T,G)-A--L or GAT. Other workers have failed to demonstrate genetic restrictions in primary responses (14) suggesting that such restrictions only occur in secondary responses. We are trying to investigate this phenomenon by looking at receptors for GRF derived from various strains - if the receptors of unprimed T cells already show self preference, it seems likely that there is genetic restriction of the primary response. The mechanism of genetic restriction of T macrophage interaction is not known - we have found no evidence of suppression, but further experiments to exclude this possibility are needed, using antisera which kill suppressor cells, and using bone marrow chimaeric mice.

C. T helper factor (HF)

This material replaces the antigen specific action of helper T cells. While the first reports of this material were published five years ago (4), there are still great gaps in our knowledge of its nature, and its mode of action. Table 2 summarises some of the published data on HF, together with some of our current results to be published (2,4,3,9,15).

HF has been obtained from *in vitro* activated T cells (2) or *in vivo* educated T cells (3,9). The latter populations contain many (mostly) dead cells, macrophages and radio-resistant cells, which have all been exposed to mouse serum, which yields problems with respect to biosynthesis of the factors.

For these various reasons, we have decided to investigate HF using preparations obtained from *in vitro* inactivated cells. With these preparations, data has been obtained which may reconcile the discrepancy between HF results previously reported (Table 2). HF_{VITRO} is antigen specific, Ia^+, Ig^+ (using chicken anti-IgM), induces IgM and IgG responses *in vivo* and *in vitro*, and acts on B cells only in the presence of macrophages. The key differences from HF_{VIVO} of Munro, Taussig, Mozes, etc. (3,9) are (a) the Ig^+ (that type of serum has not yet been tested). However, the recent report of Mozes that HF to (T,G)-A--L shares idiotypic specificities with Ig suggests a reconciliation of the previously contrasting Ia^+ and Ig^+ helper factors; (b) requirement for macrophages, and absorption by macrophages of responder or non-responder type (15); (c) potency - HF_{VITRO} can be diluted out to 10^{-5} or so (undiluted is S/N of 15 x 10^6 cells/ml), when assayed either *in vitro* (2) or *in vivo* using hapten primed spleen cells (16). In contrast, HF_{VIVO} only dilutes to less than 1/10. Whether the difference is due to the method of assay or is due to differences in concentration of HF is not known. Stability of HF preparations are probably related to potency; (d) mode of action. Munro and Taussig (3) described a model of HF action, invoking an 'acceptor' a receptor for HF on B cells, with some non-responders not responding to HF due to lack of an 'acceptor site'. Using haptenated (T,G)-A--L, we have found that these strains specifically respond to the DNP, but not the (T,G)-A--L determinants. If there is a response to DNP, on DNP (T,G)-A--L, with HF_{TGAL} as the stimulus, there must be a receptor for the HF in that strain, and the above hypothsis is not correct in its simple form. As yet, we do not know the mechanism of this determinant specific Ir gene effect (15).

Despite this listing of differences between these factors it is our opinion that these reflect technical points in their preparation and assay. This opinion is chiefly based on the identical strain distribution of HF_{TGAL} producers and non-producers among non-responder mice, if HC are induced *in vitro* or *in vivo* with the optimal dose of antigen. It would be expected that if the two factors *were* different, they would be under the control of genes which may segregate differently in the various strains with independent haplotypes tested.

D. Comparison of specific suppressor factors or extracts

Suppressor extracts were reported by Tada et al. (6) and Kapp et al. (7), made by sonication of suppressor cells and centrifugation. Supernatants with suppressor activity (suppressor factors) were reported by Feldmann (17) from *in vivo* activated helper cells, Jacobson (18) and Herzenberg from allotype specific suppressor cells (19) and G. Chouat (personal

TABLE 2

COMPARISON OF SPECIFIC HELPER FACTORS

PROPERTIES	FACTOR		
	IgT (Feldmann & Basten)	HF_{VITRO} (Howie & Feldmann)	HF_{VIVO} (Munro, Taussig, Mozes)
SOURCE	Educated T cells (T,M,B + dead cells)	Vitro HC T	Educated T cells
NATURE			
Ag specificity	+	+	+
Ig	+	+	- (idiotype +)
Ia	NT	+	+ (I-A)
MW	?150,000		∿50,000
EFFECT	B cell induction of IgM and IgG responses		
Vitro	+	+	-
Vivo	NT	+	+
TARGET	Macrophage	Macrophage (?B)	?B
Genetic restrictions	-	-	-
Ir genes production	NT	+	+
Ir genes response	NT	+	+
POTENCY	10^{-3}	10^{-4}-10^{-5}	< 1/4
STABILITY	+	+	-

(data from refs. 2,3,4, and unpublished)

communication) from tolerant cells. We have investigated supernatants of suppressor cells activated *in vitro* with high doses of antigen (8). These were termed suppressor factor (SF). These were antigen specific, released by metabolically active cells, and acted on T cells in the helper pathway (8, and unpublished data). Characterisation of this material has not progressed far - it is heat labile, sensitive to proteases, but the MW is not yet known. The mode of action has been investigated, and preliminary results are different from ones obtained with suppressor extracts (manuscript in preparation).

(1) No H-2 genetic restrictions in their mode of action have been found.

(2) Target cell is a T cell, but not a Ly-$1^+2^+3^+$ nylon wool adherent cell. It may be the helper cell, although this is not yet proven. It does not act on HF.

(3) Strains which are not sensitive to suppressor extract are sensitive to SF or suppressor cells (20) *in vitro*.

(4) Effect is similar on IgM and IgG responses (unpublished data).

Kapp et al. (21) also found no genetic restrictions with extracts, but found they only worked on GAT non-responder strains. As yet it is not possible to reconcile these differences which may be due to assay procedures, or to real heterogeneity of materials. The possibility must be considered that SF and suppressor extracts are not identical, with the extract being a precursor molecule of the SF. Properties of SF and extracts are summarised in Table 3.

E. Conclusions

While information as to the existence, nature of mediators of the type studied here is increasing, it is still insufficient to permit generalisations as to their origin, functions or mode of action. It has been stressed by several workers that these families of factors share the crucial property of bearing Ia antigen. However, the significance of this Ia is not known, and probably varies between factors. In GRF, the Ia component appears to be important, as it is recognised by the target T cells (22). In T helper factor, the function is very unclear, as it is not related to the target strain or even species, and conceivably it may act as a transport piece to permit the factor to find the right cell. There is as yet little information on the number of peptide chains in these factors, or the number of genes controlling these factors. This information, together with more detailed biochemical analysis will further our understanding of these mediators and their role in immune responses in intact animals.

TABLE 3

COMPARISON OF SPECIFIC SUPPRESSOR 'FACTORS'

PROPERTIES	FACTOR	
	SF_{VITRO} (Kontiainen & Feldmann)	S Extract (Tada et al.)
SOURCE	Ly-2^+3^+ T cell	Ly-2^+3^+ T cell
NATURE		
Ag specificity	+	+
Ig	-	-
Ia	+	+ I-J
MW		∿50,000
EFFECT	Suppresses antibody response	
Vitro	+	+
Vivo	+	+
IgM	+	-
IgG	+	+
TARGET	T cell, Ly-1(?),NW non-adherent	T cell, Ly-1,2,3, NW adherent
Genetic restriction	-	+ (I-J)
Ir genes		+
POTENCY	S/N of ∿10^3-10^5 cells	extract of ∿10^7 cells
STABILITY	+	+

(data from refs. 6,7,8,21, and unpublished)

ACKNOWLEDGMENTS

The work described here was supported by the ICRF, MRC, US Public Health Service Grant No. IMB R01 AI 13145-02, the Swiss National Research Foundation, the European Molecular Biology Organisation and the Finnish Academy of Sciences.

REFERENCES

1. Erb, P., Feldmann, M. and Hogg, N. (1976) *Eur. J. Immunol.* 6,365.
2. Howie, S. and Feldmann, M. (1977) *Eur. J. Immunol.* (submitted).
3. Taussig, M.J. and Munro, A.J. (1976) In: *Leukocyte Membrane Determinants Regulating Immune Reactivity.* Eds. V.P. Eijsvoogel, D. Roos, and W.P. Zeiglemaker. Academic Press, New York, p. 255.
4. Feldmann, M. and Basten, A. (1972) *J. Exp. Med.* 136,49.
5. Schimpl, A. and Wecker, E. (1972) *Nature New Biol.* 237, 15.
6. Taniguchi, M., Hayakawa, K. and Tada, T. (1976) *J. Immunol.* 116, 542. Tada, T., Taniguchi, M. and David, C.S. (1976) *J. Exp. Med.* 144,713.
7. Theze, J., Kapp, J. and Benacerraf, B. (1977) *Proc. 3rd Ir Gene Conference* (in press).
8. Kontiainen, S. and Feldmann, M. (1977) *Eur. J. Immunol.* (in press).
9. Mozes, E., Isac, R. and Taussig, M.J. (1975) *J. Exp. Med.* 141,703.
10. Waldmann, H. and Munro, A. (1974) *Immunology* 27,53.
11. Katz, D.H., Hamaoka, T. and Benacerraf, B. (1973) *J. Exp. Med.* 137,405.
12. Rosenthal, A.S., Barcinski, M. and Blake, J.T. (1977) *Nature* (in press).
13. Schwartz, R.A., David, C.S., Sachs, D.H. and Paul, W.E. (1976) *J. Immunol.* 117,531.
14. Pierce, C.W., Kapp, J.A. and Benacerraf, B. (1976) *J. Exp. Med.* 144,371.
15. Howie, S. and Feldmann, M. (1977) manuscript in preparation.
16. Woody, J., Howie, S. and Feldmann, M. (1977) manuscript in preparation.
17. Feldmann, M. (1974) *Eur. J. Immunol.* 4,667.
18. Jacobson, E.B. (1973) *Eur. J. Immunol.* 3,619.
19. Okumura, K., Herzenberg, L.A., Murphy, D.B., McDevitt, H. and Herzenberg, L.A. (1976) *J. Exp. Med.* 144,685.
20. Howie, S. (1977) *Immunology* (in press).

21. Kapp, J.A., Pierce, C.W. and Benacerraf, B. (1976) In: *The Role of Products of the Histocompatibility Gene Complex in Immune Responses.* Eds. D.H. Katz and B. Benacerraf Academic Press, New York, p. 583.
22. Erb, P., Maier, B. and Feldmann, M. (1976) *Nature*, 263, 601.

HELPFUL AND SUPPRESSIVE FACTORS

Co-Chairmen
Judith Kapp, Jewish Hospital, St. Louis, Mo. 63110
Philippa Marrack, University of Rochester, Rochester N. Y. 14642

Scribe.
Michael Katz, University of California at Los Angeles, Los Angeles, California 90024

At this workshop we discussed a number of different factors and, as expected, the properties of one factor often seemed reminiscent of another, though rarely were two factors studied in different laboratories similar enough for us to identify them as the same. Presentations were divided into those describing non-specific or specific factors. The factors were further sub-divided into helpful or suppressive, though in retrospect sub-division could as profitably been made by the presence or absence of Ia antigens on the factors, or by their being genetically restricted.

To begin with non-specific factors, these were usually classified as such because they were generated in some non-specific or quasi-non-specific way, for example by Concanavalin A(ConA) activation, or in mixed lymphocyte reactions (MLR), and then seemed to act equally well on responses to a number of non-crossreacting antigens. In some cases the possibility that these factors might really be a mixture of all possible antigen-specific factors, generated by polyclonal activation of the appropriate type of cells, was eliminated by demonstrating that the activity of a particular non-specific factor for affecting responses to a particular antigen could not be eliminated by pre-absorbing that factor with the antigen in question.

Some non-specific factors are not genetically restricted, and others are. In fact interesting comparisons were made between two such factors, non-specific mediator (NSM, Marrack/Harwell) and allogenic effect factor (AEF, Delovitch). These factors have similar molecular weights (35,000), have the same target (B cells, NSM; B cells + macrophages, AEF) and can be generated in related ways, in MLRs (NSM) or in primed MLRs (AEF), and yet they are apparently the products of different cell types, since NSM probably is made by T cells, and AEF by B cells and/or macrophages. NSM is Ia^-, and non-restricted, AEF bears Ia antigens coded by the I-A subregion and is restricted in that it can only help the response of B cells sharing the same I-A or I-B subregion as the donor B cells. AEF also resembles genetically related macrophage factor (GRF, Kontiainen for Erb and Feldmann) in many ways. GRF is made by macrophages, is genetically restricted at the I-A subregion and bears Ia antigens. GRF is thought to act, however, on Ly123+ T1 cells, causing them to stimulate Ly1+ $T2_H$ cells, in addition GRF is non-specifically associated with antigen.

Parallels were drawn between two genetically restricted, MLR-suppressive factors. Dr. Solliday Rich discussed a factor found in the supernatants of primed mouse cell MLRs which bears I-C or S determinants and which specifically suppresses the MLRs of responder cells bearing the same I-C subregion as the donor. Dr. McMichael described a factor found in the supernatants of a MLR between a multiparous HL-A homozygous woman and her HL-A homozygous (but different) husband. This factor suppressed the MLRs of the female donor against her husband and children and 25% of unrelated stimulators, thus showing partial specificity for the stimulator. It also suppressed the MLRs of responders sharing the HL-A type of the woman, and thus seems to be the human analogue of the Solliday Rich factor.

Different cell types appear to be involved in the synthesis of non-specific factors. AEF is made by B cells and/or macrophages, NSM by T cells (see above). Soluble immune response suppressor (SIRS, Pierce), a 48-60,000 mol. wt. factor which inhibits the antibody-forming responses of spleen cells to antigen is also made by T cells. SIRS acts, however, by stimulating macrophages to secrete yet another factor which is the immediate cause of the inhibition. Farrar suggested that this second factor might be lymphocyte activating factor (LAF), but this is probably not the case. Levy described a factor, found in the sera of tumor-bearing mice, which suppresses the rejection of tumors by recipient mice and which is probably an autoantibody, presumably made by B cells.

All the specific factors discussed whose source has been identified are made by T cells. As expected helpful specific factors seem to be made by Ly1+ T cells (Kontiainen for Howie and Feldmann, Tada) whereas suppressive specific factors are made by Ly23+ T cells (Kontiainen, Tada, Herzenberg and Okumura).

The two specific helpful factors discussed at the workshop and meeting (IgT, Kontiainen for Howie and Feldmann, enhancing factor, Tada) were both I-A+. IgT is though to bind to macrophages in the presence of antigen and there stimulate B cell responses, it is not genetically restricted. Tada's enhancing factor acts late in antigen responses, possibly stimulating Lyl+ T_H cells via amplifier Ly123+ T1 cells. It is genetically restricted, and presumably acts on an earlier stage of antibody responses than IgT.

Of the specific suppressive factors those described by Tada for hapten protein responses, by Theze and Benacerraf for GAT responses and by Greene and Benacerraf for contact sensitivity responses all bear determinants coded by the I-J subregion. Where tested the T cells synthesising suppressive factors are Ia+ (Kontiainen), in some cases the sub-region bourne has been identified as I-J (Herzenberg and Okumura, Tada). Some controversy exists over the problem of genetic restriction, the factor described by Tada only suppresses cells bearing the same I-J region as the donor cells whereas other factors (Kontiainen, Theze, Kapp and Benacerraf) are not genetically restricted. Theze described a factor which suppresses anti-tumor responses, studied by Fujimoto, which is genetically restricted by the K end of H-2, and Moorehead talked about a factor, made by immune lymph node cells, which suppresses delayed hypersensitivity responses in recipients to specific antigen only if donor and host share K or D.

All the specific suppressive factors described appear to act on T cells, either on Lyl+ T_H cells (Kontiainen) or on antigen-primed Ly123+ T cells (Tada), or on non-immune cells (Theze, Kapp and Benacerraf).

Some general observations can be made. All the factors which act on T cells appear to bear determinants encoded by the major histocompatibility complex, where this has been examined. All the factors discussed except one (Levy) had molecular weights between 35,000 and 70,000. Given the variations in technique between different laboratories it is not clear that molecular weight differences in this range can be used to differentiate between factors of these types. Finally, for some factors, mouse strains have been identified which either do not make the factor (C57BL/6, Solliday Rich, A/J, Tada) or cannot respond to it (B10 congenics, Tada). It should be possible to study the responses of these strains and thus establish an _in vivo_ role for the factors in question. This is an area which has yet to be explored.

CLONALLY-RESTRICTED INTERACTIONS AMONG T AND B CELL SUBCLASSES

Kathleen Ward, Harvey Cantor and Edward A. Boyse

Department of Medicine, Harvard Medical School,
Farber Cancer Center, Boston, Massachusetts 02115
and
Memorial Sloan-Kettering Cancer Center
New York, New York 10021

The subject of this discussion is the role of gene products that are selectively expressed on the surface of different sets of cells. These products can be identified by antibody and the gene or cluster of genes responsible for expression of these surface components can be precisely mapped. Drs. Shreffler and McDevitt (see this volume) have dealt with what is, so far, the most well-characterized gene cluster - the Major Histocompatibility Complex (MHC) of the mouse, located on the 17th chromosome. The linkage of many genes within this complex is beginning to be unraveled and the role of the products of these genes in immune responses is being investigated. As yet, we have little idea what significance can be attributed to the order in which these genes are situated. But the contiguous positioning of sets of genes ('gene families') may ultimately make sense, just as the ordering of genes makes sense in the operons of bacteria.

In general, the definition of MHC gene products is based upon antisera produced when an individual of one MHC 'type' is immunized with cells of an individual of another MHC type. In the same way, antisera produced by immunizing mice with MHC-identical T-cells, B-cells or tumor cells has led to the definition of cell surface components expressed selectively on T- and B-cell sets: each of these sets expresses a particular immune function. In other words, one can now subclassify cells whose genetic program combines information for cell surface phenotype and function.

I. CELL SURFACE PHENOTYPES OF SOME T- AND B-CELL SUBCLASSES:

The cell surface profiles of several different subclasses of T lymphocytes and a mature subclass of B-cells are shown in Table 1. According to this description, the only gene products that all these subsets share are V_H

TABLE 1

CORRESPONDENCE OF FUNCTION WITH SURFACE PHENOTYPE

Lymphocyte Subclasses	Surface Phenotype										Function
T_{ARC}:	Thy-1	Lyt-1	Lyt-23	Lyt-4	(V_H)						Antigen-Reactive
chromosomes	9	19	6	?	?						
T_H:	Thy-1	Lyt-1	(V_H)	?I							Helper
chromosomes	9	19	?	17							
T_S:	Thy-1	Lyt-23	(V_H)	'I-J'							Suppressor
chromosomes	9	6	?	17							
Mature B Cells:	I	FcR'	Lybl	Lyb2	Lyb3	Ig	((V_H)	C_H	V_L	C_L)	Antibody Secretion
chromosomes	17	?	?	4	'X'	?	?	?	?6	?6	

structures: the chromosomal linkage of this set of genes has not yet been determined. We can focus for a moment on two mature subclasses of T lymphocytes, those with helper (T_H) and cytotoxic ($T_{C/S}$) suppressive activity. Judged by their different functional potentials and by their cell surface phenotypes, these two subclasses obey different sets of genetic instructions. Isolation of these two T-cell subclasses in mice depleted of T-cells ('B-mice') has shown that they are stable, both in terms of their respective functional commitments and surface phenotypes (1). So these findings indicate that these two T-cell subclasses belong to independent lines of differentiation and are not sequential stages of a single progression.

This and other information have suggested that thymus-dependent differentiation gives rise to at least two separate sublines of mature T-cells: one programmed to amplify a number of immune responses, and a second programmed for killer/suppressor activity. A number of experiments, in which cells belonging to the third major T-cell set (Ly123 cells) are stimulated by antigen or mitogen, suggest that this latter subclass probably contains a precursor set that has acquired receptors for antigen (antigen-reactive T-cells = T_{ARC}) but is not yet committed to H or C/S functions. Thus, at least a portion of these cells may represent the reservoir from which H and C/S cells are supplied.

II. INTERACTIONS AMONG LYMPHOCYTE SETS:

Perhaps the most striking feature to emerge from studies of the functional commitments of these T-cell sets concerns the following: these different cell sets do not carry out their separate functions independently of one another. Their functions are precisely interlocked. To emphasize this, several examples are shown (Table 2). T_H cells expressing the Ly1 phenotype induce B lymphocytes to secrete free antibody and thereby express its final differentiative program (2). Under appropriate circumstances, cells of the Ly1 subclass can also induce prekiller cells to generate enhanced levels of killer activity: recent studies of B. Huber suggest that this interaction may be particularly important in vivo (3). Tada and his colleagues have indicated that preformed Ly23 suppressor cells can induce immune Ly123 T_{ARC} to generate antigen-specific T_S activity and acquire the mature Ly23 surface phenotype (see Tada et al. this volume). Two points should be noted: (1) that this feedback loop involves the recruitment (by mature T_S) of additional antigen-specific T_S from a less mature T-cell

TABLE 2

FUNCTIONAL CO-OPERATION BETWEEN CELL SETS:
A PART OF THE IMMUNOLOGICAL NETWORK

Co-operating Sets:		Function
Effector	Inducer	
B-Cell	Lyl : H	↑ Antibody
Ly2 : CS	Lyl : H	↑ Cytotoxicity
Lyl2 : ARC*	Ly2 : CS*	↑ Suppression
Lyl2 : ARC*	Lyl : H*	↑ Help
Macrophage	Lyl : H	Delayed Hypersensitivity
Macrophage	Ly2 : CS	Macrophage Killer

* After antigen stimulation

pool, and (2) this recruitment depends on the fact that $Lyl2^+$ T_{ARC} have been already stimulated by antigen. This recruitment does <u>not</u> occur if the T_{ARC} are obtained from non-immune donors. Tada and his colleagues have also described a similar recruitment circuit for generation of mature T_H cells. In this case, antigen-specific T_H cells can induce antigen-stimulated T_{ARC} to differentiate to mature antigen-specific T_H (Lyl) cells. Again, T_{ARC} cells must be obtained from immune donors. In both cases, these findings suggest that T_{ARC} have already partially differentiated along the T_H or $T_{C/S}$ path after antigen stimulation.

Both of these mechanisms probably represent 'amplification' circuits where mature, antigen-specific T_H or T_S cells can 'recruit' additional T_H or T_S cells, respectively, from a <u>partially</u> differentiated T_{ARC} population. Thus, the net result of this recruitment is to enhance the T_H or T_S system after antigen stimulation, by additional generation of antigen-specific T_H or T_S cells. This recruitment mechanism cannot therefore be viewed as a likely candidate for homeostatic feedback regulation of the magnitude or duration of an immune response.

A second set of interactions between mature T_H (Lyl) cells and T_{ARC} (Lyl23) cells may represent a more promising mechanism for feedback control: activated T_H cells induce 'virgin' T_{ARC} to generate antigen-specific T_S activity. Some of this work is discussed by Eardley et al., this volume. This T-T interaction may represent a strong 'negative' feedback circuit resulting in homeostatic control of the antibody response.

Finally, it should be noted that the two mature T-cell subclasses also have different effects on macrophages. T_H (Lyl) cells can induce monocytes to display delayed-type hypersensitivity responses to both protein and erythrocyte antigens (4); while T_S (Ly23) cells can induce macrophages to destroy intracellular Listeria (5).

III. INDUCTION AND EXPRESSION OF T_S AND T_H ACTIVITY:

A. MHC Products

It is likely that the products of the MHC provide an extraordinary fund of information for the induction of co-operative T-cell sets. For example, cells of the T_H subclass may be preferentially induced after stimulation by antigenic differences associated with I-region components. These antigens are foreign by virtue of polymorphic variation (alloantigens) or as a result of association of I-region products with 'non-self' material. Similarly, $T_{C/S}$ cells respond preferentially to H-2K/D differences, either due to polymorphic variation or because self H-2K/D molecules have been modified by association with a 'non-self' antigen, such as a virus. In sum, one result of T-cell differentiation is the generation of at least two sublines of T-cells equipped to recognize and respond to products of the MHC that are foreign by virtue of polymorphic varation (alloantigens) or because they have become associated with non-self antigens. The selective induction of either screening system by altered MHC products would then activate a series of immune reactions already programmed in one or another activated T-cell subclass (2,6,7).

In the foregoing, it is implicit that T-cells are particularly well-suited to recognize and react to components displayed on the surfaces of living cells, usually in association wtih MHC products. If one considers in particular the T-cell interactions leading to the induction of B-cell programs to secrete antibody, one can ask the following question: Can T-cells recognize non-MHC components on the surface of living cells, and can this type of recognition allow for more precise interactions between regulatory T_H/T_S cells and B-cells?

B. Non-MHC Products

At the present time the most promising system of communication capable of permitting extraordinarily precise interactions between regulatory T-cells and B-cells is based on the work of Binz and Wigzell (8) and of Eichmann and Rajewsky (9) and their colleagues. These experiments indicate that part of the T-cell receptor consists of the variable portion of immunoglobulin heavy chains (V_H). This conclusion is apparent from serological, biochemical and genetic analysis. Thus, the cell partners involved in T/B interactions speak the same language; or at least, there is a considerable overlap in their respective dictionaries. These shared cell-surface V_H structures could, in theory, allow for the most precise and restricted interactions between V_H-related lymphocyte clones.

Can T-cells recognize immunoglobulin on the surface of B-cells? There is evidence that T-cells can distinguish among classes of immunoglobulin expressed on B-cells (which represent major structural differences) (10) and that T-cells may recognize immunoglobulin 'subclasses' on B-cells, (representing relatively minor structural differences) (11). As yet there is no evidence that they can discriminate among idiotypically-different immunoglobulin structures on B-cells, although Janeway et al. have shown that T-cells can be induced to respond to idiotypic determinants carried on myeloma cell products (12).

These considerations have led us to ask the following question: Is the production of a complete regulatory T_S/T_H-B-cell circuit, resulting in the production of a single antibody or set of antibodies, governed by T-cell recognition of idiotypic structures on the surface of B-cells?

There is certainly no evidence for this possibility. In fact, there is evidence to the contrary: T-cells recognizing 'carrier' molecules have been implicated in inducing B-cells, which recognize apparently different determinants called haptens, to produce antibody. Nonetheless, the question is still not answered. One can ask the question experimentally in the following way: if one selects a single B-cell clone or set of clones, which can be identified by the expression and secretion of a single V_H product, one may study whether induction of this set of B-cell clones requires interaction with unique clonal sets of T-cells, or whether any one of many T-cell clones is capable of inducing this particular B-cell. These two models of T-B interactions are outlined in Table 3.

TABLE 3

HOW PRECISE IS THE INTERACTION BETWEEN CLONES OF T_H AND B CELLS?

Model I: There is no clonal restriction of the T_H-B interaction.

Model II: Co-operation in at least some cases is *restricted* to V_H-related T_H and B-cell clones.

(A) Induction of a single B-cell clone (B-V_HX) requires interaction with V_H-identical T_H clone = *cognate interaction*

(B) Induction of a single B-cell clone (B-V_HX) requires interaction with T_H clone expressing anti-V_HX = *complementary interaction*

In the first model ('non-specific induction'), any clone of T_H cells, each marked by unique V_H products can, under appropriate circumstances, induce a particular B-cell clone to produce antibody. According to current understanding of the immune system, the panel of T_H clones that could induce a single B-cell depends upon linkage of the relevant 'haptenic' determinant to an unrelated 'carrier' protein. According to the second model, the T_H-B circuit is clonally-restricted regardless of antigen presentation: induction of a single set of B-cell clones requires interaction with T_H cells that express identical or complementary V_H structures. The particular B-cell clone we have chosen to study has been characterized over the past several years in a series of elegant experiments by Nisonoff and his colleagues (13,14). They have shown that (1) A strain mice immunized with arsonate (Ar) on a variety of carriers produce large amounts of anti-arsonate antibodies, (2) this response includes an antibody representing 20-70% of the total anti-arsonate response; this antibody is produced by every A/J mouse and has therefore been termed a 'cross-reactive idiotype' to indicate that it is distinct from other anti-arsonate antibodies, which are produced only intermittently in different individual mice, (3) the relative affinity of this idiotype is somewhat lower than the average affinity of anti-arsonate antibodies; (4) it is not present in serum from unimmunized mice, and (5) most importantly, this antibody can be identified by an appropriate rabbit 'anti-idiotypic' antiserum. This idiotype has been partially sequenced by Capra and shows homogeneity through position #38, which includes the first hyper-variable region. N-terminal sequencing of the heavy chain also shows identity at 288 positions. In other words, this antibody is a product of a single clone or at most perhaps several clones of B-cells that are programmed to secrete antibodies carrying this V_H structure after appropriate induction by T_H cells. Since the clonal product can be specifically identified by an appropriate antiserum, the system permits analysis of the cellular basis of induction of a single set of B-cell clones which are marked by the expression and production of a single V_H product, which we shall call for the purpose of discussion 'Idiotype-A' (Id-A).

We can now go on to analyze the rules that govern the induction of a single B-cell clone, simply by monitoring the formation of the clonal product. The first experiments were directed at distinguishing between the two mechanisms of induction of the Id-A$^+$ B-cell clone outlined in Table 3.

A stringent test that might disprove the first model is based on the following argument: Every A strain mouse immunized with KLH-Ar produces Id-A. If a non-specific induction mechanism were operative for the Id-A system, Id-A$^+$ B-cell clones should be induced by all T_H cells immunized to the carrier protein alone. Specifically, T_H cells from KLH-primed mice should induce Id-A$^+$ memory B-cells from mice primed to arsonate conjugated to another carrier protein (BGG). This can be done by co-transfer of carrier- and hapten-primed cells into irradiated hosts according to the classical protocol of Mitchison (see Figure 1). The results of this approach are summarized in Table 4.

Two points should be noted: (1) isolated T_H cells and B-cells were functionally pure and did not produce detectable levels of anti-Ar antibody to KLH-Ar in irradiated hosts and (2) recipients of mixtures of graded doses of KLH-primed T_H cells and BGG-Ar-primed B-cells were able to induce Id-A$^+$ B memory cells to secrete their product.

This experiment was, on the face of it, consistent with the non-specific induction model and, at least for the Id-A

FIGURE 1

Will T_H cells primed to a protein 'carrier' help Id-A^+ B-cells that have been primed to Arsonate on another protein?

PROTOCOL:

1. Isolate T_H cells (anti-Ly2 + C) from donors primed to a protein carrier (KLH).
2. Isolate B cells (anti-Thyl.2 + C) from donors primed to Arsonate conjugated to an unrelated protein (BGG-Ar).
3. Establish that BGG-Ar-primed donors are producing substantial levels of Id-A.
4. Co-transfer graded doses of KLH-primed T_H cells with 10^7 BGG-Ar-primed B-cells to irradiated (500r) hosts.
5. Challenge with KLH-Ar (10 mcg in saline).
6. Measure total anti-Arsonate and Id-A levels.

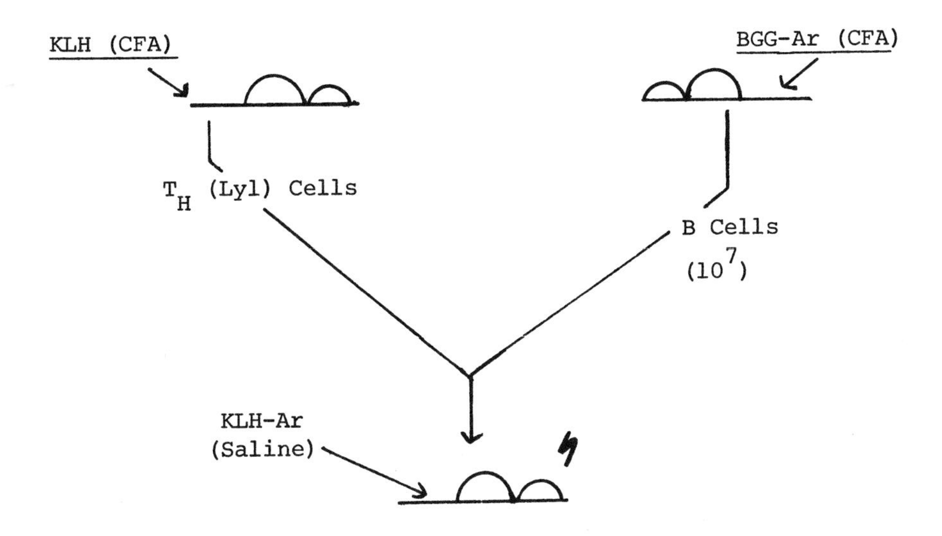

TABLE 4

B-Cells ($x10^7$)	T_H Cells* ($x10^7$)	Antibody Response Anti-Ar (mcg/ml)	Id-A (mcg/ml)
1	-	<.5	<.5
-	1	<.5	<.5
1.5	1.5	41	20

*T_H Cell Donor = KLH (CFA) immunized
B-Cell Donor = BGG-Ar (CFA) immunized

system, tended to disprove the hypothesis that the Id-A^+ B-cells were inducible mainly by T_H cells bearing idiotypically-related V_H structures.

However, it was formally possible that stimulation of mice with 250 mcg of KLH in CFA might directly or indirectly stimulate T_H clones carrying Id-A-related V_H structures. We attempted to rule out this possibility by testing whether administration of anti-Id-A before priming with KLH would eliminate the ability of these T_H cells to induce Id-A^+ B-cell clones. Therefore this experiment was repeated, this time comparing the T_H activity of a pair of donors: one group pretreated with normal rabbit serum and a second group pretreated with rabbit anti-Id-A (120 mcg Id-A binding capacity). Two weeks later both groups were immunized with 250 mcg KLH in CFA and T_H function was tested 3-5 weeks later by co-transfer with BGG-Ar-primed B-cells according to the protocol in Figure 1. The results of this experiment are summarized in Table 5.

T_H (Lyl) cells from NRS-pretreated, KLH-primed donors induced a strong anti-arsonate response, and a substantial portion was Id-A^+. In contrast, although graded doses of T_H (Lyl) cells from donors given anti-Id-A 2 weeks before KLH-priming induced similar amounts of anti-Ar, they did not induce significant levels of Id-A^+ anti-Ar antibody. To be sure that anti-Ly2 + C treatment removed idiotype-specific T_S cells (which could accound for this finding), we tested the anti-Ly2 + C-treated cells for Id-A-specific T_S activity.

TABLE 5

A. Ability of Ly1 cells from anti-Id-A→KLH - primed donors to induce Id-A:

B-Cells ($x10^7$)	Donor of T_H Cells: NRS→KLH T_H-Cells ($x10^7$)	Donor of T_H Cells: αId-A→KLH T_H-Cells ($x10^7$)	Antibody Response: Anti-Ar (mcg/ml)	Antibody Response: Id-A (mcg/ml)
1	-	-	<.5	<.5
1	0.4	-	19	17
1	-	0.4	12	0.8
1	0.8	-	43	31
1	-	0.8	33	3.5

B. Test for completeness of elimination of $Ly2^+$ Id-A-specific T_S:

KLH-Ar' Spleen Cells (10^7)	Source of T Cells (10^7): NRS→KLH' (αLy2+C)	Source of T Cells (10^7): αId-A→KLH (αLy2+C)	Antibody Response: Anti-Ar (mcg/ml)	Antibody Response: Id-A (mcg/ml)
+	-	-	23	16
+	+	-	72	59
+	-	+	56	40

That these Id-A-specific T_S cells expressed the $Ly2^+$ phenotype was first indicated by experiments performed in collaboration with A. Nisonoff: lightly irradiated (200r) hosts receiving as few as 0.5×10^6 Ly23 cells from Id-A-suppressed donors were subsequently unable to form Id-A upon repeated immunization with KLH-Ar (Ward, Nisonoff and Cantor, in preparation).

To test for completeness of T_S elimination with anti-Ly2 + C in the experiment outlined in Table 5, these same anti-Ly2 + C-treated cells were adoptively transferred with spleen cells from donors primed to KLH-Ar: no suppression of the Id-A response was seen even when large numbers of anti-Ly2 + C-treated cells were co-transferred (Table 5).

IV. COMMENT:

We have tested the proposition that $Id\text{-}A^+$ B-cells can be non-specifically induced by carrier-reactive T_H cells. We have shown that when $Id\text{-}A^+$ B-cells are induced by T_H cells from 'carrier-primed' donors, these T_H cells can be eliminated by pretreatment of the donor with anti-Id-A. We have not yet investigated the mechanism of T_H elimination after the administration of anti-Id-A, but we suspect that it reflects preferential induction of Id-A-specific T_S cells, rather than a direct effect of anti-Id-A on T_H cells. In either case, these experiments indicate that, at least for the induction of the $Id\text{-}A^+$ B-cell clones, model I is not valid and model II is indicated (Table 3).

In sum, these experiments are consistent with the following:

Induction of $Id\text{-}A^+$ B-cell clone(s) appears to be governed by idiotypically-related T_H and T_S cells. The precision of this circuit may depend on complementary or cognate V_H surface structures. It is possible that induction of B-cell clones programmed to express dominant or 'germ-line' V_H structures may be similarly regulated.

We do not yet know how much of the immune response is governed by clonally-restricted collaboration. There is no doubt that, under certain circumstances, a portion of B-cell induction is unspecific. Perhaps, the more relevant point is that this precise cellular interaction, based on idiotypically-associated cell-surface molecules, does occur. It is unlikely that this complementary dictionary is employed simply for the benefit of the immune system. People have wondered how individual retinal neuronal cells find their correct partners in the tectum, as shown by Sperry. These experiments indicate molecular mechanisms which could allow such clonally-restricted cell-interactions. It would be interesting if the existence of the presumed complementary set of markers on the connecting retinal and tectal neurons could be tested by analysis of the so-called 'tumor-associated antigens' of retinoblastoma(s), i.e., surface structures which may be unique for each tumor. It would be provocative if these were using a common ancestral dictionary and so showed homology with lymphocyte V-region structures.

ACKNOWLEDGEMENTS

We are grateful to Ms. Debra Garrigan for expert secretarial assistance. This work was supported in part by grants CA-16889, CA-08748, AI-12184 and AI-13600.

REFERENCES

1. Huber, B., Cantor, H., Shen, F.W. and Boyse,E.A. (1976) *J. Exp. Med.* 144, 1128.
2. Cantor, H. and Boyse, E.A. (1977) *Immunol. Rev.* 33, 105.
3. Huber, B., Shen, F.W. and Cantor, H. (1977), *in preparation.*
4a. Huber, B., Devinski, O., Gershon, R.K. and Cantor, H. (1976) *J. Exp. Med.* 143, 1424.
b. Vadas, M.A., Miller, J.F.A.P., McKenzie, I., Chism, S.E., Shen, F.W., Boyse, E.A., Gamble, J.R. and Whitelow, A.M. (1976) *J. Exp. Med.* 144, 10.
5a. Yakura, H., Shen, F.W. and Cantor, H., *unpublished observations.*
b. Zinkernagel, R.M., *personal communication.*
6. Zinkernagel, R.M. and Doherty, P.C. (1975) *J. Exp. Med.* 141, 1427.
7. Paul, W.E. and Benacerraf, B. (1977) *Science* 195, 1293.

8. Binz, H. and Wigzell, H. (1977) *Cold Spring Harbor Symp. Quant. Biol., in press.*
9. Eichmann, K., Berek, C., Hammerling, G., Black, S. and Rajewsky, K. (1977) *Cold Spring Harbor Symp. Quant. Biol., in press.*
10. Kashimoto, T. and Ishizaka, K. (1972) *J. Immunol.* 109, 1163.
11. Herzenberg, L.A., Okumura, K., Cantor, H., Sato, V., Shen, F.W., Boyse, E.A. and Herzenberg, L.A. (1976) *J. Exp. Med.* 144, 330.
12. Janeway, C.A., Sakato, N. and Eisen, H.N. (1975) *Proc. Nat. Acad. Sci. (USA)* 72, 2357.
13. Hart, D.A., Wang, A.L., Pawlak, L.L. and Nisonoff, A. (1972) *J. Exp. Med.* 135, 1293.
14. Nisonoff, A. and Bangasser, S.A. (1975) *Transpl. Rev.* 27, 100.

GENETIC REGULATION OF MACROPHAGE-T LYMPHOCYTE INTERACTION

Ethan M. Shevach and David W. Thomas

Laboratory of Immunology, National Institute of Allergy and Infectious Diseases, NIH, Bethesda, Maryland 20014

ABSTRACT. The products of the major histocompatibility complex (MHC) play a critical role in the regulation of the interactions of immunocompetent cells. In the guinea pig effective interaction between antigen-pulsed macrophages and immune T cells requires that the macrophage and T cell be syngeneic. Genetic studies have demonstrated that this interaction is mediated by the Ia antigens of the guinea pig MHC and not by the products of the guinea pig equivalent of the murine K or D region genes. These observations were consistent with the hypothesis that I-region genes code for specific cellular interaction structures and that homology between these structures was necessary for effective cellular interactions. However, as these studies were performed with T cells from animals primed _in vivo_, the results are also compatible with the view that T cells do not recognize antigen _per se_, but can only be sensitized to antigen-modified membrane components or to complexes of antigen with certain membrane molecules. Thus, allogeneic macrophage-associated antigen failed to activate immune T cells _in vitro_ because the T cell had been primed _in vivo_ only to antigen associated with syngeneic macrophages.

We have recently developed an _in vitro_ assay for the generation of a primary response to soluble protein antigens in which non-immune T cells can be sensitized and subsequently challenged in tissue culture with antigen-pulsed macrophages. This technique provides a method by which the mode of antigen presentation can be easily manipulated and experiments can be performed in which any combination of allogeneic or syngeneic lymphocytes and macrophages can be tested. T cells primed with syngeneic antigen-pulsed macrophages were only restimulated by syngeneic and not allogeneic antigen-puled macrophages. When F_1 T cells were primed with parental macrophages they were only restimulated by the parental macrophage used for initial sensitization and not by those of the other parent. When alloreactive guinea pig T cells were rendered unresponsive to allogeneic macrophages by treatment with bromodeoxyuridine and light, the remaining T cells could be subsequently primed and rechallenged in culture with antigen-treated syngeneic or allogeneic macrophages. T cells primed with antigen-treated allogeneic macrophages could be restimulated only with antigen-treated allogeneic, but not syngeneic macrophages. The results of this experiment strongly support the concept that Ia homology is not required for efficient T cell-macrophage collaboration in response to antigen and that the genetic restriction of this interaction is imposed only by the histocompatibility type of the macrophage used for initial sensitization.

INTRODUCTION

The major histocompatibility complex (MHC) encodes a series of genes that control or regulate a number of immune functions (1). In particular, the products of the MHC appear to be critically important in the regulation of the interaction of immunocompetent cells. In this report we will review our recent studies which have attempted to further analyze the role of the specific immune response (Ir) genes and the closely linked (or perhaps identical) I-region-associated (Ia) antigens in the interaction of antigen-pulsed macrophages and thymus derived (T) lymphocytes.

THE INTERACTION OF ANTIGEN-PULSED MACROPHAGES WITH T LYMPHOCYTES PRIMED *IN VIVO*

One of the first demonstrations of the close functional association of certain MHC alloantigens and the products of the Ir genes was that alloantisera raised by cross-immunization of inbred strain 2 and strain 13 guinea pigs would specifically inhibit antigen-induced T cell proliferative responses controlled by Ir genes in a haplotype specific manner (2). Thus, the response of primed $(2x13)F_1$ T lymphocytes to the 2,4-dinitrophenyl derivative of a copolymer of L-glutamic acid and L-lysine (DNP-GL, an antigen the response to which is controlled by an Ir gene linked to strain 2 MHC) could be inhibited by 13 anti-2 serum, while the response of these same cells to the copolymer L-glutamic acid and L-tyrosine (GT, an antigen the response to which is controlled by an Ir gene linked to the strain 13 MHC) was unaffected by culture in anti-2 serum, but could be markedly inhibited by 2 anti-13 serum.

It is now clear that sera prepared by reciprocal immunization of strain 2 and strain 13 guinea pigs exclusively recognize membrane glycoproteins homologous to the Ia antigens of mice (3). It was initially assumed that the inhibitory activity of anti-Ia sera on antigen-induced T cell proliferation was directed solely against the proliferating T lymphocyte. However, the experiments of Rosenthal and co-workers which demonstrated the critical role of the macrophage in antigen-induced T cell proliferation raised the possibility that the inhibitory activity of anti-Ia sera might be directed against both the macrophage and the T cell or even exclusively against the macrophage (4,5)

Experiments were therefore designed to define the cellular site of action of anti-Ia sera using combinations of allogeneic macrophages and T cells. In general, strain 2 antigen-pulsed macrophages activated DNA synthesis in immune strain 2 lymphocytes quite efficiently, while the same macrophages mixed with immune strain 13 lymphocytes stimulated little DNA synthesis. Similarly, antigen-pulsed strain 13 macrophages activated DNA synthesis in immune strain 13 lymphocytes, but were unable to activate immune strain 2 T lymphocytes (6). Combinations of parental macrophages and (2x13)F_1 lymphocytes did lead to T cell activation when the macrophages were pulsed with the purified protein derivative of tuberculin (PPD), an antigen not known to be under unigenic control. It was therefore of interest to evaluate whether macrophages obtained from an animal that lacked a given Ir gene (a non-responder) were capable of activating immune (non-responder x responder)F_1 T cell proliferation (Fig. 1).

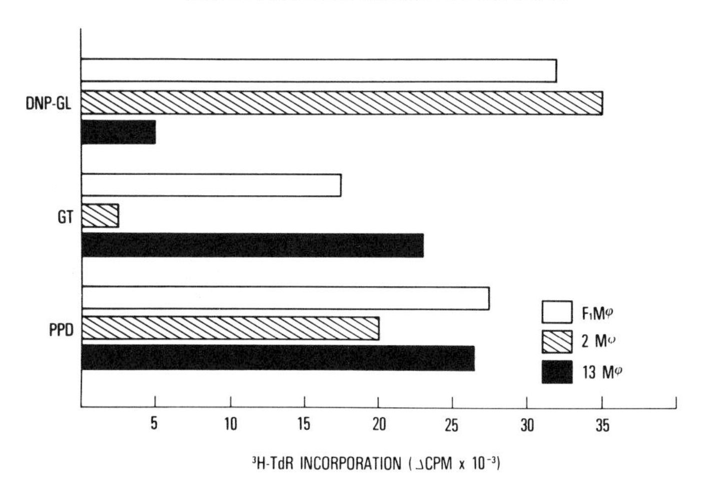

Figure 1. $(2 \times 13)F_1$, strain 2, or strain 13 macrophages were pulsed with DNP-GL, GT, or PPD for 60 minutes at 37°C, washed and then mixed with immune $(2 \times 13)F_1$ T cells. T cell proliferation was measured by the incorporation of tritiated thymidine (^{3}H-TdR) on day 3 of culture and the results are expressed as ΔCPM/culture.

When strain 2 or $(2 \times 13)F_1$ DNP-GL pulsed macrophages were mixed with immune F_1 T cells, a significant T cell proliferative response was observed; however, when non-responder strain 13 DNP-GL pulsed macrophages were used the magnitude of stimulation was approximately 1/10 that seen when responder strain 2 DNP-GL pulsed macrophages were used. A similar pattern was observed when GT pulsed macrophages were tested. Non-responder strain 2 GT pulsed macrophages were markedly inefficient in inducing T cell activation in immune $(2 \times 13)F_1$ T cells when compared to GT pulsed responder strain 13 macrophages (7). These results raise the possibility that the product of the Ir genes is functionally expressed in the macrophage and macrophages from non-responder animals are incompetent to deal with antigen, and for this reason fail to activate T cell proliferation when mixed with (non-responder x responder)F_1 T lymphocytes.

The initial experiments on the interaction of antigen-pulsed macrophages and T lymphocytes were performed several years ago before a clear understanding of the fine structure of the guinea pig MHC (the GPLA complex) was known. Recent genetic, serologic, biochemical, and tissue distribution studies have defined two major genetic regions of the GPLA complex (8). The B region contains two loci, the B and S loci, which encode a series of antigens which are homologous to the products of the K and/or D genes of the murine H-2 complex. The second region has been termed the I-region and encodes a series of antigens which resemble the products of the I-region of the murine H-2 complex. Strain 2 and strain 13 guinea pigs both bear the same antigens in the B region (B.1 and S), but appear to differ in the entire I-region. The I-region of strain 13 guinea pigs has been studied in greatest detail and has been subdivided into three subregions on the basis of chemical criteria (9) (Fig. 2). The Ia.3 and Ia.5 determinants are found on molecules composed of a 25,000 dalton chain and a 33,000 dalton chain which are probably in non-covalent association. The Ia.7 determinant is borne on a 58,000 dalton molecule in which the two chains are linked by disulfide bonds. In contrast, the Ia.1 and Ia.6 determinants are found on a molecule of 26-27,000 daltons. Sequential precipitation studies have demonstrated that Ia.3,5, Ia.7, and Ia.1,6 are borne on three distinct molecules. We have also assigned an Ir gene to a given subregion based on an association of that Ir gene with a specific Ia antigen in outbred populations and on the ability of monospecific anti-Ia sera to block T cell responses controlled by that Ir gene (10). Thus, the Ir gene controlling responsiveness to low dose immunization with DNP-guinea pig albumin (Ir-DNP-GPA) is associated with the Ia.3,5 subregion, while Ir-GT is associated with the Ia.1,6 subregion.

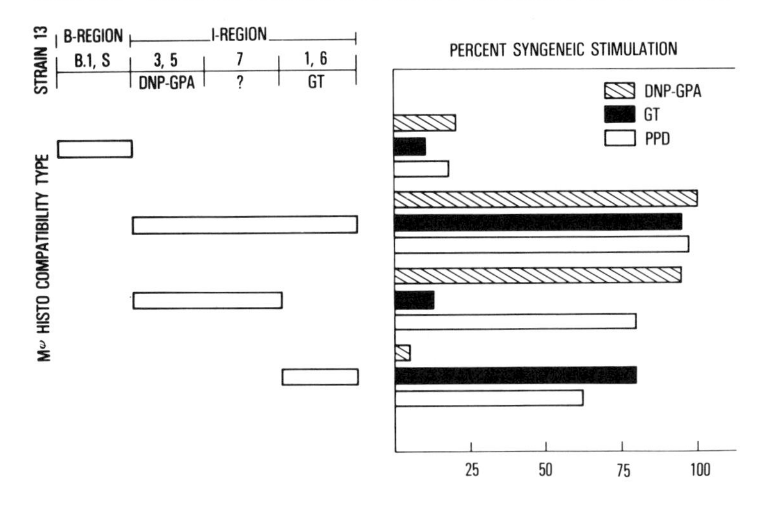

Fig. 2. Macrophages from outbred animals with serologically defined B-region and I-region determinants were pulsed with DNP-GPA, GT, or PPD and mixed with immune strain 13 T lymphocytes. The bars on the left side indicate the homology with the strain 13 MHC. The percent syngeneic stimulation is calculated as follows:

$$100 \text{ X } \frac{(\Delta\text{CPM with allogeneic macrophages})}{(\Delta\text{CPM with syngeneic macrophages})}$$

In order to more fully evaluate the role of the B-region and the I-subregions in macrophage-T cell interaction, macrophages from outbred animals typed for B-region antigens and Ia antigens were pulsed with DNP-GPA, GT, or PPD and mixed with immune strain 13 T lymphocytes (11). Macrophages from donors homologous with strain 13 animals for only the B-region antigens were markedly inefficient in the induction of T cell proliferation in strain 13 cells. However, PPD pulsed macrophages from donors homologous with strain 13 animals for at least one I-subregion were capable of activating strain 13 T cell proliferation. A different result was observed if both the Ir gene product and the linked Ia antigen were deficient in the macrophage donor; macrophages pulsed with an antigen the response to which is controlled by that Ir gene failed to activate T cell proliferation in strain 13 cells which bear both the Ia antigen and the Ir gene. For example, as shown in figure 2, macrophages from donors which have the Ia.1,6 subregion in common with strain 13 animals were capable of inducing significant T cell proliferation when pulsed with GT, but were poor activators of strain 13 T cells when pulsed with DNP-GPA. Macrophages from donors homologous with strain 13 for the Ia.3, 5 subregion functioned as well as syngeneic donors when pulsed with DNP-GPA, but were markedly inefficient when pulsed with GT. This experimental result is further evidence of specific associations between Ir genes and Ia antigens. Furthermore, it strongly suggests that the Ir gene product itself must be functionally expressed in the antigen presenting macrophage.

The model originally proposed to explain the histocompatibility restrictions on the interaction of antigen-pulsed macrophages and primed T lymphocytes was termed a "cellular interaction structure" model (7). Essentially, it proposed that the I-region genes code for specific cellular interaction structures and that homology between these structures is necessary for effective cellular interactions. A second model is derived from the observations of Zinkernagel and Doherty (12) and Shearer (13) who demonstrated that mouse T cells sensitized to hapten or virus modified cells are primarily cytotoxic for similarly modified target cells which are H-2D or H-2K compatible. This model implies that T cells do not recognize antigens *per se* but can only be sensitized to complexes of antigen and certain membrane molecules. This model can be termed the "complex antigenic determinant" model. Although this model has evolved from studies of effector T cell functions in which macrophages play no role it is also consistent with our observations on the histocompatibility requirements for macrophage-T lymphocyte interaction in the afferent arm of the immune response. As all our studies had been performed with T cells derived from immune animals, the explanation for the failure of allogeneic antigen-pulsed macrophages to activate immune T cells in vitro is that the T cells had been primed in vivo only to antigen associated with syngeneic macrophages.

THE INTERACTION OF ANTIGEN-PULSED MACROPHAGES WITH T LYMPHOCYTES PRIMED IN VITRO

In order to directly test the "complex antigenic determinant" model it was necessary to develop an assay in which non-immune guinea pig T lymphocytes can be primed and challenged in tissue culture with antigen-pulsed macrophages (14). In brief, T lymphocytes were primed by co-culturing them with antigen-pulsed or trinitrobenzene sulfonate (TNP)-treated macrophages for one week. The T cells recovered from the initial culture were washed and then restimulated with fresh normal or antigen-pulsed macrophages. Antigen-specific T cell proliferation was measured after an additional three or four day culture. Experiments were therefore designed in which any combination of allogeneic or syngeneic lymphocytes and macrophages could be tested.

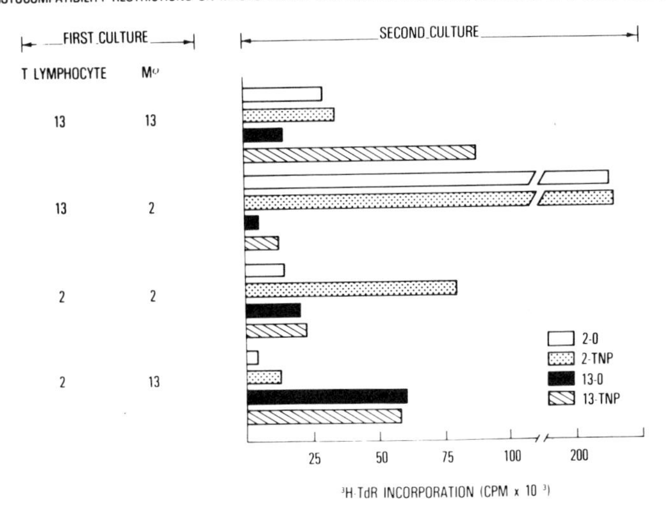

Fig. 3. In the first culture strain 13 or strain 2 adherence column purified lymph node T cells were incubated with syngeneic or allogeneic TNP-treated macrophages. T cell proliferation was measured three days after challenging the primed T cells with fresh untreated or TNP-treated syngeneic or allogeneic macrophages.

When strain 13 T cells were primed with strain 13 TNP-treated macrophages, they could be restimulated in the second culture with TNP-treated strain 13, but not strain 2 macrophages (Fig. 3). Similar results were observed when strain 2 T cells were primed with strain 2 macrophages. We were thus able to duplicate the observations made on T lymphocytes primed *in vivo* with our *in vitro* priming techniques. In order to directly test the hypothesis that T cells would respond only to the type of macrophage with which they were initially sensitized irrespective of histocompatibility type, T cells were also primed with TNP-treated allogeneic macrophages. However, with these experimental conditions a substantial secondary mixed leukocyte reaction (MLR) was seen and no TNP-specific stimulation was observed above the ongoing MLR.

In order to further explore the requirements for the interaction of allogeneic macrophages and T lymphocytes in the *in vitro* primary cultures, it was necessary to develop techniques to eliminate the MLR. The first approach used was to study the requirements for priming (2x13)F_1 T lymphocytes with parental antigen-pulsed macrophages (14). In these experiments we demonstrated that F_1 T cells initially primed with antigen-pulsed macrophages from one parent could be restimulated in the second culture only with the antigen-pulsed parental macrophage used for initial sensitization, and not with those of the other parent. If only I-region identity was required for efficient T cell-macrophage interaction, then (2x13)F_1 T cells which express the Ia antigens of both parental haplotypes, when primed with one parental antigen-pulsed macrophage should have been restimulated with antigen-pulsed macrophages from either parent. Our results are consistent with the view that the restriction on the (2x13)F_1 T cell response is imposed by the type of macrophage used for initial sensitiziation. Nevertheless, the "cellular interaction model" for macrophage-T cell interaction cannot be ruled out because F_1 T cells and parental macrophages share Ia antigens. However, in order for our experimental results to be consistent with such a model one must postulate allelic exclusion of cellular interaction structures at the T cell level.

We have recently developed an experimental system to directly demonstrate T cell sensitization to antigen-treated allogeneic macrophages (15). The experiments are based on the hypothesis that allogeneic macrophages can efficiently prime T cells if the alloreactive T cells can be removed prior to antigen priming. Strain 13 T cells were cultured for 48 hours with untreated strain 2 macrophages. Bromodeoxyuridine (BUdR) was then added to the cultures and at 72 hours the cultures were illuminated by a fluorescent light source to eliminate the alloreactive cells which had been stimulated to synthesize DNA during the 3 day culture. Following this treatment the T cells were primed for 7 days with TNP-treated syngeneic or allogeneic macrophages and T cell proliferation to normal or TNP-treated macrophages measured during a 4 day second culture.

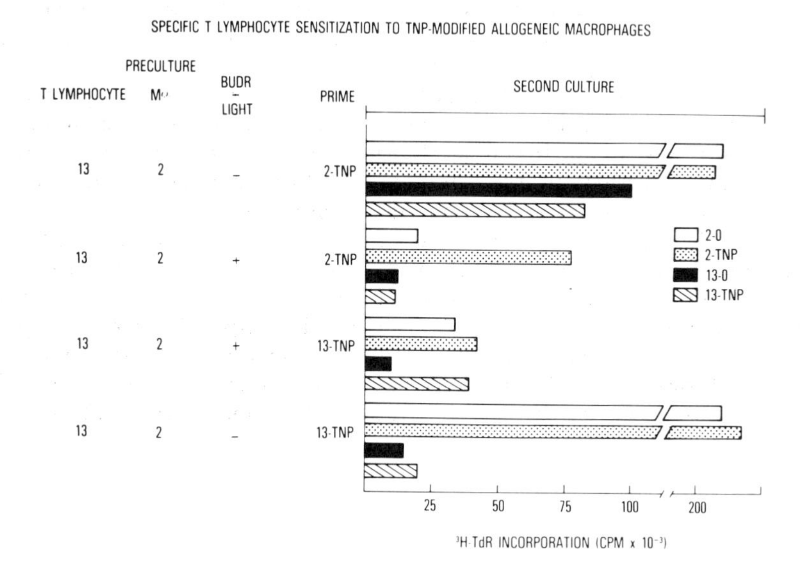

Figure 4. Strain 13 T cells were cultured for 3 days (preculture) with strain 2 macrophages and then treated with BUdR and light (+). The remaining cells were then primed for 7 days with TNP-treated strain 2 or strain 13 macrophages. The primed T cells were then stimulated in the second culture with normal or TNP-treated strain 2 or strain 13 macrophages. The results are expressed as CPM ^{3}H-TdR incorporation.

In most experiments BUdR and light treatment of alloreactive strain 13 T cells generated during the 3 day preculture with strain 2 macrophages reduced the MLR response against strain 2 macrophages in the second culture by approximately 80% (Fig. 4). BUdR and light treated strain 13 T cells could then be primed with either syngeneic or allogeneic TNP-treated macrophages. Strain 13 T cells primed with TNP-treated strain 2 macrophages responded in the second culture only to TNP-treated strain 2, but not to syngeneic TNP-treated strain 13 macrophages. Strain 13 T cells precultured with strain 2 macrophages and then treated with BUdR and light could easily be sensitized to TNP-treated syngeneic strain 13 macrophages and showed a good response when challenged with TNP-treated strain 13, but not strain 2, macrophages in the second culture (Fig. 4, group 3). However, if the MLR was not reduced by treatment of the T cells with BUdR and light, strain 13 T cells precultured with strain 2 macrophages could not be sensitized upon culture with TNP-treated syngeneic strain 13 macrophages (Fig. 4, group 4). It thus appears that the failure to obtain sensitization to TNP-modified allogeneic macrophages in the presence of an MLR may be secondary to suppression of antigen-specific priming by the ongoing MLR.

Our observation that T cells can be primed *in vitro* with TNP-treated allogeneic macrophages and then be restimulated only by antigen associated with allogeneic but not syngeneic macrophages allowed us to study the cellular site of action of anti-Ia sera in the inhibition of an antigen-specific T cell proliferative response (16). We could thereby determine if the Ia antigens of either or both macrophage or T cell are involved in an antigen-specific response. BUdR and light treated strain 13 T cells were primed with TNP-treated syngeneic macrophages and could be restimulated in the second culture by syngeneic, but not allogeneic TNP-treated macrophages in the presence of normal guinea pig serum (NGPS) (Fig. 5). When these cells were

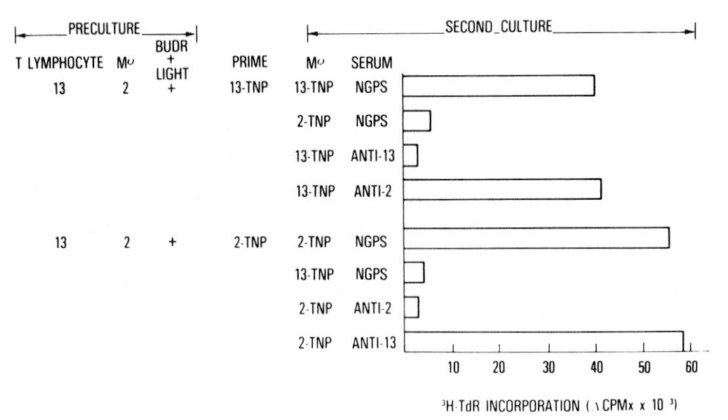

Fig. 5. Strain 13 T cells were cultured for 3 days with strain 2 macrophages and treated with BUdR and light to eliminate the alloreactive cells. The cells remaining from the preculture were primed by culture for 7 days with TNP-treated strain 2 or strain 13 macrophages. The primed T cells were then restimulated with fresh untreated or TNP-treated strain 2 or strain 13 macrophages in the presence of either NGPS, 13 anti-2 serum, or 2 anti-13 serum. After an additional three day incubation the incorporation of ^{3}H-TdR was determined and the results expressed as ΔCPM (CPM with TNP-treated macrophages - CPM with untreated macrophages).

---

cultured in the presence of 2 anti-13 serum (directed against both cell types) the TNP-specific response was markedly inhibited; 13 anti-2 serum had no effect on the response of these cells. When the BUdR and light treated strain 13 T cells were primed with allogeneic TNP-treated strain 2 macrophages and restimulated in the second culture, the T cells could only be restimulated by TNP-treated strain 2, but not strain 13, macrophages in the presence of NGPS. When 13 anti-2 serum directed solely against the stimulator TNP-modified strain 2 macrophages was incorporated into the second culture, the TNP-specific response was completely inhibited. On the other hand, 2 anti-13 serum directed solely against the responder strain 13 T cells had no effect on the TNP-specific response. These findings clearly demonstrate that antigen-specific T cell proliferation can be completely inhibited by anti-Ia sera directed against the antigen-presenting macrophage.

SPECIFIC T CELL SENSITIZATION TO ANTI-Ia SERUM MODIFIED MACROPHAGE Ia ANTIGENS

Since it had been previously shown that in the guinea pig the MLR induced by allogeneic macrophages can be inhibited by anti-Ia sera directed against the stimulator macrophage (17), we attempted to use anti-Ia sera to eliminate the MLR in the *in vitro* T cell priming culture with antigen-pulsed allogeneic macrophages. However, during the course of these experiments we made the fortuitous observation that macrophages bearing anti-Ia antibodies stimulated T cell proliferation (18)(Fig. 6). For example, when strain 13 T cells were incubated in the presence of 2 anti-13 serum in the first culture and then transferred to fresh strain 13 macrophages in the presence of 2 anti-13 serum in the second culture substantial T cell activation was observed as measured by increased DNA synthesis (Table 1). This stimulation appeared to be specific for 2 anti-13 serum induced T cell priming as no activation occurred when 2 anti-13 serum was included in only the first or second cultures.

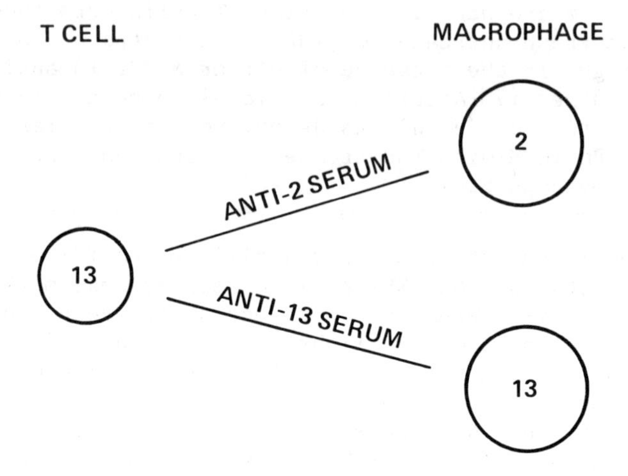

Figure 6. Strain 13 T cells can be primed to an anti-Ia induced modification of either syngeneic or allogeneic macrophage Ia antigens.

TABLE 1*

SENSITIZATION OF STRAIN 13 T CELLS TO ANTI-Ia TREATED SYNGENEIC MACROPHAGES

First Culture			Second Culture		
T Lymphocte	mφ	Serum	mφ	Serum	^{3}H-TdR
13	13	Anti-13	13	Anti-13	23,800
13	13	Anti-13	13	NGPS	700
13	13	Anti-13	2	Anti-13	3,390
13	13	Anti-2	13	Anti-2	1,540
13	13	Anti-2	2	Anti-2	1,160
13	13	NGPS	13	NGPS	1,580
13	13	NGPS	13	Anti-13	2,370

*Strain 13 T lymphocytes were cultured for 7 days with strain 13 macrophages in the presence of NGPS, 2 anti-13 serum, or 13 anti-2 serum. The incorporation of ^{3}H-TdR was determined 4 days after restimulating the cells recovered from the first culture with fresh strain 13 or strain 2 macrophages in the presence of NGPS, 13 anti-2, or 2 anti-13 serum.

The presence of 13 anti-2 serum in the first or second cultures did not lead to T cell activation indicating that the activation produced by 2 anti-13 serum is secondary to anti-Ia binding specifically to the T cells or macrophages and not to non-specific absorption of alloantibodies. The failure of 2 anti-13 serum to activate T cells in the presence of strain 2 macrophages in the second culture is a strong indication that anti-Ia induced stimulation results from antibody binding to the macrophage rather than to the T cell.

We were also able to demonstrate T lymphocyte activation to anti-Ia bound to allogeneic macrophages (Table 2). When strain 13 T cells were stimulated with allogeneic strain 2 macrophages in the first and second cultures a substantial secondary MLR was observed (Table 2, line 2). If 13 anti-2 serum was included in the first culture but replaced with NGPS in the second culture, the secondary MLR was reduced approximately 80% (Table 2, line 3). However, if 13 anti-2 serum was included in both the first and second cultures a substantial increase in T cell proliferation was observed which approximated that seen in the uninhibited secondary MLR (Table 2, line 4).

We interpret these results to indicate that at the same time anti-Ia sera directed against allogeneic macrophages blocks the MLR, the T cells also become sensitized to a new antigen created by anti-Ia serum bound to allogeneic macrophages. This finding provides further support for the concept that macrophage-T cell Ia homology is not required for effective collaboration. Although we could not determine the nature of the antibody induced stimulation, two possibilities are suggested. One possibility is that the binding of antibody to the Ia antigen induces a change in the Ia which T cells recognize as a neoantigen. Alternatively, the T cells may react against idiotypic determinants of the anti-Ia antibody which are focused on the macrophage Ia antigens.

TABLE 2*

SENSITIZATION OF STRAIN 13 T CELLS TO ANTI-Ia TREATED ALLOGENEIC MACROPHAGES

First Culture			Second Culture		
T Lymphocyte	mϕ	Serum	mϕ	Serum	^{3}H-TdR
13	2	NGPS	13	NGPS	268
13	2	NGPS	2	NGPS	181,180
13	2	Anti-2	2	NGPS	45,780
13	2	Anti-2	2	Anti-2	188,000

*Strain 13 T lymphocytes were incubated for 7 days with strain 2 macrophages in the presence of 5% NGPS or 13 anti-2 serum. The incorporation of ^{3}H-TdR was determined 4 days after restimulating the cells recovered from the first culture with fresh strain 13 or strain 2 macrophages in the presence of NGPS or 13 anti-2 serum.

DISCUSSION

The experiments presented in this report have clearly shown by two different techniques that Ia homology is not required for efficient macrophage-T cell collaboration and that T cells may be readily sensitized to antigens associated with allogeneic macrophages. Although our experiments have been performed entirely _in vitro_ a number of other studies have demonstrated allogeneic macrophage-T cell collaboration in _in vivo_ experimental models (19). One proposal put forward to explain histoincompatible cell collaboration _in vivo_ in tetraparental or radiation chimeric mice is that immature lymphoid cells undergo adaptive changes during their development _in vivo_ that allows successful collaboration of the mature cells (20). Our observations that mature T cells can be specifically sensitized to antigen-treated allogeneic macrophages _in vitro_ is against this hypothesis, unless such an adaptive process can occur with mature T cells during a 7 day _in vitro_ culture.

We have concluded from our studies that T cells can be as readily sensitized to antigen associated with allogeneic macrophages as to antigen associated with syngeneic macrophages.

We have seen no evidence for "syngeneic preference" in that in a large series of experiments the magnitude of T cell proliferation to TNP-treated allogeneic macrophages was as great as the proliferative response seen when T cells were primed with TNP-modified syngeneic macrophages. It is thus possible that the clone of T cells capable of recognizing allogeneic macrophage associated antigen is as large as the clone of T cells which recognize syngeneic macrophage associated antigen although we have not directly tested this hypothesis.

We must also integrate these observations on the histocompatibility restrictions on macrophage-T cell interaction with the possible role of the Ir gene products in this interaction. Our studies have demonstrated the existence of two distinct populations of (2x13)F_1 T cells which respond to one or the other parental macrophage; it might therefore be predicted that for antigens the response to which is regulated by unigenic Ir genes that (non-responder x responder)F_1 T cells should respond to antigen associated with either responder or non-responder parental macrophages. However, as we have reviewed in this report (Figs. 1 and 2) immune (non-responder x responder)F_1 T cells could not be activated by antigens associated with macrophages of the non-responder parent. The simplest explanation of this finding is that the Ia antigens are identical to the Ir gene products and are functionally expressed in macrophages. Non-responder macrophages would thus lack the I-region product necessary to process or present the antigen. If this is the case, then one might predict that nonresponder T cells may be able to be sensitized to antigen-pulsed responder macrophages expressing the appropriate Ia antigen. Experiments are now in progress to examine this possibility.

A number of possible roles can be postulated for the function of I-region gene products in the macrophage. The simplest concept is that there is a specific interaction between the nominal antigen and macrophage Ia antigen. The T cell would then recognize the nominal antigen-Ia antigen complex. Alternative possibilities are that the antigen induces an alteration in Ia molecules which T cells recognize as an "altered self" immunogen or that the Ia antigens themselves induce a specific modification in the nominal antigen which the T cells recognize as "altered antigen". It is still possible that no association exists between I-region gene products and nominal antigens at the level of the antigen presenting macrophage. However, in order to then explain the failure of non-responder macrophages to activate immune (non-responder x responder)F_1 T cells, one must postulate a linked dual recognition mechanism in the T cell. That is, in the F_1 T cell the receptor capable of recognizing the non-responder Ia haplotype would never be associated with the specific receptor for antigen, the response to which is controlled by the responder haplotype (21).

REFERENCES

1. Benacerraf, B., and McDevitt, H.O. (1972) Science 175: 273.

2. Shevach, E.M., Paul, W.E., and Green, I. (1972) J. Exp. Med. 136: 1207.

3. Geczy, A.F., deWeck, A.L., Schwartz, B.D., and Shevach, E.M. (1975) J. Immunol. 115: 1704.

4. Waldron, J.A., Horn, R.G., and Rosenthal, A.S. (1973) J. Immunol. 111: 58.

5. Rosenstreich, D.L., and Rosenthal, A.S. (1973) J. Immunol. 110: 934.

6. Rosenthal, A.S., and Shevach, E.M. (1973) J. Exp. Med. 138: 1194.

7. Shevach, E.M., and Rosenthal, A.S. (1973) J. Exp. Med. 138: 1213.

8. Schwartz, B.D., Kask, A.M., Paul, W.E., and Shevach, E.M. (1976) J. Exp. Med. 143: 541.

9. Schwartz, B.D., Kask, A.M., Paul, W.E., Geczy, A.F., and Shevach, E.M. (1977) Submitted for publication.

10. Shevach, E.M., Lundquist, M.L., Geczy, A.F., and Schwartz, B.D. (1977) Submitted for publication.

11. Shevach, E.M. (1976) J. Immunol. 116: 1482.

12. Zinkernagel, R.M., and Doherty, P.C. (1975) J. Exp. Med. 141: 1427.

13. Shearer, G.M., Rehn, T.G., and Garbarino, C.A. (1975) J. Exp. Med. 141: 1427.

14. Thomas, D.W., and Shevach, E.M. (1976) J. Exp. Med. 144: 1263.

15. Thomas, D.W., and Shevach, E.M. (1977) Proc. Nat. Acad. Sci. USA. In press.

16. Thomas, D.W., Yamashita, U., and Shevach, E.M. (1977) Submitted for publication.

17. Greineder, D.K., Shevach, E.M., and Rosenthal, A.S. (1976) J. Immunol. 117: 1261.

18. Thomas, D.W., and Shevach, E.M. (1977) J. Exp. Med. In press.

19. Pierce, C.W., Kapp, J.A., and Benacerraf, B. (1976) J. Exp. Med. 144: 371.

20. Katz, D.H., Chiorazzi, N., McDonald, J. and Katz, L.R. (1976) J. Immunol. 117: 1853.

21. Thomas, D.W., Yamashita, U., and Shevach, E.M. (1977) Transplantation Reviews. In press.

ROLES OF ADHERENT CELLS IN MURINE T CELL ANTIGEN RECOGNITION

Lanny J. Rosenwasser and Alan S. Rosenthal

Laboratory of Clinical Investigation
National Institute of Allergy and Infectious Diseases
National Institutes of Health
Bethesda, Maryland 20014

ABSTRACT. The cellular requirements for murine T cell antigen recognition were evaluated in an adherent cell dependent, *in vitro* proliferative assay. The immune response of sensitized T cells to complex multideterminant soluble protein antigens, such as dinitrophenylated ovalbumin, is dependent on the production of a soluble factor which is not antigen specific nor H-2 related, in that adherent cells, regardless of haptotype produce factor independent of the presence of antigen. Mouse lymphocytes do not, however, respond to supernates of either guinea pig macrophages nor murine fibroblasts, indicating the existence of species and cell type specificity. By contrast, the adherent cell requirement for T cell proliferation to the synthetic terpolymer L-glutamic acid, L-lysine, L-tyrosine (GLT^{15}), an antigen, the response to which is under specific Ir gene control, necessitated identity or partial major histocompatibility identity between macrophage and primed T cell.

INTRODUCTION

The essential role played by the macrophage (MØ) in the activation of T cells and in the expression of delayed hypersensitivity has been examined *in vivo* in the mouse (1). Macrophages function in T cell activation have been previously shown in the guinea pig and mouse to range from genetically restricted, MHC linked, physical interactions between metabolically active cells (2,3) to the secretion of various immunologically specific and non-specific substances which act both *in vivo* and *in vitro* (4,5). The recent development of reliable methods to gauge antigen specific murine T cell function *in vitro* (6,7) has allowed us to examine the precise function(s) of adherent cells in these systems. This article will concern itself with the description of an adherent cell dependent assay for antigen specific T cell proliferation and how with various probes such as alloantisera, two distinct macrophage functions can be differentiated in the mouse cell antigen recognition process.

METHODS

Antigen recognition by immune T cells in vitro - was assessed by measurement of the incorporation of tritiated Thymidine into DNA.

Immunization. 100 μg of antigen/CFA injected into the hind foot pads of the mice. Boost immunization given 2-3 weeks after primary.

Indicator lymphocytes. Lymph nodes (LN) were dissected 2-3 weeks after primary and 1-5 weeks after boost. LN were teased and passed through #200 wire mesh screen. LN cells were then placed on a nylon adherence column in 20% FCS RPMI 1640 media and incubated for 45-60 minutes. T cell enriched, adherent and B cell depleted LNL's were eluted at 1 ml/minute.

Adherent cells. Macrophage rich populations were obtained 3-10 days after injecting 1 ml sterile oil Ip, by lavaging peritoneal cavity with Heparinized HBSS. MØ were purified from peritoneal exudates (PEC) by overnight monolayer culture on petri dishes. PEC or MØ were treated with Mitomycin C 40 μg/ml at $37^{\circ}C$ for 60 minutes and washed prior to addition to culture.

OPTIMAL CELL CULTURE CONDITIONS

Media. 7.5% FBS, RPMI 1640 supplemented with Penicillin 200 U/ml, Gentamicin 10 μg/ml, Glutamine 0.3 mg/ml, 10 mM Hepes Buffer, and 2.5×10^{-5} M 2ME.

LNL density. 1.0×10^{6} LNL/ml.

PEC/MØ density. Variable - usually 0.25×10^{6} adherent cell/ml.

Duration. Cells were cultured in flat bottom microtiter plates, 200 μl/well, for 96-120 hours at $37^{\circ}C$ in 5% CO_2 incubator. During the final 16-24 hours, plates are pulsed with 1 μCi ^{3}H-TdR/well, plates are then harvested and counted. Data are expressed as cpm and delta (Δ) cpm when present antigen is added continuously to cultures.

RESULTS

Lymph node lymphocytes obtained by column purification contain greater than 90% T cells and cannot be activated into DNA synthesis even with optimal doses of the immunizing antigen, unless adequate numbers of non-immune adherent cells are added back to culture (Table 1). Antigen induced proliferation was shown to require the presence of immune T cells both by elimination of the DNA synthetic response by treatment with anti-Thy 1.2 serum plus complement and by failure to reconstitute the response when the immune B cells remaining after anti-Thy 1.2 treatment were added to non-immune T cells in the presence of antigen (Figure 1). The reconstitutive role of adherent cells in this assay is species specific in that guinea pig macrophage will not support proliferation; and cell type specific in that immune fibroblasts also do not support proliferation.

TABLE 1

RECONSTITUTION OF PROLIFERATIVE RESPONSE OF BID.A LYMPH NODE T CELLS

Antigen	Media	BID.A PEC
0	755±216	2,015±366
200 µg/ml DNA-OVA	974±98	42,117±1,394

*Data expressed as mean cpm of triplicate

Figure 1

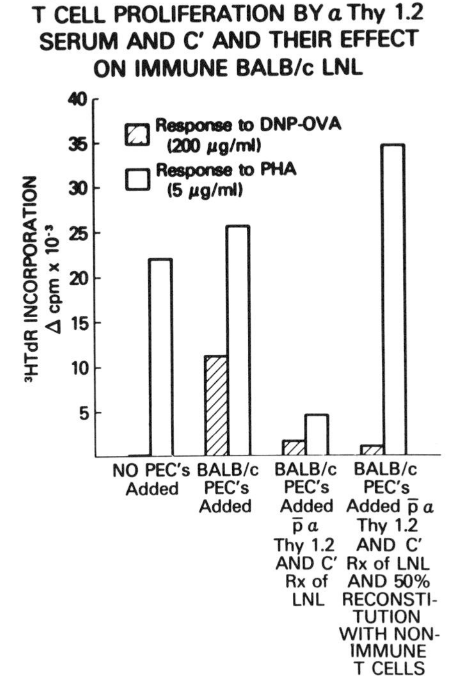

Next, the genetic restrictions on macrophage reconstitution were examined using the *in vitro* proliferative response to DNP-ovalbumin. Figure 2 illustrates the response of immune T cells as a function of antigen dose, when reconstituted with either media, syngeneic, or allogeneic peritoneal exudate cells. It is clear that at each antigen dose, both allogeneic and syngeneic PECs will restore responsiveness. Identical results can be obtained using T cells and adherent cells of other genetic backgrounds and with other multideterminant antigens such as KLH and PPD. In addition, when limiting numbers of adherent cells are added back to antigen stimulated T cells, no advantage of syngeneic cells over allogeneic cells is seen at any cell density. The possibility that the residual syngeneic adherent cell in the T cell population are functional was considered since indicator T cell populations cultured with antigen in round bottom culture dishes (thereby enhancing adherent cell attachment and density) show a ten-fold increase in stimulation over that observed with the routinely used flat bottom culture dishes.

Figure 2

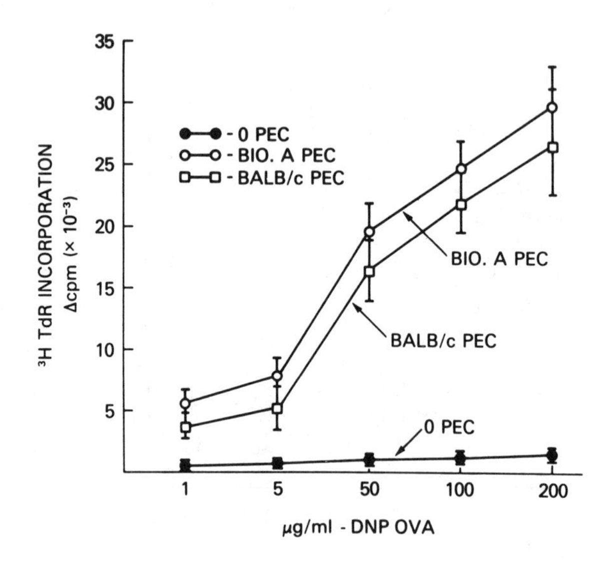

This hypothesis was further tested in experiments with adherent cell supernatants (Table 2) and alloantisera (Table 3). The data in Table 2 indicate that supernatants from cultured PECs can in the absence of added macrophages reconstitute, at least partially, the T cell response to DNP-OVA in the assay system described. Interestingly both allogeneic and syngeneic PEC supernates reconstitute the T cell response to DNP-OVA.

TABLE 2

SUPERNATANTS FROM CULTURED PEC ENHANCE THE RESPONSE OF IMMUNE T CELLS TO ANTIGEN

BALB/c LNL Immune to DNA-OVA
Date Expressed as (cpm) x 10^{-3}

DNP-OVA (200 μg/ml)	PEC/supe	(cpm) x 10^{-3}	Δcpm
0	0	2.30	-
+	0	4.03	1.73
0	BALB/c PEC	2.00	-
+	BALB/c PEC	28.81	26.81
0	B6 PEC	12.24	-
+	B6 PEC	43.78	31.54
0	BALB/c supe	4.08	-
+	BALB/c supe	15.84	11.76
0	B6 supe	2.49	-
+	B6 supe	15.07	12.58

Data in Table 3 shows that alloantisera directed against the responder T cell as well as any potential residual syngeneic macrophages inhibit proliferation even when allogeneic PEC are used to reconstitute the cultures. However, alloantisera directed against the allogeneic PEC had no effect on antigen specific proliferation, although a significant reduction in MLR is noted. Finally, Table 4, shows the response of LNLs immune to both DNP-OVA and GLT[15] reconstituted with PEC of various genetic backgrounds. As before, both allogeneic and syngeneic PEC restore the response to DNP-OVA, while reconstitution of response to GLT[15] required identity or partial identity at the H-2 complex between macrophage and responding lymphocytes.

TABLE 3

EFFECT OF ALLOANTISERA ON MOUSE T CELL PROLIFERATION IN THE PRESENCE OF SYNGENEIC AND ALLOGENEIC ADHERENT CELLS

DNP-OVA Immune LNL Reconstituted with B10.A or B10 PEC

		^{3}H-TdR Incorporation (cpm and *Δ cpm)					
Antigen†	PEC	1% NMS		1% a I^k		1% a K^bI^b	
0	B10.A	2,165	-	614	-	655	-
DNP-OVA	B10.A	34,076	*31,911	6,588	*5,974	47,611	*46,956
0	B10	3,360	-	7,651	-	1,384	-
DNP-OVA	B10	17,755	*14,395	13,407	*5,756	16,439	*15,055

† 200 μg/ml

TABLE 4

PROLIFERATIVE RESPONSE OF BALB/c LNL PRIMED WITH 100 μg GLT^{15} AND 10 μg DNP-OVA IN CFA

^{3}H-TdR Incorporation (Δ cpm) x 10^{-3}				
PEC	Haplotype	Responder GLT^{15}	200 μg/ml GLT^{15}	200 μg/ml DNP-OVA
0	-	-	0.67±0.43	2.97±0.91
BALB/c	d	+	12.97±2.28	25.39±4.94
B10.D2	d	+	11.62±1.17	21.26±4.49
(BALB/c x A/J)F_1	dxa	+	9.01±1.29	24.20±2.11
A/J	a	-	0.18±0.01	38.44±5.98
C57Bl/6J	b	-	0.47±0.34	23.51±3.39
(C57Bl/6J x A/J)F_1	bxa	+	1.35±0.89	36.20±6.15
R111	r	+	0.54±0.22	25.61±4.62

Mean ± SE of 5 Experiments.

DISCUSSION

The data presented in this paper illustrate an adherent cell requirement for antigen specific proliferation of murine T cells. Using the assay described, the immune response to complex antigens such as DNP-OVA is at least partially dependent on the production of a factor by adherent cells. In this in vitro model, proliferation of T cells immune to GLT^{15} occurred only in the presence of PEC that are syngeneic or semisyngeneic. Certain aspects of the macrophage requirement for antigen specific T cell proliferation differ from the guinea pig (2).

The apparent failure to demonstrate allogeneic restriction in antigen specific LNL proliferation following immunization with complex antigens may be interpreted in either of three general ways. The first and most trivial explanation in the case of the recognition of DNP-OVA is that sufficient residual syngeneic MØ are present to support antigen-specific immune T cell proliferation, as long as a non H-2 related factor is supplied by added PEC. A second possible explanation of the GLT^{15}-DNP-OVA paradox is that a "weak" or restricted antigen, such as GLT^{15} either elicits fewer T cell during priming in vivo such that T interaction is limiting or that the repretoire of Ir genes available in the residual macrophages in the lymphocyte population is inadequate to present an antigen of restricted heterogeneity such as GLT^{15}. A final possibility is that antigens with relatively few determinants such as GLT^{15} can only be seen in the context of gene products linked to the MHC, and would hence show restriction of cell cooperation, while for multideterminant antigens, such as DNP-OVA, an alternate pathway to activate certain subclasses of T classes (such as suppressor cells) not in context of the macrophage MHC is available. Experiments to explore these possibilities are now in progress.

REFERENCES

1. Miller, J.F.A.P., Vadas, M.A., Whitelaw, A., and Gamble, J. (1976) *Proc. Natl. Acad. Sci. USA* 73, 2486.
2. Rosenthal, A.S., and Shevach, E.M. (1973) *J. Exp. Med.* 138, 1194.
3. Lipsky, P.E., and Rosenthal, A.S. (1975) *J. Exp. Med.* 141, 138.
4. Calderon, J., Kiely, J.M., Lefko, J.L., and Unanue, E.R. (1975) *J. Exp. Med.* 142, 151.
5. Erb, P., and Feldmann, M. (1975) *J. Exp. Med.* 142, 460.

6. Schwartz, R.H., Jackson, L., and Paul, W.E. (1975) *J. Immunol.* 115, 1330.
7. Lonai, P., and McDevitt, H.O. (1974) *J. Exp. Med.* 140, 977.
8. Rosenwasser, L.J., and Rosenthal, A.S. Manuscript in preparation.

SIMULTANEOUS RECOGNITION OF CARRIER ANTIGENS AND PRODUCTS OF THE H-2 COMPLEX BY HELPER CELLS

John W. Kappler and Philippa Marrack

Department of Microbiology, Division of Immunology,
Cancer Center, University of Rochester,
Rochester, New York 14642

ABSTRACT. In F_1 mice whose parents differed at H-2, helper T cells produced by priming with keyhole limpet hemocyanin (KLH) bound to parental macrophages (MΦ) were subsequently most active when assayed *in vitro* with parental B cells and MΦ of the same parent. Using H-2 congenic mice of the B10 or C3H background this restriction of the F_1 cells was shown to be controlled by genes within the H-2 complex. Although the restriction was determined by the H-2 type of the priming MΦ *in vivo*, the phenomenon was not totally determined by the H-2 type of the MΦ present in the *in vitro* phase of the experiment, since the restriction could still be demonstrated in cultures containing optimal numbers of MΦ of both parental H-2 types.

In a second series of experiments the ability of *in vivo* sheep red blood cell (SRBC) primed helper T cells to crossreact *in vitro* with a determinant(s) present on burro RBC (BRBC) was shown to be under the control of a gene(s) in the H-2 complex (tentatively mapped to the K, I-A - I-E region). Helpers produced by SRBC priming of F1 mice of high x intermediate or low responder parents were tested separately on B cells and MΦ expressing one or the other parental H-2 haplotypes. In each case the H-2 type of the B cells and MΦ determined to what extent the SRBC-primed F_1 cells could respond to the crossreacting determinant(s). The inclusion of MΦ of the high responder parental H-2 type did not increase the ability of the F_1 cells to respond when tested on B cells and MΦ of the low responder parental H-2 type.

Our findings support the conclusion that helper T cells simultaneously recognize both antigen bound to B cells or MΦ and H-2 gene products expressed on these cells.

INTRODUCTION

We have previously reported (1) that helper T cells primed in C57BL/6 ($H\text{-}2^b$) x DBA/2 ($H\text{-}2^d$) F_1 mice (BDF_1) by immunization with keyhole limpet hemocyanin (KLH) bound to parental macrophages (MΦ) cooperate most efficiently *in vitro*

TABLE 1

THE INVOLVEMENT OF H-2 LINKED GENES

IN THE RESTRICTION OF F1 T CELLS DURING PRIMING WITH KLH-PULSED MΦ

Source of T cells	KLH-pulsed MΦ used for priming	Source of B cells and MΦ *in vitro*	Helper activity
C3H.SW x CBA F_1	C3H.SW (H-2^b)	C3H.SW	396 ± 50
"	C3H/He (H-2^k)	"	109 ± 50
"	CBA (H-2^k)	"	70 ± 50
BDF_1	B10 (H-2^b)	B10	199 ± 15
"	B10.D2 (H-2^d)	"	49 ± 10
"	DBA/2 (H-2^d)	"	34 ± 10

See ref.(1) for preparation of cells. Helper activities are slopes of titration lines (anti-TNP PFC/culture/10^6 T cells ± SEM). *In vitro* antigen was 0.1 μg TNP-KLH/ml.

with parental B cells and MΦ of the same type as the priming MΦ. Thus, when tested on BDF_1 B cells and MΦ, BDF_1 T cells primed with KLH on either C57BL/6, DBA/2, or BDF_1 MΦ had about the same activity. However, when those same T cells were tested on C57BL/6 B cells and MΦ, those primed with C57BL/6 MΦ had the greatest activity, those primed with DBA/2 MΦ had the lowest activity, and those primed with BDF_1 MΦ an intermediate activity. Preliminary experiments established the importance of H-2 linked genes in the phenomenon, although some contribution from non-H-2 linked genes was also a possibility.

In the present paper we present further evidence that this restriction in the activity of F_1 T cells is controlled primarily by H-2 linked genes. In addition we demonstrate a similar phenomenon using an antigen the response to which is controlled by an immune response (Ir) gene(s) within the H-2 complex. Finally, we demonstrate that after priming the H-2 genes of at least the B cells are important in the restriction.

METHODS

Methods for the preparation of reagents, antigens, cultures, various cell types, and antigen pulsed MΦ have been previously published in detail (1-3).

The method used for the quantitative assay of helper T cell activity has also been described in detail elsewhere (4,5). Briefly, a series of cultures containing a constant number of B cells and MΦ and a varying number of antigen (KLH or SRBC) primed helper T cells were cultured for four days with the appropriate trinitrophenylated (TNP) antigen. The number of anti-TNP PFC per culture was then determined and plotted vs. the number of primed cells added. This plot yielded a linear initial slope (units = anti-TNP PFC/culture/10^6 primed cells) which was used as an indication of the relative activity of the primed helper cells.

RESULTS

Experiments with KLH-pulsed MΦ. The results shown in Table 1 are of experiments designed to test the relative contribution of H-2 and non-H-2 linked genes in the restriction induced in F_1 T cells primed with KLH on parental MΦ. C3H.SW ($H-2^b$) x CBA ($H-2^k$) F_1 mice were primed with KLH-pulsed MΦ for helper activity. As expected the best activity was seen when priming with C3H.SW MΦ and the poorest activity with CBA MΦ. The important T cells were those primed with C3H MΦ, since this strain had the same background as C3H.SW, but the H-2 type of CBA. These T cells were no better than those primed with CBA MΦ demonstrating that the H-2 type of the immunizing MΦ and the test B cells and MΦ must be the same for optimal helper activity. Identical results were obtained in a similar experiment using BDF_1 mice primed with KLH on C57BL/10 (B10) ($H\text{-}2^b$), B10.D2 ($H\text{-}2^d$) or DBA/2 ($H\text{-}2^d$) MΦ. The primed T cells were tested on B10 B cells and MΦ. Again identity between the priming MΦ and the test B cells and MΦ at the H-2 complex was essential for optimal helper activity.

An additional important question in these experiments was whether the observed restriction was simply a manifestation of the interaction of T cells and MΦ as suggested by the work of Pierce and co-workers (6), or whether the restriction was manifest in addition at the level of T cell interactions with B cells as suggested by Katz and co-workers (7).

To begin to approach this question we have attempted to reverse the restriction *in vitro* by including MΦ of the appropriate H-2 type in cultures. The results of two such experiments are shown in Table 2.

TABLE 2

MΦ/T CELL INTERACTION CANNOT BE THE ONLY SITE OF RESTRICTION

KLH-pulsed MΦ used for priming	Source of B cells and MΦ *in vitro*	1×10^5 Additional MΦ *in vitro*	Helper activity
B10	B10	B10	831 ± 36
B10	B10	B10.D2	795 ± 48
B10.D2	B10	B10	380 ± 36
B10.D2	B10	B10.D2	385 ± 47
B10	B10.D2	B10	157 ± 46
B10	B10.D2	B10.D2	138 ± 22
B10.D2	B10.D2	B10	319 ± 41
B10.D2	B10.D2	B10.D2	293 ± 28

Conditions as in Table 1. BDF_1 T cells used throughout.

BDF_1 mice were primed with KLH on either B10 or B10.D2 MΦ. The resultant T cells were tested on B10 B cells and MΦ. Additional B10 or B10.D2 MΦ were added to the test cultures. The inclusion of additional B10.D2 MΦ in the cultures did not improve the activity of T cells primed with B10.D2 MΦ. An improvement of activity would have been expected if the lower activity of these T cells would simply be due to restricted MΦ/T cell interactions. In a separate experiment the two types of T cell preparations were tested on B10.D2 B cells and MΦ. In this case the inclusion of B10 MΦ in the cultures did not improve the lower activity of the T cells primed with B10 MΦ. These experiments indicate that although the restriction was induced by the H-2 type of the priming MΦ, the H-2 type of the test B cell at least must be important in the restriction. They do not of course eliminate the possibility that the H-2 type of both the B cell and the MΦ *in vitro* are important.

Experiments with RBC antigens. We wished to perform similar experiments with an antigen the response to which was known to be under the control of an H-2 linked Ir gene. In the course of these attempts we discovered that the response of helper T cells to a crossreacting determinant on SRBC and burro RBC (BRBC) was under such control. Originally it was found that SRBC-primed helper T cells when assayed *in vitro* would help in responses to TNP-SRBC, to a lesser extent to TNP-BRBC, and not at all to TNP-toad RBC (5). This finding was somewhat unusual at the time since anti-SRBC antibody was shown by a variety of techniques not to crossreact with BRBC. We have recently reexamined this observation and found that the ability of SRBC-primed T cells to recognize the shared determinant on BRBC was under the control of an H-2 linked Ir gene(s). For instance, Table 3 presents data from experiments in which the activity of SRBC primed helper T cells were assayed using TNP-SRBC and TNP-BRBC.

TABLE 3

H-2 LINKED Ir GENE CONTROLLING THE RESPONSE OF HELPER T CELLS TO A SHARED DETERMINANT(S) ON SRBC AND BRBC

Strain	H-2	K	I-A	I-B	I-J	I-E	I-C	S	G	D	Helper cell cross-reactivity(%)
B10.A	a	k	k	k	k	k	d	d	d	d	12 ± 7
B10.A(2R)	h2	k	k	k	k	k	d	d	?	b	5
B10	b	b	b	b	b	b	b	b	b	b	21 ± 3
B10.D2	d	d	d	d	d	d	d	d	d	d	67 ± 1
B10.AM(S)	s	s	s	s	s	s	s	s	s	s	65 ± 18

Helper cells primed on day minus four with 8 x 10^5 SRBC. SRBC-primed helper cells titrated using TNP-SRBC and TNP-BRBC as the *in vitro* antigens as in ref.(5). Crossreactivity calculated as helper activity with TNP-BRBC divided by activity with TNP-SRBC x 100. SEM shown for strains in which three or more separate determinations have been made.

Various cogenic strains of the B10 background were compared. Whereas, B10.A and B10.A(2R) mice were found to be low responders to the crossreacting determinant(s), B10 mice were intermediate responders and B10.D2 and B10.AM(S) were high responders. Other strains identified as low responders were C3H/He and CBA, both H-2^k. Other intermediate responder strains were DBA/1 (H-2^q), C3H.SW (H-2^b), and C57BL/6 (H-2^b). Other high responder strains were Balb/c and DBA/2, both H-2^d. The gene(s) controlling high responsiveness in H-2^d strains was tentatively mapped in the K, IA - IE region on the basis of the low response in B10.A and B10.A(2R) mice compared to the high response in B10.D2. Care must be taken in mapping on the basis of such limited distribution data, since recent studies showing the possibility of Ir-gene interaction in high responders prevent simple mapping studies (8).

In order to perform experiments similar to those above with KLH, BDF_1 (intermediate, H-2^b x high, H-2^d) and Balb/c x A/J F_1 (CAF_1) (high, H-2^d x low, H-2^a) mice were primed with SRBC. Their T cells were then tested on either parental B cells and MΦ or on B cells and MΦ of congenic mice carrying the parental H-2 haplotypes. Helper activities were compared using TNP-SRBC and TNP-BRBC as antigens. The results are summarized in Table 4. In each case the extent of cross-

TABLE 4

EXPRESSION OF THE Ir GENE(S) IN B CELLS AND/OR MΦ

Source of T cells	Source of B cells and MΦ	Primed helper cell crossreactivity(%)
BDF_1	C57BL/6 (H-2^b)	21
BDF_1	DBA/2 (H-2^d)	44
BDF_1	B10 (H-2^b)	29
BDF_1	B10.D2 (H-2^d)	76
CAF_1	B10.D2 (H-2^d)	60 ± 30
CAF_1	B10.A (H-2^a)	0 ± 3

Cells prepared as in ref.(1) except that B cells were not hapten-primed. Parallel titrations performed with SRBC primed and unprimed F1 T cells. Crossreactivity calculated as (activity of primed T cells minus activity of unprimed T cells using TNP-BRBC as antigen) divided by (activity of primed T cells minus the activity of unprimed T cells using TNP-SRBC as antigen) x 100. Average of two separate determinations with BDF_1 cells. Average $\pm$ SEM of three determinations with CAF_1 cells.

reaction seen was predicted by the Ir-type of the B cells and MΦ used. Thus, BDF_1 SRBC-primed T cells crossreacted highly with BRBC when tested with DBA/2 or B10.D2 B cells and MΦ, but intermediately when tested with C57BL/6 or B10 B cells and MΦ. CAF_1 SRBC-primed T cells showed very low crossreaction when tested with B10.A B cells and MΦ, but high crossreaction when tested with B10.D2 B cells and MΦ.

In order to determine which cell type was expressing the Ir gene(s) *in vitro* CAF_1 SRBC-primed T cells were tested for crossreactivity with BRBC in cultures of B10.A (low responder) B cells and MΦ to which additional B10.D2 (high responder) MΦ were added. The results are shown in Table 5.

TABLE 5

Ir GENE(S) MUST BE EXPRESSED AT LEAST IN B CELLS

Source of T cells	Source of B cells and MΦ	1×10^5 additional MΦ *in vitro*	Primed helper cell crossreactivity
CAF_1	B10.A	B10.A	8
CAF_1	B10.A	B10.D2	12

Conditions as in Table 4. Average of two separate determinations.

Low crossreactivity was observed as expected, and the crossreactivity was not increased by the presence of MΦ carrying the high responder parental H-2 haplotype. These results indicated that the high responder Ir gene(s) must be expressed at least in the B cell for high crossreaction to be seen. Whether or not the Ir gene(s) must also be expressed in the MΦ of the culture has not yet been determined.

DISCUSSION

A number of models may be constructed to explain our observations presented here and elsewhere (1) and similar ones made by other workers (6,7,9-12). However, we feel at present that the simplest explanation for these results is that helper T cells initially respond to antigen only on the surface of MΦ. During this response clones are selected which recognize not only the antigen, but also H-2 (I region?) gene products expressed on the MΦ. In the case of complex antigens such as KLH, presumably many H-2 gene products are involved. In the case of antigens under Ir-gene control a limited number of genes may be involved. Once primed, the helper T cells become operationally restricted, in that they subsequently respond best to the antigen when presented to them in conjunction with the same set of H-2 gene products as were present on the original MΦ. This second interaction can take place on the surface of a MΦ, but must also take place on the surface of a B cell which has specifically bound the antigen. This view, similarly stated by other workers (e.g.

see 9-12), represents our working hypothesis at present.

ACKNOWLEDGEMENTS

This work was supported in part by grants from the NIH, American Cancer Society, and the American Heart Association. We thank Lee Harwell and George Berry for their expert technical assistance.

REFERENCES

1. Kappler, J. W., and Marrack, P. (1976) *Nature* 262, 797.
2. Mishell, R. I., and Dutton, R. W. (1967) *J. Exp. Med.* 126, 423.
3. Marrack(Hunter), P., Kappler, J. W., and Kettman, J. R. (1974) *J. Immunol.* 113, 830.
4. Marrack, P. and Kappler, J. W. (1976) *J. Immunol.* 116, 1373.
5. Hoffmann, M. K., and Kappler, J. W. (1973) *J. Exp. Med.* 137, 721.
6. Pierce, C. W., Kapp, J. A., and Benacerraf, B. (1976) *J. Exp. Med.* 144, 371.
7. Katz, D. H., Hanaoka, T., Dorf, M. E., and Benacerrat, B. (1973) *J. Exp. Med.* 138, 734.
8. Katz, D. H., Dorf, M. E., and Benacerraf, B. (1976) *J. Exp. Med.* 143, 906.
9. von Boehmer, H., and Sprent, J. (1976) *Transpl. Rev.* 29, 3.
10. Blanden, R. V., Hapel, A. J., and Jackson, D. C. (1975) *Immunochemistry* 13, 989.
11. Paul, W. E., Shevach, E. M., Pickeral, S., Thomas, D.W., and Rosenthal, A. S. (1977) *J. Exp. Med.* 145, 618.
12. Benacerraf, B., In: The Role of Products of the Major Histocompatibility Gene Complex in Immune Responses. Ed. D. Katz and B. Benacerraf. Academic Press, New York, p. 715 (1976).

RECOGNITION RESTRICTIONS IN LYMPHOCYTE COLLABORATIVE INTERACTIONS IN IgG_1 ANTIBODY RESPONSES

Susan K. Pierce

Department of Pathology, School of Medicine
University of Pennsylvania
Philadelphia, Pennsylvania 19104

ABSTRACT. The restrictions imposed by genes of the I region on T-B lymphocyte collaborative interactions in IgG_1 antibody responses to T-dependent antigens have been investigated. We have employed the splenic fragment culture system to analyze individual primary and secondary B-cell responses to dinitrophenylated hemocyanin (DNP-Hy) in fragment cultures derived from carrier-primed irradiated syngeneic and allogeneic recipient mice. The results of these experiments demonstrate that restrictions are imposed on allogeneic T-B cell interactions but these are not absolute since the majority of primary B cells can be stimulated in allogeneic carrier-primed recipients to form clones of antibody producing cells. However, these B-cell clones synthesize only antibody of the IgM heavy chain class while the majority of DNP-specific B-cell clones, resulting from stimulation in syngeneic carrier-primed recipients, produce IgG_1 DNP-specific antibody. By selecting the appropriate congenic inbred strains of mice, it was possible to determine that the ability of T and B lymphocytes to collaborate in IgG_1 antibody responses depends on sharing of genes in the I-A or I-B subregion of the I region. We have demonstrated that the limitations on cell collaboration in primary B-cell IgG_1 antibody responses do not apply to secondary B cells, the majority of which can be antigenically stimulated to IgG_1 antibody production in syngeneic or allogeneic carrier-primed recipients. As a further probe of the requisites of T-B lymphocyte interactions, we employed the I region genetically controlled responses to the synthetic terpolymer of L-glutamic acid, L-lysine and L-phenylalanine (GLΦ). We demonstrated that I region genetic identity between T and B lymphocytes which was sufficient for DNP-specific B-cell IgG_1 antibody responses to DNP-Hy in Hy primed recipients was not sufficient for DNP-specific B-cell IgG_1 antibody responses to DNP-GLΦ in GLΦ primed recipients. The implications these studies have for the function of the I region genes in individual T and B lymphocyte collaborative interactions are considered.

INTRODUCTION

The triggering of bone marrow derived lymphocytes (B cells) to proliferate, differentiate and form clones of antibody synthesizing cells is a complex biological event which, for the majority of antigens, involves collaborative interactions between B cells and thymus-derived lymphocytes (T cells) (1, 2). The role of T cells in the promotion and regulation of B-cell responses and the mechanism by which these functions are carried out has been extensively investigated in recent years. The genetic mapping of the control of immune responses to defined synthetic polymers in inbred strains of mice to a genetic region (the I region) within the major histocompatibility locus of the mouse has focused attention on the genes and gene products of this region as playing an integral role in the functions of the immune system (3). Since the initial description of the I region and the cell surface gene products of this genetic region (Ia antigens), several immune functions of B cells have been demonstrated to be inhibitable by antisera specific for the Ia antigens on B-cell surfaces (4). The I region has also been demonstrated to play an important role in controlling interactions between T and B lymphocytes (5).

The studies summarized in this report were designed to test the postulate that individual T and B lymphocytes are limited by genes of the I region in their ability to collaborate in antibody responses to T-dependent antigens. The genes of the I region would, in this way, serve to specifically regulate the synthesis of antibody at the level of T-cell dependent B-cell stimulation. We have tested this postulate experimentally by employing the splenic fragment culture system (6, 7) to study individual antigen specific B-cell responses of B cells stimulated in carrier-primed irradiated recipient mice which differed from the donor B cells in genes of the I region. Although studies employing allogeneic T and B lymphocytes are rather non-physiological, the limitations observed in allogeneic T-B cell collaboration may reflect physiological limitations in the intact animal where individual T and B lymphocytes may clonally express genes of the I region. Such heterogeneity in the expression of I region genes would serve to regulate T and B cell interactions and thus regulate the stimulation of antibody producing B cells.

In this report we present evidence that the restrictions imposed on allogeneic T-B cell interactions are not absolute in that allogeneic T and B cells collaborated in antibody responses but the antibody synthesized as a result of this collaboration was solely of the IgM heavy chain class whereas the majority of syngeneic T-B cell stimulatory interactions resulted in IgG_1 antibody producing B-cell clones. The ability of T and B cells to collaborate in IgG_1 antibody responses depended upon identity between the responding cells in the I-A or I-B subregion of the I region. We were further able to determine that these observed limitations in T-B cell interactions did not apply to T-dependent stimulation of secondary B-cell responses. As a further probe for the requisites of B and T cell interactions, we have employed the I region genetically controlled responses to the synthetic terpolymer of L-glutamic acid, L-lysine and L-phenylalanine (GLΦ). We were able to demonstrate that I region genetic identity between T and B lymphocytes which was sufficient for dinitrophenyl (DNP) specific B-cell IgG_1 antibody responses to dinitrophenylated hemocyanin (DNP-Hy) was not sufficient for DNP-specific B-cell antibody responses to DNP-GLΦ. These studies are discussed as they relate to an understanding of the function of I region genes in individual T and B lymphocytes.

RESULTS AND DISCUSSION

A summary of previously published results (8) employing the splenic focus fragment culture system to investigate the DNP-specific responses of individual, donor, primary B cells transferred to Hy-primed, irradiated, syngeneic and allogeneic recipient mice is presented in Table 1. As shown, approximately 70% of the donor primary B cells which were stimulated by DNP-Hy in fragment cultures after transfer to syngeneic recipients could be stimulated by DNP-Hy when transferred to allogeneic recipients. However, primary B-cell stimulation in allogeneic fragment cultures resulted in B-cell clones synthesizing only IgM antibody whereas the majority of B-cell clones resulting from primary B cells stimulated in syngeneic fragment cultures produced antibody of the IgG_1 heavy chain class. Primary B-cell stimulation in both syngeneic and allogeneic recipients was absolutely dependent upon carrier-priming of the recipient and T-cell depletion of the donor cell inoculum by treatment with anti-theta antisera and complement did not affect B-cell responses in syngeneic or allogeneic recipients (8). We have also carried out similar experiments using parent and F_1 hybrids of parental strains of mice in reciprocal cell transfers. The results of these experiments demonstrated that donor B cells of parental or F_1 strains were stimulated to synthesize IgG_1 anti-DNP antibody in both parental and F_1 recipient mice. These findings indicate that the results obtained in Table 1 using allogeneic donor and recipient strains were not attributable to nonspecific allosuppression.

TABLE 1

PRIMARY AND SECONDARY ANTI-DNP SPECIFIC MONOCLONAL B-CELL RESPONSES IN SYNGENEIC AND ALLOGENEIC HEMOCYANIN PRIMED RECIPIENT MICE*

Donor B cells†	Donor and recipient H-2 identity	No. positive foci per 10^6 cells transferred	%IgG_1**	%IgM
Primary	+	1.44	79	21
	-	1.00	2	98
Secondary	+	5.80	98	2
	-	2.80	73	27

*The methodology for obtaining monoclonal antibody responses in splenic fragment cultures and the detection of antibody in culture fluids by a radioimmunoassay have been previously described (6,8,11). The heavy chain class of anti-DNP antibody was determined using ^{125}I-labeled anti-γ_1 or anti-μ purified antibodies (8). Four x 10^6 donor cells were transferred to each recipient and fragment cultures were stimulated *in vitro* with DNP-Hy at 10^{-6} M DNP.

† Donor spleen cells were obtained from nonimmune mice (primary B cells) or mice immunized with 0.1 mg DNP-Hy in complete Freunds adjuvant 6 weeks prior to cell transfers (secondary B cells).

**These figures represent the percent of all DNP-specific B-cell clones which synthesized IgG_1 or IgM antibody.

We conclude from these studies that the genes of the H-2 complex play a critical role in primary B-cell stimulation to IgG_1 antibody production. It appears that the majority of B cells were capable of interacting with T cells which did not share genes of the I region but this interaction resulted in qualitatively different B-cell antibody-producing clones.

By selecting the appropriate donor and recipient strain combinations, we were able to demonstrate that syngeny between donor and recipient strains in the I-A or I-B subregion alone was sufficient for donor B-cell IgG_1 antibody responses to DNP-Hy. A summary of these results is shown in Table 2. Although the control of immune responses to murine IgG(γ 2a) and subunits of staphylococcal nuclease have been genetically mapped to the Ir-IB subregion (9, 10), there has been no definitive identification of an Ia antigen encoded by the I-B subregion (4). Our findings therefore enable us to assign an immune function, that of allowing T-B cell collaboration in IgG_1 antibody responses, to a region for which no Ia antigens have been identified.

The results presented in Table 2 indicate that although sharing of genes in the I-A or I-B subregion alone was sufficient for IgG_1 antibody responses, the efficiency of donor B-cell stimulation, in terms of the percentage of B-cell clones which produce IgG_1 antibody, is low compared to B-cell responses in syngeneic recipients. These findings may indicate that not all T and B lymphocytes are capable of collaborating fully when limited to interacting through genes of a single subregion of the I region. If individual T and B cells clonally express genes of the I region in vivo, the types of restriction in T-B cell interactions viewed here could serve to regulate B-cell responses in vivo.

In order to determine if the I region gene limitations of T-B lymphocytes collaborative interactions observed in T-dependent primary B-cell antibody responses applied to B cells at all stages of differentiation, the response of secondary B cells stimulated in syngeneic and allogeneic carrier-primed recipients was investigated. A summary of previously published results (11) is presented in Table 1. As shown, the majority of DNP-specific secondary B cells were stimulated by DNP-Hy in allogeneic Hy-primed recipients resulting in clones of IgG_1 antibody-forming cells. Secondary B-cell stimulation would not then, appear to be regulated by I region imposed limitations which affect primary B-cell stimulation. However, the generation of secondary B cells may well be dependent upon I region identity between T and B lymphocytes.

The results presented thus far were obtained employing a carrier protein, Hy, for which there is no known genetic polymorphism in I region genetic control. It was important to determine if the limitations observed in T-B lymphocyte interactions in antibody responses to DNP-Hy would apply to cell interactions in responses to a synthetic polymer for which there is known I region gene control. The results of experiments utilizing the Ir genes controlling responses to the linear synthetic terpolymer of GLΦ for which BALB/c mice (H-2^d) are responders, C57BL/6

TABLE 2

I REGION DEPENDENCE OF IgG_1 ANTI-DNP ANTIBODY SYNTHESIS BY PRIMARY B-CELL CLONES*

Donor†	Recipient	Non-H2 Identity	H-2 Identity K	A	B	J	E	C	S	G	D	Total clones analyzed	% IgG_1 positive clones
B10.D2	B10.D2	+	+	+	+	+	+	+	+	+	+	40	80.0
B10.D2	C57BL/10	+										26	4.0
B10.D2	BALB/c	-	+	+	+	+	+	+	+	+	+	141	82.0
B10.A(4R)	B10.A(5R)	+			+							34	35.0
B10.A(4R)	B10.A(2R)	+	+	+							+	33	58.0
B10.A(5R)	B10.A(2R)	+				+	+	+	+	+		18	< 6.0
B10.A(4R)	C57BL/10	+			+	+	+	+	+	+	+	22	55.0
AQR	A/HE	-		+	+	+	+	+	+	+	+	26	42.0

*See Table 1 for methodology employed.
†Four to 6 x 10^6 donor cells were transferred to each recipient.
Pluses appear under regions of the H-2 complex which are shared by donor and recipient pairs.

mice ($H\text{-}2^b$) are non-responders and (BALB/c x C57BL/6) F_1 hybrids (CBF_1) are responders (12) are presented in Table 3. CBF_1 mice were immunized with GLΦ and used as recipients for CBF_1, BALB/c and C57BL/6 donor cell responses to DNP-GLΦ in the splenic fragment culture system. In this system CBF_1 and BALB/c mice can be distinguished from C57BL/6 mice in their ability to offer GLΦ-specific carrier help to CBF_1 donor DNP-specific B-cell responses to DNP-GLΦ. As shown, CBF_1, Hy-primed recipients were capable of helping CBF_1, BALB/c and C57BL/6 donor B cells in DNP-specific IgG_1 antibody responses to DNP-Hy. In contrast, GLΦ primed CBF_1 recipients were able to help only GLΦ responder strain B-cells (i.e. CBF_1 and BALB/c) in DNP-specific IgG_1 antibody responses while C57BL/6 GLΦ non-responder donor B cells are helped only to synthesize IgM anti-DNP specific antibody. These results are similar to those obtained by Dr. Katz and his colleagues employing the GLT genetic system (13).

TABLE 3

PRIMARY DNP-SPECIFIC B-CELL RESPONSES IN Hy or GLΦ PRIMED CBF_1 RECIPIENT MICE*

Donor	CBF_1 recipient priming	Total clones analyzed	% clones IgG_1 positive
BALB/c	Hy	36	83
C57BL/6	Hy	34	79
CBF_1	Hy	43	95
BALB/c	GLΦ	29	69
C57BL/6	GLΦ	20	5
CBF_1	GLΦ	31	55

*Manuscript in preparation: Pierce, S.K., C. Merryman, P.H. Maurer and N.R. Klinman.

The results of these studies indicate that cell interactions are more specifically limited than one would have predicted based on the studies employing DNP-Hy as an antigen. It is clear that CBF_1 T cells responsive to GLΦ are not able to interact with DNP-specific C57BL/6 B cells in IgG_1 antibody responses although Hy-specific CBF_1 T cells are able to interact with the same DNP-specific B cells in IgG_1 antibody responses. These findings further indicate that there is a relationship between I region genetic control of responses to GLΦ and the ability of cells to collaborate in IgG_1 antibody responses to DNP-GLΦ.

The studies presented have investigated individual B-cell responses in the milieu of nonlimiting numbers of T cells. It is clear that I region gene limitations of individual T and B lymphocyte interactions cannot be viewed in such a system. This would require the isolation of individual T cells as well as B cells. Preliminary results from this laboratory indicate that such T cell "cloning" may be feasible, in which case these questions may be experimentally approachable directly in the future.

ACKNOWLEDGEMENTS

I thank Dr. Norman Klinman for his help in the preparation of this manuscript. This study was supported by U.S. Public Health Service Grants CA-15822 and CA-09140.

REFERENCES

1. Mitchison, N.A. (1971) Eur. J. Immunol. 1, 18.
2. Claman, H.N., Chaperon, E.A., and Triplett, R.F. (1966) Proc. Soc. Exp. Biol. Med. 122, 1167.
3. Benacerraf, B. and McDevitt, H.O. (1972) Science 175, 273.
4. Shreffler, D.C., Meo, T. and David, C.S. (1976) in The Role of Products of the Histocompatibility Gene Complex in Immune Responses. Ed., Katz, D.H. and Benacerraf, B. Academic Press, p. 3.
5. Katz, D.H. and Benacerraf, B. (1975) Transpl. Rev. 22, 200.
6. Klinman, N.R. (1972) J. Exp. Med. 136, 241.
7. Klinman, N.R. and Press, J.L. (1975) Transpl. Rev. 24, 41.
8. Pierce, S.K. and Klinman, N.R. (1975) J. Exp. Med. 142, 1165.
9. Lieberman, R. and Humphrey, W., Jr. (1972) J. Exp. Med. 136, 1222.
10. Lozner, E.C., Sachs, D.H., and Shearer, G.M. (1974) J. Exp. Med. 139, 1204.
11. Pierce, S.K. and Klinman, N.R. (1976) J. Exp. Med. 144, 1254.
12. Merryman, C., Maurer, P.H., and Bailey, D.W. (1972) J. Immunol. 108, 937.
13. Katz, D.H., Hamaoka, T., Dorf, M.E., Maurer, P.H., and Benacerraf, B. (1973) J. Exp. Med. 138, 734.

ALLOSUPPRESSION AND THE GENETIC RESTRICTION OF CELL INTERACTIONS*†

Susan L. Swain§ and Richard W. Dutton
Department of Biology, University of California, San Diego

ABSTRACT. The addition of allogeneic T cells from normal mice suppresses the secondary *in vitro* response of primed T cells and B cells. We have shown that this negative allogeneic effect maps predominantly to the K and D regions of the major histocompatibility complex (MHC) and that the suppression is directed against the responding B cells. Secondary responses are more sensitive to suppression than are primary.

We had previously shown that carrier primed T cells can collaborate effectively with hapten primed allogeneic B cells after the removal of Ly2 positive cells. Experiments presented here show that the response of spleen cells can be stimulated by antigen bound to adherent cells. We were unable to demonstrate any preference for antigen on syngeneic macrophages.

There was thus no evidence for genetic restrictions in T-B or T-macrophage interactions in this experimental model.

INTRODUCTION. In previous studies we looked for genetic restrictions on the collaboration between carrier primed T cells and hapten primed B cells in a secondary *in vitro* response, using mouse spleen cell suspensions. In the course of these studies we found that very small numbers of T cells from a normal mouse could totally suppress the collaborative response of primed allogeneic T cells and B cells. The suppression was mediated by $Ly1^{-}$,$Ly23^{+}$ T cells which had to be present during the first day of culture (1). It was clear that allosuppressive cells in a population of carrier primed T cells would suppress any response that might take place with allogeneic hapten primed B cells. We showed that allosuppression was reduced when the T cell population was first treated with anti Ly2 and complement. Under these conditions allogeneic and syngeneic carried primed T cells collaborated with hapten-primed B cells with equal efficiency. The responses were carrier specific and not due to positive allogeneic effects (1).

Further studies, discussed here, have proceeded along two lines. First, we have characterized the allosuppressive or negative allogeneic effect in some detail: (a) to be able to assess the role of such effects in experimental models in

---

* Running title: Allosuppression and genetic restriction.

† This work is supported by Grants: USPHS AI 08795 and ACS IM-1J.

§ Supported by USPHS CA 09174.

which genetic restriction has been demonstrated, and (b) as an experimental model to study T cell mediated suppression of the immune response.

In a second set of experiments we have looked to see whether there is any genetic restriction on macrophage interactions in the induction antibody synthesis when very small amounts of antigen are induced into the system bound to the macrophage surface.

The results of these studies will be presented in more detail elsewhere (2-5).

MATERIALS AND METHODS. The mice that we have used, the antigens, the *in vivo* immunization, the secondary *in vitro* culture system, indirect plaque assay, the preparation of T cell and B cell subpopulations mitomycin treatment of cells and the assay of β-galactosidase have been described elsewhere (6).

Macrophage enriched populations were obtained from the spleens of normal mice by adherence to plastic (7). Recovered adherent cells were incubated at 4×10^6/ml of BSS with TNP β-galactosidase (ZT) antigen at 20 μg/ml for 30 minutes on ice, then washed three times in BSS with 5% FCS. These antigen-incubated macrophages were added to cultures of primed cells as the only source of antigen. The number of macrophages in cell populations was determined by esterase staining (8).

TABLE 1

INTRA-H-2 DIFFERENCES AND NEGATIVE ALLOGENEIC EFFECTS

H-2 DIFFERENCE	SUPPRESSOR (K,IA,IB,IJ,IC,S,D)	TARGET (K,IA,IB,IJ,IC,S,D)	% SUPPRESSION
D ONLY	B10.A (k k k k d d [d])	B10.A (2R) (k k k k d d [b])	75
	B10.A (2R) (k k k k d d [b])	B10.A (k k k k d d [d])	100, 86,75
	C3H.OH (d d d d d d [k])	B10.D2 (d d d d d d [d])	68
	B10.AKM (k k k k k k [q])	B10.Br (k k k k k k [k])	84
	A.AL (k k k k k k [d])	B10.Br (k k k k k k [k])	73
	B10.T (6R) (q q q q q q [d])	B10.G (q q q q q q [q])	42
K ONLY	A.AL ([k] k k k k k d)	A.TL ([s] k k k k k d)	68
	B10.AQR ([q] k k k d d d)	B10.A ([k] k k k d d d)	82,52

Table 1. The % suppression when 1×10^5 normal nylon column passed (T) suppressor cells were added to 10^6 responding cells (targets). Taken from (2).

RESULTS

Part 1. ALLOSUPPRESSION. (a) Mapping of the genetic differences that elicit an effect. Small numbers of allogeneic nylon column passed T cells from normal mice were added to whole spleen cell suspensions from hapten-carrier primed mice at zero time. The inhibition of the secondary *in vitro* IgG anti-hapten PFC response was measured at day 5.

Suppression was observed in all cases where there was a whole haplotype difference while background differences or differences at the Mls locus were without effect (2). Equally strong effects (Table 1) were

TABLE 2

ALLOSUPPRESSION DUE TO MUTATIONS AT THE K END OF $H\text{-}2^b$

TARGET (RESPONDING)	SUPPRESSOR			
	$H\text{-}2^b$	$H\text{-}2^{ba}$	$H\text{-}2^{bh}$	$H\text{-}2^{bf}$
$H\text{-}2^b$	12, 8, 0	94,64,55	87, 45	53
$H\text{-}2^{ba}$	96,93,80	29,9		96
$H\text{-}2^{bh}$	70	64	12	
$H\text{-}2^{bf}$	94	52		-49

Table 2. (See legend of Table 1). Taken from (2)

TABLE 3

INTRA-H-2 DIFFERENCES AND NEGATIVE ALLOGENEIC EFFECTS

H-2 DIFFERENCE		SUPPRESSOR (K,IA,IB,IJ,IC,S,D)	TARGET (K,IA,IB,IJ,IC,S,D)	% SUPPRESSION
I AND I+S	WHOLE	A.TH (s [s s s s s] d)	A.TL (s [k k k k k] d)	80,71
		B10.(AQR) (q [k k k d d] d)	B10.T (6R) (q [q q q q q] d)	20
	I + S	B10.T (6R) (q [q q q q q] d)	B10.(AQR) (q [k k k d d] d)	95
	IB,IJ, IC,S	B10.A (2R) (k k [k k d d] b)	B10.A (4R) (k k [b b b b] b)	76,65,0
	IJ	B10.A (3R) (b b b [b] d d d)	B10.A (5R) (b b b [k] d d d)	8
		B10.A (5R) (b b b [k] d d d)	B10.A (3R) (b b b [b] d d d)	0
	IC,S	A.AL (k k k k [k k] d)	B10.A (k k k k [d d] d)	-7
		A.AL (k k k k [k k] d)	A/J (k k k k [d d] d)	-22

Table 3. (See legend of Table 1). Taken from (2)

observed when the allosuppressive T cells differed from their targets only at the K or D end or between the wild type and mutant K^b haplotypes (Table 2). The suppression observed when the genetic differences were restricted to I and S regions were somewhat variable and were a bit weaker than those seen for K and D alone (Table 3). For instance, quite strong inhibition of A.TL responses was obtained with A.TH T cells, and B10.T (6R) T cells suppressed B10. (AQR) responding cells. The suppression of B10.T(6R) with B10.(AQR) was marginal. Differences at IJ, or IC and S did not stimulate allosuppression. The suppression of B10.A(4R) by B10.A(2R) was variable.

(b) Evidence for direct suppression of the responding B cell. Parental T cells from normal mice suppressed the response of F_1 spleens from hapten carrier primed mice. To study the cellular target of allosuppression we varied the origin of the helper T cells and responding B cells. The response of F_1 carrier primed T cells and hapten primed parental B cells (syngeneic to the allosuppressive T cells) is not suppressed (Figure 1). This was true even when F_1 B cells from unprimed mice were added as an extra stimulator cell. It would appear that the carrier primed helper T cell was not the target of suppression, and that the responding B cell or macrophage must be the target. The addition of

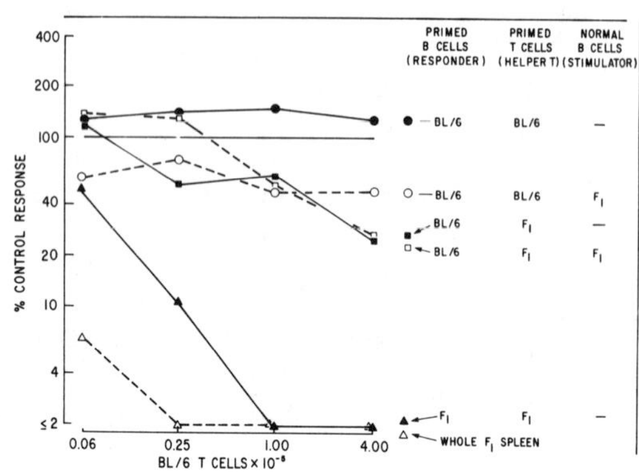

Fig. 1. The % of the appropriate control response (no T cells) obtained when different numbers of BL/6 T cells were added to responding cultures is given. The primed T and B cells were either syngeneic to these T cells or semiallogeneic (BDF_1)

macrophages syngeneic to the suppressor T cells or to the responding cells does not reverse the suppression of allogeneic B cells (data not shown). We therefore conclude that the suppression is directed against the B cell rather than the macrophage.

Further support for this comes from the finding that the response of primed B cells to the T independent antigen DNP-lys-Ficoll is also markedly suppressed (not shown).

(c) The responses of primed and unprimed cells in the presence of allogeneic T cell. We compared the addition of one population of parental T cells to BDF_1 cells responding to SRBC (primary response) and TNP-carrier secondary response)(Figure 2). The primary erythrocyte response was markedly enhanced while the secondary anti-TNP response was greatly inhibited.

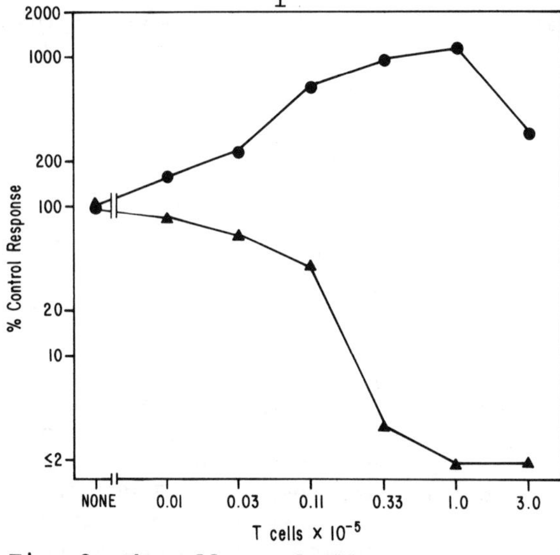

Fig. 2. The effect of allo T cells on 1° and 2° responses. Dilutions of normal BL/6 T cells were added to 10^6 normal BDF_1 cells (culture responding to SRBC (1° response)) or to ZT primed BDF_1 cells responding to ZT(▲). Direct PFCs to SRBC were determined on Day 4 and Indirect PFCs to TNP were determined on Day 5. Results are expressed as the % of the control response for no added T cells.

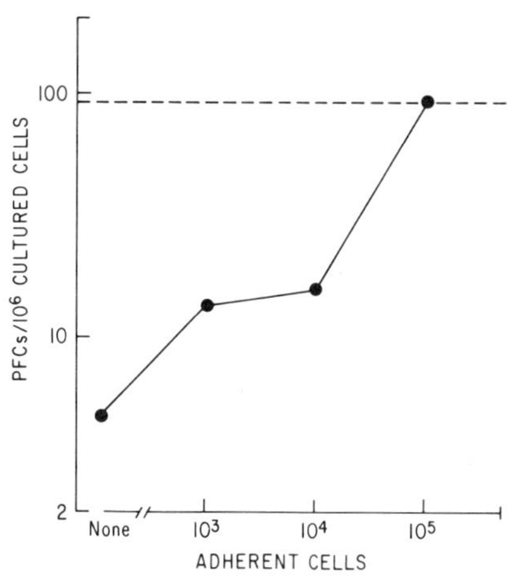

Fig. 3. Spleens of KT primed BDF_1 mice were passed over sephadex G10 and treated with anti-T cell serum + C (B cells). 10^6 B cells were mixed with 5×10^5 nylon column-passed T cells from Z-carrier primed mice. Ten-fold dilutions of BDF_1 adherent spleen cells were added to these cultures, and the response of the culture to ZT determined on Day 5.

Part 2. THE ROLE OF MACROPHAGES IN THE SECONDARY IN VITRO RESPONSE. In previous studies (1) we demonstrated that carrier primed, nylon column passed, T cells could effectively collaborate with allogeneic anti T cell-treated spleen cells from hapten primed mice if they were first treated with mitomycin and anti Ly2 sera and complement. These findings are in conflict with the observations of other investigators (9,10) who demonstrated a genetic restriction, in related experimental models. Studies of the humoral response to GAT presented on macrophages (11) and the studies which have demonstrated a genetic restriction of T cell macrophages interactions (12) suggest an alternative explanation of the requirement for genetic identity in T-B collaboration, i.e., that the antigen must be presented on macrophages compatible with macrophages used in the in vivo priming (macrophages syngeneic with the carrier primed T cells). It could be argued that our failure to demonstrate a genetic restriction was because syngeneic macrophages were introduced into the culture alone with the carrier primed T cell population.

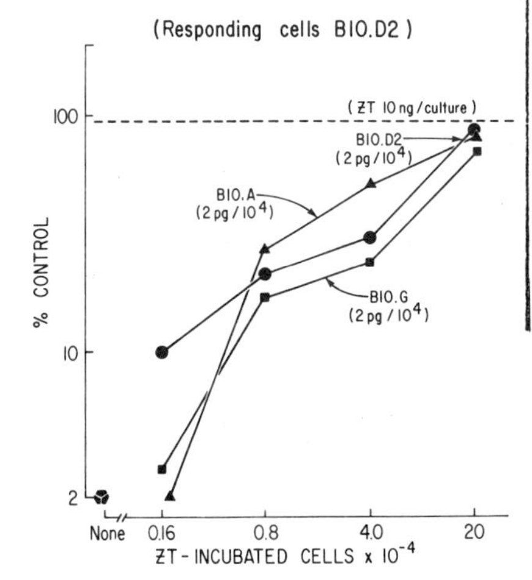

Fig. 4. Each culture contained 10^6 primed B10.D2 cells treated with anti-T cell sera + C plus 2×10^5 T_{mit} cells. Antigen (ZT) was introduced in soluble form or on Ag-incubated adherent cells from mice of different haplotypes.

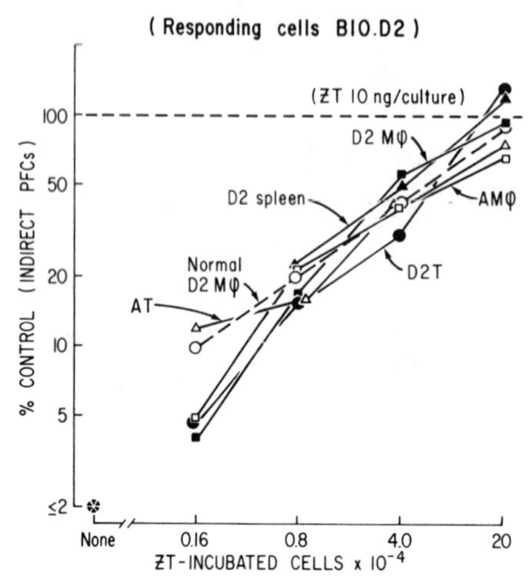

Fig. 5. Conditions the same as for Fig. 4 except whole spleen, nylon column passed T, and adherent cells from primed mice were incubated with ZT and used as a source of antigen.

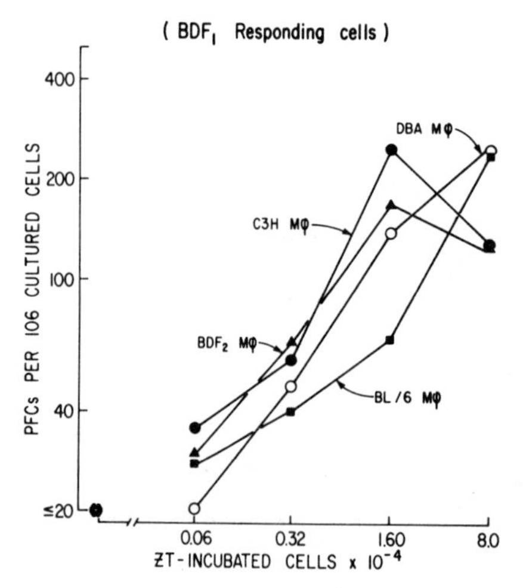

Fig. 6. Adherent cells were incubated with ZT antigen, washed and cultured overnight. The following morning they were harvested, mitomycin-treated and washed extensively before they were used as a source of antigen.

In our first experiments to test this possibility we demonstrated that removal of the macrophages from the B cell population by passage through sephadex G10 (13) alone was sufficient to abrogate the response (Figure 3). The response could be restored by the addition of syngeneic adherent spleen cells. In other experiments (data not shown) it was found that allogeneic and syngeneic adherent cells were equally effective. It was clear that the nylon column passed T cell population contains too few macrophages to support the response. This was confirmed by microscopic examination of esterase stained preparations. In several experiments an average macrophage content for whole spleen was 9%, the nylon column passed T, 2%, the G10 passed spleen contained 3%.

Thus in the experiments in which carrier primed T cells collaborated with allogeneic B cells, the bulk of the macrophages present were of the B cell haplotype and thus allogeneic to the T cells.

It could be argued that the macrophages performed two functions, one in which relatively large numbers of genetically unrestricted macrophages were required and a second in which much smaller numbers of genetically restricted macrophages were needed to present antigen to the T cells. The presence of an adequate number of such cells in the nylon column passed T cell population could not be excluded.

In the second series of experiments, very small amounts of antigen were introduced into the cultures bound to the surface of adherent spleen cells. The antigen used was TNP-β-galactosidase and the amount of antigen bound was determined by the enzyme activity present in the cell populations (6). Varying numbers of antigen incubated adherent cells were added to mixtures containing limiting numbers of carrier primed T cells and hapten primed B cells. The response could be stimulated by very small amounts of antigen introduced in this way (Figure 4). As little as 2 pg were effective. Spleen cells from B10.D2 were equally stimulated by antigen on B10.D2, B10.G, or B10.A adherent cells. There was no preference for antigen on syngeneic macrophages. Antigen was equally effective when introduced on syngeneic or allogeneic nylon passed T cell populations (not shown) or on T cells or adherent cells from primed spleen populations (Figure 5). Similar results have been obtained with another antigen (TNP-KLH) and with antigen-incubated macrophages incubated overnight before use (Figure 6). We were thus unable to obtain any evidence of genetic restriction in this model.

DISCUSSION. (1) Allosuppression. The recognition of genetic differences within the MHC elicits a strong suppression of a secondary *in vitro* antihapten response to a hapten carrier conjugate. The suppression is mediated by a T cell (1) and is directed against the responding B cell rather than the carrier specific T cell or a macrophage performing some accessory role. In this respect it differs from allotype suppression in which the suppressor T cells have been shown to act on allotype specific helper T cells (14) or the antigen specific suppressor T cells which act via an $Ly123^+$ intermediary T cell (15, and Tada personal communication). It is clear that genetic differences at the K or D region alone are able to elicit such a response. The response, however, is not exclusively to the K and D antigens since I region differences also lead to suppression although it is variable and in general somewhat weaker. It is striking that even minor differences such as the differences between K^b, K^{ba}, K^{bh}, and K^{bf} produced marked effects.

The genetic differences that lead to a negative effect can be distinguished from those that lead to a positive effect but there is considerable overlap. Strong positive effects are elicited by differences in the I region but moderate effects are also obtained when the differences are located only at K or D or are between K^b, K^{ba}, and K^{bh} (Panfili, shortly to be published). The only locus to give a clear separation of the two effects is Mls which elicits strong stimulation and no suppression. It should be noted that the positive effect is assayed in a primary response to one antigen, SRBC and the negative in a secondary response to another, a TNP-carrier conjugate.

The addition of allogeneic T cells markedly suppresses the secondary and stimulates the primary response (Figure 3). The primary response is most sensitive to positive effects while the secondary response in our hands is not enhanced even when negative effects are eliminated. On the other hand the primary response is suppressed only by the higher allogeneic T cell concentrations while the secondary response is suppressed by lower numbers of T cells.

(2) Antigen presentation on macrophages. These experiments show that the presentation of antigen on adherent spleen cells is a very effective way of introducing antigen into the system. However, syngeneic or allogeneic, adherent cells or T cells, primed cells and unprimed cells were all equally effective and no genetic preferences could be demonstrated in an extensive series of experiments. Similar findings were reported earlier by Katz and Unanue(8). It should be noted that both antigen-does response titrations and antigen-presentation on macrophages have rather shallow slopes so that small differences in efficiency might not make a noticeable difference. The possibility that antigen present on the cells that are added to the cultures is then transferred to macrophages, syngeneic to the T cell population, is not totally excluded but is considered unlikely. One would have to assume (a) that antigen initially attached to the antigen-incubated cells (adherent or T cell population) is ineffective in that site, (b) is slowly released (the rate limiting step), and (c) is then efficiently transferred to syngeneic macrophages before being presented to the T cells.

We conclude that, in this experimental model, there are no genetic restrictions on T cell-macrophage interaction.

(3) Conclusions. In our experimental model we are unable to demonstrate any genetic restrictions on cell interactions.

It is clear that in this model the presence of allosuppressive effects precludes the observation of cooperation between carrier primed T cells and hapten primed allogeneic B cells. Such collaboration can only be seen when Ly2 positive cells are removed from the T cell population. Allosuppression, however, maps preferentially to the K and D regions rather than I while the genetic restrictions in T-B and T-macrophage interaction map preferentially to I. Two further points can be made. First, differences at IA do lead to suppression (see examples in Table 2); second, the treatment with anti Ly2 sera and complement may eliminate other suppressor systems in addition to the negative effects characterized in our studies. Even so it would appear that allosuppression does not provide a general explanation for genetic restrictions in other systems and indeed allosuppression was excluded in other systems by controls in which allogeneic T cells failed to suppress syngeneic collaboration.

There is thus an unresolved discrepancy between the results obtained in this system where there is no genetic restriction and the results obtained by a number of investigators in several systems where restriction has been clearly demonstrated.

We conclude that genetic restriction may not be a mandatory requirement in cell interactions.

We speculate that: (a) later, more mature B cells can be assisted by T cells in some way that does not involve a restricted interaction and/or (b) the stimulation of T cells to give some responses, e.g., proliferation, may involve a restricted interaction while the stimulation to other responses - induction of primed helper activity, may not.

ACKNOWLEDGEMENT. We would like to express our thanks to Michele English and Sara Albanil for their skilled technical assistance and to Kathy Wong for her help in preparing the manuscript.

REFERENCES

1. Swain, S.L., Trefts, P.E., Tse, H. Y.-S. and Dutton, R.W. (1977) *Cold Spring Harb. Symp. Quant. Biol.* in press.
2. Swain, S.L. and Dutton, R.W. (1977) a *J. Immunol.* in press.
3. Swain, S.L. and Dutton, R.W. b (Submitted to *J. Immunol.*)
4. Swain, S.L. and Dutton, R.W. c (In preparation)
5. Swain, S.L. and Dutton, R.W. d (In preparation)
6. Swain, S.L., Modabber, F. and Coons, A.H. (1976) *J. Immunol.* 116, 915.
7. Pierce, C.W., Kapp, J.A., Wood, D.D. and Benacerraf, B. (1974) *J. Immunol.* 112, 1181.
8. Yam, L.T., Li, C.Y. and Crosby, W.H. (1971) *Am. J. Clin. Pathol.* 55, 283.
9. Katz, D.H. and Benacerraf, B. (1975) *Transplant. Rev.* 22, 195.
11. Pierce, C.W., Kapp, J.A. and Benacerraf, B. (1976) *J. Exp. Med.* 144, 371.
12. Rosenthal, A.S. and Shevach, E.M. (1976) In: *Contemp. Topics in Immunol.* 5, 47.
13. Ly, I. and Mishell, R.I. (1974) *J. Immunol. Methods* 5, 239.

14. Herzenberg, L.A., Okumura, K., Cantor, H., Sato, V.L., Shen, F.W., Boyse, E.A. and Herzenberg, L.A. (1976) *J. Exp. Med.* 144, 330.
15. Tada, T., Taniguchi, M. and David, C.S. (1976) *J. Exp. Med.* 144, 713.
16. Katz, D.H. and Unanue, E.R. (1973) *J. Exp. Med.* 137, 967.

CELL COMMUNICATION AND RESTRICTIONS ON COMMUNICATION

Workshop No. 13

Conveners: David H. Katz and Rolf M. Zinkernagel

Department of Cellular and Developmental Immunology
Scripps Clinic and Research Foundation
La Jolla, California 92037

The main topics discussed were: 1) what regions of the major histocompatibility gene complex are involved in immunological cell communication; 2) what cells are involved; 3) are the interactions one way or two way; 4) is communication mediated by cell-cell contact or by a factor; 5) what may be the *in vivo* relevance of the phenomena demonstrated *in vitro*.

In the domain of T-macrophage interactions and supporting the relevance of factors in T-B collaboration, Marc Feldmann (University College, London) reported that in his two-step induction assay the genetically-restricted factor (GRF) adsorbs more efficiently to syngeneic than to allogeneic unprimed T cells. This is the case even if the T cells are from von Boehmer-Sprent type chimeras. Both these findings are interpreted as indicating that genetic restriction exists *before* priming. The discussion on the relative relevance of factors versus cell-cell contact in T-B collaboration could not be settled.

This general question was briefly speculated upon by George Bell (Los Alamos, New Mexico). Assuming a frequency of 10^{-4} antigen specific B cells of average affinity about 400 specific T-B interactions occur per gram of lymphoid tissue and per hour. Since most of the assumptions cannot be tested as yet this remains a speculation.

John Niederhuber reported on the differential expression of serologically detectable I structures in macrophages. Splenic adherent cells from mice of various *H-2* haplotypes were treated with specific anti-*I*-subregion reagents + C. Anti-*I-A*, anti-*I-J* and anti-*I-E* lysed 20-40% of these cells and prevented T helper cell induction to burro erythrocytes. This data suggests that macrophages can act in induction only at certain differentiation stages that can be found in restricted organs only.

The question of the role of macrophages in expressing Ir gene functions in a delayed-type hypersensitivity assay system was addressed by Mathew Vadas (Walter and Eliza Hall, Melbourne). Adoptive transfer of DTH with F_1 cells to either parent results sometimes in high response in one, and in low response in the other parent. Analysis with lacto-dehydrogenase B as antigen revealed that cyclophosphamide treatment of the donor could abrogate the failure to transfer DTH in a nonresponder combination.

The validity of *H-2* restriction in DTH was questioned by J. Kettman (Southwestern Medical School, Dallas). Transfer of DTH to SRBC is not obviously *H-2*-restricted if effector lymphocytes and SRBC are injected directly into the footpad. However, if immune spleen cells are transferred i.v., *H-2* restriction is observed. The analysis of this phenomenon is inconclusive as yet. Additional evidence in favor of *H-2* restriction of T-macrophage interactions emerged from the *Listeria* model infection in mice (Rolf Zinkernagel, Scripps, La Jolla). Adoptive transfer of anti-Listeria protection maps predominantly to the *I*-region of *H-2*. This is compatible with the interpretation that I-structures may serve as receptors for cell-specific differentiation signals, for example, in macrophages leading to increased enzymatic activity enabling the digestion of intracellular bacteria.

Zoltan Tökes (University of Southern California, Los Angeles) reported that surface proteolytic activity of glass-adherent human leukocytes correlated with their capacity to transform T lymphocytes *in vitro*. These results suggest that surface proteolytic activity of cells may be an important factor in cell-interactions. Robert Mishell (U.C. Berkeley) found that macrophage factor can protect helper T cells against adverse effects of hydrocortisone in an *in vitro* system. The relevance of these two interesting findings is as yet unclear.

The interaction of T and B cells were discussed by John Kappler (University of Rochester), David Katz (Scripps Clinic, La Jolla) and John Sprent (University of Pennsylvania, Philadelphia). Kappler presented evidence that T cells from F_1 mice primed with antigen-pulsed parental macrophages are present in the test system. Furthermore, F_1 (low x high responder) helper T cells cooperate poorly with low responder B cells irrespective of whether the macrophage comes from a low or high responder. David Katz reported that helper T cells could be sensitized *in vivo* to antigen administered on allogeneic macrophages. However, T cells primed in this manner maintained their restrictive capacity to interact only with B cells histocompatible with the T cell donor. Additional experiments revealed that priming of F_1 thymocytes with antigen in irradiated parental recipients resulted in a population of cells that provided effective help for B cells of this same parental strain, but little or no help for B cells from either the opposite parent or F_1 strains. Evidence was presented that this phenomenon reflects the generation of *H-2*-restricted F_1 helper and suppressor activities.

J. Sprent used thoracic duct canulation to select F_1 T helper cells when transferred to irradiated, antigen-injected parents negatively (1-2 d after transfer) or positively (5 d after transfer). Negatively selected F_1 T helpers collaborated best with the other parental B cells, positively selected ones cooperated best with their syngeneic B cells. In both cases, *H-2* restriction was confirmed. Neither of the three sets of data could address the question of dual versus single recognition.

Susie Swain (UCSD, La Jolla) argued that allogeneic suppression might prevent cooperation between *H-2*-incompatible T and B cells, since elimination of anti-Ly2,3-sensitive plus mitomycin C treatment of cell populations allows successful cooperation across *H-2* barriers under all circumstances.

This workshop has clearly illustrated that of the initial questions only few can be answered reasonably well. The *H-2* regions involved in different immunological cell interactions are well known, and in many instances we know what cells are involved. Little is known on the fundamental question of the relevance of factors versus cell-cell contact, the directions of signals and, except, for speculations mainly from the models in infectious disease immunology and embryology, our knowledge on the *in vivo* relevance of the phenomena discussed is limited.

GENETIC CONTROL OF THE IMMUNE RESPONSE

M.E. Dorf, T.J. Kipps, N.K.V. Cheung and B. Benacerraf

Department of Pathology, Harvard Medical School
Boston, Massachusetts 02115

ABSTRACT. In this article, we survey several Ir gene dominated systems presenting recent data that shed light onto the possible mechanism(s) for Ir gene control. Discussed are: 1) the specificities of Ir gene recognition in contrast to the observed specificity of antibodies induced to antigens under Ir gene control; 2) the finding that responder and nonresponders to the polymer, GLØ, respond respectively to GLØ and to GLØ coupled to an immunogenic carrier by the production of antibodies which share idiotypic determinants; 3) the ability of Ir genes to regulate the immune response quantitatively in several systems; 4) the possible influence of Ir genes on the class of antibody; 5) the phenomenon of coupled complementation of Ir genes; and 6) the influence of non-H-linked genes on the regulation of several systems under Ir gene control.

INTRODUCTION

The recognition of specific antigens as immunogens by individual animals or inbred strains is governed by the products of dominant immune response (Ir) genes located within the major histocompatibility complex. This has now been verified in several mammalian species. These genes have been termed histocompatibility or H-linked Ir genes. The discovery of specific H-linked Ir genes has depended upon experiments wherein the immunological systems was presented with an antigenic challenge of highly restricted heterogeneity and specificity such as the injection of 1) synthetic polypeptides with limited number of amino acids, 2) alloantigens, or 3) limiting immunizing doses of complex multideterminant antigens. These conditions tend to limit the possible specific interactions between the antigens and the clones of immunocompetent cells. This approach has revealed that the responses of experimental animals to a wide variety of antigens are under the control of dominant H-linked Ir genes. All of these antigens share an important characteristic in that they are all thymus dependent. This may have considerable significance when an analysis is made of the process controlled by H-linked Ir genes.

In this review, we shall concentrate on the genetic regulation of the immune response. Detailed consideration of the mechanisms affected are presented elsewhere in this volume (1).

RESULTS AND DISCUSSION

Several antigenic systems have been used to identify H-linked Ir genes. In our laboratories, we primarily have used a series of linear polypeptide antigens containing two or three of the following L-amino acids:glutamic acid (G), lysine (L), proline (pro), serine (ser), alanine (A), leucine (leu), tyrosine (T) and phenylalanine (Ø). In Table 1, we have compiled our most recent available information on the responses of strains with 12 distinct H-2 haplotypes to seven linear synthetic polypeptides. One of the polymers, GL, is non-immunogenic in all inbred strains tested. However, the responses to each of the other six antigens are under dominant control by H-linked Ir genes, many of which have been mapped within the I region of the H-2 complex (2-4).

TABLE 1

H-2 CONTROL OF IMMUNOLOGICAL RESPONSIVENESS

H-2	IMMUNOGEN						
Haplotype	GL	GLpro^{10}	GLser^{7}	GLA^{10}	GLleu^{10}	GLT^{5}	GLØ^{9}
b	—	—	—	+	—	—	—
d	—	—	—	+	+	+	+
f	—	—	+	+	—	—	—
j	—	—	—	+	+	+	+
k	—	—	+	+	—	—	—
p	—	—	—	—	—	—	+
q	—	—	—	—	—	—	+
r	—	—	—		+	+	+
s	—	+	—	+	—	—	—
u	—	—	—	+	+	+	+
v	—	—	—		+	—	—
z	—	—			+		+

The most important information which can be derived from Table 1 concerns the genetically controlled differences in the responsiveness to the synthetic polypeptide antigens— GLpro, GLser, GLA, GLleu, GLT and GLØ that differ from each other in only the third amino acid which constitutes from 5 to10% of the residues of the terpolymers. The patterns of genetic response for all of these GL containing polypeptides are absolutely distinct for different H-2 haplotypes. The antibody responses to these antigens, however, are very cross-reactive, and responsiveness to each polypeptide can be accurately ascertained by measuring the serum antibodies reactive with GLT (3,4). These findings indicate that the H-linked Ir gene control of specific immune responsiveness may not be concerned with the structural genes for immunoglobulin V regions and that the specificity for individual antigens displayed by the traits controlled by individual H-linked Ir genes is not contributed by conventional immunoglobulin antibodies.

Animals unable to respond to an antigen because they lack the relevant Ir genes may, nevertheless, form antibodies against determinants on that antigen when immunized with the antigen bound to an immunogenic carrier, such as fowl gamma globulin (FγG). Thus, when GL is chemically conjugated to FγG, it is immunogenic in all inbred strains tested (Table 2). The ability to respond to the haptenic GL molecule is dependent upon the presence of T cells which recognize the FγG carrier. Thus, congenitally athymic mice (nu/nu) fail to make detectable GL responses following immunization with GL-FγG (Table 2).

We have recently compared the anti-GLØ idiotypes of antibodies produced in responder and nonresponder mice when immunized with GLØ and GLØ-FγG respectively. The anti-GLØ idiotype antiserum was prepared by immunization of guinea pigs with purified anti-GLØ antibodies from an individual 5R mouse (5). After absorbtion with normal 5R immunoglobulins, the anti-idiotype serum was used to inhibit the binding of anti-GLØ sera with radiolabeled ligand. Thus, this assay detects idiotypes primarily associated with the antibody combining site. As indicated in Table 3, following immunization with GLØ, approximately 75% of the 5R and B10.D2 produced anti-GLØ antisera which was inhibited from binding radiolabeled ligand by prior incubation with anti-idiotypic sera. Moreover, approximately half the mice produced high levels of the cross-reactive idiotype(s), and are inhibited by more than 50% in their antigen binding capacity.

Table 2
GL NONRESPONDER MICE PRODUCE SPECIFIC ANTIBODY RESPONSES TO GL-FγG

Strain	H-2 Haplotype	Immunogen GL	Immunogen GL-FγG
A/J	*a*	< 10*	52*
C57BL/10	*b*	< 10	43
C3H.NB	*p*	< 10	42
C3H.Q	*q*	< 10	54
B10.S	*s*	< 10	40
SM	*v*	< 10	29
BALB/c	*d*	< 10	43
BALB/c (nu/nu)	*d*	NT	-1

* Percent antigen binding of a 1:5 serum dilution.

Table 3
SHARED IDIOTYPIC DETERMINANTS BETWEEN RESPONDER AND NONRESPONDER MICE

Strain	Immunogen	Number tested	Percent inhibition of Ag binding by anti-idiotype* < 20	20-50	> 50
5R	GLØ	23	5	8	10
B10.D2	GLØ	12	3	3	6
B10	GLØ-FγG	15	4	4	7
BALB/c	GLØ	11	8	3	0
CB-20	GLØ	4	0	1	3

* Anti-GLØ idiotype was prepared in guinea pigs by immunization of purified anti-GLØ antibody from an individual 5R mouse.

Following immunization with GLØ-FγG, the frequency of non-responder C57BL/10 mice (B10) that produce high and intermediate amounts of idiotype bearing molecules is virtually identical to that found among responder mice. Preliminary data indicate that the GLØ idiotypic determinants are controlled by genes linked to the immunoglobulin heavy chain allotype linkage group. Thus, none of 11 BALB/c responder mice produced high levels of anti-GLØ antibody with the cross-reactive idiotype while 3 of 4 CB-20 mice, congenic with the BALB/c but possessing the C57BL, $Ig\text{-}1^b$ heavy chain allotype, generated high levels of antibody bearing these idiotypic determinants (Table 3).

While responder mouse strains possessing the relevant Ir genes always produced high levels of serum antibodies, nonresponder strains, depending upon the antigen and the strain used, were often characterized as "low responders". This was particularly the case for the branched multi-chain synthetic copolymer most extensively studied by McDevitt and Sela (6), (T,G)-A--L, to which the ability of inbred mice to make antibodies are a quantitative genetic trait. In contrast, the antibody response to the linear polypeptide antigens is generally all-or-none in most inbred strains of mice, even with hyperimmunization with antigen in complete Freund's adjuvant. However, following extensive analysis of additional linear polypeptide systems, we have become aware of the occurrence of marginal or intermediate responses following hyperimmunization with antigens GLT, GLØ and GLleu (3,7). In Table 4, we demonstrate this phenomenon with another antigen, GLA. It should be noted that three levels of responsiveness exist for these copolymers and that these levels are all under H-linked Ir gene control. First, mice carrying the $H\text{-}2^q$ haplotype or the I^q region (6R) fail to respond to GLA or DNP-GLA. The failure of nonresponder mice to make a DNP response following immunization with DNP-GLA indicates that the H-linked Ir genes are concerned with the recognition of the carrier portion of these hapten-carrier conjugates. Secondly, mice carrying the $H\text{-}2^b$ or $H\text{-}2^a$ haplotypes or the I^a region (B10.AQR) make intermediate responses to both GLA and DNP-GLA, in comparison to the high responses of mice bearing the $H\text{-}2^f$ haplotype.

Using a hemagglutination (HA) assay (8), we have compared the classes of anti-GLA antibody produced by intermediate and high responder strains. Secondary sera from B10 and B10.A mice immunized with 100 μg of GLA in complete Freund's adjuvant contained predominantly 2-mercaptoethanol sensitive antibodies.

Table 4

H-2 CONTROL OF THE MAGNITUDE OF THE IMMUNE RESPONSE

Strain	H-2 Haplotype	IMMUNE RESPONSE* GLA	DNP-GLA (DNP)	
B10.G	*q*	5 ± 10 (3)	-4 ± 3 (5)	NONRESPONDER
C3H.Q	*q*	2 ± 5 (5)	2 ± 2 (4)	
6R	*y2*	3 ± 1 (5)	2 ± 1 (4)	
A	*a*	44 ± 6 (5)	48 ± 6 (4)	INTERMEDIATE RESPONDER
B10.A	*a*	44 ± 5 (4)	58 ± 4 (4)	
B10.AQR	*y2*	41 ± 9 (4)	43 ± 12 (4)	
A.BY	*b*	59 ± 5 (4)	40 ± 7 (4)	
B10	*b*	47 ± 7 (5)	25 ± 5 (8)	
A.CA	*f*	86 ± 4 (4)	71 ± 8 (6)	HIGH RESPONDER
B10.M	*f*	72 ± 7 (5)	75 ± 5 (4)	

* Percent antigen binding ± SE, number of mice indicated in parentheses.

In contrast, nearly all the antibodies produced by B10.M high responder mice were 2-mercaptoethanol resistant. Similar observations have been reported for GLleu and the synthetic branched polypeptide (T,G)-A--L (9,10). Thus, in addition to controlling the magnitude of the immune response, the same H-linked Ir genes may also control antibody class. Alternatively, genetically linked but distinct genes may control these two properties. Resolution of these alternatives requires precise gene mapping.

It was originally assumed that a single Ir gene was required for the control of the response to a single antigen. In several systems, evidence has accumulated that two complementing Ir genes were required for responsiveness (11). The data supporting this conclusion stemmed from studies using either F_1 or recombinant mice. Table 5 illustrates the requirement of two genetically distinct genes for GLØ responsiveness. B10 and B10.A mice carrying the $H\text{-}2^b$ and $H\text{-}2^a$ haplotypes respectively, are GLØ nonresponders, while F_1 hybrids between these two strains and the 3R and 5R recombinant strains are high responders. The two interacting Ir loci have been termed α and β. The Ir-GLØ-β locus has been mapped to the I-A subregion while the Ir-GLØ-α locus can be localized between the I-J and S regions (4). Complementing Ir genes have been identified for several antigens including GLleu, GLT, lactic dehydrogenase and the H-2 alloantigen, H-2.2 (3,7,9,12). In all of the latter systems, the α and β genes map in the vicinity of the I-C and I-A regions, respectively.

Table 5

COMPLEMENTATION OF H-2 LINKED GENES

Strain	H-2 Haplotype	H-2 Region Formulae									GLØ Response
		K	I-A	I-B	I-J	I-E	I-C	S	G	D	
B10	*b*	*b*	*b*	*b*	*b*	*b*	*b*	*b*	*b*	*b*	1 ± 3 (13)
B10.A	*a*	*a*	*a*	*a*	*a*	*a*	*a*	*a*	*a*	*a*	4 ± 2 (7)
5R	*i5*	*b*	*b*	*b*	*a*	*a*	*a*	*a*	*a*	*a*	73 ± 5 (10)
3R	*i3*	*b*	*b*	*b*	*b*	*a*	*a*	*a*	*a*	*a*	59 ± 7 (5)
18R	*i18*	*b*	*b*	*b*	*b*	*b*	*b*	*b*	?	*a*	5 ± 2 (5)
$(B10 \times B10.A)F_1$	*b/a*	*b/a*	*b/a*	*b/a*	*b/a*	*b/a*	*b/a*	*b/a*	*b/a*	*b/a*	55 ± 8 (10)
		Ir-GLØ-β[b] (K, I-A, I-B)				Ir-GLØ-α[a] (I-E, I-C, S, G)					

Table 6

COUPLED COMPLEMENTATION

Strain	Ir-H-2.2 alleles		H-2.2 Response (HA titer)
	β	α	
B10.BR	*k*	*k*	100
5R	*b*	*d*	90
B10.A	*k*	*d*	< 5
18R	*b*	*b*	< 5
B10.D2	*d*	*d*	< 5

Recently, we noted the α and β alleles derived from different haplotypes are not equivalent in their ability to complement. This is best illustrated using the Ir-H-2.2

system (12), summarized in Table 6. The Ir-H-2.2-β^k allele complements with the Ir-H-2.2-α^k allele to provide high responsiveness to H-2.2 as seen in B10.BR mice. Similarly, the Ir-H-2.2-β^b allele complements with the Ir-H-2.2-α^d allele as indicated by the high level response of 5R mice. Thus, the Ir-H-2.2-β^k and α^d alleles are perfectly functional when complemented with the appropriate genes. However, the B10.A strain which possesses both these alleles is a nonresponder to the H-2.2 antigen. We refer to such asymmetric patterns of complementation as coupled complementation and believe they represent polymorphism of the complementing Ir alleles. Similar asymmetric complementation patterns exist in other complementing Ir gene systems, *i.e.*, the Ir-GLleu, Ir-GLØ and Ir-LDH systems (9,12,13). We have previously identified coupled complementation in a different genetic system, the H-linked immune suppressor (Is) genes (11). The interacting Is genes map in the same chromosomal areas as the complementing Ir genes. A further discussion of this topic is presented elsewhere (9).

Table 7

INFLUENCE OF NON-H-2 GENES ON THE IMMUNE RESPONSE

Strain	H-2 Haplotype	ANTIBODY RESPONSE H-2.2 (HA titer)	GAT (ABC-33)
B10.A	*a*	< 5	125
A	*a*	75	3,600
(A x B10.A)F_1	*a*	90	3,900

Finally, we should consider the role of non-H-2 genes in the regulation of the immune response. This area has been generally overlooked in the past and relatively little information is available. Table 7 illustrates that non-H-2 genes can have as dramatic an influence on the strength of the immune response as the H-linked Ir genes. For example, A and B10.A mice which share the same H-2 haplotype but differ at multiple non-H-2 loci have different response potentials to the antigens H-2.2, GAT, nuclease, KLH, and sheep erythrocytes (9,14-17). However, in contrast to the H-linked Ir genes, the non-H-2 linked genes may influence the immune response in a non-antigen specific fashion. For in all of the systems mentioned, genes present in the A background are associated with a higher response potential than genes in the B10 background. We have analysed the influence of non-H-2 genes in the GAT system (14), and concluded that several loci were involved including at least one locus linked to the heavy chain allotype linkage group. Biozzi and associates (18), have estimated that 7-13 genes or gene clusters are involved in the genetic regulation of the immune response to sheep red blood cells. This includes at least one H-2 and one allotype-linked gene cluster. The mechanisms by which these non-H-2 genes can influence the immune response remains to be determined.

ACKNOWLEDGMENTS

This work was supported by Grants PCM-75-22422 from the NSF and AI-13419 and AI-00152 from the NIH. The authors express their sincere appreciation to Mrs. Sharon Smith for expert secretarial services.

REFERENCES

1. Benacerraf, B., Waltenbaugh, C., Thèze, J., Kapp, J., and Dorf, M. (1977), this volume
2. Dorf, M.E., Plate, J.M.D., Stimpfling, J.H., and Benacerraf, B. (1975) *J. Immunol.* 114:602.
3. Dorf, M.E., Twigg, M.B., and Benacerraf, B. (1976) *Eur. J. Immunol.* 6:552.
4. Dorf, M.E., and Benacerraf, B. (1975) *Proc. Nat. Acad. Sci. U.S.* 73, 3671.
5. Kipps, T.J., Dorf, M.E., and Benacerraf, B. Unpublished data.
6. McDevitt, H.O. and Sela, M. (1965) *J. Exp. Med.* 122, 517.
7. Dorf, M.E., Maurer, P.H., Merryman, C.F., and Benacerraf, B. (1975) *J. Exp. Med.* 43, 889.
8. Cheung, N.K.V., Dorf, M.E., and Benacerraf, B. (1977) *Immunogenetics* 4, 163.
9. Dorf, M.E., Stimpfling, J.H., Cheung, N.K., and Benacerraf, B. In: Proceedings of the Third Ir Gene Workshop. H.O. McDevitt, ed., Academic Press, New York, New York (1977).

10. Mitchell, G.F., Grumet, F.C., and McDevitt, H.O. (1972) *J. Exp. Med.* 135, 126.
11. Benacerraf, B., Dorf, M.E. (1977) Proc. of Cold Spring Harbor Symposia on Quantitative Biology, in press.
12. Melchers, I., and Rajewsky, K. (1975) *Eur. J. Immunol.* 5, 753.
13. Dorf, M.E., Twigg, M.B., and Benacerraf, B. (1977) *Transplantation Proc.*, in press.
14. Dorf, M.E., Dunham, E.K., Johnson, J.P., and Benacerraf, B. (1974) *J. Immunol.* 112, 1329.
15. Berzofsky, J.A., Schechter, A.N., Shearer, G.M., and Sachs, D.H. (1977) *J. Exp. Med.* 145, 111.
16. Cerottini, J-C., and Unanue, E.R. (1971) *J. Immunol.* 106, 732.
17. Silver, D.M., and Winn, H.J. (1973) *Cell Immunol.* 7, 237.
18. Stiffel, C., Mouton, D., Bouthillier, Y., Heumann, A.M., Decreuseford, C., Mevel, J.C., and Biozzi, G. (1974) *Progress in Immunol.* 2, 203.

GENE COMPLEMENTATION IN THE T-LYMPHOCYTE PROLIFERATION ASSAY: A DEMONSTRATION THAT BOTH Ir GLΦ GENE PRODUCTS MUST BE EXPRESSED IN THE SAME ANTIGEN PRESENTING CELL IN ORDER TO OBTAIN AN IMMUNE RESPONSE TO GLΦ

RONALD H. SCHWARTZ, AKIHIKO YANO AND WILLIAM E. PAUL

Laboratory of Immunology, NIAID, NIH

The antibody and T-lymphocyte proliferative responses to GLΦ are known to be controlled by two major histocompatibility complex-linked immune response genes. In this paper we examine the cellular sites of expression of the two gene products through the use of subregion specific anti-Ia antisera to inhibit T-lymphocyte proliferation to GLΦ, through an analysis of the GLΦ response of radiation chimeras made in lethally irradiated responder F_1 mice by reconstitution with both types of nonresponder parental bone marrow, and through a comparison of the presenting ability of GLΦ to F_1 responder lymphocytes by F_1 responder spleen cells and either type of parental nonresponder spleen cells. The results suggest that both gene products must at least be expressed on the surface of the same antigen presenting cell in order to elicit a T-lymphocyte proliferative response to GLΦ.

The ability of inbred strains of mice to mount an immune response against the thymic-dependent antigen, GLΦ, has been shown to be under the control of two independent, major histocompatibility (MHC)-linked immune response (Ir) genes (1). This was demonstrated at both the antibody (2) and the T lymphocyte proliferation (3) levels in that responder strains could be obtained from two nonresponder strains either by complementing the genes in the trans position through the use of F_1 hybrids or by complementing the genes in the cis position through the use of strains possessing recombinant-MHC haplotypes. Although the results obtained with the T-cell assay demonstrated that B-cell function is not required in order for the Ir genes to manifest themselves, the exact cellular sites of expression of the two complementing IrGLΦ genes in this system were not clear since both T cells and macrophages are felt to play a role in generating the response (4). In this paper we will present the results from 3 separate types of experiments which suggest that both genes must be expressed in an individual cell. Furthermore, the products of both Ir genes, or of the closely linked Ia genes, must be expressed on the surface of at least the antigen presenting cell in order to generate an immune response to GLΦ.

Earlier studies from our laboratory demonstrated that antibodies directed against Ia determinants could inhibit the T-lymphocyte proliferative response to a variety of antigens (5). Since the GLϕ response of most strains of mice is controlled by two genes mapping in separate subregions of I (3), it was possible to analyze the ability of anti-Ia sera directed against each subregion to inhibit the GLΦ proliferative response. This study was carried out on the $H-2^d$ alleles for the GLϕ alpha and beta genes and the data are presented in Tables 1 and 2. The proliferative response to GLΦ of PETLES from B10.D2 mice (dddddddd)[1] was completely inhibited by a (B10xA) anti-B10.D2 serum, which could potentially contain antibodies directed against MHC products coded for by the d alleles of K, I-A, I-B, I-J and I-E (Table 1). This serum did not inhibit the GLΦ response of B10.A (5R) PETLES (bbbkkdddd), which possess the d allele of I-C, but lack the d alleles of K, I-A, I-B, I-J and I-E (Table 1). When the serum was adsorbed with spleen cells from D2.GD mice (ddbbbbbbb), which possess the d alleles of only K and I-A, the blocking activity for B10.D2 PETLES was completely removed from low concentrations of serum (1/8%), although not from higher concentrations (1%) (Table 1). Adsorption with B10.A (4R) spleen cells (kkbbbbbbb) did not remove the blocking activity, indicating that the adsorption with D2.GD cells was specific. These results suggest that antibodies in a (B10xA) anti-B10.D2 serum directed against products of the d alleles of K and/or I-A are critical for the inhibition of the GLΦ proliferative response of B10.D2 PETLES at low serum concentrations. Since earlier studies demonstrated that anti-H-2K antibodies do not inhibit T-lymphocyte proliferation (5), it seems reasonable to conclude that the antibodies directed against products of the d allele of I-A are required for the inhibition. The residual blocking activity at higher concentrations of the D2.GD adsorbed serum might represent the presence of other blocking antibodies or an incomplete adsorption of the anti-I-A antibodies.

The proliferative responses to GLΦ of PETLES from both B10.S (9R) (ss?kkdddd) and B10.HTT (ssskkkkd) mice were inhibited by an A.TH anti-A.TL serum (Table 2).

[1]Letters listed in parentheses after each mouse strain indicate the haplotype source of the K, I-A, I-B, I-J, I-E, I-C, S, G, and D alleles of the major histocompatibility complex.

TABLE 1

Inhibition of the T-Lymphocyte Proliferative Response to GLΦ by a (B10xA) Anti-B10.D2 Serum

PETLES	Serum	Adsorbed With	Proliferation (CPM ± SEM) Medium	GLΦ	%NMS Response
B10.D.2	1% NMS	-	200 ± 25	22,000 ± 1,700	-
	1% (B10 x A)α B10.D2	-	176 ± 30	900 ± 200	5
	1% (B10 x A)α B10.D2	GD spleen	125 ± 40	4,600 ± 1,200	20
	1% (B10 x A)α B10.D2	4R spleen	1,500 ± 100	1,600 ± 1,500	1
	1/8% NMS	-	160 ± 30	18,700 ± 3,300	-
	1/8% (B10 x A)α B10.D2	-	100 ± 30	1,900 ± 150	10
	1/8% (B10 x A)α B10.D2	GD spleen	120 ± 30	17,000 ± 3,100	91
	1/8% (B10 x A)α B10.D2	4R spleen	90 ± 15	2,400 ± 300	12
B10.A(5R)	1% NMS	-	3,400 ± 100	59,700 ± 4,600	-
	1% (B10 x A)α B10.D2	-	2,700 ± 500	63,100 ± 5,000	100

TABLE 2

Inhibition of the B10.HTT and B10.S(9R) T-Lymphocyte Proliferative Responses to GLΦ by an A.TH Anti-A.TL Serum

PETLES	Serum	Spleen Adsorption	Proliferation GLΦ (Δ cpm)	% NMS Response
B10.S(9R)	NMS	-	60,900	-
	A.THαA.TL	-	5,500	9
B10.HTT	NMS	-	43,900	-
	A.THαA.TL	-	3,800	9
	A.THαA.TL	B10.D2	41,400	94
	A.THαA.TL	B10	6,200	14

This antiserum could potentially contain antibodies directed against products of the k alleles of the I, S and G regions of the MHC. In the case of the B10.S (9R) cells, the blocking antibodies could only be directed against products of the k alleles of the I-B, I-J, I-E and I-C subregions (the last of these because of shared specificity, Ia.7, between the k and d allelic products of I-C). In the case of the B10.HTT cells, the blocking antibodies could only be directed against products of the k alleles of I-E, I-C, S or G. If one makes the reasonable assumption that the blocking antibodies are the same for both strains, then these antibodies would have to be directed against products of the subregions they share in common, namely: I-E and/or I-C. Adsorption of the A.TH anti-A.TL serum with B10.D2 spleen cells completely removed the blocking activity for the GLΦ response of B10.HTT PETLES (Table 2). Adsorption with B10 spleen cells did not. This suggests that antibodies directed against the $I\text{-}C^k$ subregion (Ia.7) are responsible for the inhibitory activity in the serum.

The results of the blocking experiments suggest that antibodies directed against either I-A gene products or I-C gene products can inhibit the T-lymphocyte proliferative response to GLΦ. At the antibody level, the β IrGLΦ gene has been mapped to the I-A subregion and the α IrGLΦ gene has been mapped to either I-E or I-C (2). If the Ia antigens represent determinants on Ir gene products, as has been suggested on the basis of correlative evidence (5), then one can conclude that the two IrGLΦ gene products (α and β) are both expressed on the cell surface and that they can be inhibited from functioning in a similar way by anti-Ia antisera.

The next question we asked was whether both genes have to be expressed in the same cell for complementation to occur. This was tested by the production of radiation chimeras. $[B10.A(2R)xB10]F_1$ mice were lethally irradiated (900R) and reconstituted with either anti-Thy 1-treated syngeneic $(2RxB10)F_1$ bone marrow cells or a mixture of anti-Thy 1-treated 2R and B10 parental bone marrow cells. Irradiated B10.A (5R) mice were reconstituted with B10.A (5R) bone marrow cells as an additional control. Two months after reconstitution the mice were immunized with (T,G)-A--L as a measure of the immune function of $H\text{-}2^b$ cells, (H,G)-A--L or the IgA myeloma protein, TEPC 15, as a measure of the immune function of $H\text{-}2^{h2}$ cells, and GLΦ to measure $H\text{-}2^b$ and $H\text{-}2^{h2}$ cell interactions. If both genes must be expressed in the same cell in order to generate a GLΦ response, then chimeras which possess the α and β genes in separate cells should not respond to GLΦ. The proliferation results are shown in Table 3. PETLES from $[B10.A(2R)xB10]F_1$ mice reconstituted with a mixture of parental bone marrow cells, responded to PPD, (T,G)-A--L, TEPC 15 and (H,G)-A--L (one of two experiments)

TABLE 3

FAILURE OF RADIATION CHIMERAS TO RESPOND TO GLΦ

Source of PETLES	% H - 2^b^	Proliferative response (Δ CPM)				
		(T,G) -A -L	IgA Myeloma	(H,G) -A-L	GLΦ	PPD
BIO.A(5R) in	–	20,700	–	–	20,900	33,500
5R recipient	–	54,300	–	–	69,200	47,800
[BIO.A(2R) x BIO] F_1	–	5,100	300	–	6,600	20,300
in F_1 recipient	–	21,800	800	–	36,100	53,100
BIO.A(2R) plus BIO	–	7,300	5,100	–	500	–
in F_1 recipient	70	11,300	–	12,400	<0	–
	62	4,500	–	1,200	<0	15,500

but not to GLΦ. In two of the three experiments, the PETLES were assessed for balanced chimerism by indirect immunofluorescence staining with a (B10.D2xA) anti-B10.A(5R) serum (anti-K^b, I-A^b, I-B^b). Although the populations in those two experiments were physically balanced in terms of the numbers of $H-2^b$ cells present (70% and 62%) one of the populations appeared not to be functionally balanced because it failed to give a significant proliferative response to (H,G)-A--L. Thus, the failure of this group to respond to GLΦ could be attributed to its lack of functional $H-2^{h2}$ cells. In the other two cases, however, both parental cell types are functionally intact and the failure to respond to GLΦ must be attributed to other causes.

Control PETLES from B10.A(5R) mice reconstituted with B10.A(5R) bone marrow cells responded well to GLΦ and (T,G)-A--L (Table 3). Control PETLES from $[B10.A(2R)xB10]F_1$ mice reconstituted with syngeneic F_1 bone marrow cells also responded to GLΦ and (T,G)-A--L, although they failed to proliferate in response to TEPC 15. This last result was unexpected since normal F_1 mice respond to TEPC 15 even though it is a rather weak immunogen compared to GLΦ or PPD. This suggests that lethally irradiated, bone marrow reconstituted mice are immunologically somewhat less competent than normal F_1 mice and stresses the need for functional evaluation of each parental cell type in a radiation chimera in order for a negative result to be meaningful. Thus, the most striking B10.A(2R)↔ B10 chimera experiment is the one in which the PETLES successfully responded to the weaker immunogens TEPC 15 and (T,G)-A--L, but still failed to proliferate in response to the stronger immunogen GLΦ (Table 3). These results suggest that having the α and β gene products present in separate cells is not sufficient to generate a T-cell proliferative response to GLΦ. At least one cell type involved in generating this immune response must express both gene products.

In order to investigate the possibility that both IrGLΦ genes, or I region genes closely associated with the α and β GLΦ genes, function in the antigen-presenting cell to generate a proliferative response to GLΦ, experiments using antigen-pulsed spleen cells were undertaken. In these experiments, ammonium chloride lysed spleen cells were exposed to 20-100 μg/ml of antigen for 60 min. at 37°C in the presence of 50 μg/ml of mitomycin C. The cells were washed to remove unbound antigen and mitomycin C, and then mixed with antigen-primed PETLES. Proliferative responses were measured 5 days later. As previously described in the guinea pig (6), this method of antigen presentation is subject to certain genetic restrictions. An example of this phenomenon in the mouse is shown in Table 4. DNP_6OVA presented on syngeneic spleen cells gave a substantial proliferative response, whereas the same antigen presented on

TABLE 4

Antigen Presentation by Syngeneic, Allogeneic and F_1 Spleen Cells

Mouse Strain		Proliferation (CPM ± SEM)		
PETLES	SPLEEN	Non-Pulsed	DNP_6-OVA-Pulsed	ΔCPM
B10.D2	B10.D2	2,600 ± 800	22,200 ± 2,300	19,600
B10.D2	B10	26,400 ± 2,300	27,800 ± 2,100	1,400
B10.D2	B10x B10.D2	29,900 ± 960	38,900 ± 100	9,000
B10	B10	1,600 ± 300	31,600 ± 100	30,000
B10	B10.D2	21,500 ± 100	23,700 ± 4,400	2,200
B10	B10x B10.D2	13,600 ± 1,200	28,300 ± 1,400	14,700

allogeneic cells gave a minimal response. Thus, B10.D2 spleen cells presented well to B10.D2 PETLES but poorly to B10 PETLES whereas B10 spleen cells presented well to B10 PETLES but poorly to B10.D2 PETLES. Use of semisyngeneic F_1 cells resulted in intermediate levels of stimulation for both parental types of responder cells.

The failure of allogeneic cells to present antigen did not appear to be a result of the mixed lymphocyte reaction since, as shown in Table 5, mixtures of nonpulsed allogeneic cells and antigen-pulsed syngeneic cells presented antigen much better than mixtures of the same numbers of antigen-pulsed allogeneic cells and nonpulsed syngeneic cells. Thus, in a situation where an identical MLR stimulus was present, antigen on syngeneic cells was presented much better than antigen on allogeneic cells.

The genetic analysis of antigen-presentation in the case of GLΦ involved a slightly different protocol. The MLR was avoided entirely by using $(B_6A)F_1$ responder PETLES and by presenting the GLΦ on mitomycin C treated F_1 or parental

TABLE 5
Antigen Presentation By Syngeneic and Allogeneic Spleen Cells in the Presence of the Same Mixed Lymphocyte Reaction

Number of Spleen Cells ($x10^4$)		Cells Pulsed with DNP_6OVA	Proliferative Response of B10 PETLES (CPM ± SEM)
B10.A	B10		
10	10	None	3,700 ± 100
10	10	B10	45,200 ± 700
10	10	B10.A	13,200 ± 2,500
3	3	None	3,600 ± 1,400
3	3	B10	31,900 ± 4,100
3	3	B10.A	6,500 ± 2,000

spleen cells. In this situation, for antigens such as DNP_6OVA or PPD the F_1 cells present antigen better than either parental cell type (Table 6).

TABLE 6
B6A F_1 Spleen Cells Present Antigen Better than A or B6 Spleen Cells to B6A F_1 PETLES

Presenting Spleen Cells	Proliferation (CPM ± SEM)		
	Non-Pulsed	DNP_6OVA-Pulsed	PPD-Pulsed
B6A F_1	600 ± 40	37,700 ± 2,600	10,500 ± 2,500
B6	1,100 ± 300	20,900 ± 2,600	5,400 ± 700
A	700 ± 100	18,000 ± 1,000	6,000 ± 300

Each parental strain gave one half or less of the stimulation seen with the F_1. However in the case of GLΦ (Table 7), neither parental population presented to F_1 PETLES. Note that the same parental cell populations were capable of presenting PPD. Only $(B_6A)F_1$ cells, which possess both IrGLΦ genes could present GLΦ. Mixtures of the two parental cells, which contain all the genetic material, but expressed in separate cells, were also incapable of presenting GLΦ (Table 7), although they presented PPD better than either parent alone. Thus, genes associated with the $H-2^a$ and $H-2^b$ haplotypes must

TABLE 7
Failure of Nonresponder B6 or A Parental Spleen Cells to Present GLΦ to Responder B6A F_1 PETLES

Exp. No.	Presenting Spleen Cells	Proliferative Response of B6A F_1 PETLES to	
		GLΦ (ΔCPM)	PPD (ΔCPM)
1	B6A F_1	19,200	29,000
	B6	1,000	8,000
	A/J	< 0	11,800
2	B6A F_1	6,100	34,300
	B6	700	8,100
	A/J	800	4,600
	B6 plus A/J	500	16,100

be expressed in the same antigen presenting cell in order to initiate a T-lymphocyte proliferative response to GLΦ.

In conclusion, these results suggest that both IrGLΦ gene products must be expressed on the surface of the antigen presenting cell in order to generate a secondary T cell proliferative response to GLΦ *in vitro*. The results reinforce and enrich similar findings made in the guinea pig (7) which suggested that macrophages may be one site for the expression of Ir gene products. It should be noted, however, that none of the data has ruled out the possibility that Ir gene products might also be expressed in T-lymphocytes. This possibility is currently under investigation.

ACKNOWLEDGMENTS

We wish to thank Drs. Gustavo Cudkowicz, Chella David, David Sachs, Martin Dorf, Baruj Benacerraf and Charles Janeway, Jr., for helpful contributions to various aspects of this work. The invaluable technical assistance of Mrs. Clare Horton is also gratefully acknowledged.

REFERENCES

1. Dorf, M.E., Stimpfling, J.H., and Benacerraf, B. J. Exp. Med. 141, 1459 (1975).

2. Dorf, M.E., and Benacerraf, B. Proc. Natl. Acad. Sci. U.S.A. 72, 3671 (1975).

3. Schwartz, R.H., Dorf, M.E., Benacerraf, B., and Paul, W.E. J. Exp. Med. 143, 897 (1976).

4. Rosenstreich, D.L., and Rosenthal, A.S. J. Immunol. 112, 1085 (1974).

5. Schwartz, R.H., David, C.S., Sachs, D.H., and Paul, W.E. J. Immunol. 117, 531 (1976).

6. Rosenthal, A.S. and Shevach, E.M. J. Exp. Med. 138, 1194 (1973).

7. Shevach, E.M., and Rosenthal, A.S. J. Exp. Med. 138, 1213 (1973).

INFLUENCE OF THE MAJOR HISTOCOMPATIBILITY COMPLEX ON THE TRANSFER OF DELAYED TYPE HYPERSENSITIVITY TO ANTIGENS UNDER Ir GENE CONTROL IN MICE

M.A. VADAS and J.F.A.P. MILLER

From the Robert B. Brigham Hospital,
Harvard Medical School, Boston and
the Walter and Eliza Hall Institute
for Medical Research, Melbourne,
Australia

ABSTRACT

Transfer of DTH to GAT was shown to be restricted by the I-A region of the MHC. Sensitized cells from F_1 hybrid mice between responder and nonresponder strains transferred DTH to syngeneic F_1 mice and to naive parental-strain recipients of the responder but not of the nonresponder haplotypes. These results may be interpreted to favor the postulate that the MHC linked Ir genes exert their effects by coding for components which allow interactions between particular I region gene products and the antigen to form stable structures immunogenic for T cells.

INTRODUCTION

Delayed type hypersensitivity (DTH) is a classical expression of cell mediated immunity. The specific initiation of this reaction takes place when sensitized T lymphocytes meet a local deposition of antigen. There are two important characteristics of the antigen in this reaction. First, it is a cell bound rather than a soluble form which is likely to interact with T cells (1) and, second, T cells recognize not only antigen, but also products of major histocompatibility complex (MHC) genes of the cells presenting it (2,3). Thus the initiation of DTH involves a cell to cell interaction: That of sensitized T cells with cell (probably often macrophage) associated antigen. The necessity for cell associated antigen has also been noted for T cells proliferating in vitro in response to antigen (4,5) and for the induction of helper (6) or cytotoxic (7) T cells.

The involvement of the MHC in DTH is evident from cell transfer studies. For successful transfer, the MHC of the donor of sensitized lymphocytes and that of the recipient need to be identical at certain regions (2,3,8). An interesting point arose from experiments primarily designed to show that rejection of the transferred lymphocytes was not responsible for the restriction patterns obtained. F_1 sensitized lymphocytes were used to transfer DTH into parental strains. Although transfer was possible into both parents, it was noted that using (CBA X C57BL)F_1 or (CBA X BALB/c)F_1 cells sensitized to fowl gamma globulin (FGG), the transfer to the C57BL or BALB/c parent was always better than to the CBA. That this was not due to a special property of the CBA was evident as when dinitrofluorobenzene (DNFB) rather than FGG was used no such discrepancy existed. It was also shown that genes in the MHC were again responsible for the differences noted (3). It was thought that these observations may reflect an Ir gene controlled regulation of the immunogenecity of FGG. In this paper, we show evidence that using an antigen under strict Ir gene control transfer is possible from sensitized (Responder [R] X Non-Responder [NR])F_1 only into R or F_1 recipients. These observations suggest that the expression of Ir genes is at the level of the antigen presenting cell rather than the T cell receptor.

METHODS

Mice. Two to 3 month old female mice were used for sensitization and as recipients of sensitized cells. Higly inbred strains A. TH, A.TL, A.SW, A.QR, B10.A, B10.A(4R), B10.A(5R), BALB/c, C57BL, CBA and SJL were obtained from the Walter and Eliza Hall Institute stock. Their origins and maintenance have been described elsewhere (3,8,9).

Antigens and Sensitization. Two days before antigen administration, mice were given 200 mg/kg cyclophosphamide (Endoxan, Asta; Mead Johnson, Crows Nest, Australia) subcutaneously. The random terpolymer of L-glutamic acid^{60}-L-alanine^{30}-L-tyrosine^{10} (GAT) was a generous gift from Professor Baruj Benacerraf. It was made up as described elsewhere (10), emulsified in complete Freund's adjuvant, and 10 μg in 0.1 ml injected in the 4 footpads and subcutaneously in the abdomen. For transfer of sensitized state, peripheral lymph node or spleen cells were taken from mice 5 days after sensitization and injected intravenously into naive mice.

Test for DTH. This has been described in detail elsewhere (11). Briefly, the assay measures the influx of cells labeled with ^{125}I-5-iodo-2'-deoxyuridine (^{125}I-UdR, specific activity 90-110 μCi/μg, Radiochemical Centre, Amersham, U.K.) into DTH lesions. Antigen is deposited into the left (L) pinna and the right (R) remains uninjected. 24-48 hr later, the ears are cut off the ratio of radioactivity in the L/R ears (L/R ^{125}I-UdR uptake) reflects the extent of DTH. To increase labeling efficiency, 10^{-7} mol of 5-fluorodeoxyuridine is injected intraperitoneally 20-30 min before the intraperitoneal injection of 2 μCi ^{125}I-UdR. Values in nonsensitized mice or in naive mice not given sensitized cells are usually $\lesseqgtr$ 1.2.

RESULTS

Mice of the H-$2^{n,p,q,s}$ haplotypes are "non-responders" to GAT because they produce no detectable antibodies after injection of the antigen (10,12,13). It was, therefore, of interest to determine whether DTH responsiveness to GAT could be elicited in NR strains. The results in Table 1 show that no DTH to GAT can be elicited in the NR tested mice ($H-2^s$, $H-2^q$), although it could readily be produced in other strains.

TABLE 1

Mouse strain differences in DTH responsiveness to GAT

Strain of mice sensitized to GAT	MHC*	L/R	^{125}I-dUrd uptake[+]
CBA	k k k k k k k k		3.4±0.4
BALB/c	d d d d d d d d		2.2±0.2
A.TL	s k k k k k k d		2.2±0.2
A.TH	s s s s s s s d		1.1±0.1
A.SW	s s s s s s s s		1.2±0.1
SJL	s s s s s s s s		0.8±0.2
DBA/1	q q q q q q q q		1.0±0.1

*Regions indicated are K, I-A, I-B, I-J, I-E, I-C, S and D.
+Values are arithmetic means ± 1 SEM. Five to six mice per group.

In previous studies we established that the transfer of DTH to protein antigens, such as FGG, required identity at the I-A region of the MHC (3,8). It can be seen from Table II that identity between donors and recipients at the I-A region alone (A.TL → B10.A(4R) and B10.A(4R) → A.TL) was sufficient to allow DTH transfer to GAT. It should be noted that the MHC-linked Ir gene governing responsivenss to GAT has been mapped in the region which encompasses both I-A and I-B (12). Identity at I-B alone (B10.A(4R) → B10.A(5R) or at D alone (A.TL → BALB/c) did not allow transfer. Whether K identity alone would have allowed transfer could not be determined with the strains available, since in combinations identical only at K, one member of the pari was a NR and transfer could not be achieved from R to NR recipients (e.g., A.TL → A.SW).

TABLE 2

MHC restriction of the transfer of DTH to GAT

Donor of sensitized lymphoid cells	Recipient Strain	MHC of recipients*	L/R ^{125}I-Udr uptake in recipient mice +
CBA	CBA	k k k k k k k k	1.7±0.1
	B10.A	k k k k k *d* *d* *d*	1.7±0.1
	A.TL	*s* k k k k k k *d*	1.6±0.1
	AQR	*q* k k k k *d* *d* *d*	1.8±0.2
	C57BL	*b* *b* *b* *b* *b* *b* *b* *b*	1.2±0.1
BALB/c	BALB/c	d d d d d d d d	2.6±0.2
	B10.A	*k* *k* *k* *k* *k* d d d	1.2±0.1
A.TL	A.TL	s k k k k k k d	2.1±0.2
	B10.A(4R)	*k* k *b* *b* *b* *b* *b* *b*	1.6±0.2
	BALB/c	*d* *d* *d* *d* *d* *d* *d* d	0.9±0.2
	A.SW	s *s* *s* *s* *s* *s* *s* *s*	1.1±0.1
B10.A(4R)	B10.A(4R)	k k b b b b b b	1.5±0.04
	A.TL	*s* k *k* *k* *k* *k* *k* *d*	1.6±0.2
	B10.A(5R)	*b* *b* b *k* *k* *d* *d* *d*	1.0±0.02
	CBA	k k *k* *k* *k* *k* *k* *k*	1.5±0.1

*Regions indicated are K, I-A, I-B, I-J, I-E, I-C, S and D; letters in italics point to differences in regions between donor and recipient mice.

+Arithmetic means ± SE. Five mice per group.

We have previously demonstrated that primed lymphoid cells from F_1 mice could successfully transfer DTH to both parental strains as well as to the F_1 (8). It was therefore of interest to determine whether sensitivity could be transferred to naive mice of the parental strains from F_1 mice between R and NR strains, which themselves showed high levels of sensitivity to GAT. As can be seen from Table 3, when sensitized (R X NR)F_1 lymphoid cells were transferred to naive parental-strain recipients and to F_1 sensitivity was detected only in the F_1 and in parental recipients of the R haplotype, never in parental recipients of the NR haplotype.

TABLE 3

DTH transfer to GAT by sensitized F_1 (responder x nonresponder) cells +

Naive recipients of 5 X 10^7 F_1 cells	L/R ^{125}I-UdR Uptake+
(BALB/c x SJL)F_1	2.2±0.2
BALB/c	2.6±0.1
SJL	1.2±0.1
(A.TL x A.TH)F_1	1.9±0.1
A.TL	1.8±0.2
A.TH	1.2±0.1

*Sensitivity in the donors was as follows: (BALB/c x SJL)F_1 9.3±0.9 and (A.TL x A.TH)F_1 , 2.5±0.3.

+Values are arithmetic means±SE. Five to six mice per group.

DISCUSSION

No detectable DTH responsiveness could be elicited in mice which are known to be NR to GAT even after cyclophosphamide pretreatment (Table 1). This is unlike the situation we observed with the isozyme B of lactic dehydrogenase in which a short period of sensitivity was obtained after cyclophosphamide (2). There may thus be a difference in the mechanisms of nonresponsiveness to these antigens, for example, resulting from the activation of cyclophosphamide sensitive suppressor effects which may operate exclusively or only to a certain degree in one system. This possibility is currently under investigation.

DTH transfer to GAT was possible if sensitized donors and naive recipients were identical at the I-A region of the MHC (Table 2), just as had previously been observed with FGG (3,8). One implication is that DTH transfer to all protein and polypeptide antigens requires I-A identity. Alternatively, the region of the MHC which imposes constraints on DTH transfer may conceivably also control responsiveness to that antigen. If this is the case, restrictions with certain antigens should be imposed by different I subregions corresponding to those in which have been mapped the Ir genes governing responsiveness to those antigens.

DTH to GAT was transferrable from sensitized (R x NR)F_1 mice to naive recipients of the R but not of the NR haplotypes (Table 3). The alternative possibility that Ir genes are expressed as defect at the level of T cell recognition for antigen is difficult to reconcile with these data. Thus, if one assumed that the Ir gene coded for the T cell receptor for antigen, one would expect an F_1 animal to have T cells with receptors able to recognize antigen associated with MHC products of the NR haplotype, since the Ir gene coding for such a receptor would be derived from the R parental strain. But if this were so, sensitivity from the F_1 would be transferrable to naive recipients of LR haplotypes.

There are two other possibilities for the method of action of Ir genes. One envisages mechanisms whereby Ir genes control idiotype specific suppressor T cells (expressed in a (R x NR)F_1) which block the expression of cells activated to antigen and NR I region product. A second possibility is that Ir genes are expressed at the level of antigen presenting cell by controlling the immunogenic display of antigen. In fact, the situation here would be analogous to that already observed in (CBA x BALB/c)F_1 mice sensitized to FGG-pulsed macrophages derived from one parental strain: Transfer of DTH from these F_1 mice was possible but only to naive recipients of the parental strain of the same genotype as that from which the macrophages used for sensitization were obtained (2). In the NR, therefore, the antigen under MHC linked Ir gene control may fail to associate effectively with a particular I region gene product displayed on the surface of the macrophages and thus be unable to form a product immunogenic for T cells involved in DTH. According to this view, which we expressed before (8), the MHC linked Ir genes would exert their effects by allowing the appropriate display of antigen on the surface of macrophages. There is no reason not to consider the Ir gene product itself as being the relevant I region gene molecule involved in restriction. On the other hand, one could imagine Ir genes coding for components which allow interactions between particular I region products and the antigen to form stable structures. These ideas are in general agreement with notions recently expressed by Paul and Benacerraf (14) and by Schwartz _et al._ (5) and based on other experimental systems.

ACKNOWLEDGEMENTS

We are very grateful to Professor Baruj Benacerraf for a generous gift of GAT. The original work reported in this article was performed in collaboration with Misses A. Whitelaw, J. Gamble and A. Plantinga and with Dr. C. Bernard. The secretarial assistance of Ms. Barbara Berger and Ms. Lynn White is gratefully acknowledged. Support for this was obtained from the National Health and Medical Research Council of Australia, and from the U.S. Public Health Service Research Grant No. 63992 from the National Cancer Institute.

REFERENCES

1. Oppenheim, J.J. and Seeger, R.C. (1976). Induction of cell-mediated immunity. In Immunobiology of the Macrophage. D. Nelson, editor. Academic Press, New York, p. 112.
2. Miller, J.F.A.P., Vadas, M.A., Whitelaw, A. and Gamble, J. (1976). Proc. Natl. Acad. Sci. U.S.A. 73, 2486.
3. Vadas, M.A., Miller, J.F.A.P., Whitelaw, A. and Gamble,J. (1977). Immunogenetics 4, 137.
4. Rosenthal, A.S.,Lipsky, P.E. and Shevach, E.M. (1975). Fed. Proc. 34, 1743.
5. Schwartz, R.H., David, C.S., Sachs, D.H. and Paul, W.E. (1976). J. Immunol. 117, 531.
6. Erb, P. and Feldmann, M. (1975). J. Exp. Med. 142, 460.
7. Cerottini, J.C. and Brunner, K.T. (1974). Adv. Immunol. 18, 67.
8. Miller, J.F.A.P., Vadas, M.A., Whitelaw, A. and Gamble, J. (1975). Proc. Natl. Acad. Sci. U.S.A. 72, 5095.
9. Vadas, M.A., Miller, J.F.A.P., McKenzie, I.F.C., Chism, S.E., Shen, F.W., Boyse, E.A., Gamble, J and Whitelaw, A. (1976). J. Exp. Med. 144, 10.
10. Kapp, J.A., Pierce, C.W. and Benacerraf, B. (1973). J. Exp. Med. 138, 1107.
11. Vadas, M.A., Miller, J.F.A.P., Gamble, J. and Whitelaw, A. (1975). Intern. Arch. Allergy Appl. Immunol. 49, 670.
12. Benacerraf, B. and Katz, D.H. (1975). Adv. Cancer Res. 21, 121.
13. Kapp, J.A., Pierce, C.W., Schlossman, S. and Benacerraf, B. (1974). J. Exp. Med. 140, 468.
14. Paul, W.E. and Benacerraf, B. (1977). Science, in press.

IMMUNE RESPONSE GENES CONTROL THE HELPER-SUPPRESSOR BALANCE

Eli E. Sercarz, Robert L. Yowell, and Luciano Adorini

Department of Bacteriology, University of California, Los Angeles, California 90024

ABSTRACT

Control over the immune response to a small antigen such as lysozyme, determined by H-2 linked Ir genes, seems to depend on a single suppressive determinant (SD)* present on non-immunogenic lysozymes and absent on immunogenic lysozymes. The suppressor cell receptor complexed with antigen at the SD will nullify helper cells directed against most other determinants on the molecule except those which are so close to the SD as to be sterically shielded from suppression. Such protected helper determinants will then be responsible for any evidence of positive T-cell proliferation or antibody production arising from an antigen bearing the SD.

INTRODUCTION

During the past decade, there has been an information explosion regarding the variety and function of immune response genes. Awareness of I-region genes controlling the existence of a particular immune response, the interaction of lymphocytes, or the presence of antigenic determinants on factors or factor-acceptor sites, has contributed to increased understanding. Nevertheless, essential issues remain unclarified.

In the study of two different regulatory systems (1,2), one reflecting H-2 genetic control, we have found that a single suppressive determinant on an antigen can drastically influence the course of the immune response. Logical consideration plus some extrapolation flowing from this fact allow us to address several of these unresolved issues, e.g. the number and specificity of Ir genes, and the role of T helper and T suppressor cells.

*See Table of Abbreviations at end of article.

PROPOSITIONS

The following assumptions will serve as viewpoints from which to examine the data.

re: Suppressor vs. Helper Specificity

1. In the response to most T-dependent antigens, T helper cells are directed against certain antigenic determinants and T suppressor cells against others.

2. The repertoire of T suppressor-inducing epitopes is very small, so that for the most part, a small protein molecule under obvious Ir control will have only one suppressor determinant (SD).

3. The activity of helper cells directed against most other epitopes on the antigen will be vulnerable to suppressor cell action.

4. However, some helper cells will be protected from suppressor effects because they are directed against epitopes very close to the SD.

These relationships are portrayed in Figure 1.

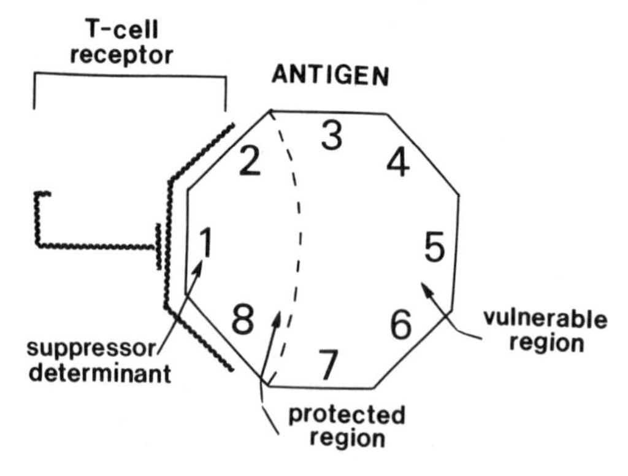

Figure 1. Antigen under genetic control is complexed to T-suppressor cell receptor which is specific for suppressor determinant (=SD) #1. The T-cell receptor is comprised of an idiotypic portion in association with an I-J region product. Helper cells directed against vulnerable determinants (=VD) #3,4,5,6, or 7 will be inhibited after interaction with the receptor-antigen complex. Helper cells directed against protected helper determinants (=PHD) #2 or 8 will be shielded from suppressive effects, since portions of the suppressor receptor will prevent access by helper receptors to nearby determinants.

re: Ir Gene Control

5. Whether one or two H-2 sub-regions are apparently involved in control of the B-cell response to a multideterminant monomeric antigen will depend on the balance of helper cells directed against PHDs versus the suppressor cells directed against SD.

6. Closely related, slightly different Ia molecules are coded for by multiple cistrons in any one I-subregion (such as I-A or I-J); each of these Ia molecules is restricted to associating with only a subset of all idiotypic receptors.

EXPERIMENTAL

We will now examine these postulates as they apply to a particular situation, the lysozyme immune response genes. The response to chicken (hen) egg-white lysozyme, HEL, is controlled by two H-2 linked genes, one presumably in or near I-A and the other in or near I-J (3). In accord with recent evidence relating I-A molecules to helper effects and I-J molecules to suppressor effects, e.g., see Tada (4), we would like to present evidence for both suppressor and helper function versus HEL in the $H\text{-}2^b$ non-responder strain, C57BL/10 (B10).

Suppressor Effect in B10, but not B10.A

Priming of B10 and its congenic, responsive partner, B10.A, with HEL-CFA leads to opposite results when spleen cells are subsequently cultured with HEL-TNP. Table I demonstrates that the anti-TNP response obtained with HEL-TNP in unprimed B10 spleen cells can be suppressed by the addition of T-cells from the 3-week primed spleen. On the other hand, this priming regime provides help to the B10.A response (not shown). The suppressor cells are sensitive to 600r; in these experiments, suppression can be reinstated by addition of unirradiated, HEL-CFA primed cells.

TABLE I

HEL-CFA Priming Induces Suppressor T Cells

HEL-CFA cells ($x 10^6$)	Treatment	Unprimed cells ($x 10^6$)	Net direct anti-TNP PFC/10^6)
-	-	10	304
10	-	-	17
5	-	5	41
5	anti-T + C	5	478
5	C only	5	0

10^6 spleen cells were cultured under Mishell-Dutton conditions with 5 x 10^{-5} 2-ME for 5 days. The antigen culture was HEL-TNP, 1 μg/ml. 10^6 spleen cells were incubated for 30 minutes at 37° C with 1 ml of rabbit-anti-mouse T-cell serum (final conc. 1:100) and 1 ml of spleen cell-absorbed guinea pig complement (final conc 1:30).

Helper Cells Raised by HEL in Non-Responder Mice

Despite the presence of suppressor activity, several lines of evidence favor the concept that HEL-specific T helper cells also are primed by HEL in H-2^b mice. First, stimulation of HEL-CFA in the footpads of B10, B6 or A.BY mice leads to IgG formation in the local, popliteal nodes (5, and Yowell, R., unpublished). Second, HEL primes for a response to the immunogenic lysozyme, JEL (Hayden, B. and Hill, S.W., unpublished). Immunogenic lysozymes for the H-2^b strain, such as JEL, REL AND TEL are presumed to lack an SD. The cross-reactive helper cells primed by HEL can then be utilized by any of the immunogenic lysozymes, but not by HEL itself, because its SD will nullify the helper cells directed against VD.

Protected Helper Determinants

We would like to discuss a third line of evidence *in extenso* since it includes consideration of protected helper determinants. In experiments on the stimulation of T-cell proliferation by HEL in H-2^b mice (Yowell, R., Schwartz, R. and Sercarz, E., unpublished) using the system developed by Schwartz et al (6), it was found that

HEL clearly stimulated a small but reproducible response. A histogram showing the relative stimulation of the HEL-primed T cells by related, non-immunogenic NEL and immunogenic REL and TEL is shown in Figure 2. T cells from the B10.A cross-react nicely with each of these lysozymes, but HEL stimulates the maximal response. However, in the B10 mouse, NEL (and OEL not shown in Figure 1) are almost completely non-cross-reactive! When the amino acid sequences were examined, the only amino acid which differed jointly in OEL and NEL from all other lysozymes was position 84 (leucine) where OEL bears a methionine and NEL a glutamine. Remarkably, the molecular model of lysozyme shows that amino acid residue 84 lies very close on the molecular surface to the position of the presumed SD. This is exactly where a PHD should be found! Our interpretation is that the PHD(s) on HEL which are responsible for the small, HEL-stimulated response in $H-2^b$ does not cross-react with OEL or NEL.

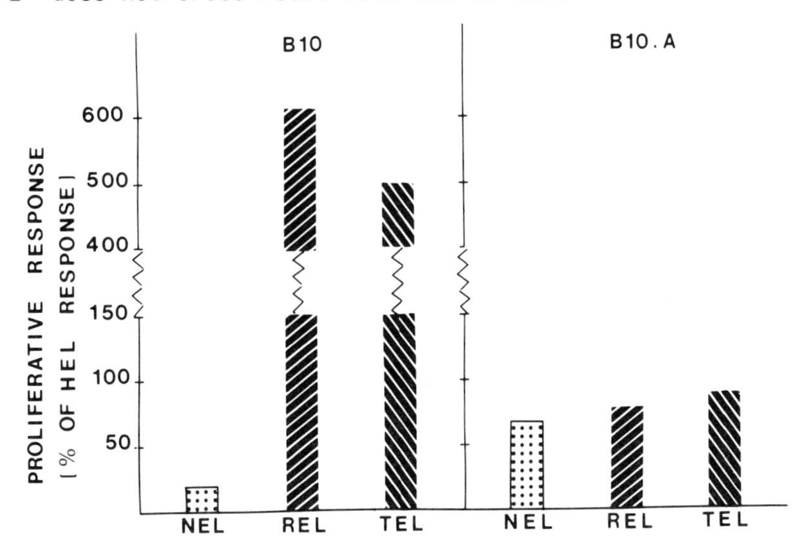

Figure 2. HEL-primed T-cell proliferation in PETLES from B10 and B10.A mice primed 3 weeks earlier in the footpads with 25 μg HEL-CFA. The values for the in vitro stimulation by the optimal concentrations of NEL, REL and TEL, are expressed as the ratio to the stimulation by HEL itself. (In the B10 strain, HEL-induced a net incorporation of ^{3}H-thymidine on the fifth day of culture of 1270 CPM; the value for HEL in B10.A was higher, 3450 CPM.)

DISCUSSION

Evidence has been presented which indicated that both suppressor and helper cells can be stimulated by HEL in the genetically non-responsive H-2^b strain. We would now like to consider some of the propositions stated earlier in light of the evidence and the relationships shown in the model (Figure 1).

Helper Cells are Directed Against Different Determinants than Suppressor Cells in Systems Showing Genetic Control

The major reason that non-responsiveness is apparent in the first place is owing to the masking of positive helper effects in the non-responder, which can be shown to exist if properly provoked. We assume that suppressor cells interact directly with helper cells via a receptor-antigen complex, as pictured in Figure 1, to nullify helper activity. This may occur by direct cell interaction or via a soluble factor. In any interactive event, the Ts receptor is directed against a determinant different from the Th receptor. The "lesion" in non-responder strains lies in the vulnerability of the great majority of their helper cells to suppressor cell action: few or weak Th cells exist with specificity for PHDs. In the H-2 mouse, we assume that the only help available is this rather weak help from the PHDs. On the contrary, in the HEL-responsive H-$2^{a/b}$ F1, there should be strong helper cells directed against determinant 1 which will be invulnerable to suppression. Alternatively, the vigorous anti-HEL response of the F1 may be owing to complementation by non-H2 genes and/or H-2 genes that vitiates the suppression.

Were two suppressive determinants to exist on the same molecule, e.g., 1 and 5 above, it would follow that _no_ helpers would be protected from the combined action of suppressor T cells of both specificities. In such systems, no sign of T-cell of B-cell activity should be evident.

It is still early in the description of idiotypes and antigen-binding entities on T and B cell subpopulations to hazard generalities about their specificity libraries. Nevertheless, we can imagine several possibilities. (a) Th, Ts and B cells possess identical libraries; (b) Th and Ts libraries are overlapping but not identical; (c) Th and Ts have complementary libraries and Th + Ts = B. Evidence that idiotypes are shared on T and B cells suggests alternatives (a) or (b). However, it remains to be seen whether idiotypes are in fact completely private. Our findings in the β-galactosidase system (2), that a peptide derived from GZ seems to be able to induce only suppressors and not helpers seems to favor possibility (b) or (c).

Are Suppressor Cells Directed Against a Variety of Epitopes on the HEL Molecule?

Although the complete answer still remains to be learned, our evidence points to a very limited number of "suppressive epitopes." Several closely related lysozymes are immunogenic in $H\text{-}2^b$ mice, such as JEL, REL, TEL. JEL, in particular, differs from HEL by only six amino acid residues, all of which happen to cluster on one quadrant of the small lysozyme molecule. We would suggest that JEL bears no suppressive determinant for the $H\text{-}2^b$ mouse and that the SD on HEL involves one or more of these 6 different amino acids. Phenylalanine at position 3 seems to be crucial in the SD: tyrosine is in this position on the immunogenic lysozymes. REL immunogenicity has been universal in a large number of haplotypes, (D. Kipp and S. Hill, unpublished), suggesting that molecules of this size are at the borderline of being too small to possess an SD. Larger antigens may possess several SDs; however, it is empirically difficult to demonstrate Ir genes in these cases, possibly because the topological relationships may be less restrictive with regard to VDs and PHDs.

The Course of Responsiveness at the T and B-Cell Levels

Despite the presence of triggerable B cells in non-responsive strains, as demonstrable by using extrinsic helpers, the animals cannot make an antibody response. Nevertheless, in several systems now (7; Hill, Miller and Sercarz, unpublished) repeated immunization leads to some antibody production in an increasing number of animals. We postulate that initially, Th of many specificities (2-8 on HEL in Figure 1) are expanded. The obvious priming by HEL for a proliferative response elicited by REL (see Figure 2) supports the notion that helper cells expand but are controlled by the late accession of suppressors.

Subsequently, Th 3-7 are prevented from expressing themselves due to suppressor T cells of specificity 1.

However, why don't the PHDs directed against 2 + 8 induce a fine immune response at the B cell level? Possibly, Ts cells interact directly with virgin B cells and prevent their expansion. If we consider PHD-2, the most appropriate B cells for collaboration should be of specificities 6, and then 5 or 7 (8, 9).

Topological and steric arguments dictate that B cells of specificities 4, 5, and 6 should be the most easily inhibited by Ts combining with HEL at determinant 1. Occasionally, B cells of specificity 7 should be available that are resistant to such suppressors. However, interaction of these two rare cell types (Th versus PHD-2 and B versus 7) may be a limiting factor, leading to the extremely sluggish anti-HEL response in the B10 strain. The fact that any anti-HEL which does arise depends on helper cells specific for the PHD, receives support from evidence that NEL, which lacks an HEL-NEL cross-reactive PHD, is completely non-immunogenic in B10 mice (Hill, S.W., unpublished). A further prediction we are testing is whether Th directed against VD 3-7 are actually prevented from helping B-cells in B10 mice. If so, we should never see antibodies specific for determinant 1.

Location and Number of H-2 Linked Genes Controlling the Immune Response

We will assume that a single sub-region controls helper cell interactions (I-A) and another sub-region controls suppressor cell interactions (I-J). Evidence obtained in our laboratory (3; D. E. Kipp, unpublished) has shown that two genes within H-2 are involved in control over the anti-HEL and the anti-HUL responses.

An interesting facet of the gene mapping is that the position of the telomeric gene, in or near I-J, is different for HEL than for HUL, despite the close evolutionary and structural similarity of the two lysozymes.

One way to accommodate this result is shown in Figure 3. Three I-J bearing T-cell receptors are shown. Each of these receptors possesses an idiotypic portion, an

I-J-coded antigenic portion, and an adapter area at the site of association of these 2 elements. The two I-J molecules shown have different adapters which will only fit with those idiotypic portions of the correct configuration. Thus, each I-J constant region will only be able to associate with a subset of the idiotypes available in Ly 23^{+} cells. A similar arrangement should exist in the I-A subregion. The number of adapter families is unknown. Figure 3 also suggests that different haplotypes may use identical adapter regions. It can be appreciated that a distinct adapter gene is not a <u>sine</u> <u>qua</u> <u>non</u> in this model.

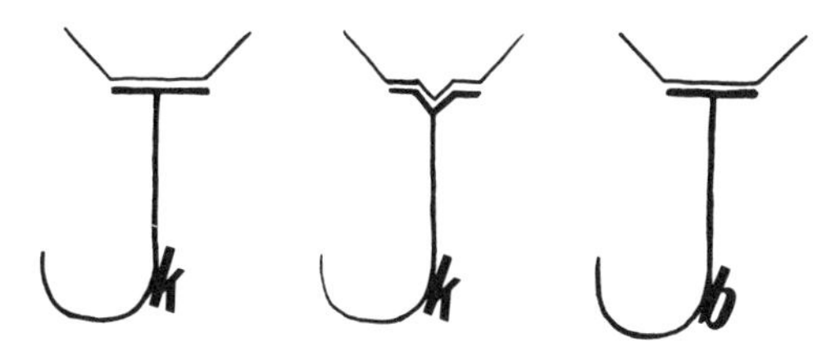

Figure 3. Three I-J bearing T-cell receptors are shown. See text for details.

What is essential for mapping in two I-regions is the presence of helper cells directed against protected helper determinants, invulnerable to the action of suppressor cells. Haplotypes lacking PHD receptors should appear to map solely in the I-J region. Likewise, in the absence of a suppressive mechanism, one-region mapping will arise when helper factor or its acceptor are not produced (10).

The foregoing model still is in large part hypothetical, although most of its postulates are readily testable.

ACKNOWLEDGEMENTS

This work was supported by NIH grants AI-08198 and AI-11183 and NCI contract CB-43972. We thank Sandy Barschi and Shelley Parker for their expert secretarial assistance. Luciano Adorini is a Fellow of the Cancer Research Institute, Inc.

Table of Abbreviations: Lysozymes: HEL = chicken (Hen); HUL = human; NEL = guinea hen (Numida); OEL = chachalaca (Ortalis); REL = ringed-neck pheasant; TEL = turkey. Antigenic determinants: PHD = protected helper determinant; SD = suppressor determinant; VD = vulnerable determinant. Mouse strains: B6 = C57BL/6J; B10 = C57BL/10J. T cells: Th = helper T cells; Ts = suppressor T cells.

REFERENCES

1. Hill, S. W. and Sercarz, E. E. (1975) Eur. J. Immunol. 5, 317.
2. Sercarz, E. E., Corenzwit, D. T., Eardley, D. D., and Morris, K. M. in "Suppressor Cells in Immunity," S. K. Singhal and N. R. St. C. Sinclair, Eds., The University of Western Ontario Press, Canada, p. 19, (1975).
3. Hill, S. W., Melchers, I., Klein, J., and Sercarz, E.E. Immunogenetics (submitted for publication).
4. Tada, T. Present Volume, Article 43.
5. Hill, S. W., Yowell, R. L., Kipp, D. E., Scibienski, R. J., Miller A., and Sercarz, E. E. in "Advances in Experimental Medicine and Biology," M. Feldman and A. Globerson, Eds., Plenum Press, New York, London (1975).
6. Schwartz, R. H., Jackson, L., Paul, W. E. (1975) J. Immunol. 115, 1330.
7. Berzofsky, J.A., Schechter, A. N., Shearer, G. M., and Sachs, D. H. (1977) J. Exp. Med. 145, 111.
8. Cecka, J. M., Stratton, J. A., Miller, A., and Sercarz, E. E. (1976) Eur. J. Immunol. 6, 639.
9. Sercarz, E. E., Cecka, J. M., Kipp, D., and Miller, A. (1977) Ann. Immunol. (Inst. Pasteur) 128C, 599.
10. Munro, A. J., and Taussig, M. J. (1975) Nature 256, 103.

GENETIC CONTROL OF THE IMMUNE RESPONSE TO H-2.32

Nobukata Shinohara and David H. Sachs

Immunology Branch, National Cancer Institute
National Institutes of Health
Bethesda, Maryland 20014

ABSTRACT. Strains of congenic mice of various H-2 haplotypes on the B10 background were all found to be low humoral responders to H-2.32, the H-2D private specificity of $H\text{-}2^k$, irrespective of their H-2 haplotypes. Low responsiveness appeared to be antigen-specific and attributable to non-H-2 linked gene(s) present in the B10 background. In an analysis of backcross progeny of the type (AKR.MxB10.AKM)F_1 x B10.AKM, the anti-H-2.32 antibody response was found to be predominantly controlled by a single locus. This locus was found not to be linked to the Ig heavy chain allotype nor to the Ly2 (hence kappa light chain) locus. Despite the observed difference in antibody production, no significant difference between AKR.M and B10.AKM mice was observed in induction of H-2D-specific killer cells. Thus, this non H-2 linked Ir gene appears to specifically control B cell reactivity and not T cell reactivity to H-2.32.

INTRODUCTION

Extensive immunogenetic studies on mice have revealed important roles played by the immune response (Ir) genes located within the I region of the major histocompatibility complex (MHC) in the regulation of immune responses to a variety of antigens. On the other hand, several Ir genes not linked to the MHC have also been reported, indicating the complex nature of genetic regulation of immune responses (1-4). Recently, McKenzie reported a genetically determined unresponsiveness of B10.AKM ($H\text{-}2^m$) mice to the $H\text{-}2D^k$ private specificity H-2.32 (5). The AKR.M mice which presumably share the same $H\text{-}2^m$ haplotype did produce a reasonable amount of anti-H-2.32, thus indicating control of anti-H-2.32 antibody response by non-H-2-linked gene(s). In this report, we have used several congenic strains of mice and appropriate backcross animals to study the specificity and possible linkage relationships of this Ir gene(s).

METHODS

Immunization of animals, absorption of antisera with lymphocytes, and cytotoxicity assays were performed by previously described methods(6) as were mixed leukocyte cultures (MLC) (7). The *in vitro* induction of and assays for cytotoxic cells were performed according to Neefe *et al*. (8). Methodological details will be presented elsewhere (manuscript in preparation.)

RESULTS

In order to confirm the interpretation that the low responsiveness of B10.AKM mice to H-2.32 is determined by non-*H-2* linked gene(s) present in the B10 background, the pooled antisera of several B10 congenic strains immunized with tissues of *H-2*^{k} mice were subjected to an analysis for anti-H-2.32 cytotoxic activity (Table 1). In the combinations *H-2*^{m} anti-*H-2*^{k} , and *H-2*^{a} anti-*H-2*^{k}. the only defined specificity which should be recognized by the responder strain is H-2.32. In other combinations, antibodies against multiple H-2 or Ia specificites besides H-2.32 are present. Therefore, such sera were tested for cytotoxicity on B10.BR cells before and after complete absorption with B10.AKM cells which share all H-2 and Ia specificities of H-2^k except H-2.32.

As shown in Table 1, none of the antisera raised in the B10 congenic strains showed high anti-H-2.32 cytotoxic activities. On the contrary, AKR.M anti-AKR, (AKR.MxB10. AKM)F_1 anti-B10.BR and C3H.SW anti-C3H had reasonable titers of anti-H-2.32 cytotoxic activity. This result is consistent with the interpretation that the low anti-H-2.32 responsiveness of the B10.AKM mouse is attributable to a recessive gene(s) present in the B10 background. It should be noted, however, that low but definite positive anti-H-2.32 cytotoxicity was detected in the antisera of B10 congenic mice except for the B10.D2 anti-B10.BR. The specificity of these reactions was checked by testing the antisera on B10.AKM cells and none of them showed any cytotoxicity.

The B10.D2 anti-C3H.OL serum had cytotoxicity of 1:128 on B10.BR cells and comparable cytotoxicity on B10.AKM cells (data not shown), indicating the presence of anti-public specificity antibodies which crossreact with both H-2K^k and and H-2D^k products. Since the MHC difference between B10.D2 and C3H.OL is restricted to *S* and *D* regions, these anti-public specificity antibodies must have been produced in response to antigenic stimulation by H-2D^k products. This result indicates that H-2D^k molecules are immunogenic to

TABLE 1. Anti-H-2.32 Cytotoxic Titers of Various Antisera Raised in B10 Congenic Strains

Antiserum	Specificity	Cytotoxic Titer	
		H-2 k*	H-2.32 **
B10.A anti-B10.BR (H-2^{a}) (H-2^{k})	kkkddd[+] anti-kkkkkk		1:4
B10.AKM anti-B10.BR (H-2^{m}) (H-2^{k})	kkkkkq anti-kkkkkk		1:8
AKR.M anti-AKR (H-2^{m}) (H-2^{k})	kkkkkq anti-kkkkkk		1:128
(AKR.MxB10.AKM)F_1 anti-B10.BR (H-2^{m}) (H-2^{k})	kkkkkq anti-kkkkkk		1:64
B10.D2 anti-B10.BR (H-2^{d}) (H-2^{k})	dddddd anti-kkkkkk	1:128	1:2
B10 anti-B10.BR (H-2^{b}) (H-2^{k})	bbbbbb anti-kkkkkk	1:64	1:4
C3H.SW anti-C3H (H-2^{b}) (H-2^{k})	bbbbbb anti-kkkkkk	1:2048	1:256
B10.D2 anti-C3H.OL (H-2^{d}) (H-2^{ol})	dddddd anti-ddddkk	1:128	1:2

The cytotoxicity tests were done on B10.BR cells.
* Cytotoxic titer of multi-specific antiserum before the absorption with B10.AKM cells.
**In case of multi-specific sera, cytotoxic titer after the complete absorption with B10.AKM cells.
+ Letters refer to the haplotypes of origin of the H-2 regions and subregions K, I-A, I-B, I-C, S, D, respectively.

B10 congenic mice. Thus the low responsiveness of B10 congenic mice to H-2.32 appears to be specific to this determinant rather than to the whole H-2D molecule.

Backcross analysis of the gene(s) controlling the anti-H-2.32 antibody response. (AKR.MxB10.AKM)xB10.AKM backcross offspring were immunized with B10.BR tissues and the cytotoxic antibody responses of individual mice were measured. Serum immunoglobulin allotype was determined by Ouchterlony immunodiffusion technique. As shown in Fig. 1, the backcross mice showed a bimodal distribution with respect to anti-H-2.32 antibody titer with a roughly 1:1 ratio of high and low responders. This indicates that a single gene (or a group of genes) segregating in the backcross generation plays a predominant role in determining the responsive status of the animals. In contrast to the preliminary results of McKenzie (5), no linkage was found between the Ig-1 locus and the gene(s) which determines anti-H-2.32 responsiveness. A collaborative study of this relationship is in progress and will be published elsewhere (Shinohara, Sachs, McKenzie, manuscript in preparation).

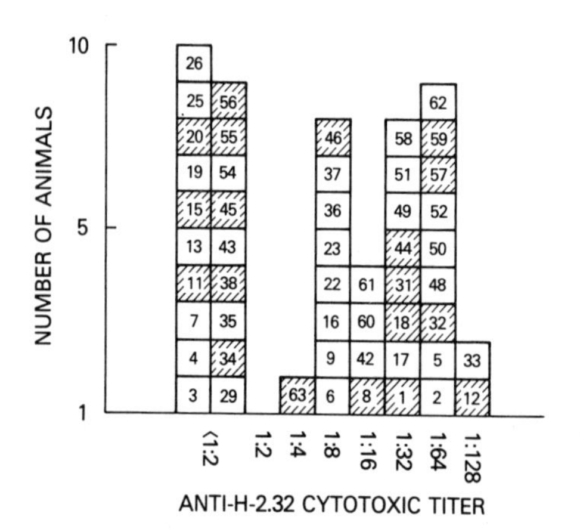

Fig. 1. Anti-H-2.32 cytotoxic antibody responses of (AKR.Mx B10.AKM)xB10.AKM backcross mice. The mice were immunized with a skingraft followed by 4 injections of B10.BR lymphocytes. Each numbered square represents an individual mouse. Open squares represent Ig-1 d/b and hatched squares Ig-1 b/b mice.

The linkage of this Ir gene to another immunoglobulin structural gene, the k light chain locus, was studied taking advantage of the fact that Ly2 locus and k chain locus are genetically linked (9). Thymocytes from the backcross animals which had been typed for anti-H-2.32 antibody response were typed for Ly2.1 antigen by cytotoxicity assay using anti-Ly2.1 serum with the kind help of Ms. Liz Schurko and Dr. E. Boyse. The results showed independent segregation.

Sixteen backcross animals typed for anti-H-2.32 response were subsequently immunized with B10.D2 tissues, in order to study whether this *Ir* gene controls antibody response in general or not. After 3 boosts, anti-H-2 cytotoxicity of individual sera was determined (Table 2). There was no significant difference in the anti-H-2^d cytotoxic antibody response between the low and the high responders to H-2.32. This result indicates that the segregating *Ir* gene controlling anti-H-2.32 responsiveness does not control humoral antibody responses in general. Thus the function of this *Ir* gene appears to be antigen-specific.

TABLE 2
Responses of High and Low Responders to the Second Antigen H-2^d

anti-H-2.32 response	No. of mice	$\log_2$ anti-H-2 cytotoxic titer
high responder	9	8.22 $\pm$ 1.78
low responder	7	8.28 $\pm$ 1.38

Cell mediated immune response to H-$2D^k$. Although the B10.AKM and the AKR.M mice showed a difference in anti-H-2.32 humoral antibody production, both strains rejected the immunizing skin grafts of H-2^k mice in 18 to 30 days. Therefore, the ability of these mice to generate H-$2D^k$ specific cytotoxic cells in vitro was studied (Table 3).

TABLE 3. The Induction of H-2D Specific Cytotoxic Cells

Responder	Stimulator	% Cr Release from Target	
		AKR	B10.BR
AKR.M	AKR	15.8	19.6
AKR.M	B10.BR	41.4	41.3
B10.AKM	AKR	13.9	17.4
B10.AKM	B10.BR	38.6	47.6

Both AKR.M and B10.AKM cells generated H-2D^k specific cytotoxic cells after in vitro sensitization with AKR or B10.BR cells. There was no significant difference in the cytotoxic activity of sensitized cells between the AKR.M and the B10.AKM responder cells, indicating that at least by this criteria, T cell recognition of H-2.32 is not affected by the defect in the B10 background.

DISCUSSION

These data suggest that this Ir gene (tentatively termed (Ir-H-2.32) controls antibody production in an antigen-specific manner. An attractive explanation for the mechanism of such a genetically determined, antigen-specific B cell low responsiveness might be a defect of the B cell clones of the relevant specificity, perhaps due to the lack of variable region information. Indeed, McKenzie's preliminary data suggested linkage between this Ir gene and the Ig heavy chain allotype locus (5). However, in the present strain combination, it was found that this Ir gene is not linked to the Ig heavy chain allotype locus nor to the k light chain locus. In addition, the fact that the B10 congenic mice did produce a small amount of anti-H-2.32 antibody might indicate that the function of this Ir gene is regulatory rather than structural. Thus a precise understanding of the nature of this Ir gene awaits further studies.

The genetically determined antigen-specific low responsiveness of B10.BR mice to the public specificity H-2."28" has recently been reported by Hansen et al. (4). This responsiveness was also shown to be under the dominant control of a single non-H-2 locus not linked to the Ig heavy chain locus nor to the k light chain locus. Although these two Ir genes share similar characteristics, the genetic relationship between them is not yet known.

REFERENCES

1. Moses, E., McDevitt, H. O., Jaton, J. C., and Sela, M. (1969) *J. Exp. Med.* 130, 1264.
2. Blomberg, B., Greckler, W. R., and Weigert, M. (1972) *Science* 177, 178.
3. Biozzi, B. (1972) in *Genetic Control of Immune Responsiveness* McDevitt, H. O. and Landy, M. (eds.) Academic Press, New York.
4. Hansen, T. H., Cullen, S. E., Shinohara, N., Schurko, E., and Sachs, D. H. (1977) *J. Immunol* ., in press.
5. McKenzie, I.F.A. (1975) *Immunogenetics* 1, 529.

6. Sachs, D. H., Winn, H. J., and Russell, P. S. (1971) *J. Immunol.* 107, 481.
7. Schwartz, R. H., Fathman, C. G. and Sachs, D. H. (1976) *J. Immunol.* 116, 929.
8. Neefe, J. R. and Sachs, D. H. (1976) *J. Exp. Med.* 144,996.
9. Gottlieb, P. D. (1976) *J. Exp. Med.* 140, 1432.

IN VITRO ANTIBODY RESPONSE OF SPLEEN CELLS FROM BIOZZI MICE

G. Doria and G. Agarossi

CNEN-Euratom Immunogenetics Group
Laboratory of Radiopathology
C.S.N. Casaccia (Rome) Italy

ABSTRACT. The aim of this study was the identification of the cell type in which genes selected for high or low response to SRBC express their functions. Spleen cells from high (H) and low (L) responder mice were immunized with SRBC in the Mishell and Dutton system. An antibody response of different magnitude was found in cultures of H and L spleen cells, the difference being at least as great as that observed in vivo. This finding under experimental conditions allowing the exclusion of any influence of the animal milieu during the immune response, suggest macrophages, B, and T lymphocytes as possible target cells of gene action. In vitro cell separation and recombination experiments with SRBC, TNP-LPS, or TNP-HRBC indicate that the differences between H and L responders brought about by genetic selection are expressed in lymphocytes rather than in macrophages. Among lymphocytes, B cells but not helper T cells were found more responsive in cultures of spleen cells from H than from L mice.

INTRODUCTION

H and L responder mice genetically selected by G. Biozzi for maximal and minimal agglutinin response to SRBC have the ability to mount high and low responses also to many other unrelated immunogens. This general antibody responsiveness is a polygenic character determined by a group of about 10 independent genes (1). Since from previous studies it is unclear whether the genetic selection has direct effects on lymphocytes or macrophages, the present work was undertaken with the following aims: 1) To establish whether H and L spleen cells display in vitro different antibody responses to T-dependent and -independent immunogens. 2) To determine the relative contribution of lymphocytes and macrophages to the difference in responsiveness between H and L mice. 3) To investigate whether the antibody responsiveness of H and L mice reflects differences in B and/or helper T cells.

IN VITRO ANTIBODY RESPONSE

Spleen cells were prepared and cultured in vitro according to Mishell and Dutton (2). In a series of experiments 1.5×10^7 nucleated spleen cells in 1 ml medium were immunized in vitro with 1×10^7 SRBC and assayed by the Jerne technique (3) for direct PFC anti-SRBC from day 3 on. The results of Fig.1 show that the peak response at day 4 of H spleen cells was

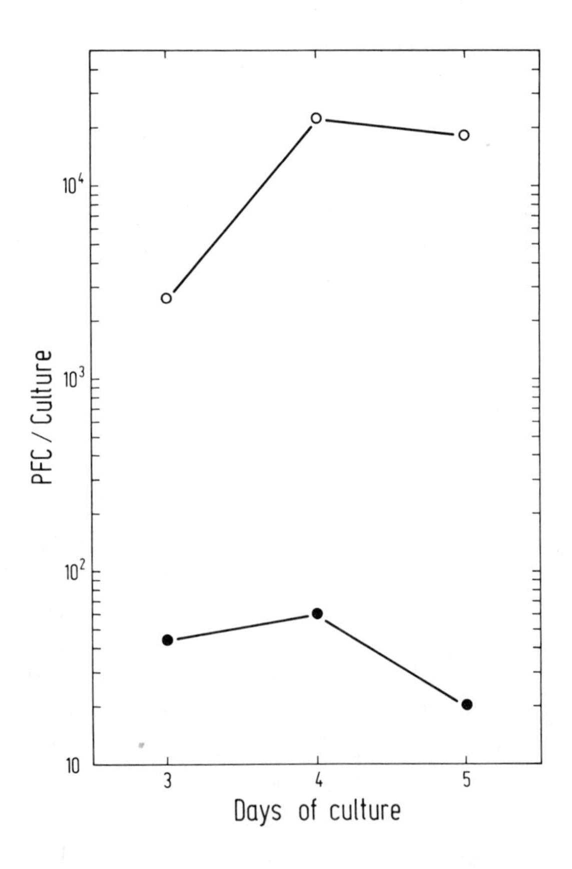

Fig.1. In vitro anti-SRBC primary response of spleen cells from H (open circles) or L (closed circles) mice.

about 400 fold higher than that of L spleen cells, the difference being at least as great as that usually observed in vivo (1). These and other findings obtained in similar experiments indicate that genetic differences between H and L mice were expressed at the cellular level under conditions allowing the exclusion of any influence of the animal milieu during the immune response. The different responsiveness of H and L spleen cells to SRBC could reflect differences in lymphocytes and macrophages.

LYMPHOCYTES VERSUS MACROPHAGES

Spleen cells from H or L mice were separated by plastic adherence in macrophages (adherent) and lymphocytes (non-adherent) according to Mosier (4). Cultures of $1.5x10^7$ unseparated or non-adherent cells in 1 ml medium, without or with $5x10^{-5}$ M mercaptoethanol (ME), were immunized with $1x10^7$ SRBC and assayed for direct PFC anti-SRBC from day 3 on. The responses at day 4 are reported in Fig.2.

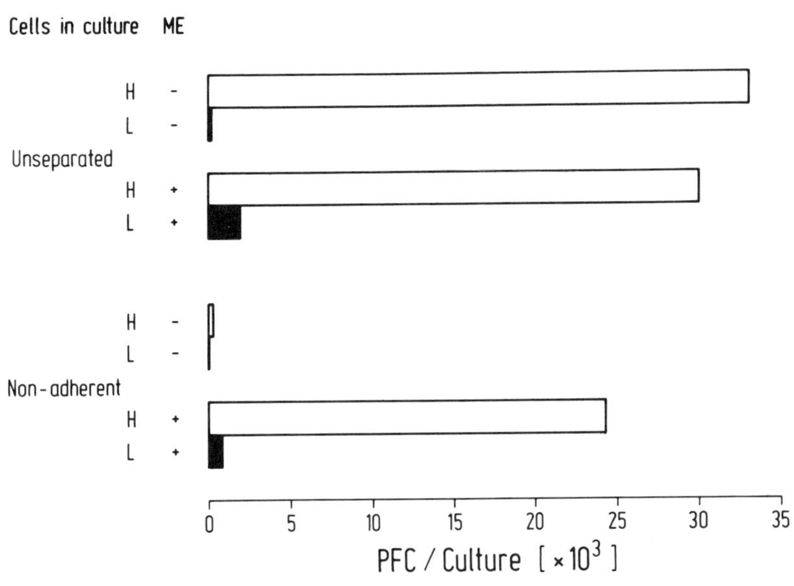

Fig.2. In vitro anti-SRBC primary response of unseparated or non-adherent spleen cells from H or L mice.
ME: mercaptoethanol.

Removal of macrophages from spleen cells drastically reduced the antibody response of H and abolished that of L lymphocytes. The addition of ME increased the responses of both H and L lymphocytes, but the difference in their responses remained at least as great as that between H and L unseparated cells. Thus, differences in lymphocytes could entirely account for the different responses of H and L unseparated spleen cells whereas differences in macrophages seem to play a negligeable role.

B VERSUS HELPER T LYMPHOCYTES

Spleen cells from H or L mice were separated in macrophages and lymphocytes as described above, and then immunized in vitro with 1 µg TNP-LPS (5). Separated cells were also recombined syngenically (e.g. $1x10^7$ non-adherent H cells were added to the adherent H cells obtained from $1x10^7$ spleen cells) and immunized in vitro. Fig.3 shows the responses at day 4 in

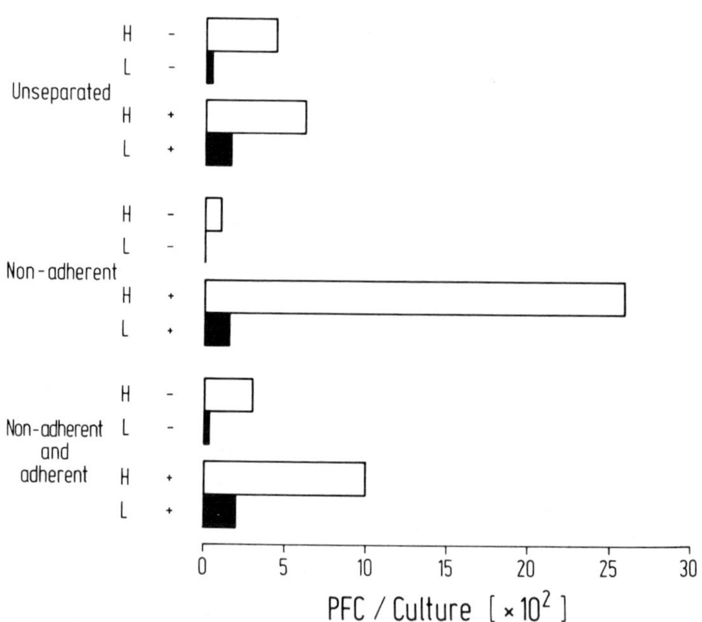

Fig.3. In vitro anti-TNP primary response of unseparated, non-adherent, or syngenically recombined adherent and non-adherent cells from H or L mice. Antigen in vitro: TNP-LPS.

terms of direct PFC anti-TNP detected by TNP-SRBC (6) in the Jerne technique. H and L spleen cells displayed different responsiveness also to TNP-LPS, a macrophage-dependent and T-independent antigen (7). The anti-TNP responses of lymphocytes stimulated after the removal of macrophages indicate that the different responsiveness of H and L spleen cells could totally depend on differences in lymphocytes rather than in macrophages, as already shown for the selection antigen SRBC. Furthermore, while the different antibody responses to SRBC may reflect differences at both T and B cell levels the data obtained with TNP-LPS point to differences in B cells.

Whether the different antibody response of H and L mice to T-dependent antigens could be attributed to differences in helper T cells was investigated by the following experiment. 1×10^7 spleen cells from (HxL)F1 hybrid mice were immunized in vitro with 2×10^5 TNP-HRBC (8) after the addition of no cells or of graded numbers of carrier-primed spleen cells from H or L mice, which had been injected i.v. with 2×10^5 HRBC 3 days before the sacrifice. Immediately prior to the sacrifice carrier-primed mice were exposed to 2000 R of X-rays to kill unprimed B and T cells in order to prevent their spleen cells from producing in vitro anti-TNP antibodies and anti-F1 allogeneic effects. The anti-TNP responses at day 4 are

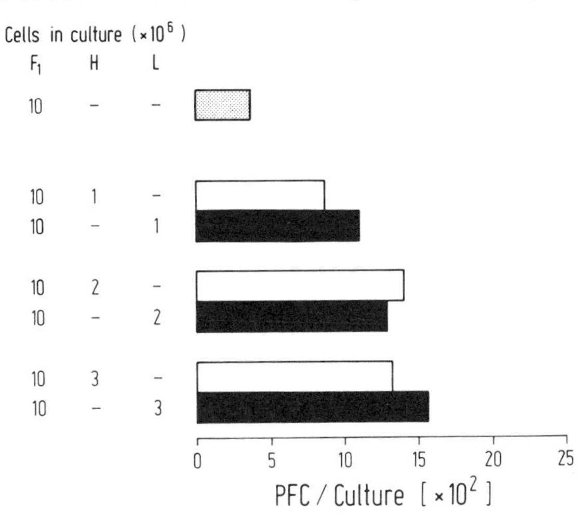

Fig.4. In vitro anti-TNP response of (HxL)F1 spleen cells alone or with spleen cells from HRBC-primed and irradiated H or L mice. Antigen in vitro: TNP-HRBC.

reported in Fig.4. The data illustrate that the addition of 1-$3x10^6$ carrier-primed cells helped F1 cells to produce anti-TNP antibodies. However, it was irrelevant whether helper cells had been provided by H or L mice. Thus, under the present experimental conditions, spleen cells from H and L mice displayed no difference in helper T cells.

It can be concluded that the different responsiveness of H and L mice to T-dependent and -independent antigens is well expressed in spleen cells immunized in vitro and can be accounted for by differences in B cells rather than in helper T cells and macrophages.

ACKNOWLEDGMENTS

Work supported by CNEN-Euratom Contract. Publication No. 1457 of the Euratom Biology Division.

REFERENCES

1. Biozzi, G., Stiffel, C., Mouton, D. and Bouthillier, Y. (1975) In: Immunogenetics and Immunodeficiency, p.180. Ed. Benacerraf, B. Publ. MTP, Lancaster, England.
2. Mishell, R.I. and Dutton, R.W. (1967) J. Exp. Med. 126, 423.
3. Jerne, N.K. and Nordin, A.A. (1963) Science 140, 405.
4. Mosier, D.E. (1967) Science 158, 1575.
5. Jacobs, D.M. and Morrison, D.C. (1975) J. Immun. 114, 360.
6. Rittenberg, M.B. and Pratt, K.L. (1969) Proc. Soc. Exp. Biol. Med. 132, 575.
7. Lee, K.C., Shiozawa, C., Shaw, A. and Diener, E. (1976) Eur. J. Immun. 6, 63.
8. Kettman, J. and Dutton, R.W. (1970) J. Immun. 104, 1558.

WORKSHOP NO. 6 - Ir CONTROL

Summary prepared by M. Bevan and J. Forman

Participants in the workshop presented data relating to the genetic control of the immune response to several different antigenic molecules.

The immune response to synthetic polymers, including L-glutamic, L-lysine, L-phenylalanine (GLϕ) and G,L-leucine was discussed. The response is under control of two genes that map in the IA and IC region of the *H-2* complex termed α and β respectively. Two non-responder strains, each lacking one of these genes, complement each other. The fact that not all α and β genes complement suggests that Ir genes are polymorphic. Non-responder strains to GLϕ were shown to produce antibody (following immunization with GLϕ coupled to FGG) bearing the same idiotype as responder strain antibody to GLϕ.

The immune response to the polymer (Glu^{80}-Phe^{20}) is controlled by at least two genes, one of which maps in the IA region of the *H-2* complex. Four phenotypes are observed: high-, intermediate-, low-, and non-responders. Furthermore only 50% of $H\text{-}2^k$ mice of one strain respond. T cells from high-responder strains give secondary responses in *in vitro* T cell proliferation assays.

The genetic control of the response to the sequential polypeptides $(T\text{-}A\text{-}G\text{-}Gly)_n$, $(T\text{-}G\text{-}A\text{-}Gly)_n$, $(Phe\text{-}G\text{-}A\text{-}Gly)_n$, and $(Phe\text{-}A\text{-}G\text{-}Gly)_n$ was studied. Only $H\text{-}2^b$ mice respond to $(T\text{-}G\text{-}A\text{-}Gly)_n$, whereas, mice of the $H\text{-}2^b$, $H\text{-}2^f$ and $H\text{-}2^r$ haplotypes respond to $(T\text{-}A\text{-}G\text{-}Gly)_n$. No responders were observed against $(Phe\text{-}G\text{-}A\text{-}Gly)_n$, and only mice of the $H\text{-}2^f$ haplotype responded against $(Phe\text{-}A\text{-}G\text{-}Gly)_n$. The genes controlling the response to $(T\text{-}G\text{-}A\text{-}Gly)_n$ and $(T\text{-}A\text{-}G\text{-}Gly)_n$ were mapped to the IA region. It was noted that only the helical form of the polymer is immunogenic. Some non-uniformity in antibody responses, as measured by a modified Farr assay, is observed with $(T\text{-}G\text{-}A\text{-}Gly)_n$ and $(T\text{-}A\text{-}G\text{-}Gly)_n$. More uniform responses were detected using an *in vitro* T cell proliferation assay. It was also noted that several of these polymers are mitogenic for B cells of all strains although they do not act as polyclonal activators. Since these materials can affect B cells directly, the possibility was raised that the non-uniformity in antibody responsiveness may be due to this property of these antigens.

Experiments were performed to test whether both the α and β genes, required for the response to GLϕ, have to be expressed in the same cell. One attempt to answer this question was the use of lethally irradiated F_1 animals repopulated with bone marrow cells from two complementing non-responder strains. The radiation chimeras failed to respond to GLϕ, indicating that both genes have to be expressed in the same cell. Another approach was to sensitize an F_1 responder animal derived from two non-responder complementing parents with GLϕ and then rechallenging *in vitro* with antigen pulsed splenic macrophages from each parent. The results indicate that the F_1 T cells did not respond unless antigen was presented on the surface of F_1 cells, indicating that both Ir genes need be expressed at the level of the antigen presenting cell.

Models were discussed relating to how Ir genes may function at the stimulating cell level. One possibility is that antigen interacts with Ia molecules on the surface of the antigen presenting cell. This interaction would account for a limited type of specificity that would differ from antibody recognition. When two *H-2* Ir genes are involved, Ia antigens might form a trimolecular complex with the antigen. If the T cell repertoire can recognize the resulting antigenic complex, then an immune response would occur. An alternative is that the T cell expresses receptors for the two Ir gene products, both of which must be recognized by the T cell which in addition bears a receptor for antigen. Complementing genes might serve to increase the specificity of Ir genes by combinatorial product interaction. No data was presented to indicate whether Ir genes were expressed at the T cell level.

The genetic control of the immune response to the Thy-1 antigen was presented. So far, three genes control the response, one about 17 cM to the right of $H\text{-}2^d$ (Ir 5) and two genes localized within the *H-2* complex. Evidence was presented to suggest that one of the intra-*H-2* genes controlling the response to Thy-1 maps within the *K* region. The evidence is based on the response in *H-2* recombinant strains and the observations that two *H-2 K*-region mutants (M523 and $H\text{-}2^{bf}$) have an altered response to the Thy-1 antigen compared to the wild type.

The T-cell mediated cytotoxic response to TNP-modified $H\text{-}2D^d$ determinants was shown to be controlled by two *H-2* genes, one of which maps to the left of the IA region. The rate of recovery from the disease caused by Friend virus is also influenced by genes mapping at the *D* end of *H-2*. Various F_1 mice derived from crosses between A.BY and the $H\text{-}2^b\text{-}H\text{-}2^a$ recombinants were used in this study. The severity of thyroiditis induced by injecting thyroglobulin is influenced by genes mapping in the *K* and *D* regions. David suggested that some of the Ir genes which map in the peripheral regions of *H-2* may in fact control immune suppression. It was also pointed out Thy-1, TNP-H-2^d, viral antigens and autoantigens all may represent true cell membrane antigens.

Evidence was presented to indicate that T cell mediated cytotoxic crossreactivity between haplotypes using TNP-modified target cells is under *H-2* control. B10.BR animals, sensitized to their own TNP-modified cells, do not crossreact with *H-2* unrelated TNP-modified targets; whereas, other strains sensitized against their own modified cells crossreact extensively with unrelated TNP-modified targets, including B10.BR.

Ir genes controlling the delayed-type-hypersensitivity (DTH) and antibody response to limiting doses of ovalbumin were examined. In addition to an *H-2* Ir gene, a non-*H-2* gene also influenced the antibody and DTH response to this antigen. The immune response to other soluble proteins, hen and human egg white lysozyme, (HEL) and (HUL) was shown to be controlled by three genes, two within the *H-2* complex, and one non-*H-2*-linked. The background gene could be distinguished in the C57BL/6 Jackson strain as compared to the C57BL/6 Bailey strain.

Several immune response were controlled by non-*H-2* non-allotype linked genes. Whether their role in control of the immune response is similar to *H-2* genetic control is presently not known. Little is understood about their antigen specificity.

FEEDBACK INDUCTION OF SUPPRESSION BY IN VITRO EDUCATED LY 1 T HELPER CELLS

Diane D. Eardley[a], *Fung W. Shen*[b], *Harvey Cantor*[c], *and Richard K. Gershon*[a].

ABSTRACT. T cells purified from mouse spleen by passage over nylon wool can be educated in vitro to mediate potent specific suppressor activity when added to assay cultures of normal spleen cells or T + B combinations. Using Ly antisera to fractionate subpopulations of T cells in the educated and assay populations, we have traced some of the cellular interactions that occur during the induction and effector phases of suppression. Activated Ly 1 cells are unable to act directly as suppressor cells but can induce suppressor activity in Ly 2$^+$ populations. In addition a non-T cell population, most likely B cells, in the assay culture can regulate Ly 1 helper activity.

INTRODUCTION

The use of Ly antisera to subdivide T cells into separate, functionally distinct subclasses has resulted in the passage of the suppressor T cell from a concept to an entity (1). Thus, cells which do not express the Ly 1 differentiation antigen but do express the Ly 2 and Ly 3 antigens (Ly 23 cells) can function as suppressor (or cytotoxic effector) cells ($T_{C/S}$) but have not been shown to act as helper cells (T_H). Both their function and Ly phenotype appear to be stable over long periods of observation (2) suggesting that these cells have reached a final differentiated state in which they are specialized to perform a single function. The only decision they have to make is whether to act or not. This interpretation is supported by studies showing that purified Ly 23 cells can be activated by a mitogen, such as Con A, and when activated act as obligatory suppressor cells independent of the dose of Con A used for activation or the

(a) Pathology Dept., Yale Medical School, New Haven, Ct.,
(b) Sloan-Kettering Memorial Cancer Research Center, New York NY
(c) Sidney Farber Cancer Center, Harvard University, Boston, MA.

number of Con A activated cells used in assay cultures (3).

Ly 1 positive, 23 negative cells (Ly 1 cells) show similar behavior in a converse fashion; that is, these cells appear to be programmed to act as helper cells and show the same stability and single function after mitogen activation as do Ly 23 cells.

Cells which bear all three of the Ly alloantigens, Ly 123 cells, (T_e) have not been shown to have the same stability and appear to be a heterogeneous cell population containing helper/suppressor potential, and/or helper/suppressor amplification. This cell subclass appears to be the one which ultimately determines the level at which the immune response is set.

We have used Ly antisera to decipher some of the regulatory T cell interactions which result in the generation and activation of antigen-specific suppressor T cells. We have used an in vitro culture system where nylon wool purified splenic T cells can be induced, by relatively high doses of sheep red blood cells (SRBC), to act as potent specific suppressor cells after transfer to fresh assay cultures of normal spleen cells (4). We have found that education of Ly 2^- cells, by a method which produces potent suppressor cell activity when Ly 2^+ cells are present, leads to; (a) the appearance of significant helper activity when the educated Ly 2^- cells are added to purified B cells and (b) the induction of suppressor cell activity when small number of Ly 2^+ cells are added to the system, either during the education procedure, or in the assay cultures.

MATERIALS AND METHODS

Animals: C57BL/6 mice were obtained from Jackson Labs, Bar Harbor, ME.

T and B cell fractionation: Spleens from normal mice were removed and single cell suspensions were prepared in Hank's balanced salt solution + 5% fetal calf serum. T cells were purified by the nylon fiber technique of Julius et al (5). Two techniques were employed to obtain Ly 2^- cells. (a) The effluent cells from nylon wool columns were treated once with an anti-Ly 2 serum and rabbit complement before culture. We refer to these cells as Ly 1 cells below; or (b) to insure the removal of all Ly 2 cells, we treated nylon purified T cells twice with both anti-Ly 2 and anti-Ly 3 sera plus complement. After these purified Ly 1 cells were educated in vitro we treated them a third time with both anti-Ly 2 and anti-Ly 3 sera plus complement, at the end of the education period. We refer to these cells as purified Ly 1 cells below.

Three methods were used to prepare B cell populations. Spleen cells were either treated once with an anti-theta serum plus rabbit complement, or twice with an anti-theta serum plus complement or once with an anti-theta serum and once with anti-Ly 2 and anti-Ly 3 sera plus complement. We refer to the latter population as *Ly 2^- B cells.*

The production and use of anti-Ly 1.2 (C3H anti-CE thymocyte antiserum and anti-Ly 2.2 (C3H Beta anti-ERLD) have been described previously (6). The Ly antisera were diluted 1:30 and absorbed with C3H thymus cells to remove autoantibody.

Tissue culture: *The tissue culture techniques for education and assay cultures has been described (4,7). Briefly, T cells (10^7 cells/ml) were cultured (educated) with approximately 2 x 10^6 SRBC for 4 days using Mishell-Dutton culture conditions. After 4 days, the educated cells were recovered from culture, washed, counted for viability by the trypan blue exclusion method, and added to assay cultures. Assay cultures were whole spleen cells or T + B combinations, as indicated, and 1 ml cultures were immunized with 2 x 10^6 SRBC. Cultures were maintained for 5 days and then tested for plaque forming cell (PFC) response by the Cunningham plaque assay (8).*

Results and Discussion

Educated, purified Ly 1 cells expressed no demonstrable suppressor activity when added to assay cultures containing purified non-immune Ly 1 cells and Ly 2^- B cells (Table 1).

Table 1: Comparison of the suppression effected by educated T cells treated 3 times with anti-Ly 2 and anti-Ly 3 sera with the suppression effected by NMS treated controls in assay cultures of purified Ly 1 T and Ly 2^- B cells.

Educated T cells	Degree of Suppression
NMS	++++
Ly 1	+

The addition of identical numbers of unselected educated T cells resulted in greater than 90% suppression. In these experiments the educated Ly 1 cells expressed substantial levels of T_H activity when added to Ly 2^- B cells alone (see below). Under the same conditions very little T_H activity was expressed by the unselected educated T cells. Thus, even after in vitro education using optimal suppression inducing *conditions, purified Ly 1 cells can only act as helper cells and cannot directly suppress immune responses.*

However, if we used less rigorous techniques to deplete Ly 2^+ cells from either the educated T cells or the assay cultures, substantial amounts of suppressor activity could be demonstrated (Table 2).

Table 2: Suppression can be mediated by residual Thy-1^+ Ly 2^+ cells in assay cultures of B cells plus normal Ly 1 T cells.

Treatment of B cells	*Degree of suppression produced by educated Ly 1 cells*
anti-Thy 1.2, 1 treatment	*+ to +++*
anti-Thy 1.2 + anti-Ly 2 and anti-Ly 3	*0*

Thus, the presence of very small numbers of Ly 2^+ cells in either the in vitro educated T cell population or in the assay population resulted in substantial inhibitory effects, which appear to be disproportionate to the number of Ly 2^+ cells that remained after a single treatment with an anti-Ly 2 serum plus complement. We now routinely test our anti-theta treated spleen cells for Con A responsiveness and have consistently found that when some Con A responsiveness remains the "B cells" do not respond by antibody formation as well as do those without any residual Con A activity, after the addition of educated purified Ly 1 T cells.

Our findings are analogous to those of Cantor and Boyse, in their studies on the generation of allo-killer cells (9). They showed that extremely small numbers of Ly 23 cells (0.5×10^5) could produce substantial levels of allo-reactive cytotoxic effector activity when relatively large numbers of Ly 1 cells were present. It therefore seems that small numbers of residual Ly 2^+ cells, in the presence of large numbers of Ly 1 cells, can be induced to express substantial levels of Ly 2^+ $T_{c/s}$ activity.

Perhaps the most intriguing finding to come from our experiments is that inclusion of small numbers of antigenically naive Ly 2^+ cells (which had persumably escaped elimination after a single treatment with anti-theta serum plus complement) in the assay cultures could exert substantial levels of feedback suppressor activity. The fact that treatment with an anti-Ly 2 and an anti-Ly 3 serum after anti-theta serum treatment eliminated this suppressive effect substantiates this conclusion. The feedback nature of the effect is emphasized by the observation that the addition of increasing numbers of activated purified Ly 1 cells to cultures containing antigenically naive Ly 2^+ cells, resulted in decreasing formation of anti-SRBC (data not shown).

These findings indicate the potential importance of T-T interactions in the homeostatic regulation of the antibody response. At present we have no evidence as to whether the induced Ly 2^+ suppressor cell comes from an Ly 23 or an Ly 123 subclass. Our preliminary evidence, however, indicates that there may be a subpopulation of Ly 123 cells which is particularly resistent to killing with antiserum plus complement.

Whatever the phenotype the induced suppressor cell has, the fact that we have shown that Ly 2^+ cells, in the presence of large numbers of Ly 1 cells, can exert suppressor effects which are far greater than would be predicted by their small numbers indicates that caution must be taken in interpreting results which suggest that Ly 1 cells can act directly as suppressor cells. Our results indicate that even under the most optimal conditions for generating suppressor cells, no direct suppressor activity can be demonstrated in purified Ly 1 cell populations although these cells can be potent inducers of suppressor activity in Ly 2^+ cells.

In addition to the homeostatic T-T interactions discussed above, our results also suggest an important regulatory feedback role for a non-T cell population. This conclusion is based on our observation that the tempo of the anti-SRBC response in the presence of antigen activated purified Ly 1 T_H cells is considerably different from that noted after the addition of non-immune purified Ly 1 T_H cells (Table 3).

Table 3: Educated Ly 1 cells produce a precocious antibody response when added to Ly 2^- B cells.

Educated T cells	PFC/Culture	
	day 3	day 5
NMS	±	±
Ly 1	+++	±

In the latter case, the PFC response peaks at day 5, while in the former there is an accelerated formation of PFC resulting in a peak response on day 3 and a greatly reduced response on day 5. Thus, there is a precocious antibody response and a more rapid shutoff. We are presently examining the class of antibody produced when activated purified Ly 1 T_H cells are added to purified B cells. It would be interesting if these educated T_H cells, added to purified Ly 2^- B cell assay cultures at day 2 or 3 after the addition *of antigen, would result in a substantial increase in the response on day 5. We are presently testing this prediction.*

From the results presented above, we suggest that the following cellular events ensue after antigen stimulation:

Activation of T_H *cells induces an accelerated formation of* $T_{c/s}$ *(Ly 23) effector cells from both* T_E *(Ly 123) cells and resting* $T_{c/s}$ *(Ly 23) precursors. The induced* $T_{c/s}$ *effectors in turn inhibit both the delivery of* T_H *activity to the B cell and perhaps also the generation of* $T_{c/s}$ *effector cells. It is possible that the latter suppression is an indirect one, in that the generation of* $T_{c/s}$ *effector cells is dependent to a large extent upon* T_H *activity. Thus,* $T_{c/s}$ *suppression of* T_H *activity can remove the signal responsible for the continued generation of* $T_{c/s}$ *effector cells, and thus when all* T_H *activity is gone,* $T_{c/s}$ *activity also rapidly disappears.*

According to this notion the following predictions can be made:

(a) T_H *activity should always be accompanied by* $T_{c/s}$ *activity after immunization. (b) the generation of optimal* $T_{c/s}$ *activity requires the presence of Ly* 1^+ *population during induction and (c) soluble mediators will be identified which can induce resting Ly* 2^+ *cells to express* $T_{c/s}$ *activity. Evidence in support of prediction 1 has already been published (10). In addition, both we (unpublished observations) and Feldmann, et al (11) have found that the second prediction appears to be true, in that the generation of optimal* $T_{c/s}$ *activity has been shown to require the presence of Ly* 1^+ *cells. We do not know of any evidence in support of the third proposition. It will be interesting to determine whether T helper factors which are responsible for the induction of the B cell program can also be shown to be responsible for the induction of feedback suppression from Ly* 2^+ *cells or whether this regulatory T-T interaction is mediated by a distinct family of molecules.*

In summary, our results are consistent with the current immunological paradigm that Ly $1^+ 23^-$ *immunoregulatory T cells act as obligatory helper* (T_H) *cells and cannot be induced to act directly as suppressor cells by any mode of immunization. In addition the data indicate that Ly* 2^+ $T_{c/s}$ *cells can be activated or induced by high levels of* T_H *activity to produce considerably more suppression under these circumstances than would have been predicted by stoichiometric considerations. We suggest that Ly 1* T_H *activation of Ly* 2^+ $T_{c/s}$ *is an important homeostatic immunoregulatory mechanism.*

ACKNOWLEDGMENTS

These studies were supported by grants CA-08593, CA-14216, AI-10497, AI-12184, AI-13600 and contract CB-43994 from the National Institutes of Health. DDE is an NIH fellow. We gratefully acknowledge the assistance of Astrid Swanson and John Barberia.

REFERENCES

1. Cantor, H. Contemp. Topics in Immunol. (1977) 6, in press.

2. Huber, B., and Cantor, H., Shen, F.W., and Boyse, E.A. (1976) J. Exp. Med. 144, 1128.

3. Jandinski, J. and Cantor, H., Tadakuma, T., Peavy, D.L., and Pierce, C.W. J. Exp. Med. 143, 1382.

4. Eardley, D.D. and Gershon, R.K. (1976) J. Immunol. 117, 313.

5. Julius, M., Simpson, E., and Herzenberg L. (1973) Eur. J. Immunol. 3,645.

6. Shen, F.W., Boyse, E.A. and Cantor, H. (1975) Immunogenetics 2,591.

7. Eardley, D.D., Staskawicz, M.O., and Gershon, R.K. (1976) J. Exp. Med. 143, 1211.

8. Cunningham, A.J. and Szenberg A. (1968) Immunology 14, 599.

9. Cantor, H. and Boyse, E.A. (1975) J. Exp. Med. 141, 1390.

10. Weizman, S., Shen, F.W. and Cantor, H. (1976) J. Immunol. 117, 2209.

11. Feldmann, M., Beverley, P.C.L., Woody, J. and McKenzie, I.F.C. (1977) J. Exp. Med. 145,793.

SPECIFIC ENRICHMENT OF SUPPRESSOR T CELLS BEARING THE PRODUCTS OF I-J SUBREGION

Ko Okumura, Toshitada Takemori and Tomio Tada

Laboratories for Immunology, School of Medicine,
Chiba University, Chiba, Japan

ABSTRACT. The expression of I-J subregion gene on T cells was directly demonstrated by cytotoxic killing of suppressor T cells which were specifically separated by adsorption to and elution from an antigen-coated column. Serological analysis showed that more than 30% of these enriched suppressor T cells were lysed by an antiserum directed at I-J subregion of the haplotype in the presence of complement, while the lysis of original splenic T cells by the same antiserum was insignificant. The functional assay using *in vivo* and *in vitro* secondary antibody response indicated that this procedure enriched suppressor T cells about one hundred fold. Sequential cytotoxic tests by the combination of the anti-Ly and anti-I-J alloantisera showed that about 30% of the fraction consisted of $I\text{-}J^+$, Ly $2^+_, 3^+$ T cells.

Analysis of the enriched suppressor T cell fraction by a fluorescence activated cell sorter (FACS) demonstrated that about 30% of the cells bear I-J subregion product on their surface, whereas such cells are undetectable in unseparated spleen cell suspensions. No Fc receptor (FcR) was detectable on the I-J bearing suppressor T cell.

INTRODUCTION

The Ia-4 locus which defines I-J subregion has been noted mainly by functional assays in which activities of suppressor T cell and suppressive T cell factor were removed by antisera reactive with I-J subregion products (1-3). Selective expression of I-J subregion genes on allotype specific suppressor T cell and carrier-specific suppressor factor has been proved by the successful absorption of anti-I-J activity with nylon wool purified T cells but not with B cells (2,4). No direct demonstration of the I-J subregion product as surface Ia antigen has been reported using usual cytotoxic assays, probably due to the minute number of T cells expressing this gene product among lymphoid cells. It is an obvious importance to devise a method for detecting the I-J subregion product by serological procedure in order to establish the specificity and distribution of this newly described Ia antigen among T cells of various haplotype, and to study the relationship between known phenotypic expressions and I-J product on T cells. Thus, we have tried to enrich carrier-specific suppressor T cells by adsorption to and elution from an antigen-coated column.

Method to enrich suppressor T cells. Spleen cells from mice (C57BL/6J) which had been immunized with keyhole limpet hemocyanin (KLH) were applied to the anti-mouse immunoglobulin (anti-MIg)-coated Sephadex G-200 column (Schlossman's column) for the depletion of B cells. This T cell-enriched spleen cell suspension (3×10^8 cells) was then incubated for 30 min in an KLH-coated column, which was prepared by coupling KLH to cyanogen bromide-activated Sephadex G-200. In general, 20 mg of KLH was coupled to 20 ml of Sephadex G-200, and the KLH coated Sephadex was packed in a 20 ml disposal syringe. After this incubation period, the column was thoroughly washed with warm (37°C) medium (Dulbecco's PBS) until the effluent contained less than 10^4 cells/ml. The retained cells were eluted by flushing the column with cold (0°C) medium. The elution pattern of KLH-primed spleen cells is depicted in Fig. 1. The majority of applied cells passed through the column with the warm medium, and only less than 1% of the original cells were recovered by elution with the cold buffer.

Fig. 1.

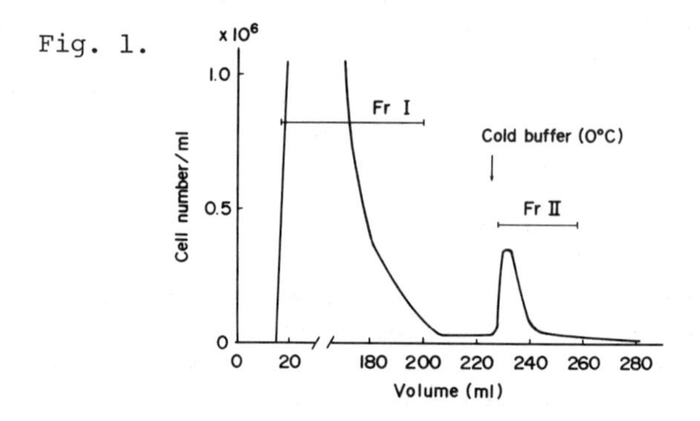

Functional analysis of the fractionated T cell population. Each cell fraction eluted from the KLH-coated column was assayed in *in vivo* or *in vitro* secondary antibody response. As shown in Table 1, the cells eluted with the cold medium (Fr. II) showed a striking suppression in the adoptive secondary antibody response at a dose of 2×10^5. On the other hand, suppressive activity contained in the unseparated cell fraction had become almost totally non-existent in Fr. I. To assay the activity of the fractionated cell in an *in vitro*

TABLE 1

ENRICHMENT OF SUPPRESSOR T CELL FUNCTION
BY ANTIGEN-COATED COLUMN

Cells transferred ($x10^6$)				Indirect anti-DNP-PFC/10^6
T depleted DNP-KLH 1°	KLH 1° helper (nylon passed)	KLH 2° suppressor		
		Fraction	Cell No.	
5	0	——	0	<10
5	1	——	0	1,500
5	1	Unseparated	1	600
5	1	Fr. I	1	1,200
5	1	Fr. II	0.2	<10

* Cells from C57BL/6J were transferred to 600R irradiated C57BL/6J with 10 μg of DNP-KLH,and PFC assay was done on day 7.
* T cells were depleted with anti-BAT (brain associated T antigen) plus complement.
* The number of direct PFC (less than 15% of total PFC) was subtracted.

secondary antibody response, the sonicated extract from each cell fraction was prepared as described elsewhere (3). As expected, the extract of cells in Fr. II showed a strong suppressive activity, whereas no suppression was observed with the extract of Fr. I (data not shown) (4).

Cytotoxic analysis of the cell fractions separated. The separated cell fraction was analysed by cytotoxic ^{51}Cr release with various alloantisera and rabbit complement. As shown in Table 2, both Fr. I and Fr. II from C57BL/6J ($H-2^b$) contained about 70% of Thy 1.2 positive cells. The killing by anti-I-J^b of Fr. II was about 25%. Anti-I-J^k antiserum as a control did not show detectable cytotoxicity. No cytotoxicity was detected with anti-I-J^b when it was tested with Fr. I and unfractionated spleen cells.

Since it became possible to detect I-J bearing cells by the conventional serological assay using the enriched cell population as a target, we tried to titrate the cytotoxic activity of the anti-I-J^b antiserum by two-step killing. As we expected from the removal of suppressor T cell activity

Fig. 2

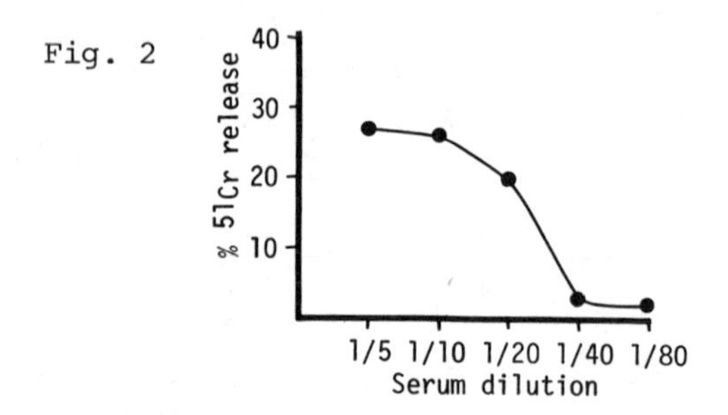

only with a low dilution of antiserum, the direct cytotoxic titer was not so high in which the cytotoxic curve reached a plateau at the dilution of 1:10 (Fig. 2).

Another interesting observation was the increase in cytotoxicity of Fr.II by anti-Ly 2,3 as compared with that of Fr. I. The results are consistent with the finding that carrier-specific as well as allotype-specific suppressor T cells bear Ly 2,3 antigen on their surface. To confirm the evidence serologically that this procedure, in fact, enriched $I\text{-}J^+$, Ly 2^+, 3^+ cells, we have attempted to analyse the fraction by the modified sequential killing method as described by Cantor *et al.* (5).

TABLE 2

CYTOTOXIC ANALYSIS OF CELL FRACTIONS SEPARATED BY ANTIGEN-COATED COLUMN

Cell fraction	Antisera	^{51}Cr release
Fr. I	Anti-Thy 1	70 %
	Anti-I-J^b*	<5
	Anti-Ly 1	45
	Anti-Ly 2,3	20
Fr. II	Anti-Thy 1	75
	Anti-I-J^b	25
	Anti-I-J^k**	<5
	Anti-Ly 1	20
	Anti-Ly 2,3	40

* B10.A(5R) anti-B10.A(3R)
** B10.A(3R) anti-B10.A(5R)

Analysis of phenotypic expressions of specifically enriched suppressor T cells by the sequential cytotoxic assay. ^{51}Cr labelled Fr. II cells were distributed to two sets of microtiter trays, and both trays were treated with alloantisera (1st killing) as indicated in Table 3, after which one of these trays was further treated with other alloantisera (2nd killing). About 30% of the cells were killed by anti-I-J^b at the 1st killing. If the same cells were first treated with the anti-Thy 1.2 antiserum, no detectable killing was observed by anti-I-J^b at the 2nd killing, indicating that all I-J bearing cells are Thy 1 antigen positive T cells. It was further found that 10% of Fr. I cells were sensitive to anti-Ly 1 antiserum and about 40% consisted of anti-Ly 2,3 sensitive cells. If the cells were first treated with anti-Ly 2,3, the killing of anti-Ly 1 at 2nd killing was almost completely diminished. The result indicated that the cells sensitive to anti-Ly 1 at the first killing are, in fact, Ly 1,2,3 positive cells.

Similar analysis was performed with the combination of anti-Ly and anti-I-J antisera as shown in Table 3. It should be noted that the treatment with anti-Ly at the 1st killing resulted in the decrease of anti-I-J^b cytotoxicity at the 2nd killing by about 10%. The treatment with anti-Ly 2,3 beforehand completely eliminated the killing by anti-I-J at the 2nd killing. From these results, which were repetitively confirmed, we can conclude that the specifically enriched suppressor T cell fraction by antigen coated column consists of two disdinct T cell subsets, i.e., I-J^+, Ly 2^+, 3^+ T cells (about 30%) and I-J^+, Ly 1^+, 2^+, 3^+ T cells (about 10%).

TABLE 3

PHENOTYPIC EXPRESSIONS OF SPECIFICALLY ENRICHED SUPPRESSOR T CELL

1st killing		2nd killing	
Serum	^{51}Cr release	Serum	^{51}Cr release
Anti-I-J^b	33 %		%
Anti-I-J^k	0		
Anti-Thy 1.2	70	Anti-I-J^b	<5
Anti-Ly 1	12		
Anti-Ly 2,3	38	Anti-Ly 1	<5
NMS	0	Anti-Ly 1	12
Anti-Ly 1	12	Anti-I-J^b	25
Anti-Ly 2,3	40	Anti-I-J^b	0

Fig. 3

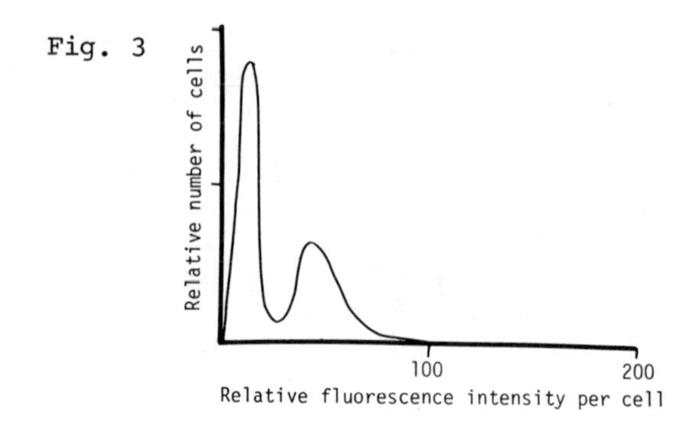

A trial to separate I-J bearing cells by fluorescence activated cell sorter (FACS). We also tried to analyse and separate I-J bearing T cells from the enriched suppressor T cell fraction by FACS. Cells from each fraction were treated with the anti-I-J antiserum and then stained with rhodaminated specific anti-mouse IgG. These stained cells were then analysed by FACS after gating the live cells by the size scattering. Fig. 3 shows the fluorescence profile of Fr. II stained with anti-I-J^b. As shown in this figure, a definite fluorescence positive fraction was sorted, whereas no such a peak was seen when the cells were stained with anti-I-J^k. Thus it became possible to separate the suppressor T cell as a relatively pure fraction by the combination of poor man's cell sorter and FACS together with anti-I-J alloantiserum.

ACKNOWLEDGEMENT

We wish to thank Mr. H. Takahashi for his assistance in performing the experiments and Ms. Yoko Yamaguchi for her excellent secretarial help.

REFERENCES

1. Okumura, K., Herzenberg, L.A., Murphy, D.B., McDevitt, H.O. and Herzenberg, L.A. (1976) *J. Exp. Med.* 144, 685.
2. Murphy, D.B., Herzenberg, L.A., Okumura, K., Herzenberg, L.A. and McDevitt, H.O. (1976) *J. Exp. Med.* 144, 699.
3. Tada, T., Taniguchi, M. and David, C.S. (1976) *J. Exp. Med.* 144, 713.
4. Okumura, K., Tokuhisa, T. and Tada, T. (1977) In: Proceedings of IIIrd Ir Gene Workshop. edited by H.O. McDevitt. Academic Press. in press.
5. Cantor, H. and Boyse, E.A. (1975) *J. Exp. Med.* 141, 1376.

TWO RECENTLY ACTIVATED T-CELLS NECESSARY IN THE GENERATION OF SPECIFIC SUPPRESSOR CELLS

Diane Turkin and Eli E. Sercarz

Department of Bacteriology
University of California
Los Angeles, California 90024

ABSTRACT

The appearance of suppressor cell activity was studied in vitro by the addition of nylon wool-separated fractions of primed CBA/J spleen cells to normal or β-galactosidase primed cells in culture. Both the nylon wool-retained (NWR) and the nylon wool-passed (NWP)fractions were necessary for the expression of suppressors. Each of the partners in this collaborative event had to be recently primed. Antiserum studies showed that they each were of the T-cell lineage. A homeostatic model was favored in which the disruption of equilibrium induced by antigen administration is restored by the interaction of antagonistic T cells.

INTRODUCTION

Interactions between T cell subpopulations have been described in the generation of cytotoxic cells (1,2), of suppressor cell activity (3,4), and in graft vs. host reactivity (5). Our own studies on the T-cell priming to β-galactosidase (GZ), demonstrating overlapping waves of help and suppression, have been suggestive of T-cell interaction in generating these two subclasses of T cells and in achieving homeostasis (6-8). The work of Gershon and his colleagues (9-11), demonstrating helper-suppressor interaction was fundamental in the development of such a notion. Other collaborative suppressive circuits are implied by the work of Hodes and Hathcock (12), who showed a synergy between two cells in their non-specific suppressor cell system. Likewise, in this volume, Tada (13) describes studies on the Ly 123^+ T-cell target of Ly 23^+ T-suppressor cells in a system showing positive feedback.

We shall here consider the interaction of two T cells with different adherence properties in the expression of a suppressive signal.

EXPERIMENTAL

Suppressive populations contain hidden helpers

Recent work from our laboratory has shown that 7 days after priming CBA/J mice with 100 μg GZ in CFA, spleen cells from these animals are unable to respond to haptens conjugated to GZ when the hapten-GZ molecule was used as antigen in an in vitro response (7,8). This non-response was a dominant suppression capable of inhibiting responsiveness of normal or carrier primed cells. In order to investigate which cell type was responsible for the suppression, various cell fractionation procedures were attempted.

Aliquots of the 7-day GZ primed spleen cell preparation were treated with (a) anti-Thy 1.2 plus complement to remove T cells, or (b) anti Ly-2.1 plus complement (the kind gift of F. Shen, H. Cantor and E. Boyse), to eliminate Ly 2-bearing T cells (14), or (c) passaged over nylon wool to remove B cells and macrophages (15). In addition to collecting the cells which passed through the nylon wool (NWP), we also collected those partially adherent cells which could be eluted by additional repeated washing and mechanical compression, the nylon wool retained-eluted (NWR) fraction. These two fractions comprised approximately 75-85% of the cells originally applied to the column; the remaining cells were strongly adherent to nylon wool and could not be recovered.

As can be seen in Table 1, Anti-Thy 1.2 plus complement treatment completely abolished the suppression, indicating that T cells were involved in the generation or expression of the suppression. However, when 7-day GZ primed spleen cells were fractionated on nylon wool, and the NWP fraction was mixed with normal <u>or</u> 30-day GZ-primed spleen cells the NWP cells were not suppressive. In fact, they enhanced the subsequent response of the normal and carrier primed cells, revealing a population of helpful cells which had heretofore been masked.

In view of this interesting finding, the Ly phenotype of the suppressor cell was investigated. After treatment of the 7-day GZ-primed spleen with anti-Ly 2.1 serum and complement, the suppressive activity was again abolished and the response restored to the level observed for normal cells. It was concluded that an Ly 2+, Thy-1+

TABLE 1

Suppressive Populations Contain Hidden Helpers

GZ' 7-day cells	Treatment	Normal cells	Anti-FL PFC/10^6
10^7	-	-	10
-	-	10^7	120
	-	7×10^6	92
3×10^6	-	7×10^6	14
3×10^6	NWP	-	0
3×10^6	NWP	7×10^6	210
3×10^6	anti-Thy 1.2	-	6
3×10^6	anti-Thy 1.2	7×10^6	125

6 to 8 week old CBA/J mice were injected with 100 μg GZ in CFA. Seven days later, spleen cell suspensions were prepared, aliquots of which were anti-Thy 1.2 + C treated, or passed over nylon wool columns. These cells were added to normal cells under Mishell-Dutton culture conditions, in the presence of 1 μg GZ-FL/ml. 4 days later, direct anti-FL PFC were enumerated.

nylon-adherent cell was involved in the GZ-specific suppression.

<u>Nylon wool separates two populations needed for suppresor cell expression</u>

The inability to recover the suppression in the 7-day NWP cells suggested that either the T cell which was causing the suppression was adhering to the column, or alternatively, that the suppression was mediated by more than one cell type, one of which adhered to the nylon wool. To distinguish between these two possibilities, 7-day GZ-primed NWP and NWR cells were prepared and added either alone or together to normal or primed helper cells. The data in Table 2 show that whereas neither cell type was suppressive when added to normal or primed cells, the suppressive activity returned if the NWP and NWR populations were mixed. It should be noted that the design of the experiment allows us to conclude that the appropriate collaborating cell is neither present in the normal population nor in the 30-day GZ-primed population[1] This suggested that there was a synergy between two recently activated cell populations

Table 2

Nylon Wool Separates Two Populations Needed for Suppressor Cell Expression

GZ' 7-day cells	Normal cells	Primed cells*	Anti-FL PFC/10^6
10^7			5
	10^7		150
		10^7	523
3×10^6 NWP			0
3×10^6 NWP	7×10^6		517
3×10^6 NWP		7×10^6	740
3×10^6 NWR			609
3×10^6 NWR	7×10^6		519
3×10^6 NWR		7×10^6	385
1.5×10^6 NWP + 1.5×10^6 NWR			0
1.5×10^6 NWP + 1.5×10^6 NWR	7×10^6		1
1.5×10^6 NWP + 1.5×10^6 NWR		7×10^6	1

Same conditions as for Table 1. *"Primed cells" were from spleens of mice injected 4 weeks earlier with GZ in CFA.

for the expression of suppression.

B cells and macrophages are not involved in the suppressive interaction

An alloantigen coded for by the Mls locus (16) appears on the surface of B cells and macrophages, but not T cells, which can stimulate a mixed lymphocyte reaction. Antiserum directed against antigens encoded by genes inseparably linked to this locus in the CBA/J mouse, has recently been described. The antiserum is designated AST-101, and was the kind gift of S. Tonkonogy and H. Winn. Antiserum and complement treatment was performed as described (17).

GZ 7-day primed spleen cells were treated with either anti-Thy 1.2 or AST-101 plus complement before mixing with either normal or carrier primed helper cells. Table 3 shows that elimination of the T cells completely relieves the suppression; AST-101 treatment, which eliminates both B cells and macrophages, has no effect on the suppressive nature of the 7-day GZ primed spleen.

TABLE 3

B Cells and Macrophages are not Involved in the Suppressive Interaction

GZ' 7-day cells	Treatment	Normal cells	Primed cells	Anti-FL PFC/10^6
10^7	-	-	-	28
-	-	10^7	-	322
-	-	-	10^7	370
3×10^6	anti-Thy 1.2	-	-	0
3×10^6	anti-Thy 1.2	7×10^6	-	194
3×10^6	anti-Thy 1.2	-	7×10^6	222
3×10^6	AST-101*	-	-	0
3×10^6	AST-101	7×10^6	-	3
3×10^6	AST-101	-	7×10^6	22

Same conditions as for Tables 1 and 2.
* AST-101 is an antiserum cytotoxic for B cells and macrophages (for details see text).

It appears, therefore, that the suppression expressed by 7-day GZ primed spleen requires the concomitant functioning of more than one cell, both of which are of the T lineage. This view was supported in an experiment where it was shown that NWP and NWR populations were each sensitive to anti-Thy 1.2 plus complement treatment.

DISCUSSION

The evidence in this report suggests that a collaborative event is required between T cell subpopulations with different adherence properties in the generaton of specific suppression. Each of the partners has had to have been recently activated by antigen.

Activated cells of either helper or suppressor subpopulations are retained by nylon wool.

It is clear that "adherence" is not a fixed characteristic of a lymphocyte subpopulation but depends on a particular stage after antigen activation. Three days after GZ-priming, unfractionated spleen cells show a predominant helper effect for the in vitro anti-fluorescein (FL) response to GZ-FL. A majority of the activated helper cells are adherent to nylon wool, because the passed fraction at this time is actually suppressive! Likewise, at seven days, when the whole spleen population displays a predominant suppressive effect, it appears that the activated suppressor cells at this time are also NW adherent, since the passed fraction contains helpers.We assume that the early helpful population matures through an activated, sticky phase to now acquire the ability to elude the fibrous nylon network.

The nature of the synergizing cells

It is remarkable that neither the NWP nor NWR cell type could find its collaborative partner among the normal spleen cell population. The data in Table 2 therefore point to antigen-activation as a necessary step for each partner in the cell collaboration. More surprising is the deficiency of the long-termed primed, helper cell population in each of these required cell types. The evidence therefore points to a recently activated T-cell in each of the NWP and NWR fractions as effective partners in a collaborative event leading to specific suppression.

The specificity of the acute suppressive phase in the GZ system has been demonstrated elsewhere (8); suffice it to say that the anti-fluorescein in vitro response to KLH-FL is not interfered with by GZ suppressor cells even in the presence of GZ.

Two different sorts of T-cell collaboration have been envisioned in the generation of suppressors. It is probable that each of these regulatory circuits plays some part in any overall Grand Design. Tada has presented evidence (13) for a regulatory circuit in which a suppressive T cell factor is extracted from the membrane of specifically primed Ly 23+ cells that can act upon an acceptor Ly 123^{+} cell which then matures into a non-specific suppressive cell. Tada's exhausted suppressor cell seems to be long-lived, but reactivation of suppression requires the Ts to awaken an earlier Ly 123+ precursor via a positive feedback loop. This activated Ly 123+ cell has to be antigen-primed, implying that Ly 123+ cells can exist in both virgin and activated states. Thus, antigen must act on both the exhausted Ly 23+ suppressor and the virgin Ly 123+ cell prior to the synergistic event.

Gershon and his colleagues have described interactions between helper T cells and suppressor T cells in facilitating the full development and subsequent decline of suppression (11). These mechanisms, in which an excess of positive effectors (i.e. helper cells) act to shorten their own hegemony by facilitating the induction of suppressors, must be of primary significance for homeostatic control. We have shown earlier (7) that a short-term wave of suppression, appearing at about one week following antigen priming in CFA, is sandwiched between an early and a later helper stage. Our studies, using antigen in adjuvant, can be expected to lead to quite a different time frame for the evolution of helper and suppressor phases.

The findings reported here favor the notion that two recently activated T-cells, one presumably from a helper subpopulation and the second from a suppressive subpopulation, antagonistically regulate each other's activities and guide the system via an oscillating path towards homeostatic equilibrium. The exact nature of these T cell subpopulations awaits examination.

ACKNOWLEDGEMENTS

These studies were supported by NIH grant AI 11183.

REFERENCES

1. Wagner, H. J. Exp. Med. 138:1379, 1973.

2. Cantor, H. and E.A. Boyse. J. Exp. Med. 141:1390, 1975.

3. Gershon, R.K. In The Immune System: Genes, Receptors, Signals, E.E. Sercarz, A.R. Williamson and C.F. Fox, eds., Academic Press, New York, p. 471, 1974.

4. Taniguchi, M., T. Tada, T. Tokuhisa. J. Exp. Med. 144: 20, 1976.

5. Asofsky, R., H. Cantor and R.E. Tigelaar. Progress in Immunology 1:369, 1971.

6. Sercarz, E.E., D.T. Corenzwit, D.D. Eardley and K. Morris. In Suppressor Cells in Immunity. University of Western Ontario Press, p. 19, 1975.

7. Eardley, D. and E. Sercarz. J. Immunol. 116:600, 1976.

8. Turkin, D. and E. Sercarz. Submitted, 1977.

9. Gershon, R.K. Contemp. Topics Immunobiol. 3:1, 1974.

10. Eardley, D.D. and R.K. Gershon. J. Exp. Med. 142:524, 1975.

11. Eardley, D.D., F.W. Shen, H. Cantor and R.K. Gershon. This Symposium, article 61.

12. Hodes, R.J. and K.S. Hathcock. J. Immunol. 116:167, 1976.

13. Tada, T. This Symposium, article 43.

14. Shen, F.W., E.A. Boyse and H. Cantor. Immunogenetics 2: 591, 1975.

15. Handwerger, B.S. and R.H. Schwartz. Transplantation 18: 544, 1974.

16. Festenstein, H. Transplantation 18:555, 1974.

17. Tonkonogy, S.L. and H.J. Winn. J. Immunol. 116:835, 1976.

GENETIC AND ANTIGENIC CONTROL OF SUPPRESSOR CELL ACTIVITY FOR CELL-MEDIATED IMMUNE RESPONSES

Susan Solliday Rich, Gary A. Truitt,
Frank M. Orson and Robert R. Rich

Department of Microbiology and Immunology and
The Institute of Comparative Medicine
Baylor College of Medicine and
The Methodist Hospital, Houston, Texas 77030

ABSTRACT. Alloantigen-activated splenic T cells and their soluble products suppress in vitro proliferative responses in mixed leukocyte reactions (MLR). The interaction between suppressor and responder cells is controlled by genes of the *H-2* complex. Two aspects of this suppressive regulation are reviewed in this report. First, factor produced from a strain exceptional in this system, C57BL/6, failed to demonstrate suppressor activity. The defect was not *H-2* linked and did not affect expression of a receptor for suppressor factor on responding cells. Thus interaction between *H-2* and non-*H-2* genes appears to be required for effective MLR suppressor factor production. Second, the MLR suppressor cell system has been extended to a second model of cell mediated immune function, the in vitro generation of cytotoxic lymphocytes (CL). Inclusion of alloantigen-activated splenic T cells suppressed CL generation in MLR in an antigen-nonspecific fashion, provided a second antigen-specific stimulus was presented to the primed suppressor cells in CL culture.

INTRODUCTION

Attempts to dissect critical mechanisms of regulatory cell interactions in the immune response have revealed the complexity of such interactions, if not the essence of the mechanisms. Particularly, recent efforts to understand suppressor T cell activities (rev.1,2,3) suggest interactions of multiple T cell subpopulations in generation of suppressor T cells, various immediate and secondary cell targets, and a spectrum of antigen and target cell specificity. Moreover, similar, yet subtly differing, modes of genetic control of various suppressor activities have been identified.

As a model for suppressor cell functions in cell-mediated immune responses, we have investigated the activities of alloantigen-activated cells and their soluble products in regulation of in vitro lymphocyte proliferation and subsequent generation of cytotoxic lymphocytes. Our previous studies have demonstrated antigen-nonspecific suppression of in vitro proliferative responses in mixed leukocyte reactions (MLR) by splenic T cells previously allosensitized in vivo (4). Sonicates of primed cells or supernatant fluids of these cells restimulated in vitro suppressed MLR responses in a similar nonspecific fashion (5,6). However, although nonspecific with regard to stimulating alloantigens, MLR suppressor factor exhibited marked *H-2* linked genetic restriction in its capacity to interact with MLR responder cells and thereby effect suppressor activity (6). Recent identification of a strain unique in its lack of suppressive activity in this system suggested more complex genetic control of regulatory cell activities than previously identified. Therefore, to dissect critical genetic influences in suppressor cell activities and to examine concepts developed for regulation of MLR in other models of cell-mediated immune functions, two distinct but related studies have been undertaken and will be briefly reviewed in this communication.

EFFECT OF NON-*H-2* GENES IN SUPPRESSION OF MLR

The preparation and biological characterization of MLR suppressor factor has been described previously (5,6). Briefly, MLR suppressor factor is prepared from spleen cells of mice injected by footpad with allogeneic cells; sonication or 24 h culture with cells of the original sensitizing strain releases soluble factors with suppressor activity. Antigen-nonspecific suppression results from the addition of suppressor factor to MLR, i.e., proliferative responses are suppressed to stimulating alloantigens unrelated to the original sensitizing antigen (5,6). The critical element of specificity appears rather to be syngenicity between the responder and suppressor cells for the *I-C* subregion of the *H-2* complex (6). In addition, preliminary data suggest that the suppressive T cell factor is, or is associated with, a product of the *H-2* complex, as indicated by removal of suppressor activity upon exposure to immunoadsorbents with specificity for the right hand (*I-C*, *S*,) portion of the complex (7). Past studies indicated that non-*H-2* gene background had no apparent role in this *H-2* directed regulatory cell interaction. Recently, however, we observed that a single strain of mouse, C57BL/6(B6) did not demonstrate suppressor factor activity in MLR (Table 1). Factor produced by B6 spleen cells in response to a variety of allogeneic stimuli in multiple experiments failed to suppress the response of syngeneic B6, *H-2* compatible C57BL/10(B10), semisyngeneic (BALB/c x C57BL/6)F_1 or allogeneic BALB/c responder cells. In order to determine if the defect of suppressor activity was linked to the *H-2*^{b} haplotype, factors prepared from other *H-2*^{b} strains were tested. In contrast to B6, factors of alloactivated B10 and A.BY suppressed responses of syngeneic and *H-2* matched responder cells, including those of B6 responder cells. The same factors did not affect allogeneic BALB/c responses. Two conclusions may be drawn from these observations. First, *H-2* linkage of this defect is not apparent since other *H-2*^{b} strains produce factors with suppressive activity. Second, B6 MLR responses are effectively suppressed by active factors of *H-2* homologous (B10, A.BY) or semi-syngeneic (BALB/c x B6)F_1 strains. Thus the defect in B6 suppressor activity appears to be related to inability to produce an active factor, while the capacity to respond to suppressive signals is unimpaired.

TABLE 1

SUPPRESSIVE ACTIVITY OF VARIOUS *H-2*^{b} FACTORS

Suppressor Strain	Responder Strain	Percent Suppression[1]
B6 (*H-2*^{b})	B6 (*H-2*^{b})	-19
	B10 (*H-2*^{b})	10
	(BALB x B6)F_1 (*H-2*$^{d/b}$)	15
	BALB (*H-2*^{d})	15
B10 (*H-2*^{b})	B10 (*H-2*^{b})	*48* [2]
	B6 (*H-2*^{b})	*47*
	A.BY (*H-2*^{b})	*31*
	BALB (*H-2*^{d})	- 7
A.BY (*H-2*^{b})	A.BY (*H-2*^{b})	*61*
	B6 (*H-2*^{b})	*64*
	B10 (*H-2*^{b})	*49*
	BALB (*H-2*^{d})	10
(BALB x B6)F_1 (*H-2*$^{d/b}$)	(BALB x B6)F_1 (*H-2*$^{d/b}$)	*55*
	BALB (*H-2*^{d})[1]	*58*
	B6 (*H-2*^{b})	*59*
	CBA (*H-2*^{k})	-11

[1]Supernate factors tested at 20% final concentration in MLR. Data derived from 3-20 experiments.

[2]Significantly different from relevant control values by two-tailed Wilcoxon rank-sum test ($p<0.01$) (ital.nos.).

The T cell factor obtained from (BALB/c x B6)F_1, representing a suppressor x nonsuppressor hybrid, suppressed responses of syngeneic F_1, as well as those of both parental strains (Table 1). Studies, not shown, using factors derived from an F_1 hybrid of two suppressor strains, BALB/c (H-2^d) and CBA (H-2^k), have demonstrated that suppressor molecules reactive to both parental strains are codominantly expressed in F_1 suppressor cells. Furthermore, adsorption of a (BALB/c x CBA)F_1 factor by Con A-activated thymocytes of either parent left residual suppressive activity against the opposite parental responder cell, implying that F_1 suppressive activity for parental cells is associated with separate, *H-2* specific molecules. Consequently, results of the (BALB x B6)F_1 study suggest that B6-reactive suppressor molecules are produced by the F_1 suppressor cell. Preliminary adsorption studies with Con A-activated parental thymocytes support this suggestion. Since the B6 suppressor defect is not *H-2* linked, non-*H-2* gene(s) may be required to allow expression of suppressor activity; in the (BALB/c x B6)F_1, production of B6 reactive factor may be promoted by dominant regulatory genes provided by the BALB/c parental haplotype.

To verify the role of non-*H-2* genetic background in defective B6 factor production, suppressor activity was tested in a B6 congenic strain, B6.C-H-2^d(B6.C)(Table 2). Supernates of alloactivated B6.C spleen cells failed to suppress responses of syngeneic B6.C, H-2^d identical BALB/c or non-*H-2* background identical B6. In contrast, the proliferative response of B6.C cells was suppressed by active H-2^d identical BALB/c factor, but not by background-matched B6 factor. From these experiments it was concluded that congenic strains of B6 background are incapable of producing suppressor factor but respond normally to *H-2* compatible suppressor signals. Therefore, it appears that *H-2* and non-*H-2* genes interact in the elaboration of suppressive regulatory factors. Furthermore, while restrictions to syngenicity for the *I-C* subregion suggests an homology interaction of products of the same or closely linked *H-2* coded genes expressed in the suppressor and target moieties, the non-*H-2* gene here identified appears to affect only the suppressor molecule.

TABLE 2

FAILURE OF SUPPRESSOR ACTIVITY OF B6.C-*H-2*^{d} FACTOR

Suppressor Strain	Responder Strain	Shared Regions *H-2*	Shared Regions Non-*H-2*	Percent Suppression[1]
B6.C (*H-2*^{d})	B6.C (*H-2*^{d})	All	All	-28
	BALB (*H-2*^{d})	All	Dissimilar	-11
	B6 (*H-2*^{b})	Dissimilar	All	4
BALB (*H-2*^{d})	BALB (*H-2*^{d})	All	All	*42*[2]
	B6.C (*H-2*^{d})	All	Dissimilar	*44*
	B6 (*H-2*^{b})	Dissimilar	Dissimilar	-12
B6	B6 (*H-2*^{b})	All	All	- 5
	B6.C (*H-2*^{d})	Dissimilar	All	-13
B10	B10 (*H-2*^{b})	All	All	*45*
	B6 (*H-2*^{b})	All	Similar	*52*
	B6.C (*H-2*^{d})	Dissimilar	Similar	7

[1]Supernate factors tested at 12.5% final concentration in MLR. Data derived from 2-7 experiments.

[2]Significantly different from relevant control values by two-tailed Wilcoxon rank-sum test ($p < 0.01$)(italized numbers).

SUPPRESSION OF CYTOTOXIC LYMPHOCYTE GENERATION BY ALLOSENSITIZED SPLEEN CELLS

As previously described, the inclusion of allosensitized spleen T cells in MLR results in significantly reduced proliferative responses. Under appropriate culture conditions, extended MLR cultures generate cytotoxic lymphocytes (CL) with specificity for stimulator cell alloantigens. Collaboration of subpopulations of T lymphocytes, variously characterized, has been identified in the generation of this cell-mediated immune response (8,9). Therefore, it has been of great interest to determine whether suppression of the MLR phase of CL generation leads to depressed cytotoxic responses, and subsequently to characterize the immediate target and possible genetic controls of allosensitized suppressor cells in CL generation.

CL responses developed during a six-day incubation of 5 x 10^{6} BALB/c spleen cells with 1 x 10^{6} irradiated (1500 r) C57BL/6 spleen cells. The influence of regulatory cells was assessed as the ability to suppress the CL response when

5 x 10^6 irradiated allosensitized BALB/c spleen cells were added to the cultures (addition of 5 x 10^6 irradiated normal BALB/c spleen cells were added to replicate cultures to serve as control). With this protocol, a highly reproducible suppression of CL responses has been obtained (10). Because we have shown that allosensitized spleen cells demonstrate a radioresistant cytotoxicity, albeit weak and transient (10), it was particularly critical to identify and dissociate possible cytotoxic activity from suppressor function. Antigenic specificity of the putative suppressor effect was characterized, since suppression consequent to possible cytotoxic removal of stimulator cells would be largely antigen-specific. Therefore, BALB/c mice were sensitized with C57BL/6 or C3H spleen cells and tested for their ability to suppress the generation of CL against $H\text{-}2^b$ stimulator cells or $H\text{-}2^k$ stimulator cells or a mixture of the two (Table 3). CL responses generated against a single alloantigen were suppressed only by regulator cells activated by that haplotype, e.g., BALB/c cells activated by C57BL/6 ($BALB_{B6}$) suppressed the CL response generated only against $H\text{-}2^b$ targets. However, when CL were generated against both alloantigens simultaneously, cytotoxic responses toward both targets were equivalently suppressed by each regulator cell population. Thus allosensitized spleen cells required antigen-specific restimulation, which was provided to either regulatory population by the mixed target cell population. After restimulation, however, suppression was effected in an antigen-nonspecific fashion. These characteristics follow closely the requirement for specific antigenic challenge of alloactivated splenic suppressor cells prior to release of MLR suppressor factor, although subsequent suppression is antigen nonspecific (4). It will now be of great interest to determine at what level of cell interaction in CL generation this suppressive effect is manifest.

TABLE 3

SUPPRESSIVE INFLUENCE OF ALLOSENSITIZED SPLEEN CELLS ON CYTOTOXIC LYMPHOCYTE GENERATION IN VITRO

Responder	Regulator[1]	Stimulator[1]	Target Cells: EL-4 (*H-2^b*) Specific Release cpm[2]	EL-4 (*H-2^b*) Specific Release percent[3]	AKR SNTL^{-13} (*H-2^k*) Specific Release cpm	AKR SNTL^{-13} (*H-2^k*) Specific Release percent
BALB	BALB	B6	3402±209	42	395± 42	3
	$BALB_{B6}$		1304±117	16	58± 61	1
	$BALB_{C3H}$		4095±109	51	396± 26	3
BALB	BALB	C3H	335±138	4	3376±126	30
	$BALB_{B6}$		476± 69	6	3984±125	35
	$BALB_{C3H}$		28± 28	1	652± 61	6
BALB	BALB	B6 + C3H	3195± 86	39	3207± 77	28
	$BALB_{B6}$		1525± 24	19	2188± 89	19
	$BALB_{C3H}$		289± 67	4	368± 8	3

[1]Inactivated by 1500 r gamma irradiation. Subscripts designate regulator presensitization.
[2]Shown are mean cpm SEM of triplicate determinations at an effector-to-target cell ratio of 20:1. Data are corrected to specific release by subtraction of spontaneous release values.
[3]Percent specific release was obtained by dividing the cpm data shown by the maximal (freeze-thaw) release minus the spontaneous release (x 100).

ACKNOWLEDGMENTS

This research was supported by USPHS RO1-AI-13810, NO1-AI-42529, HL-17269, NASA NAS 9-14368, and the Kelsey-Leary Foundation. RRR is in receipt of USPHS Research Career Development Award 1-KO4-AI00006. We wish to thank Chris Arhelger and Kathleen Cowen for technical assistance and Mrs. Elaine Bowers for preparation of this manuscript. We also wish to thank Dr. Donald Bailey for providing the B6.C-H-2^d mice.

REFERENCES

1. Pierce, C.W., and Kapp, J.A. (1976) *Contemp.Top. Immunobiol.* 5, 91.
2. Asherson, G.L., and Zembala, M. (1976) *Brit.Med.Bull.* 32, 158.
3. Rich, S.S., and Rich, R.R. (1977) *J.Pharmacol.Exp. Therapeut.* In press.
4. Rich, S.S., and Rich, R.R. (1974) *J.Exp.Med.* 140, 1588.
5. Rich, S.S., and Rich, R.R. (1975) *J.Exp.Med.* 142, 1391.
6. Rich, S.S., and Rich, R.R. (1976) *J.Exp.Med.* 143, 672.
7. Rich, S.S., Orson, F.M., and Rich, R.R. (1977) in *Proc. Third Ir Gene Workshop,* ed. H.O.McDevitt. Academic Press, New York. In press.
8. Wagner, H. (1973) *J.Exp.Med.* 138, 1379.
9. Cantor, H., and Boyse, E.A. (1975) *J.Exp.Med.* 141, 1390.
10. Truitt, G.A., Rich, R.R., and Rich, S.S. (1977) Submitted for publication.

LATENT HELP

N.A. Mitchison and P. Lake

IRCF Tumour Immunology Unit, Department of Zoology,
University College London

ABSTRACT. An antigen may be unable to evoke a primary response without the help of other antigens. It may nevertheless be able to evoke a later response by itself. If it does so, and if the later response still depends on helper activity, that activity was previously latent. H-2D and H-2K, in certain response systems, provide examples of latent help.

Cooperation between cells plays an important part in regulating the immune response to cell surface antigens. Our analysis of mechanisms of cooperation depends on (i) exploiting defined allo-antigens by the use of congenic, recombinant and inbred-recombinant mouse strains, (ii) separation of cell populations by means of their allo-antigenic surface markers or by physical means, (iii) the use of irradiated hosts or modified Marbrook culture vessels to cocultivate populations of cells, and (iv) assay of the response of B cells by means of a plaque test (for the response to Thy-1) or a computer-calculated cytotoxicity assay (for the response to H-2D, H-2K or Thy-1) (1-3). With these methods we have been able to distinguish two forms of helper activity, intra- and inter-molecular. In the former case, helper T cells (T_H) help the response of B cells by recognising determinants carried on the same molecule as is recognised by the B cells; in the latter case they react with another molecule. Intermolecular help may nevertheless require that the two molecules be carried on the same cell, thus defining the category of inter-molecular, intrastructural help. We believe that T cell receptors operate by presenting antigen to receptors of B cells, probably via binding to the surface of an intermediate cell.

In the course of a study of intermolecular, intrastructural help between I and K or D antigens, we encountered the following new phenomenon, which we now term latent help. An antigen may be unable to initiate a response without the help of other antigens. Once immunity has been established with the aid of intermolecular help, the same antigen may on its own be able to maintain or recall the response. If this occurs, and if the recall depends on T helper cell activity, that activity must originally have been latent. Latent help was first encountered while carrying further the work of Wernet and Lilly on the response of $H\text{-}2^b$ mice to $H\text{-}2D^d$, and has been examined in detail in the activation of latent help to H-2K^s and H-2K^q. We belive that the phenomenon is a general one, and that it may underlie some or all forms of auto-immunity.

THE RESPONSE OF H-2^b MICE TO D^d

C57 mice (H-2^b) respond to repeated immunization with Balb/c (H-2^d) or B10.D2 (H-2^d) spleen cells by producing anti-H-$2D^d$ IgG antibody, but do not so respond to B10.A(5R) cells. B10.A(5R) cells are bbb/ddd, and thus lack foreign IA and IB antigens. This finding confirms the work of Wernet and Lilly, and is attributed to a need for help from IA or IB (or other antigens) in generating the response. We have not found it necessary to distinguish between IgG and IgM antibodies as did Wernet and Lilly since essentially all the cytotoxic antibody present in hyperimmune mice is IgG.

Spleen cells from donor C57 mice hyperimmunized with Balb/c cells can be stimulated to make anti-D^d antibody upon adoptive transfer, but cells from donors which received similar numbers of 5R cells cannot. Experiments in which cells from both types of donor are mixed indicate that the 5R-treated cells are not suppressive. Since Wernet and Lilly find that administration of 5R cells inhibits the capacity to generate subsequently a response to D^d, it seems likely that this form of treatment inactivates B cells.

Adoptively transferred cells require antigenic stimulation in order to make a response. 5R cells are just as effective as Balb/c or B10.D2 cells in boosting cells taken from hyper-immunized (6 injections) donors. Thus, the D^d antigen when presented in relative isolation on 5R cells is able to recall but not to initiate a response.

THE RESPONSE OF H-$2^{b/k}$ MICE TO K^q AND K^s

T and B cells of F_1(B10xCBA) mice (H-$2^{b/k}$) were immunized separately, respectively against B10.G (H-2^q) and Balb/c (H-2^d). The B cells were prepared by treatment of spleen cells with anti-Thy-1.2 plus complement, while the T cells were left unseparated. The two cell populations were combined in syngeneic adoptive transfers, and boosted with either B10.T(6R) cells (qqqqqd) as a '+I' treatment or AQR cells (qkkddd) as a '-I' treatment. Thus with +I treatment the anti-D^d response could be helped by T cells primed to H-2$(K+I)^q$, but with -I treatment only by T cells primed to H-2K^q. Groups of mice receiving +I treatment showed helper activity on the part of T cells collected after a single donor immunization, and showed no significant improvement of helper activity as the number of donor immunizations increased. In contrast, the -I groups showed no initial helper activity, and showed activity only after five successive immunizations of the T cell donors. Essentially similar results were obtained with T cell donors immunized against B10.S ($H-2^s$); here, +I treatment was obtained by boosting with B10.S(7R) cells (sssssd) and -I with A.TL cells (skkkkd).

The early activation of T helper cells directed at I antigens had been expected (2). The late activation of T helper cells directed at K^q or K^s provides further examples of latent help.

MECHANISM OF LATENT HELP

Three hypotheses can be proposed, illustrated in Fig.1. Consider for example, latent help for D^d.

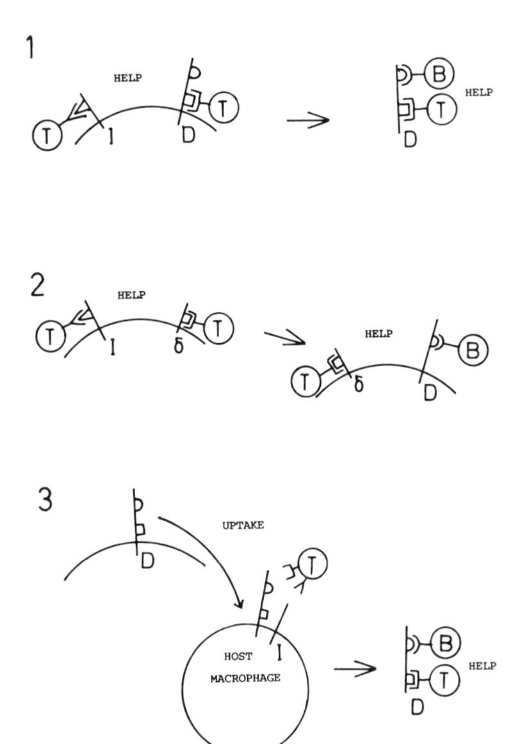

Fig. 1. Hypothetical mechanisms of latent help.

(1) As a result of T-T cooperation, T helper cells directed at determinants on the D^d molecule become activated. These D^d molecule determinants are inadequate to stimulate T_H cells without additional help.

(2) As a result of T-T cooperation, T helper cells become activated against a weak antigen ϕ, closely linked to D^d (i.e. not separated by the 5R recombination).

(3) During prolonged immunization the D^d antigen becomes separated from its own I region products, is taken up by host macrophages, and thus associates with the host I region products. In this location, it can elicit a T helper cell response.

The first hypothesis is unattractive in so far as it breaks the rules of dual recognition, by asking a T helper cell to recognise a foreign D molecule not associated with host-I. The second and third keep within the rules. Of the two, the third seems least extravagant - and is open to test using antigen-pulsed macrophages.

GENERALISING

All late allo-antibody responses that have been adequately examined seem to require help, but show no signs of needing intermolecular help. The position over initiating responses is entirely different, for intermolecular, intrastructural help seems often to be essential or at least useful (2). This suggests that latent help may be quite common, at least as regards major histocompatibility complex components and Thy-1. Auto-antibody responses can be interpreted in the same way, particularly those induced by concomitant immunization with foreign antigens but which persist in the absence of the original antigenic stimulus (4). The response to rat platelets, an AGB^+ I^- cell type, can also be interpreted in terms of latent help. These cells can generate a secondary, but not a primary antibody response (5).

REFERENCES

1. Lake, P. (1976) *Nature* 262,297.
2. Mitchison, N.A. and Lake, P. (1977) *Cold Spring Harbor Symp. Quant. Biol.* (in press).
3. Mitchison, N.A. and Lake P. (1977) *Immunological Communications* (in press).
4. Baechtel, F.S. and Prager, M.D. (1977) *J. Immunol.* 118,175.
5. Batchelor, J.R. (1977) *Eur. J. Immunol.* (in press).

ANTI-DNP IgE PRODUCTION AND SUPPRESSION IN SJL MICE[1]

Zoltan Ovary, Takaki Itaya[2], Judith Levinson, Steven S. Caiazza and Naohiro Watanabe[2,3]

Department of Pathology, NYU School of Medicine, New York, New York 10016

Preferential production of IgE antibody can be obtained in certain strains of mice if the adjuvant is $Al(OH)_3$ (1). It is also known that helminth infections easily elicit IgE antibody responses in both man and animals (2-5).

Mice, like guinea pigs (6-10), may produce two kinds of antibody capable of sensitizing the homologous species. One, IgG1 (11-13), may be produced in mg amounts. The other, IgE (1,14-16) is generally produced in very small amounts and detectable only by passive cutaneous anaphylaxis (PCA)[4] (17, 18).

It has also been shown that in the mouse, carrier specific thymus derived (T) lymphocytes collaborate with hapten specific bone marrow derived (B) lymphocytes in the primary and secondary anti-hapten IgE antibody production (19-21).

Taking advantage of the above known facts, a reproducible model of immunization was developed (Fig. 1) to obtain high and generally persistent anti-hapten IgE antibody (ahIgE) production in different strains of mice (22). This model consists of injecting i.p. 1 μg dinitrophenylated keyhole limpet hemocyanin (DNP-KLH) with 1 mg $Al(OH)_3$. On day 21 the mice are infected with 750 3-rd stage larvae of *Nippostrongylus brasiliensis* (N.Br) and on day 35 injected i.p. with 1 or 10 μg of dinitrophenylated *Nippostrongylus brasiliensis* worm extract (DNP-Nb) with 1 mg of $Al(OH)_3$. On day 42, and every 7 days thereafter the mice are bled from the retroorbital sinus (23). IgG1 and IgE antibody titers are

[1] Supported by grants from the NCI, 5-P01 CA16247-03 and from NIH AI 03075-18.

[2] Recipient of fellowship from the National Cancer Institute Inc., New York.

[3] Present address: Department of Parasitology, Jikei Univ. School of Medicine, Tokyo, Japan.

[4] Abbreviations: ahIgE = anti-hapten IgE antibody, B = bone marrow derived lymphocytes, DNP-BSA = dinitrophenylated bovine serum albumin, DNP-KLH = dinitrophenylated keyhole limpet hymocyanin, Nb = Nippostrongylus brasiliensis worm protein, N.Br = Nippostrongylus brasiliensis, PCA = passive cutaneous anaphylaxis, T = thymus derived lymphocytes.

determined by PCA in mice (17,18) and rats (18,24) and IgM and IgG2 antibody titers by passive lysis (8).

Fig. 1

Day 0	21	35	42 - - -
1 μg DNP-KLH 1 mg $Al(OH)_3$ i.p.	Infection 750 3rd stage larvae of N.Br	1 μg DNP-Nb 1 mg $Al(OH)_3$ i.p.	

Low doses of antigen (1 μg DNP-KLH) are used to obtain high titer IgE antibody as recommended by Vaz et al. (25). Similar results were obtained in adoptive transfer experiments (26).

In most strains of mice ahIgE produced is of high and persistent titer. A notable exception is the SJL strain of mice (Table 1) in which the titer of ahIgE is lower and, more importantly, is not persistent.

TABLE I
Strain Difference of Antihapten Antibody Response

Strain	H-2 type	Anti-DNP PCA titer*			
		Day 42‡		Day 56‡	
		IgE	IgG_1	IgE	IgG_1
A/J	a	2,560	640	1,280	1,280
C57BL/6	b	2,560	320	2,560	80
BALB/c	d	2,560	640	2,560	320
AKR	k	1,280	640	160	80
C3H	k	1,280	640	10,240	160
CBA	k	1,280	160	10,240	80
DBA/1	q	2,560	1,280	2,560	2,560
ASW	s	2,560	160	640	1,280
SJL	s	320	160	20	160

* Titer determined by pooled serum from five mice of each group.
‡ Days after DNP-KLH immunization.

One of the important factors in the regulation of antibody production is the action of suppressor T cells (27-29).

Helper T cells are generally more radioresistent (30) than suppressor T cells (29-31). Therefore, SJL mice were irradiated with 540 R on day 36, one day after injection of DNP-Nb. The lethal dose in this strain is much higher (32). Though lower doses of irradiation (360 R) were sufficient to eliminate suppressor T cells and even 180 R was almost as effective a dose as 540 R, 540 R was chosen because at this dose mostly ahIgE is obtained. With this high dose of x-irradiation, IgG1 anti-hapten antibody production is hampered as IgG antibody producing B cells in this strain of mice are more radiosensitive than IgE anti-hapten antibody producing B cells (26). The exquisite radiosensitivity of IgG producing B cells has also been confirmed in other strains as well (33).

When SJL mice were immunized as described above and irradiated on day 36 with 540 R, ahIgE production persisted. However if in these irradiated mice, $5x10^7$ spleen cells from normal untreated SJL mice were injected one day after irradiation (on day 37), ahIgE production fell as in non-irradiated mice (Fig. 2). These results were subsequently confirmed by Chiorazzi et al. (34). In addition, these authors further showed that treatment with cyclophosphamide also produces an elimination of cells capable of suppression IgE antibody production. Other strains, such as AKR, BALB/c, A/J and (BALB/c x SJL)F1 hybrids, gave similar results (34,35).

The cells responsible for this ahIgE suppression are T cells, as anti-Thy-1.2 antibody and complement treated spleen cells did not produce this suppression (26).

The class specific (IgE) non antigen specific T cells belong to the $Ly-1^+$ subset of T cells (36) (Table 2). They

TABLE 2

Anti-hapten antibody response of immunized-irradiated SJL mice injected with normal SJL spleen cells from which $Ly-1^+$ or $Ly-2^+$ T cell subclasses had been eliminated

Group	Transferred cells pre-treated with complement and:	PCA Titer of Recipients[b]	
		IgE	IgG1
I	αLy-1.2 absorbed with Ly-1.1 cells	800	400
II	αLy-1.2 absorbed with Ly-1.2 cells[a]	200	200
III	αLy-2.2	100	200

[a]Control for specificity of Ly1.2 antiserum; compare with Group I.

[b]on day 56

therefore differ in this respect from the antigen specific suppressor T cells demonstrated in other strains of mice (37-41) and from the specific allotype suppressor T cells demonstrated in SJL mice (42) which all have the phenotype $Ly-1^-2^+$.

It is well known that antigen-specific helper T cells investigated in other strains of mice belong to the Ly-1+ subset of lymphocytes (37-41). We confirmed this observation in the SJL strain of mice too (Table 3).

TABLE 3

Helper effect of antibody response after treatment with anti-Ly antibody in SJL mice.

Group	Transferred Cells[a] treated with ($5x10^7$)	Anti-hapten antibody titers (14 days after cell transfer)[b]		
		PCA Titer		Complement fixing antibody titer[c]
		IgE	IgG1	
I	αLy-1.2 absorbed with Ly-1.1 cells	50	10	100
II	αLy-1.2 absorbed with Ly-1.2 cells	400	100	400
III	αLy-2.2	400	100	400

[a]Spleen cells from infected and boosted mice were transferred to DNP-KLH primed (3 weeks) irradiated (540 R) recipients and challènged with 10 μg DNP-Nb and 1 mg Al(OH)3 after cell transfer.

[b]Titers were determined by sera pooled from 5 mice.

[c]Passive lysis.

TABLE 4

Anti-DNP titer of different antibody classes in non-irradiated SJL mice.

Group	Normal Spleen cells injected	Day 42			Day 56		
		IgE	IgG1	IgG2	IgE	IgG1	IgG2
I	0	800	800	1600	400	800	800
II	5×10^7	400	800	800	100	1600	800
III	7.5×10^7	200	800	1600	<10	800	400

For antibody determinations see text.

It is remarkable that these antigen non-specific suppressor cells do not seem to suppress IgG production. IgG1 and IgG2 antibody production persisted for a long period of time in non-irradiated mice, well beyond the time when ahIgE production is terminated. Injection of normal spleen cells in non-irradiated animals, where the IgG production is not hampered, does not show inhibition of IgG production (Table 4).

The substance producing ah IgE suppression can be extracted from normal SJL spleen cells. It is a high molecular weight (> 300,000 daltons, as determined by the Sephadex G-200 column), heat labile protein. As it is not sedimented by ultracentfifugation at 100,000 g for 90 minutes, it is unlikely that it is a virus (43). The site of action of this suppressor substance is currently under investigation.

All of the progeny of the cross between SJL and BALB/c mice (F1) immunized as indicated above produced persistent ahIgE of high titer (26). The backcross of the F1 mice to the SJL parent were then examined for persistence of a high titer

TABLE 5

Antibody Response in Backcross Mice

	PCA titer								Passive Lysis Titer	
	Day 42[a]				Day 56[a]				Day 42	Day 56
	anti-DNP		anti-Nb		Anti-DNP		anti-Nb	anti H-2[d]	Anti-DNP	Anti-DNP
Mice	IgE	IgG1	IgE	IgG1	IgE	IgG1	IgE	HA[b]		
♀ 1	50	800	0[c]	ND[d]	0	1600	ND	+	1600	800
2	200	1600	50	''	50	400	''	+	1600	400
3	50	100	0	''	50	50	''	+	600	200
4	50	800	0	0	0	800	0	-	1600	800
5	50	100	0	ND	0	800	ND	-	1200	800
6	0	100	0	0	0	200	0	-	600	200
♂ 1	50	400	0	0	0	400	0	+	800	1600
2	0	100	0	ND	0	100	ND	+	800	400
3	0	200	0	''	0	400	''	-	800	600
4	200	800	0	0	0	400	0	+	800	800
♀ 7	3200	1600	100	0	1600	1600	100	+	2400	800
8	3200	1600	100	ND	400	1600	ND	-	2400	800
♂ 5	3200	3200	0	0	3200	3200	50	+	2400	1600
6	800	1600	0	0	3200	1600	50	+	1600	1600
7	3200	1600	100	ND	3200	1600	ND	+	1600	1600
8	1600	400	100	''	3200	400	''	+	1600	1600
9	1600	800	100	0	3200	800	100	+	1600	800
10	1600	400	100	ND	3200	800	ND	-	800	800

[a] Days after DNP-KLH immunization.

[b] (+) positive hemagglutination reaction with anti $H-2^d$ sera diluted 1/800.

[c] o: No PCA reactions with sera diluted 1/50 except for anti-Nb IgG_1 where the dilution was 1/10.

[d] ND: not done

of ahIgE. About half of the progeny produced persistent ahIgE of high titer. The segregation is not sex linked and the persistence of ahIgE is not linked to the H-2 gene complex. These facts are consistent with the hypothesis that inheritence of this recessive autosomal gene is not linked to the H-2 gene complex (Table 5).

It has been suggested that helper T cells for IgE producing B cells are different from those which collaborate with the IgG producing B cells (44,45). However, these results were obtained in rabbits. It has also been suggested by Katz et al. (46) from results obtained in BALB/c and A/J mice that the same helper T cell is able to collaborate with both B cells producing IgE or IgG anti-hapten antibodies. On the other hand, Herzenberg et al. demonstrated that "in the (SJL-BALB/c)F1 and (SJAxBALB/c)F1 congenic pair, T helper cells show specificity for the immunoglobulin commitment of the B cells which they help" (42). In view of the fact that the carrier protein plays an important role in the respective amount of IgE or IgG produced and because of the important role of adjuvants such as $Al(OH)_3$ or pertussis vaccine in IgE production, and, finally, because IgE production is generally higher with parasitic infections and hapten coupled to worm extracts, it has been pointed out (22) that it is more likely that different T cells collaborate with IgE and IgG producing B cells. However, until precise methods can be developed to examine helper T cells, it is extremely difficult to decide if the same T cell may or may not collaborate effectively with IgG and IgE producing B cells.

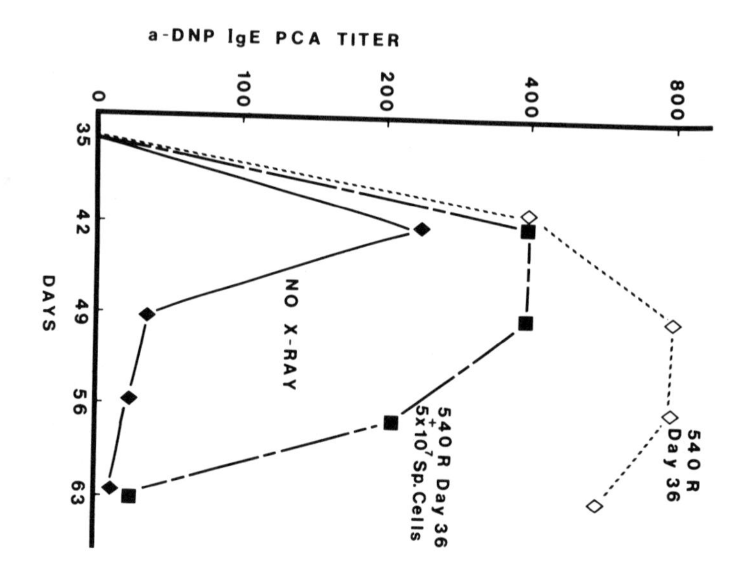

Fig. 2

REFERENCES

1. Revoltella, R. and Ovary, Z. (1969) Immunol. 17,45.
2. Bloch, K.J. (1967) Prog. Allergy 10,84. S. Karger, Basel.
3. Smithers, S.R. (1967) "Immunologic Aspects of Parasitic Infections" (O. Bier, Ed.) 43. Pan Am H. Org. Washington.
4. Ogilvie, B.M. and Jones, V.E. (1969) "Cellular and Humoral Mechanisms in Anaphylaxis and Allergy" (H.Z. Movat, Ed.) 13. S. Karger, Basel. Publ.
5. Sadun, E.H. (1972) "Immunity to Animal Parasites" (E.J.L. Soulsby, Ed.) 97.
6. Benacerraf, B., Ovary, Z., Bloch, K.J. and Franklin, E.C. (1963) J. Exp. Med. 117, 937.
7. Ovary, Z., Benacerraf, B. and Bloch, K.J. (1963) J. Exp. Med. 117, 951.
8. Bloch, K.J., Kourilsky, F.M., Ovary, Z. and Benacerraf, B. (1963) J. Exp. Med. 117, 965.
9. Catty, D. (1969) The immunology of nematode infections. Trichinosis in guinea pigs as a model. Monogr. Allergy, vol. 5 (Karger, Basel, 1969).
10. Ovary, Z., Kaplan, B. and Kojima, S. (1976) Int. Archs. Allergy appl. Immun. 51, 416.
11. Nussenzweig, R.S., Merryman, C. and Benacerraf, B. (1964) J. Exp. Med. 120, 315.
12. Fahey, J.L. and Barth, W.F. (1965) Nature 206, 730.
13. Ovary, Z., Barth, W.F. and Fahey, J.L. (1965) J. Immunol. 94, 410.
14. Mota, I. and Peixoto, J.M. (1966) Life Sci. 5, 1723.
15. McCamish, J. (1967) Nature 214, 1228.
16. Prouvost-Danon, A., Peixoto, J.M. and Queiroz-Javierre, M. (1968) Immunology 15, 271.
17. Ovary, Z. (1964) Passive cutaneous anaphylaxis. (Ackroyd,J. F.Ed.) 259. Immunological methods. Oxford Sci. Publ.
18. Ovary, Z., Caiazza, S.S. and Kojima, S. (1975) Intern. Archs. Allergy appl. Immun. 48, 16.
19. Ishizaka, K. and Okudaira, H. (1973) J. Immunol. 110,1067.
20. Okudaira, H. and Ishizaka, K. (1973) J. Immunol. 111,1420.
21. Hamaoka, T., Katz, D.H. and Benacerraf, B. (1973) J. Exp. Med. 138, 538.
22. Kojima, S. and Ovary, Z. (1975) Cell.Immunol. 15, 274.
23. Pettit, A. (1913) Comptes Rendus de la Societe de Biologie, 74, 11.
24. Mota, I. and Wong, D. (1969) Life Sci. 8, 813.
25. Vaz, E.M., Vaz, N.M. and Levine, B.B. (1971) Immunology 21, 11.
26. Watanabe, N., Kojima, S., Ovary, Z.(1976) J. Exp. Med. 143, 833.
27. Gershon, R.K. (1974) Contemp. Top. Immunobiol. 3, 1. (M.J. Henna, Ed.). Plenum Press, New York.

28. Basten, A., Miller, J.F.A.P., Sprent, J. and Cheers, C. (1974) J. Exp. Med. 140, 1685.
29. Tada, T., Taniguchi, M., Okumura, K. (1971) J. Immunol. 106, 1012.
30. Hamaoka, T., Katz, D.H. and Benacerraf, B. (1972) Proc. Natl. Acad. Sci. USA 69, 3453.
31. Kapp, J.A., Pierce, C.W., Schlossman, S. and Benacerraf, B. (1974) J. Exp. Med. 140, 648.
32. Lerman, S.P., Chapman, J., Carswell, E.A. and Thorbecke, G.J. (1974) Fed. Proc. 33, 616.
33. Fox, D.A., Chiorazzi, N. and Katz, D.H. (1976) J. Immunol. 117, 1622.
34. Chiorazzi, N., Fox, D.A. and Katz, D.H. (1977) J. Immunol. 118, 48.
35. Chiorazzi, N., Fox, D.A. and Katz, D.H. (1976) J. Immunol. 117, 1629.
36. Watanabe, N., Kojima, S., Shen, F. W. and Ovary, Z. (1977) J. Immunol. 118, 485.
37. Feldmann, M., Beverley, P.C.L., Dunkley, M. and Kontiainen, S. (1975) Nature, 258, 614.
38. Kisielow, P., Hirst, J.A., Shiku, H., Beverley, P.C.L., Hoffman, M.K., Boyse, E.A. and Oettgen, H.F. (1975) Nature 253, 219.
39. Cantor, H. and Boyse, E.A. (1975) J. Exp. Med. 141, 1376.
40. Cantor, H., Shen, F.W. and Boyse, E.A. (1976) J. Exp. Med. 143, 1391.
41. Jandinski, J., Cantor, H., Tadakuma, T., Peavy, D.L. and Pierce, C.W. (1976) J. Exp. Med. 143, 1382.
42. Herzenberg, L., Okumura, K., Cantor, H., Sato, V.L., Shen, F.W., Boyse, E.A. and Herzenberg, L.A. (1976) J. Exp. Med. 144, 330.
43. Watanabe, N. and Ovary, Z. (1977) Submitted for publication.
44. Kishimoto, T. and Ishizaka, K. (1973) J. Immunol. 111, 1.
45. Kishimoto, T. and Ishizaka, K. (1973) J. Immunol. 111,720.
46. Katz, D.H., Hamaoka, T., Newburger, P.E., Benacerraf, B. (1974) J. Immunol. 113, 974.

REGULATORY DETERMINANT WORKSHOP

J. W. Goodman and N. A. Mitchison

This workshop addressed itself to mechanisms of T cell activation by antigens, progressing from simple synthetic model antigens to more complex cell surface markers.

S. Fong discussed T cell responses to azobenzenearsonate derivatives of tyrosine (RAT) and an analog of tyrosine in which the side chain is replaced by a more hydrophobic propyl group (RAN). The former is immunogenic in guinea pigs whereas the latter is not, despite the close structural similarity of the two and their indistinguishable reactivity with anti-ABA antibody. In order to investigate the basis for the immunologic discrimination between the two compounds, macrophages where pulsed with 14C-labelled molecules; uptake and retention of the compounds were determined, as well as the capacity of pulsed macrophages to form clusters with RAT-sensitive lymphocytes by the Lipsky-Rosenthal technique. Although RAN was taken up and retained by macrophages to at least as great a degree as RAT, RAN-pulsed macrophages were less efficient at inducing clusters. Thus, the presentation of antigen on macrophage surfaces may be a critical step which distinguishes between these two molecules. The possible importance of association of antigens with Ia molecules on accessory cell surfaces for provoking T cell responses was raised, but this information for RAT and RAN was not yet available. It was pointed out that only 15-55% of murine macrophages bear detectable Ia, the highest proportion being found in spleen. Macrophages could well be functionally heterogeneous, with only those bearing Ia involved in successful antigen presentation to T cells.

This led to a discussion of the nature of Ia association with antigens. S. Leskowitz hypothesized that antigens become covalently bound to Ia molecules by enzymatic action. He speculated that this might explain the non-immunogenicity of RAN (due to its altered tyrosyl side chain, which might hinder enzyme action), of D-amino acid polymers and of ABA coupled to polysaccharides. However, it would be difficult to explain how enzymes distinguish between foreign and autologous proteins to prevent Ia saturation by the latter. In addition, soluble products of macrophages which can replace intact cells and do not consist of covalent Ia-antigen complexes have been described.

F. Celada described his comparative studies of T cell and antibody specificity to beta galactosidase in rabbits, using small peptides derived from the parent protein. He found differences between antibody-binding and T cell reactivity with the battery of peptides, providing another example of dichotomy in antigen recognition by T and B cells.

The induction of suppressor T cells was considered next. RAN, though non-immunogenic, was capable of inducing unresponsiveness to RAT in guinea pigs, although suppressor T cells were not yet identified in this system by transfer experiments. The work of W. Bullock and his associates was mentioned, in which active suppressors were demonstrated in guinea pigs immunized with RAT in incomplete Freund's adjuvant. D. Turkin reported that a cyanogen bromide fragment of beta galactosidase induced suppression of an _in vitro_ response of murine cells to a flourescein conjugate of the protein. The effector cell was sensitive to anti-Thy-1 and anti-Ly-2 antisera, consistent with expectations concerning properties of suppressor cells. R. Yowell described preliminary, as yet inconclusive, evidence for suppressor cells specific for a determinant in the N-terminal region of chicken egg lysozyme. Mice of $H\text{-}2^b$ haplotype are normally non-responders to the intact protein, but respond to the partially degraded protein which lacks an N-terminal peptide. However, direct evidence for suppression in this system is lacking. Finally, B. Benecerraf related how the addition of tyrosine sequences to copolymers of glutamic acid and alanine produced antigens which generated suppression in certain strains of mice. Thus, the cumulative evidence supports T suppressor induction by defined determinants, a point which should be conclusively resolved in the next few years. The question of what determines whether the net response to a given epitope will be help or suppression is less clear, although adjuvants appear to play a significant role in the outcome.

Discussion next turned to suppression in systems in which the antigens were less well defined. Z. Ovary has found that in SJL mice the IGE response to _Nippostrongylus brasiliensis_ protein coupled with DNP groups in transient, but can be prolonged by irradiating the animals one day after challenge with antigen. Injection of the irradiated mice with normal syngeneic spleen, lymph node or thymus cells abruptly terminated the IgE antibody response. The suppressor cells were eliminated by anti-Thy-1 and anti-Ly-1 sera, but not by anti-Ly-2 serum, an unexpected finding. Breeding experiments showed that suppression was determined by a single autosomal, recessive gene which was not linked to the MHC. A soluble suppressor substance has also been extracted from normal spleen cells. Elucidation of this suppressor system and reconciliation with other models of suppression will be awaited with great interest.

F. Bach discussed suppressor cell induction in allogeneic systems, in which incompatibilities at K and D loci appear to outweigh I-region differences. Suppressor cells can be triggered by targets which do not express Ia or LD markers. S. Swain has been studying allogeneic suppression directed against K or D products on TNP-specific B cells. The phenomenon has only been observed when the B cell and the suppressor differ at K or D and is reversible (the B cell is not destroyed).

An interesting observation in a tumor system was related by F. Bach, who found that a specific anti-leukemic proliferative response could be mounted by normal cells exposed to leukemic sibling cells, provided the siblings differed at LD or CD loci. This may represent an example of what N.A. Mitchison refers to as "latent help," and he described other examples. One such is the difficulty of inducing anti-Thy-1 responses in strains of mice congenic for that marker. However, sensitization will take place if I-region differences are also present. Once the anti-Thy-1 response has been initiated, it can be sustained by congenic immunization alone. Thus, weak antigenic differences require additional help initially, but once the response is in full swing, latent help can be dispensed with. Another example proffered by Mitchison is the inability of $H\text{-}2^b$ mice injected with cells which bear the K^s, K^q or D^b markers to make primary responses to these antigens. In contrast, animals which differ from the challenge cells across the H-2 complex will make responses to the K and D-determined markers, and will subsequently respond to challenge with these markers in isolation. Latent help could be an important factor in the induction of autoimmunity.

The workshop concluded on that note.

Workshop #8

T CELL SUBPOPULATIONS: R.K. Gershon, Yale University School of Medicine and F.H. Bach, University of Wisconsin.

The primary focus of this workshop was to discuss the different T cell subclasses involved in a number of in vitro and in vivo systems as well as the different types of histocompatibility antigens that elicit responses in the antigen-specific clones within each subpopulation.

Generation of Cytotoxic Cells

The basic findings regarding the generation of cytotoxic T lymphocytes in a mixed leukocyte culture were summarized. The primary target determinants for CTLs are coded for by genes of the H-2 K and D regions and the serologically defined antigens of the H-2K and D loci serve as excellent marker s for these cytotoxic targets. However, since it is not certain that the target determinants for the CTLs and for antibody are the same, the cytotoxic T lymphocyte target determinants are referred to as CD. In addition to the K and D region CD determinants, there are CD determinants coded for by genes in the I region although for the most part the reaction to these I region coded CD determinants is much weaker than the reaction to the K and D region determinants.

To generate a maximal response against K or D region CD antigens, it is desirable to stimulate the responding cells not only with the antigens coded for by the K and/or D region but also by I region determined LD antigens. Whether the CD determinants by themselves can generate a cytotoxic response in normal allogeneic MLCs is not clear, since uv or heat-treated stimulating cells, which present their CD antigens but do not functionally express LD, do not generate a cytotoxic response by themselves but do when the responding cells are simultaneously stimulated with x-irradiated stimulating cells that differ from the responding cells by I region determined LD antigens. K or D region different stimulating cells that are x-irradiated do generate a relatively weak cytotoxic response either because the CD antigens themselves can elicit that response as present on x-irradiated cells or because they are weak "LD-like" antigens coded for by genes in the K and D regions.

With respect to the T cell subpopulations responding to LD and CD antigens, it appears that Ly 1 helper cells respond primarily to LD-like antigens and Ly 23 cytotoxic T lymphocytes respond primarily to CD antigens.

In a secondary in vitro response (after sensitization in vivo) there is some controversy whether CD antigens alone can elicit a response. Evidence presented from some laboratories suggest that uv-treated stimulating cells that do not generate a primary response, can cause significant restimulation of the secondary cytotoxic response by themselves; evidence from other laboratories suggest that such uv-treated cells can only cause a secondary response when in fact they present some level of LD-like stimulus. This is an important question to be resolved since it addresses one of the major questions in this entire area: namely, the T helper cell dependence of various effector reactions, in this case including the reaction of the cytotoxic T lymphocyte.

With respect to the chemical modification of self, evidence has been presented which suggests that chemical modification results in both altered CD-like and LD-like antigens. It has not been clearly established that these LD-like and CD-like altered-self antigens activate the same cell populations that are activated in an allogeneic system. In fact it seems that Ly 123 cells play a more important role in reactions against altered self than they do in alloreactions.

Under certain circumstances cytotoxicity can be generated *in vitro* without a proliferative response, such as with glutaraldehyde fixed stimulating cells. Several explanations were offered to explain these findings. First, it may be that the glutaraldehyde fixed cells present only the CD antigen and this antigen by itself can stimulate the entire response. Second, it may be that the glutaraldehyde fixed cells secondary to the modification of the cell surface presents signal 2 to the cytotoxic T lymphocyte precursor as well as presenting the CD antigen as signal 1. Third, it was emphasized that the lack of proliferation in such a system does not rule out the activation of helper cells by the glutaraldehyde fixed stimulating cells, i.e. helper cells may not proliferate or at least may not be responsible for more than a very small amount of the proliferation in an MLC. Thus it is not critical to equate the presence of proliferation in the MLC with the activation of helper cells.

Asking some of these same questions from a cellular perspective, it has been demonstrated that a responding cell population depleted of Ly 1^+ cells can still generate cytotoxic cells in response to an allogeneic stimulus. The degree of the cytotoxic response can be increased by adding Ly 1 cells. Whether the response of the population that has been depleted of Ly 1^+ cells is due to a few remaining Ly 1^+ cells that can provide sufficient help to generate a significant response, or whether there are "helper-type" cells within the Ly 2 3 population, or whether Ly 2 3 cytotoxic T lymphocyte precursors can directly respond to an allogeneic haplotype has not been resolved.

The question was raised as to whether presentation of LD

and CD determinants on a single cell is a more effective stimulus than presenting these two types of determinants on two different cells. In the past protocols presenting LD and CD determinants on separate stimulating cells in an MLC has been referred to as 3-cell protocol since the responding cell population as well as two separate stimulating cell populations were involved. The suggestion was made that this is probably a 4-cell experiment in the sense that at least two separate subpopulations of T lymphocytes react within the responding cells: the Ly 1 helper cell and the Ly 2 3 cytotoxic cell. The evidence presented suggests that the presence of the LD and CD determinants on the same cell is a more effective way of stimulating the responding cells.

Further, evidence supporting the need for a helper stimulus was presented using a system of stimulating with glutaraldehyde fixed or otherwise metabolically inactivated stimulator cells. In this case helper cells that had been previously activated *in vitro* were added to mixtures of responding cells and glutaraldehyde fixed stimulating cells which, without the addition of the helper cells, would not yield a significant (or at most a very weak) response. The addition of the helper cells allowed the generation of highly significant cytotoxicity. These results also provided evidence that glutaraldehyde does not destroy determinants recognized by helper T cells.

The cell populations responsive to *K* or *D* region different stimulating cells alone were investigated with the use of the anti-Ly antisera. Evidence was presented demonstrating that treatment of the responding cell population with either the anti-Ly 1 or the anti-Ly 2 antiserum markedly decreased the proliferative response to the *K* or *D* region different stimulating cells. In addition pretreatment of the precursor population with either anti-Ly 1 or anti-Ly 2 markedly decreased the level of cytotoxicity generated. These results, presented as preliminary data, were interpreted to show that either the cell responding to a *K* or *D* region difference alone had the Ly 1 2 3 phenotype or that both the Ly 1 and the Ly 2 3 cells responded to a *K* or *D* region difference with the Ly 1 population providing help for the generation of cytotoxicity and the Ly 2 3 population contributing a significantly greater proportion of the proliferative response than when an entire H-2 difference is used.

Further, evidence concerning markers on responding subpopulations *in vitro* was presented using anti-Ia antisera raised against concanavalin A activated thymocytes. One such antiserum could be demonstrated to reduce the proliferative response in an MLC by approximately 50%, without decreasing the cytotoxic response. Since at least a part of the proliferative response could be removed with these antisera without affecting the cytotoxic response, it was postulated that either the antisera were not strong enough to remove all the proliferating cells, or that the antisera may be removing a subpopulation of the proliferating cells that was different from the subpopulation remaining, or that the antisera were affecting a cell other than the proliferating lymphocytes (e.g. a Ia+ macrophage) that affected the proliferative response. No conclusion could be reached about the need for the proliferating response to allow a cytotoxic response to develop until further studies are done.

In Vivo Graft Rejection

Data was presented from two perspectives regarding the rejection of an allograft. First, with respect to the antigens that are involved in leading to rejection, results using thyroid lobe transplantation under the kidney capsule were discussed. If the thyroid is from a donor who is only *K* or *D* region different from the recipient then the thyroid is rejected relatively slowly. If at the time as thyroid transplantation the recipient animal receives *I* region determined LD different lymphoid cells intraperitoneally then the *K* region different thyroid transplanted under the kidney capsule is rejected much more rapidly. These results are interpreted to demonstrate LD-CD collaboration *in vivo,* i.e. that i.e. cytotoxic T lymphocytes activated by CD antigens in the thyroid receive help from helper T cells activated by LD antigens on the lymphoid cells injected intraperitoneally and thus increase the rejection response. Preliminary data was also presented to indicate that a cultured thyroid, which shows very prolonged or indefinite survival when transplanted under the kidney capsule, can be rejected if the recipient is appropriately stimulated with *I* region determined LD different cells at the time of thyroid transplantation. These results could be explained, however, either on the basis of a helper T lymphocyte--cytotoxic T lymphocyte interaction or on the basis of rejection of the thyroid allograft due to a delayed type hypersensitivity reaction alone.

With respect to the cell populations responding to a skin allograft it would appear that both Ly 1^+ and Ly 2 3^+ cells are involved. B mice that are reconstituted with either immune Ly 1 T cells or with immune Ly 2 3 T cells reject allografts slowly; rapid rejection ensues when reconstitution is done with both T cell populations. Thus *in vivo* as *in vitro* studies delineating the antigens which are responsible for graft rejection parallel studies with different T cell subclasses. Overall, it would appear that the helper T cells respond primarily to LD-like antigens and that cytotoxic T lymphocytes respond primarily to the CD-like antigens and that help allows the development of a maximal cytotoxic response. To what extent the precursor CTLs can respond to CD antigens alone, in vitro or in vivo, has yet to be determined.

Delayed Type Hypersensitivity (DTH)

Studies on DTH responses gives similar results. Although some DTH responses can be produced in mice depleted of all Ly 2$^+$ cells, evidence was presented that Ly 2$^+$ cells play a significant role in contact hypersensitivity reactions and in anti-viral DTH reactions. The DTH reactions which involve Ly 2$^+$ cells also appear to require K and D homology between donor and recipient for efficient adoptive transfer. The presentation of these findings elicited a lively debate as to whether T cell functions segregate completely with Ly antigen. The general cocensus was that there was insufficient evidence to reach any firm conclusion. A study on vasoactive amine dependence of different types of DTH indicated that those forms of DTH which do not require Ly 2$^+$ cells are initiated by Ly 1 cells which recruit non-lymphocytic effector cells - e.g. monocytes and polymorphonuclear leukocytes. These types of effector cells cannot leave the blood and participate in DTH reactions without the help of vasoactive amines. On the other hand at least one of the effector cells in contact hypersensitivity may be an Ly 2$^+$ lymphocyte (?cytotoxic cell) which can migrate out of the blood in vasoactive amine depleted animals. Thus, the latter type of DTH reaction (like acute graft rejection) may be brought about by interactions of 2 functionally distinct T cell subclasses; an Ly 1 initiator cell and an Ly 2 3 effector cell. Other DTH responses may mimic chronic graft rejection and be due to Ly 1 initiator cells interacting with non-lymphocytic effector cells.

Suppressor T Cells

Studies of suppressor T cells have indicated that these cells are remarkably similar to the cytotoxic T cell. Both are Ly 2$^+$, both react preferentially with CD type alloantigens, and both require significant Ly 1 helper activity for optimal activity. Some evidence was presented indicating that Ly 1 cells could act as suppressor cells for DTH responses. This view was challenged by studies showing that Ly 1 cells could act as inducers of suppressor cells in the Ly 1 2 3 T cell population and thus suppression effected by Ly 1 cells may be indirect and require the presence of Ly 2$^+$ cells. We were reminded not to interpret the fact that T cells can reconstitute the antibody response of B mice, to indicate that T cells make antibody. This admonition suppressed further discussion and thus closed the session.

THE ROLE OF MHC GENES IN T-CELL MEDIATED RESPONSES TO SYNGENEIC MODIFIED CELLS

Gene M. Shearer, Anne-Marie Schmitt-Verhulst, Stephen Shaw, Carla Pettinelli, Pierre A. Henkart and Terry G. Rehn

Immunology Branch, National Cancer Institute
Bethesda, Maryland 20014

ABSTRACT. Cell-mediated lympholysis (CML) has been generated in vitro against syngeneic murine spleen cells modified with trinitrobenzene sulfonate (TNBS). The specificity of the effectors generated is such that the stimulator and target cells must both be modified by the same agent and must also express the same H-2K and/or H-2D haplotype. Further analysis of the specificity indicated that these effector cells against TNBS-modified autologous cells could not lyse: (a) H-2-matched targets modified with the trinitrophneyl group (TNP) separated from the cell surface proteins by a tripeptide spacer; nor (b) H-2-matched targets in which the TNP group was presented on the cell surface by the insertion of a TNP-fatty acid-dextran conjugate into the lipid bilayer of the cell membrane. Furthermore, cells presenting TNP either by (a) or (b) were ineffective inhibitors of the lysis of H-2-matched TNBS-modified targets by effector cells sensitized with TNBS-self. These findings indicate that the immunogenic unit recognized by these cytotoxic effectors includes more than the modifying agent and involves the cell surface proteins (possibly H-2-coded products). However, it was unexpectedly observed that cytotoxic effectors could be generated by culturing spleen cells with TNBS coupled to soluble proteins. These effector cells were also restricted to lyse only TNBS-modified H-2K and/or H-2D matched target cells.

Cultures primed in vitro with TNBS-modified autologous cells were able to generate secondary effectors only if restimulated with TNBS-modified stimulators sharing H-2K and/or H-2D with the primary stimulators. Secondary proliferative responses were induced only with TNBS-modified cells sharing either the K, I-A or I or D regions with the primary stimulating cells. These results extend the requirements for K and/or D homology at the effector phase of the CML to include homology at K and/or D between primary and secondary stimulators. Furthermore, they implicate a role for I region products when proliferative responses are studied in the modified self system. Primary and secondary proliferative responses have been recently obtained against TNBS-modified human lymphocytes. Preferential proliferative responses were obtained when the primary and secondary modified stimulating cells were from the same donor.

In addition to the role of H-2-coded products in the specificity of the TNBS-modified self system, the MHC also influences the ability to generate cytotoxic effectors against TNP-H-$2D^d$. The phenomenon appears to be controlled by multiple Ir genes, one of which maps to the left of I-A, and the other between the I-A and I-J subregions of the H-2 complex.

INTRODUCTION

It has been recently discovered that self-restricted cell-mediated lympholysis (CML) reactions are among the numerous immunologic functions associated with products of the murine major histocompatibility (H-2) complex (MHC). Such CML reactions have been generated against virally infected or chemically modified cells only when the target cells shared H-2K and/or H-2D region products with the infected or modified stimulating or responding cells or both (2-5, 7, 9, 14-16). Similar requirements for H-2 restriction have been demonstrated for sensitization and lysis of cells expressing weak transplantation antigens (1, 6). The investigations that have been carried out in this laboratory using the chemically modified autologous system will be presented, including a brief review of earlier published work, as well as a summary of new results.

H-2 RESTRICTION IN PRIMARY IN VITRO CML

Thymus-derived cytotoxic effector cells have been generated in vitro in primary 5-day cultures for sensitization of responding mouse splenic lymphocytes against TNBS-modified autologous stimulating spleen cells (14). (TNBS is a reagent that reacts covalently with cell surface amino groups.) The specificity requirements that have been observed for detecting lysis of target cells is summarized in Table 1. The target cells have to be modified by the same agent as the stimulating cells (9), and also have to express the same H-2K modified or H-2D region haplotypes on the modified stimulating and/or responding cells (3, 15). Responding cells for F_1 hybrid donors sensitized against modified parental stimulating cells generated effectors capable of lysing only modified targets of the same parental haplotype as that expressed by the stimulators (3, 15). These results indicate that H-2 homology only between effector and modified target cells is not sufficient to account for this phenomenon, and suggest that either H-2 homology between modified stimulators and modified targets or among effectors, stimulators, and targets

TABLE 1

SUMMARY OF SPECIFICITY REQUIREMENTS IN CML REACTIONS GENERATED BY MODIFIED AUTOLOGOUS CELLS

Responding Cells	Stimulating Cells	Target Cells	Lysis of Target by Effector Cells
A	A-M	A	No
		A-M	Yes
		B-M	No*
		B-m	No
	A-m	A-m	Yes
		A-M	No
$(AxB)_1$	A-M	A-M	Yes
		B-M	No
		$(AxB)F_1$-M	Yes
	B-M	B-M	Yes
		A-M	No

*Except if A and B share K and/or D H-2 regions.
M and m indicate two distinct modifying agents.

is required. These results indicate that both the modifying agent and H-2-coded cell surface products are involved in the recognition of T-effector cells. The recognition of the modifying agent as well as H-2-coded products could be accounted for by one of at least two basic models (16). These two models essentially differ in that one involves a single receptor recognizing self H-2 products (either altered or unaltered in close association with the infecting or modifying agent), whereas the other involves two distinct receptors--one specific for the agent and the second specific for self H-2K- or H-2D-coded cell surface products. In the single receptor model, the agent could be modifying cell-surface antigenic structures (altered self). For the two receptor or dual recognition model, one receptor would be specific for the "hapten" or infecting agent. The second receptor would be a responder or effector cell receptor which would function as a responder-stimulator and/or effector-target interaction structure recognizing syngeneic K or D region products. Attempts to block the lytic interaction between effector and target cells with unlabelled cells presenting one of the two components involved in the recognition alone (TNBS-modified allogeneic cells or syngeneic unmodified cells) were unsuccessful (3, 15). These results could be accounted for by a single receptor recognizing the two moities of the immunogenic complex or by two low affinity receptors, each of which would be insufficient alone for binding. Blocking of either the TNP groups or the specific H-2K- or H-2D- coded products (involved in the sensitization) on the target cells with antiserum resulted in the inhibition of cytolysis by the effector cells (10).

FINE SPECIFICITY OF THE CYTOTOXIC EFFECTOR CELLS

One of the advantages of the in vitro chemically-modified syngeneic CML response is that this system can be utilized to investigate the fine specificity of the "haptenic" moiety. No cross reactivity was detected between effectors generated against TNBS-modified stimulating cells when asayed on ^{51}Cr-labelled N-(3-nitro-4-hydroxy-5-iodophenylacetyl)-β-alanylglycylglycylazide-(N-) modified target cells and vice versa (9). Furthermore, effectors generated by sensitization against TNP-modified syngeneic cells did not lyse H-2-matched target cells modified by the TNP moiety separated from the cell surface by a β-alanylglycylglycyl tripeptide (AGG) (see Table 2). (Similar to NTBS, both N and TNP-AGG react with cell surface amino groups.) These results not only demonstrate the high degree of fine specificity contributed by the "haptenic" or modifying agent, but they also tend to argue against dual recognition in its simplest form involving recognition of the TNP moiety and self H-2 by two independent receptors on the T-effector cell (8).

TABLE 2

IN VITRO INDUCTION OF CYTOTOXICITY OF C57BL/10 SPLEEN CELLS BY TNBS-MODIFIED OR TNP-AGG-MODIFIED SPLEEN CELLS TO TNP-MODIFIED AND TNP-AGG-MODIFIED $H\text{-}2^b$ TUMOR TARGET CELLS

Stimulating C57BL/10 Spleen Cells Modified With:	$H\text{-}2^b$ Tumor Target Cells Modified With:	% Specific Lysis±SE in exp. # 1	2	3
TNBS	TNBS	45.2±1.7	45.6±2.1	39.6±2.9
TNBS	TNP-AGG	6.4±2.3	--	2.0±2.9
TNP-AGG	TNP-AGG	11.5±2.7	22.2±2.3	13.0±2.3
TNP-AGG	TNBS	8.6±1.7	11.9±2.0	0.8±0.4

The fine specificity of the TNBS-modified syngeneic CML system has been further investigated by TNP-modification of cells either with TNBS or with TNP-stearoyl-dextran (TSD), an amphipathic molecule, which can be inserted into the lipid bilayer of the cell membrane and which binds to cells by non-covalent forces. Quantitatively equivalent amounts of TNP were detected on the surface of cells modified by both reagents as measured by fluorescein anti-TNP antibody and the fluorescence activated cell sorter. These cell preparations were compared for their ability to: (a) sensitize syngeneic splenic lymphocytes leading to the generation of cytotoxic effector cells; (b) serve as lysable targets in a 4-hour ^{51}Cr-release assay for effector cells generated in (a); and (c) act as blocking cells in the lysis of TNBS-modified targets lysed by TNP-self effector cells generated in (a). In none of these three experimental systems did TSD-modified syngeneic spleen or H-2-matched tumor cells act either as a sensitizing immunogen or as a target antigen. In contrast, TNBS-modified spleen cells sensitized syngeneic lymphocytes to generate effectors against TNBS-modified syngeneic targets. Furthermore, TNBS-modified, H-2-matched cells served as specific lysable targets and as inhibiting cells for such effectors (see Tables 3 and 4). These results indicate that

TABLE 3

COMPARISON OF TNBS AND TSD MODIFIED CELLS AS CML STIMULATORS AND TARGETS

Responding Cells	Stimulating Cells	% Specific Lysis±SE Assayed on:		
		RDM-4-TNBS	RDM-4-TSD	RDM-4
B10.BR	B10.BR-TNBS(10mM)	42.6±2.0	-8.0±1.7	-5.5±1.6
	B10.BR-TNBS(1mM)	39.5±3.1	-1.3±2.7	-4.5±2.0
	B10.BR-TSD	-8.0±1.9	-6.0±2.8	-3.7±1.7
	B10.BR-TSD(X2)*	-7.0±2.0	-9.2±1.9	-5.3±1.9

Effector:target cell ratio = 20:1.
*Responding cells restimulated with B10.BR-TSD spleen cells after 24 hours of culture.

the manner in which the TNP group is associated with the cell surface is important in the immunogenicity and antigenicity of hapten-modified syngeneic stimulating cells in generating H-2-associated CML reactions. These findings raise the possibility that a covalent or at least a stable linkage with cell surface proteins (possibly H-2-controlled products) is important for immunological function. Furthermore, these observations do not favor the dual receptor model for H-2-restricted syngeneic CML if it is assumed in such a model that one receptor is specific for the TNP moiety and the second for unmodified self major histocompatibility products.

TABLE 4

COMPARISON OF THE BLOCKING OF CYTOTOXICITY AGAINST TNBS-MODIFIED TARGETS WITH TNBS-MODIFIED AND TSD-MODIFIED INHIBITING CELLS

Blocking Cells	% Specific Lysis of RDM-4-TNBS Target Cells at Blocking:Effector Cell Ratio of:		
	(2:1)	(1:1)	(1:2)
RDM-4	45.9±2.7	47.1±4.1	51.6±3.7
RDM-4-TNBS(10mM)	6.2±0.7	12.5±1.0	21.0±1.6
RDM-4-TNBS(1mM)	24.7±1.4	32.7±1.5	34.2±2.6
RDM-4-TSD	39.0±1.0	41.6±1.6	45.9±2.1

Effector:target cell ratio = 20:1.
Specific lysis in the absence of blocking cells was 53.3+1.2.
Responding and TNBS-modified stimulating cells were from B10.BR donors.

The results summarized thusfar indicate that the T-cell recognition of TNP occurs when the TNP group is presented on the H-2 matched target cells in the same way that it was presented on the modified sensitizing cells. Furthermore, non-covalent binding of TNP to cell surfaces did not produce an immunogenic complex. (For the remainder of this report TNP-modified cells will indicate cells that have been modified with TNBS.) In contrast to the above findings, it has been recently observed that spleen cells can be sensitized in vitro to TNP-BGG (bovine gamma globulin) or TNP-BSA (bovine serum albumin). As shown in Table 5, the effectors generated lysed TNBS-modified target cells H-2 matched with the responding cells, but not TNBS-modified allogeneic target cells. Although the mechanism by which this H-2-associated sensitization occurs is presently not understood it suggests that physical modification of cell surface proteins including H-2-coded products is not an absolute requirement--even though the specificity is H-2 restricted. These latter results tend not to favor altered H-2 coded structures as an absolute requirement for immunogenicity, but leave open either an associative recognition model, in which TNP groups and unaltered H-2-coded products (13) are recognized via a single receptor, or the dual recognition model, in which one receptor recognizes TNP and the other recognizes self H-2-coded structures (16).

TABLE 5

IN VITRO GENERATION OF H-2 RESTRICTED CML EFFECTOR CELLS WITH TNP-MODIFIED SYNGENEIC CELLS AND SOLUBLE TNP-BCG(*) OR TNP-BSA(#)

Responder Cells Strain	Immunogen	% Specific Lysis ± S.E. on Target Cells RDM4 (H-2^k)	RDM4-TNP	LSTRA (H-2^d)	LSTRA-TNP	EL-4 (H-2^b)	EL-4-TNP
B10.BR	--	0.0	0.0	N.T.	N.T.	0.0	0.0
kkkkkkkk	B10.BR-TNP	-0.2±3.4	50.6±2.3			-5.7±4.4	9.2±3.7
	BGG	-1.5±3.8	-3.3±2.0			-3.0±3.4	0.4±2.5
	BGG-TNP	-0.9±3.5	36.5±1.0			1.9±5.5	6.8±5.6
B10.A	--	0.0	0.0	0.0	0.0	N.T.	N.T.
kkkkkddd	B10.A-TNP	-6.7±1.6	35.6±7.1	-7.6±2.2	4.7±3.5		
	BGG	-4.3±2.8	-3.4±5.5	-0.8±3.9	-1.7±1.2		
	BGG-TNP	-1.2±2.4	41.3±5.8	-5.5±3.2	0.9±3.5		
	BSA	-3.8±1.1	-11.7±5.3	-7.9±4.1	-5.8±1.3		
	BSA-TNP	-6.0±1.3	32.7±11.9	-10.3±1.8	-4.1±1.4		

(*) BGG = bovine gamma globulin; (#) BSA = bovine serum albumin.

The TNP-BGG and TNP-BSA were at 100 μg/1 ml at initiation of the 5-day culture.

Effector to target ratio = 40:1.

N.T., not tested.

SECONDARY CYTOTOXIC AND PROLIFERATIVE RESPONSES TO TNP MODIFIED SELF

We have employed one further approach to define the specificity of lymphocyte stimulation by TNP-modified syngeneic cells. Secondary stimulation of in vitro primed cultures was performed, and both secondary CML and proliferative (MLR) responses were monitored. In these experiments attention was focused on the requirement for H-2 homology between primary and secondary stimulators. Table 6 shows a secondary CML experiment which indicates that lymphocytes sensitized by culturing TNP-syngeneic cells in vitro for 7 days could be restimulated by stimulator cells presenting the TNP on the same type of cells, but not by TNP-modified cells from a congenic resistant strain of mice differing in the expression of the H-2 haplotype. The inability of the cultures sensitized by TNP-syngeneic cells to generate a primary allogeneic reaction after restimulation with allogeneic cells is probably due to the development of nonspecific suppressor cells of precursors of primary CML and MLR effectors in unsensitized or sensitized spleen cell cultures. Such conditions are favorable for the analysis of the H-2 homology requirements for secondary CML and MLR allowing the use of modified allogeneic stimulator cells in the absence of an allogeneic reaction.

TABLE 6

H-2 HOMOLOGY REQUIREMENT BETWEEN PRIMARY AND SECONDARY TNP-MODIFIED STIMULATOR CELLS

Responding Cells	Primary Stimulators	Secondary Stimulators	% Specific Lysis±SE on: RDM4-TNP	LSTRA-TNP
B10.BR	B10.BR-TNP	B10.BR	0	0
		B10.BR-TNP	39.3±5.0	5.3±0.6
		B10.D2	-5.5±2.7	0.4±0.6
		B10.D2-TNP	-7.3±1.3	1.1±0.5
B10.BR	B10.D2-TNP	B10.BR	0	0
		B10.BR-TNP	11.0±1.5	13.4±1.1
		B10.D2	17.3±0.9	88.2±3.5
		B10.D2-TNP	6.6±0.9	56.7±1.4

Effector:target = 10:1

Although the generation of a primary CML reaction against TNP-syngeneic cells is accompanied by only marginal cellular proliferation as measured by thymidine incorporation

TABLE 7

SECONDARY PROLIFERATIVE RESPONSE TO TNP-MODIFIED SELF AFTER IN VITRO PRIMING WITH TNP-SYNGENEIC OR TNP-ALLOGENEIC CELLS

Responding Cells	Primary Stimulating Cells	Secondary Stimulating Cells	^{3}H-Thymidine CPM ± S.E.	Ratio	CPM
B10.BR	B10.BR-TNP	B10.BR	2890 ± 279	1.0	0
		B10.BR-TNP	42900 ± 1050	14.9	40000
		B10.D2	12200 ± 222	4.2	9300
		B10.D2-TNP	9920 ± 419	3.4	7030
	B10.D2-TNP	B10.BR	5900 ± 401	1.0	0
		B10.BR-TNP	9300 ± 382	1.6	3400
		B10.D2	27600 ± 932	4.7	21700
		B10.D2-TNP	23800 ± 1130	4.1	17900

TABLE 8

MAPPING THE REGIONS WITHIN THE H-2 COMPLEX RESPONSIBLE FOR THE SECONDARY PROLIFERATIVE RESPONSE AGAINST TNP-MODIFIED STIMULATING CELLS

Primary Stimulating Cells	Secondary Stimulating Cells	H-2 Haplotype at: K	A	B	J	E	C	S	D	^{3}H-Thymidine CPM ± S.E.	Ratio	CPM
B10.BR-TNP	B10.BR									2890. ± 279.	1.0	0.
	B10.BR-TNP	$\underline{\underline{k}}$	$\underline{\underline{k}}$	$\underline{\underline{k}}$	$\underline{\underline{k}}$	$\underline{\underline{k}}$	$\underline{\underline{k}}$	$\underline{\underline{k}}$	$\underline{\underline{k}}$	42900. ± 1050.	14.9	40000.
	C3H/H3J									1640. ± 174.	0.6	-1240.
	C3H/HeJ-TNP	$\underline{\underline{k}}$	$\underline{\underline{k}}$	$\underline{\underline{k}}$	$\underline{\underline{k}}$	$\underline{\underline{k}}$	$\underline{\underline{k}}$	$\underline{\underline{k}}$	$\underline{\underline{k}}$	47000. ± 2620.	16.3	44100.
	C3H.OH									6370. ± 345.	2.2	3480.
	C3H.OH-TNP	d	d	d	d	d	d	$\underline{\underline{d}}$	$\underline{\underline{k}}$	26000. ± 1360.	9.0	23100.
	B10.D2									12200. ± 222.	4.2	9300.
	B10.D2-TNP	d	d	d	d	d	d	d	d	9920. ± 419.	3.4	7040.
	A.TL									5700. ± 504.	2.0	2800.
	A.TL-TNP	s	$\underline{\underline{k}}$	$\underline{\underline{k}}$	$\underline{\underline{k}}$	$\underline{\underline{k}}$	$\underline{\underline{k}}$	$\underline{\underline{k}}$	d	26900. ± 1330.	9.3	24000.
	A.TH									3320. ± 188.	1.2	432.
	A.TH-TNP	s	s	s	s	s	s	s	d	4160. ± 327.	1.4	1280.

Responding lymphocytes from B10.BR donors were restimulated 7 days after the initiation of the primary cultures.

H-2 subregions common to primary and secondary stimulating cells are indicated by $\underline{\underline{\ \ }}$.

into the cells, a strong proliferative response could be measured 2 to 3 days after secondary stimulation with TNP-modified cells syngeneic to the primary TNP-modified stimulating cells. The results shown in Tables 7 and 8 indicate that optimal TNP-dependent restimulation was obtained when the secondary stimulating cells shared the whole H-2 (irrespective of the non-H-2 background) with the primary TNP-stimulating cells. When only the D region (C3H.OH) or the I region (A.TL) haplotype were shared between primary and secondary TNP-modified stimulating cells (Table 8), a lower but significant TNP-dependent proliferative response was observed. These results suggest: (a) that the specificity of lymphocytes sensitized *in vitro* against TNP-modified syngeneic cells is for TNP and H-2-coded products not only as measured by the specificity of the cytotoxic effector cells, but also for the restimulation of primary sensitized precursors of CML and MLR responsive cells; and (b) that TNP modification can be recognized in association with I region products by strongly proliferative cells, whereas this recognition did not occur by the cytotoxic effectors.

Recently, we have shown that human peripheral blood lymphocytes undergo a proliferative response to TNP-modified autologous cells. There was a weak primary response which characteristically peaked at day 7 with a stimulation index of 3 fold or less. An accelerated secondary response peaked 3-4 days following restimulation of primed cells with TNP-modified autologous cells. Stimulation indices of 10 to 20 fold were often observed in such secondary responses. As in the murine system, proliferation in the secondary response was generally greater when primed cells were restimulated with TNP-modified autologous cells than with TNP-modified unrelated cells. This self specificity is illustrated by the experiment shown in Table 9. Two individuals were primed by self-TNP and then restimulated with autologous and unrelated cells modified with TNP. Each primed responding population proliferated maximally in response to TNP modified autologous cells; response to TNP-modified unrelated cells was less than 40% of this maximal response. This data suggests that the secondary proliferation is not a response to the TNP moiety alone, but reflects recognition of some additional determinant(s) on the TNP-modified cells. Extrapolation from the murine system suggests that MHC determinants may be involved. Mapping studies in humans is currently in progress to determine whether recognition is indeed HLA-linked.

IMMUNE RESPONSE GENES FOR H-2-RESTRICTED CYTOTOXICITY

It was observed in the initial studies of the CML response against TNP-modified syngeneic cells that effector

TABLE 9

SELF SPECIFICITY IN HUMAN SECONDARY PROLIFERATIVE RESPONSE TO TNP-MODIFIED AUTOLOGOUS CELLS

Restimulating Cells:	Responding Cells Primed as Shown Below: ^{3}H-Thymidine CPMx10^{-3}		Stimulation Index		ΔCPMx10^{-3}	
	LαL-TNP	PαP-TNP	LαL-TNP	PαP-TNP	LαL-TNP	PαP-TNP
Autologous	3.0±0.3	3.3±0.4	--	--	--	--
L-TNP	*26.3±4.0	12.5±1.1	*8.8±1.7	3.8±0.6	*23.4±4.0	4.2±4.2
P-TNP	8.1±1.2	*50.3±5.1	2.7±0.5	*15.3±1.3	5.1±1.2	*47.0±5.1
S-TNP	5.9±0.6	19.4±2.9	2.0±0.3	5.9±1.1	2.9±0.7	16.1±2.9

*Indicates restimulation with modified autologous cells.

L, P, and S indicate lymphocytes from 3 unrelated individuals.

cells from certain inbred mouse strains lysed TNP-modified targets expressing either the K or D H-2 haplotype of the responding and modified stimulating cells, whereas other mouse strains generated effectors capable of lysing TNP-modified targets expressing the K haplotype, but not those expressing the D haplotype of the responder and stimulating cells (15). Mice which exhibited these differences in D-end-restricted CML differed in the H-2 haplotypes in the left part of the H-2 complex (which included K and a portion of the I region), although they expressed the same haplotype at H-2D. When the lytic potential of effectors generated by B10.A and B10.D2 responding cells were compared, the B10.D2 effectors were observed to lyse TNP-modified target cells matched at either the K- or D-end with the cells of the sensitizing phase, whereas B10.A effectors did not lyse TNP-modified D-end matched targets although they did lyse TNP-modified K-end matched targets (10-12, 15).

These strain-dependent preferential response patterns to TNP-modified syngeneic cells restricted at the D region appear to be regulated by H-2-linked immune response (Ir) genes, and the ability to respond seems to be a dominant trait, since (B10.AxB10.D2)F_1 mice are high responders to TNP-H-$2D^d$ (11). That responding cells from the B10.A strain are defective in their response potential to TNP-H-$2D^d$ has been verified by CML inhibition experiments in which region-specific anti-H-2 sera were used to block the interaction between effector cells and TNP-modified syngeneic target cells (10). The results indicated that A.TL anti-A.AL sera (directed against H-$2K^k$) severely reduced the lysis by B10.A effectors, whereas B10.BR anti-B10.D2 sera (directed against H-$2D^d$) had no effect on the lysis of B10.A-TNP targets by B10.A effectors. In contrast, the B10.BR anti-B10.D2 sera (directed against H-$2D^d$) abolished the lysis of B10.A-TNP targets by B10.D2 effectors. These results indicate that the TNP-self cultures from B10.D2 donors generated effectors specific for H-$2D^d$-TNP, whereas cultures from B10.A donors did not generate effector cells specific for H-$2D^d$-TNP.

Mapping studies for the Ir control of effector cells specific for H-$2D^d$-TNP have been performed using recombinant mice on the C57BL/10 and A strain backgrounds (12). As summarized in Table 10, a higher level of responsiveness was found when the s haplotype was expressed to the left of the I-E subregion as compared with the K haplotype. The A.TH and B10.HTT were the highest CML responders to H-$2D^d$-TNP, whereas B10.A and A.AL were the poorest responders to H-$2D^d$-TNP, although all four of these strains were high CML responders to their respective TNP-modified H-2K restricted specificities. The A.TL strain, which expresses the high responder s haplotype at K and the low responder k haplotype

TABLE 10

IN VITRO CYTOTOXIC RESPONSES TO TNP-MODIFIED AUTOLOGOUS CELLS IN DIFFERENT INBRED MOUSE STRAINS ON THE C57BL/10 AND GENETIC BACKGROUNDS

Responding Spleen Cells	Stimulating Spleen Cells	H-2 Haplotype at: K A B J E C S D	TNP-Modified Target Cells*	% Specific Lysis ± S.E.	Range of Lysis
B10.A	B10.A-TNP	k k k k k d d d	$H\text{-}2^d$	2.2 ± 1.1	1-10
			$H\text{-}2^s$	5.2 ± 1.0	2-16
			$H\text{-}2^k$	30.6 ± 1.7	22-46
B10.HTT	B10.HTT-TNP	s s s s k k k d	$H\text{-}2^d$	38.0 ± 1.4	38-53
			$H\text{-}2^s$	49.9 ± 2.6	50-55
			$H\text{-}2^k$	27.3 ± 2.1	
A.AL	A.AL-TNP	k k k k k k k d	$H\text{-}2^d$	8.2 ± 1.6	2-8
			$H\text{-}2^s$	3.1 ± 1.9	0-6
			$H\text{-}2^k$	32.5 ± 4.7	23-33
A.TL	A.TL-TNP	s k k k k k k d	$H\text{-}2^d$	19.9 ± 1.1	20-23
			$H\text{-}2^s$	45.6 ± 0.9	10-46
			$H\text{-}2^k$	11.1 ± 2.5	0-11
A.TH	A.TH-TNP	s s s s s s s d	$H\text{-}2^d$	33.3 ± 1.4	33-44
			$H\text{-}2^s$	51.8 ± 0.6	16-52
			$H\text{-}2^k$	11.4 ± 1.7	5-11

EFFECTOR:TARGET RATIO = 40:1

*$H\text{-}2^d$, $H\text{-}2^s$, and $H\text{-}2^k$ target cells were 48 hour PHA-stimulated blast spleen cells from B10.D2, SJL/J or B10.S, and B10.BR donors, respectively.

throughout the I region was an intermediate responder to H-2D^d-TNP, but a high responder to its respective H-2K^s-TNP restricted specificity (12). These results are compatible with the regulation of CML responsiveness specific for H-2D^d-TNP being under the control of at least two genes, one mapping to the left of the crossover between K and I-A (in the A.TL strain) - possibly in the K region, and the second mapping inside the I-A through I-J subregions.

The finding that at least one of the genes controlling responsiveness to H-2D^d-TNP maps to the left of I-A is the first published example of an H-2-linked Ir gene which appears outside of a known I subregion. This raises the possibilities that: (a) Ir genes might be randomly distributed so that not all of them would map in the I region; (b) some Ir genes that are associated with the I-A subregion are situated to the left of the crossover that occurred in the A.TL strain as a result of this particular recombinant event having actually taken place within I-A; or (c) that functionally distinct Ir genes map in different regions of the MHC.

CONCLUSIONS

The results summarized in this report can be most readily interpreted either by the altered self model or by a "modified" dual recognition model--since if two receptors do exist, they appear not to be independent from each other. For the TNP-self system a modified dual recognition model could be valid if it is assumed that one of the receptors recognizes the haptenic moiety plus a portion of the adjacent amino acids. The second receptor would recognize the unaltered H-2 product(s). The development of the proliferative response by human lymphocytes against TNP-modified autologous cells permits an investigation of the role of HLA-coded products as recognition structures in human cellular immune responses, as well as the possible role of Ir genes in such reactions. The fact that the murine TNP-modified syngeneic CML system is under the control of H-2-linked Ir genes as well as being restricted for K and D region products, implies multiple functional roles for H-2-coded gene products in cell-mediated immunity. If the immune phenomena reviewed here are relevant for natural immunity, MHC products may be simultaneously important for recognition by T-lymphocytes (under the control of Ir genes) and for the H-2-restricted or modified products they recognize (the K, I, or D region products associated with the infectious agent).

ACKNOWLEDGEMENTS

We gratefully acknowledge the support of Dr. William D. Terry throughout these studies and thank Pam Gilheany and Matthew Miller for technical assistance and Mrs. Marilyn Schoenfelder for preparation of the manuscript.

REFERENCES

1. Bevan, M.J. (1975) *Nature (Lond)* 256, 419.
2. Blank, K.J., Freedman, H.A., and Lilly, F. (1976) *Nature (Lond)* 260, 250.
3. Forman, J. (1975) *J. Exp. Med.* 142, 403.
4. Gardner, I.D., Bowern, N.A., and Blanden, R.V. (1975) *Eur. J. Immunol.* 5, 122.
5. Gomard, E.V., Dreprez, U., Henire, Y., and Levy, J.P. (1976) *Nature (Lond)* 260, 707.
6. Gordon, R.D., Simpson, E., and Samelson, L.E. (1975) *J. Exp. Med.* 142, 1108.
7. Koszinowski, V., and Thomssen, R. (1975) *Eur. J. Immunol.* 5, 245.
8. Rehn, T.G., Inman, J.K., and Shearer, G.M. (1976) *J. Exp. Med.* 144, 1134.
9. Rehn, T.G., Shearer, G.M., Koren, H.S., and Inman, J.K. (1976) *J. Exp. Med.* 143, 127.
10. Schmitt-Verhulst, A.-M., Sachs, D.H., and Shearer, G.M. (1976) *J. Exp. Med.* 143, 211.
11. Schmitt-Verhulst, A.-M., and Shearer, G.M. (1975) *J. Exp. Med.* 142, 914.
12. Schmitt-Verhulst, A.-M., and Shearer, G.M. (1976) *J. Exp. Med.* 144, 1701.
13. Schrader, J.W., Cunningham, B.A. and Edelman, G.M. (1976) *Proc. Natl. Acad. Sci. USA* 72, 5066.
14. Shearer, G.M. (1974) *Eur. J. Immunol.* 4, 527.
15. Shearer, G.M., Rehn, T.G., and Garbarino, C.A. (1975) *J. Exp. Med.* 141, 1348.
16. Zinkernagel, R.M., and Doherty, P.C. (1974) *Nature (Lond)* 248, 701.

POSSIBLE BIOLOGICAL FUNCTION OF CELL SURFACE STRUCTURES RECOGNIZED BY H-2 RESTRICTED T CELLS

Rolf M. Zinkernagel

Dept. of Cellular & Developmental Immunology, Scripps Clinic & Research Foundation, La Jolla, Calif. 92037

ABSTRACT. Virus-specific cytotoxic T cells are also specific for cell-surface structures coded by the major histocompatibility gene complex in mice and also in rats. In mice *Listeria*-immune T cells are similarly H-2 restricted. However, whereas virus-specific cytotoxic T cells are specific for *K* or *D* coded self markers, *Listeria*-immune T cells are predominantly specific for *I* structures. This differential specificity of effector T cells for distinct cell-surface markers suggests that 1) the ubiquitously expressed *K* and *D* structures are receptors for cytolytic signals to destroy "active" and "ubiquitous" antigens such as viruses and cells expressing foreign transplantation antigens such as tumor antigens; 2) selectively expressed *I* structures are receptors for cell specific differentiation signals that result in macrophage activation, release of lymphokines, antibody production, etc., that can destroy intra- and/or extracellularly "inert" and exclusively with phagocytes associated antigens such as intracellular bacteria or toxins.

Introduction

Many viruses have been tested for generation of cytotoxic T cells in infected mice. For all viruses that produce *in vitro* measurable cytotoxicity, the H-2 restriction has been documented (1,2) similar results are now also available for the rat (3). The H-2 restriction indicates that the immune T cell receptor has two specificities, one for self plus one for antigen, or alternatively it recognizes a complex of antigen plus self or altered self (1-2). At this time none of the experimental data prove or disprove either possibility. Nevertheless, independent of which model is more likely to be correct, some fundamental concepts have emerged recently that may explain T cell specificity for self in mechanistic terms and possibly relate this immunological phenomenon to a universal concept of cellular recognition.

A Mechanistic Function for H-2 Coded Self-Markers

What is the biological role of major transplantation antigen or self-cell-surface structures coded in the major histocompatibility gene complex? Why are virus-specific cytotoxic T cells specific for virus plus H-2K or D and not for virus plus I coded structures? We have speculated earlier on the possible biological role of gene duplication and the genetic polymorphism of major transplantation antigens. More recent results from *in vivo* and *in vitro* studies suggest that the cell surface structures coded in H-2 may be receptors for or fulfill distinct functions.

If the generation of high levels of virus-specific cytotoxic T cells reflects the fact that these T cells play an important role in early and rapid virus elimination, then cytolysis itself must be antiviral. In fact, *in vitro* virus-specific cytotoxic T cells can lyse acutely infected cells during the eclipse phase before virus-progeny assemble and before virus is released from the cells and thus destroy the virus. Once virus has assembled cytotoxic T cells cannot act antivirally any more directly (5, Fig. 1-upper part). In the case of noncytopathogenic viruses which release virus continually, cytotoxic T cells stop further virus production by destroying the virus producing cell.

The evidence from adoptive transfer experiments *in vivo* (6-7), circumstantial evidence from studies on early appearance of target antigens after infection (8), and this direct demonstration *in vitro* strongly suggests that cytolysis of acutely infected cells is a very effective antiviral defense mechanism. If T cells react to virus plus self then the self-marker involved must fulfill two conditions, and this is independent of the two basic T cell-recognition models mentioned previously. First, viruses infect actively phagocytic and, more important, also nonphagocytic host cells within the range of host cells that are susceptible to them. Therefore, the self-marker that is involved in T cell recognition of viruses must be ubiquitous. Second, if cytolytic T cell action is a crucial early defense mechanism against virus then the self-marker must serve as a punch-hole opener of all cells (9). Both these conditions seem to be exquisitely fulfilled by the K and D products of H-2. Considerations similar to events that follow when viruses infect cells can be applied to cellular changes by mutations, i.e. tumor cells, allogeneic cells (reviewed in 11) or cells that were chemically modified (12).

However, unlike these "active" antigens (such as viruses), "inert" antigens, such as intracellular bacteria, toxins, etc., cannot be destroyed by cell lysis since this would simply result in their release

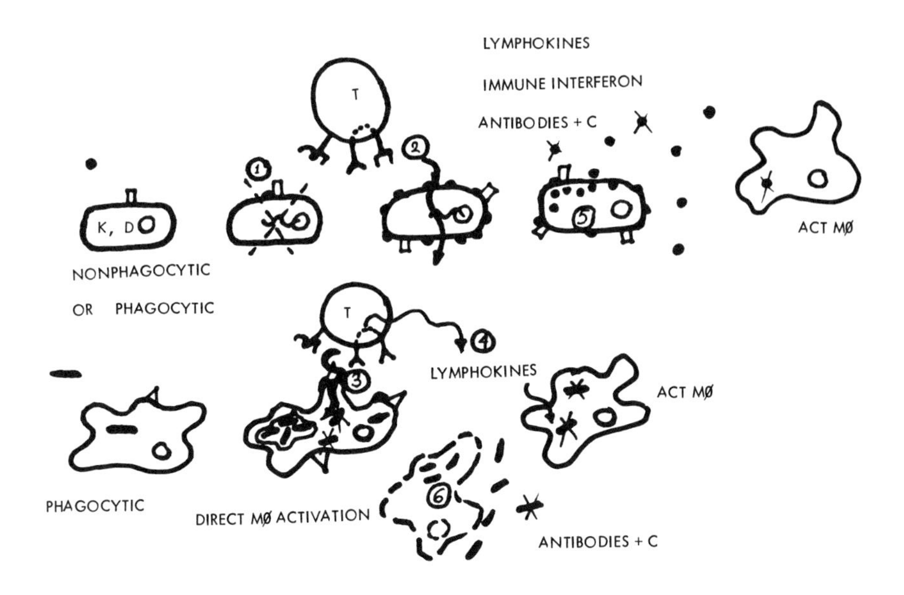

Fig. 1. Explanation of the possible mechanistic role of H-2 coded self-markers in T cell-mediated immunity. T cells can control virus production and/or spreading by nonphagocytic or phagocytic cells in several ways (upper half of Fig.): Virus and H-2K or H-2D specific cytotoxic T cells can lyse acutely infected target cells during the eclipse phase before virus progeny assemble and thus destroy the virus ①. Once virus progeny is assembled and released, cytotoxic T cells cannot act antivirally directly ②. Other antiviral mechanisms include immune interferon, neutralizing antiviral antibodies ⑤ and activated macrophages (15). Lytic interactions between phagocytic cells that have ingested inert antigens such as intracellular bacteria or toxins, would result in release of infectious or toxic antigens ⑥. Such inert antigens could be inactivated either by direct probably I-specific T cell-mediated ③ activation (via the I-structure) of enzymatic activities of the phagocytic cell that has ingested the antigens, or by I-specific triggering of lymphokines that mediate activation ④ of recruited macrophages, which inactivate phagocytosed bacteria or, alternatively, by antibody and complement. Since it is not known whether T cells recognize antigen plus self by a single or dual recognition mechanism, both possibilities are indicated on K, D specific cytotoxic or I specific noncytolytic T cells.

from cells (13, Fig. 1-lower part). In the case of bacteria, the released bacteria would infect surrounding phagocytic cells and the infection would continue. If intracellular bacteria or other inert antigens were to be handled by T cells similar to how virus is handled, then the self-markers recognized by the former T cells would have the following characteristics: First, the self-marker is expressed selectively on phagocytic cells or other restricted cell populations handling such antigens and second, the self-marker would be the receptor for signals that improve the digestive capacity of phagocytes and/or trigger production of antigen-neutralizing substances such as antibodies. Self-markers coded in the H-2I region appear to fulfill these conditions (14). In fact, the adoptive transfer of T helper cells promoting antibody production by B cells and of T cell-mediated immunity to Listeria monocytogenes an intracellular bacterium is restricted predominantly by the H-2I region (13,14).

H-2 coded cell surface structures may, therefore, constitute a variety of partially cell-specific receptors for differential functions that are triggered specifically by T cells. The functions that are triggered through these structures may be regarded as final differentiation steps resulting in activation of enzyme production in phagocytes or production of antibodies (15). In the case of K and D structures it is conceivable that these structures are identical with or linked to ion pumps (K+, Na+, Ca++, etc.) which control the osmolar equilibrium within the cell; triggering of these structures by T cells could thus result in an osmodysregulation that results in cell death. Although unknown as yet, the respective functions of H-2 coded structures in cell-mediated immunity and during phylogeny and/or during differentiation in ontogeny may be related. A better understanding of the differentiation processes may thus well result in the analysis of lymphocyte interactions.

ACKNOWLEDGMENTS

I thank Drs. P.C. Doherty, R.E. Langman, R.V. Blanden and D.H. Katz for many stimulating discussions and Dr. F.J. Dixon for continuous support. The secretarial assistance of Mrs. J. Gouveia in preparing this manuscript is greatly acknowledged. These studies were supported in part by U.S.P.H.S. Grants AI-07007 and AI-12734. This is publication no. 1293 from the Immunology Departments and publication no. 21 from the Department of Cellular & Developmental Immunology, Scripps Clinic and Research Foundation, La Jolla, Calif. 92037.

REFERENCES

1. Zinkernagel, R.M. and Doherty, P.C. (1974) Nature 251, 547.
2. Doherty, P.C., Blanden, R.V. and Zinkernagel, R.M. (1976) Transplant. Rev. 29, 89.
3. Zinkernagel, R.M. and Jensen, F. (1977) Nature, submitted.
4. Doherty, P.C. and Zinkernagel, R.M. (1975) Lancet i, 1406.
5. Zinkernagel, R.M. and Althage, A. (1977) J. Exp. Med. 145, 644.
6. Kees, U. and Blanden, R.V. (1976) J. Exp. Med. 143, 450.
7. Zinkernagel, R.M. and Welsh, R.M. (1976) J. Immunol. 117, 1495.
8. Ada, G.A., Jackson, D.C., Blanden, R.V., Thalha, R. and Bowern, N.A. (1976) Scand. J. Immunol. 5, 23.
9. Forman, J. and Vitetta, E.S. (1975) Proc. Nat. Acad. Sci. USA 72, 3661.
10. Zinkernagel, R.M. and Oldstone, M.B.A. (1976) Proc. Nat. Acad. Sci. USA 73, 3666.
11. Cerottini, J.C. and Brunner, K.T. (1974) Adv. Immunol. 19, 67.
12. Shearer, G.N.,Rehn, T.G. and Schmitt-Verhaust, A. (1976) Transplant. Rev. 29, 222.
13. Zinkernagel, R.M., Althage, A., Adler, B., Blanden, R.V. Davidson, W.F., Kees, U., Dunlop, M.B.C. and Shreffler, D.C. (1977) J. Exp. Med., in press.
14. Katz, D.H. and Benacerraf, B. (1975) Transplant. Rev. 22, 175.
15. Blanden, R.V. (1974) Transplant. Rev. 19, 56.

POSSIBLE IMPLICATIONS OF THE INFLUENZA MODEL FOR T CELL RECOGNITION

Peter C. Doherty, Jack Bennink, Rita B. Effros and William E. Biddison

The Wistar Institute of Anatomy and Biology
36th Street at Spruce
Philadelphia, Pennsylvania 19104

ABSTRACT. Mice primed with serologically distinct type A influenza viruses generate at least two populations of cytotoxic T cells--the one specific for the virus used to immunize, the other highly cross-reactive for target cells infected with other type A viruses. These cross-reactive T cells may be recognizing a shared determinant. A possible candidate is the virion matrix protein, which is very hydrophobic and is the major structural component of the virus particle. Current serological studies indicate that the matrix protein is also expressed on the surface of the virus-infected cell, though this was thought previously not to be the case. The possible consequences of cytotoxic T cells preferentially recognizing a virus determinant presented as a minor antigen on cell membrane are discussed in the context of the altered-self hypothesis.

INTRODUCTION

We originally proposed two alternative models to explain our finding that recognition of virally modified cells by virus-immune cytotoxic T cells involves the H-2 gene complex (1). These were designated "intimacy" and "altered-self." The basis of the first idea was that interaction between like H-2 antigens (or structures coded for by closely linked genes) is essential for the functional association of lymphocyte and target. The second considered that the T cell receptor is specific for a complex of viral and H-2 antigens, or recognizes some neoantigen whose expression is determined by both host (H-2) and viral genomes.

Subsequent experiments have tended to exclude the intimacy concept (2). Studies using radiation chimeras (3-5) and negative selection protocols (D. Wilson and J. Sprent, personal communication) have established beyond reasonable doubt that the T cell and target cell need not phenotypically express the same H-2 gene products. The requirement for H-2 identity is between stimulator environment and target cell. Effector T cells are specific for the spectrum of H-2 (self) and non-self encountered during sensitization.

The altered-self model has always suffered, however, from the difficulty of proposing a molecular basis for the association of virus [or minor H antigens (6, 7)] and H-2 on the cell membrane. Further problems for explanatory models which propose the generation of novel antigenic determinants were raised by the experiments of Binz and Wigzell (8), Eichmann and Rajewsky, and their colleagues which indicate that the T cell receptor expresses the same idiotype as the corresponding B cell (9). Hypotheses were thus advanced to accomodate the evidence about idiotypes with the H-2 restriction phenomenon (10-12). Is specificity for self and non-self mediated via two V_H gene products, the one reactive to H-2, the other to virus? Could these operate as two distinct recognition units, or is there a need for assembly into a single, high-affinity structure which binds to a complex of virus and H-2 antigen?

These ideas should not, however, be treated as anything more than working hypotheses. There is an obvious need for further information about T cell receptor idiotype in other experimental systems, especially those which involve surveillance of self. Binz and Wigzell (8) are studying alloreactive T cells, which may comprise a functionally distinct subset. Eichmann and Rajewsky are considering helper activity: evidence for antigen binding is found for both helper and suppressor factors which contain H-2I-region determinants (13, 14). The attractions of a unifying concept are obvious, but other possibilities are not yet excluded.

Our current work is concerned with trying to establish which viral components are involved in T cell-mediated cytotoxicity. Analysis using influenza virus has provided interesting, and somewhat unexpected, results (15, 16). The intention here is to discuss T cell recognition in the context of these findings.

THE INFLUENZA VIRUSES

Nature provides, in the influenza viruses, a unique system for the analysis of T cell specificity. The influenza A viruses share almost identical internal components (17), the matrix (M) protein and ribonucleoprotein (RNP), but express serologically different hemagglutinin and neuraminidase antigens. No cross-neutralization is recognized between the various major subtypes (designated H0N1, H2N2 and H3N2) which have been isolated from man over the past 48 years. Influenza A and B viruses grown in the chick allantois express a common carbohydrate moiety of host origin, but are otherwise serologically distinct. It is thus possible to study a range of viruses with similar molecular biology, but with differing spectra of antigenicity.

HETEROGENEITY OF THE CYTOTOXIC T CELL RESPONSE

Mice which are primed either intraperitoneally with a large dose of influenza virus or intranasally with a much smaller dose (which causes pneumonia) generate potent cytotoxic T cell populations (15, 16). Maximal cell-mediated lysis (CML) is recognized at 5 to 7 days after inoculation, depending on the route of exposure and the dose of virus given. Reciprocal specificity of CML is found between influenza A virus, influenza B virus, rabies virus, and lymphocyte choriomeningitis virus. There is, however, extensive cross-reactivity in the cytotoxic T cell response to different influenza A viruses (15, 16, 18), even though the surface antigens of these viruses are quite distinct serologically.

Use of "cold target" competitive inhibition protocols has indicated that at least two populations of influenza A immune effector T cells are generated in the primary response (15, 16), the one specific for the virus which was used to immunize, the other highly cross-reactive with other type A viruses. Secondary stimulation (after one month) with a different type A virus results in restimulation of the cross-reactive T cells, CML being demonstrable within 24 hours of challenge.

The effector T cell response thus shows, when analyzed using very similar viruses, a spectrum of specificity which would not have been predicted on the basis of serum antibody patterns. Mice primed by the technique used to develop CML do not generate antibody populations which bind to target cells infected with heterologous type A influenza viruses (15). This is the first time that such a divergence has been demonstrated. What are these T cells recognizing?

T CELL RECOGNITION AND THE VIRUS-INFECTED TARGET CELLS

The virus-specific T cells. Defining T cell recognition other than by idiotypic analysis of the receptor depends essentially on generating useful correlations with antigen expression. We know (by serological criteria) that the major virus component present on the surface of the infected target cell is the hemagglutinin, variations in which essentially define the virus subtype. The existence of T cell populations which are apparently specific for the immunizing virus might be thought to indicate that the hemagglutin is being recognized. However, we have not, to date, been able to support this idea by comparing viruses expressing closely related, or very distant, hemagglutinins.

Attempts at blocking T cell recognition with antibody to hemagglutinin have not yet been successful. These blocking experiments were done using target cells that had been lightly fixed with glutaraldehyde (19), so that virus expression is presumably constant and no new antigen is synthesized. Also, no diminution of CML is seen when cytotoxic T cell assays are incubated in the presence of excess tumor cells secreting large amounts of antibody to the virus hemaglutinin antigen (a somatic cell hybrid developed by Drs. H. Koprowski, C. Croce, and W. Gerhard). Furthermore, addition of excess erythrocytes (100:1) to fixed, virus-infected target cells does not inhibit T cell effector function (Y. Goldfeld, unpublished data), even though the hemagglutinin binds to sialic acid residues on the red cells [hemadsorption phenomenon (20)]. Negative findings are obviously unsatisfactory, but it is evident that we have not succeeeded so far in establishing a role for the viral hemagglutinin in T cell recognition.

The cross-reactive T cells. Most of our current work is concerned with the cross-reactive T cell populations. There are two reasons for this preoccupation: the divergence between specificity patterns determined by CML and antibody is surprising and needs to be explained; very potent cross-reactive lymphocyte populations are readily generated (both *in vitro* and *in vivo*) by secondary stimulation with heterologous influenza A viruses.

We know from the experiments of J. Forman (21), G. Shearer, and A. Schmitt-Verhulst (personal communication) that T cells distinguish between target cells modified with TNP and DNP. Could T cell recognition in influenza be directed against shared "haptenic" determinants on the different hemagglutinin antigens? The problem with this idea is that there is no obvious reason for the failure to generate antibody against such structures, as they would always be presented in the context of a non-self carrier (i.e., other components of the hemagglutinin). Furthermore, findings from base sequence analysis of genes coding for the hemagglutinin tend not to support this possibility (Dr. C. Scholtissek, personal communication). The demonstration that there is specificity for TNP and DNP also argues against the concept (22) that binding sites of high affinity for one hemagglutinin but low affinity for another might, when arranged as multiple units on the surface of a lymphocyte, form a recognition complex which is much more cross-reactive (for repeated virus determinants on a target cell) than free antibody.

Do these T cells recognize shared antigens, expressed on the surface of cells infected with all type A influenza viruses? Obvious candidates are the internal M protein and RNP components. Both are highly hydrophobic molecules and have been thought not to be expressed on the outer surface of the virus, or of the virus-infected cell (17). Recent experiments indicate, however, that RNP is presented externally on the cell membrane (23). Furthermore, hyperimmune goat antisera prepared by Dr. R. G. Webster against highly purified M protein of an influenza A virus are, in our laboratory, showing high titers (by complement-mediated lysis) against target cells infected with a range of influenza A viruses, but not for cells infected with a type B virus (B/Lee). Present experiments are concerned with attempting secondary stimulation of virus-immune T cells with purified M protein and hemagglutin (supplied by Dr. Webster), and with trying to block effector T cells with antisera to the various components of the virus. No clear pattern is yet emerging.

SPECULATIONS ABOUT THE ALTERED-SELF HYPOTHESIS

Studies with the oncornavirus model suggest that T cell recognition may be directed against a group-specific protein (p30), which is an internal component of the virion (24). Recent analysis has shown that a glycoprotein with the antigenicity of p30 is also present on the cell surface (25). We cannot yet say that an analogous situation exists for the influenza A viruses, but the cross-reactivity in the cytotoxic T cell response indicates that such may be the case. If so, why should a major component of the cytotoxic T cell response be directed at what is apparently (at least so far as the antibody response during infection is concerned) a minor antigen on the cell membrane? The implications of this possibility are worth considering.

It is useful to summarize current ideas of how influenza viruses are constituted (17). The hemagglutinin spikes on the virus are trimers, made up of three identical subunits (20). Each of these subunits has a heavy chain (HA1) and a light chain (HA2) joined by disulphide bonds, and is thought to orient in the cell membrane via the hydrophobic region of HA2. During virus synthesis the HA spikes assemble on the outer aspect of the cell plasma membrane. A layer of M protein accumulates beneath this region, forming a site for localization of RNP. The virus then buds from the cell surface, and the plasma membrane fuses where the released virion pinches off. Host proteins are excluded from the intact virus particle, even though the viral membrane and cell plasma membrane are contiguous throughout. Choppin and Compans (17) have speculated that an essential function of the M protein may be to maintain a localized viral domain within the plasma membrane. This implies that the M protein is able to affect the re-distribution of host membrane proteins, including H-2 antigens. There is thus some conceptual basis (other than the H-2 restriction phenomenon) for considering that the M protein and H-2 antigen may interact.

Analogies exist between H-2 antigens, the Ig molecule, and the influenza virus hemagglutin. Each is presented as an independent entity within the cell plasma membrane. All have well recognized binding functions: to β_2 microglobulin, antigens, and sialic acid, respectively. The primary antibody response to the virus hemagglutinin apparently requires helper T cell function (26). Delayed-type hypersensitivity to Ig (fowl γG) shows a requirement for association with H-2I (27). Recognition of this class of molecules by T cells may thus involve the I-region of the H-2 complex.

Cytotoxic T cell activity against hapten-modified cells is associated with H-2K and H-2D (28). The TNP molecule binds directly to amino groups on the H-2 molecule (29). There is thus a direct physical attachment of non-self to self. Antibody to hapten is not readily generated in the context of this self-carrier (30). The same may be true for some virus components, such as the M protein, which elicit little (if any) antibody response during viral infection. Association with H-2K or H-2D molecules may cause proliferation of cytotoxic rather than helper T cells. The mechanism would, at least from a teleological aspect, be advantageous in that the possibility of T cell surveillance being blocked by antibody (with resultant immunological enhancement) should be minimized.

ACKNOWLEDGEMENTS

We thank Dr. W. Gerhard and Dr. M. Halpern for useful discussions. This work was funded by grants from the National Institutes of Health and the National Multiple Sclerosis Society.

REFERENCES

1. Zinkernagel, R. M. and Doherty, P. C. (1974) Nature (Lond.) 251, 547.
2. Doherty, P. C., Blanden, R. V. and Zinkernagel, R. M. (1976) Transplant. Rev. 29, 89.
3. Pfizenmaier, K., Starzinski-Powitz, A., Rodt, H., Röllinghoff, M. and Wagner, H. (1976) J. Exp. Med. 143, 999.
4. Zinkernagel, R. M. (1976) Nature (Lond.) 261, 139.
5. Von Boehmer, H. and Haas, W. (1976) Nature (Lond.) 261, 141.
6. Bevan, M. J. (1975) Nature (Lond.) 256, 419.
7. Gordon, R. D., Simpson, E. and Samelson, L. E. (1975) J. Exp. Med. 142, 1108.
8. Binz, H. and Wigzell, H. (1975) J. Exp. Med. 142, 1218.
9. Black, S. J., Hämmerling, G. J., Berek, C., Rajewsky, K. and Eichmann, K. (1976) J. Exp. Med. 143, 846.
10. Doherty, P. C., Götze, D., Trinchieri, G. and Zinkernagel, R. M. (1976) Immunogenetics 3, 517.
11. Janeway, C. A., Jr., Wigzell, H. and Binz, H. (1976) Scand. J. Immunol. 5, 993.
12. Zinkernagel, R. M. and Doherty, P. C. (1976) Cold Spring Harbor Symp. Quant. Biol. 41, in press.
13. Erb, P., Feldmann, M. and Hogg, N. (1976) Eur. J. Immunol. 6, 365.
14. Taniguchi, M., Hayakawa, K. and Tada, T. (1976) J. Immunol. 116, 542.
15. Effros, R. B., Doherty, P. C., Gerhard, W. and Bennink, J. (1977) J. Exp. Med. 145, 557.
16. Doherty, P. C., Effros, R. B. and Bennink, J. (1977) Proc. Natl. Acad. Sci. USA, in press.
17. Choppin, P. W. and Compans, R. W. in "The Influenza Viruses and Influenza," E. D. Kilbourne, ed., Academic Press, N.Y., 1975, p. 15.
18. Zweerink, H. A., Courtneidge, S. A., Skehel, J. J., Crumpton, M. J. and Askonas, B. A. (1977) Manuscript submitted for publication.
19. Bubbers, J. E. and Henney, C. S. (1975) J. Immunol. 114, 1126.
20. Schulze, I. T. in "The Influenza Viruses and Influenza," E. D. Kilbourne, ed., Academic Press, N.Y., 1975, p. 53.
21. Forman, J. (1976) Transplant Rev. 29, 146.
22. Doherty, P. C., Effros, R. B. and Bennink, J. (1977) Perspectives in Virology 10, in press.
23. Virelizier, J. L., Allison, A. C., Oxford, J. S. and Schild, G. C. (1977) Nature (Lond.) 226, 52.

24. Bruce, J., Mitchison, N. A. and Shellam, G. R. (1976) Int. J. Cancer 17, 342.
25. Nowinski, R. C. and Watson, A. (1976) J. Immunol. 117, 693.
26. Burns, W. H., Billups, L. C. and Notkins, A. L. (1975) Nature (Lond.) 256, 655.
27. Miller, J. F. A. P., Vadas, M. A., Whitelaw, A. and Gamble, J. (1976) Proc. Natl. Acad. Sci. USA 73, 2486.
28. Shearer, G. M., Rehn, T. G. and Schmitt-Verhulst, A. -M. (1976) Transplant. Rev. 29, 223.
29. Forman, J., Vitteta, E. S., Hart, D. A. and Klein, J. (1977) J. Immunol., in press.
30. Scott, D. W. and Long, C. A. (1976) J. Exp. Med. 144, 1369.

H-2/VIRAL PROTEIN INTERACTION AT THE CELL MEMBRANE AS THE BASIS FOR *H-2*-RESTRICTED T-LYMPHOCYTE IMMUNITY

Kenneth J. Blank, J. Eric Bubbers and Frank Lilly

Department of Genetics, Albert Einstein College of Medicine
Bronx, New York 10461

ABSTRACT. Cytotoxic T lymphocytes from congenic BALB.B *(H-2^b)* and BALB.G *(H-2^g)* mice immunized with syngeneic Friend virus (FV)-induced tumor cells displayed *H-2* restriction of their cytotoxic activity. For FV-induced target tumor cells to be killed, it was sufficient that they express the *H-2D^b* molecules. In addition, FV particles from the serum of infected (BALB.B x BALB/c)F_1 mice, were found to have incorporated only *H-2D^b* (and not *H-2D^d*, *H-2K^b* or *H-2K^d*) molecules from the membranes of cells from which they had matured by budding. To explain these results, we propose that a molecular complex is formed at the membrane surface between *H-2D^b* molecules and certain virus molecules, and that formation of this complex (1) is responsible for the incorporation of *H-2D^b* antigenic specificities into Friend virions, and (2) creates antigenic specificities which are recognized by cytotoxic T lymphocytes.

INTRODUCTION

In mice a relationship exists between *H-2* type and resistance to disease induced by Gross (1), Friend (2) and lymphocytic choriomeningitis viruses (3). Although no mechanism for this association has been firmly identified, it has been suggested that resistance is mediated by the expression of *H-2*-linked immune response genes which promote the generation of effective immune responses to virus-associated antigens present on the surfaces of infected cells (3,4). Recently there has been speculation that T lymphocytes from mice immunized with virus-infected cells do not respond to virus antigens per se but rather to antigenic specificities expressed on "altered self" molecules (5). One version of this hypothesis is that the "altered self" molecule is generated on the surface of a virus-infected cell by an interaction between a virus molecule and an *H-2* molecule. This idea arose from the observation that cytotoxic T lymphocytes generated in a mouse during an immune response to virus displayed *H-2* restriction with respect to their ability to kill virus-infected target cells from allogeneic mice, i.e., these immune T cells killed only those virus-infected target cells which shared *H-2K* and/or *H-2D* regions with the immunized mice (5,6).

Here we shall summarize recent evidence obtained in the Friend virus (FV) system in support of the hypothesis that *H-2* restriction is a consequence of *H-2*/viral glycoprotein interactions on the cell surface. The existence of such interactions is suggested by our finding that *H-2* molecules are incorporated in a nonrandom fashion into virions collected from the serum of FV-infected mice (7). Moreover, since we find a correlation between the incorporation of specific *H-2* molecules into Friend virions, on the one hand, and the specificity of the *H-2* restriction of T killer cells generated in the system, on the other hand, we suggest that the common factor is the formation on cell surfaces of *H-2*/viral protein molecular complexes.

RESULTS

H-2-restricted, T cell-mediated killing of cells of FV-induced tumor lines. We have previously described the *H-2*-restricted, T cell-mediated killing of FV-induced tumor cells derived from mouse strains congenic with respect to the *H-2* complex (8). Briefly, these experiments have shown that in a lymphocyte-mediated cytotoxicity (LMC) assay *in vitro*, T cells from mice immunized with cells from syngeneic FV-induced tumor lines of the HFL series (9) kill only those cells used as targets which express the same *H-2* haplotype as those FV-induced tumor cells used for immunization.

Table 1. H-2 Restriction of T-Cell Mediated Killing

Source of immune PEC	H-2 haplotype				Syngeneic immunizing cells	Target cells (% ^{51}Cr release)			
	K	I	S	D		HFL/b	HFL/g	HFL/d	HFL/k
BALB.B	b	b	b	b	HFL/b	18	15	0	0
BALB.G	d	d	d	b	HFL/g	20	37	1	0
BALB/c	d	d	d	d	HFL/d	0	0	0	0

* For methods see reference 8

Thus, in this entirely congenic system, PEC obtained from either BALB.B or BALB.G mice after secondary immunization with syngeneic FV-induced HFL/b or HFL/g tumor cells, respectively, kill syngeneic ^{51}Cr-labelled cells of both the HFL/b and HFL/g but not the HFL/d lines (Table 1). Cells of all these tumor cell lines express both the FMR antigen and the virus envelope antigens associated with the gp70 molecule.

Pretreatment of immune PEC with anti-Thy-1.2 and complement abolishes the specific killing of syngeneic tumor cells.

Cross reactive antigens recognized by immune T cells on the surface of FV-infected tumor cells and their relation to the *H-2* complex. In the present studies were concerned with defining the specific region(s) of the *H-2* complex responsible for the observed *H-2* restriction. As shown in Table 1, PEC from BALB.G mice, whose $H\text{-}2^g$ haplotype is a product of recombination between the $H\text{-}2^d$ and $H\text{-}2^b$ haplotypes, immunized with syngeneic HFL/g tumor cells were able to kill both syngeneic HFL/g target cells as well as HFL/b target cells which share only the $H\text{-}2D^b$ region with HFL/g. However, these BALB.G anti-HFL/g PEC were not able to kill HFL/d tumor cells which share the *K, I,* and *S* regions with HFL/g. Therefore, it appears that identity with respect to the $H\text{-}2D^b$ but not the $H\text{-}2K^d$ region between the immunizing and the target cells was sufficient for the occurrence of T cell-mediated killing.

Similarly,since PEC from BALB.B mice immunized with HFL/b tumor cells were able to kill HFL/g target cells, it also appeared that in this system identity between the cells used for immunization and the cells used as targets with respect to the *H-2D* region was sufficient for T-cell killing to occur. However, since no FV-infected cell line was available which expressed antigens associated with the $H\text{-}2K^b$ region but not the $H\text{-}2D^b$ region, this experimental approach could not be employed to determine if $H\text{-}2K^b$ region identity between the immunizing and target cells would be sufficient for T cell killing to occur. No T-cell cytotoxicity has been detected among PEC from BALB/c mice immunized with syngeneic HFL/d cells.

It has previously been shown that T cell-mediated killing of virus infected cells could be blocked by adding high concentrations of specific *H-2* antiserum to the reaction medium (10). Furthermore, the inhibitory effects of the antiserum were found to be exerted upon the target cells and not the T-killer cells. We thus attempted to block the killing of HFL/b target cells by immune BALB.B PEC with the addition of *H-2* antisera or antisera against various virus components to the LMC assay reaction medium. The results, shown in Fig. 1, indicate that an antiserum raised against antigenic specificities associated with the $H\text{-}2D^b$ region inhibited the lysis of HFL/b cells whereas antiserum raised against specificities associated with the $H\text{-}2K^b$ region had no greater inhibitory effect on killing than did an antiserum specific for irrelevant *H-2* specificities. We also attempted to block cytolysis with antisera raised against various virus

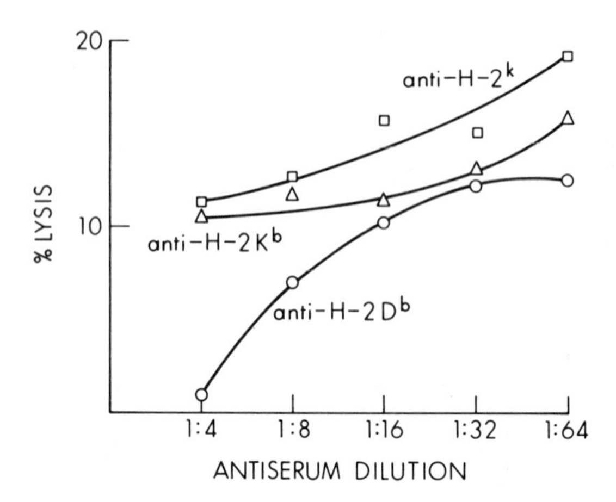

Fig. 1 Lysis of HFL/b target cells by immune BALB.B PEC in the presence of anti-H-2 sera.

components. A xenoantiserum raised against the gp70 molecule significantly inhibited T-cell killing, whereas antiserum to the p30 molecule did not block this activity. However, only one pool of goat anti-mouse gp70 out of three tested demonstrated this inhibitory activity (data not shown).

From these experiments, therefore, it appears that T-cell recognition of foreign determinants on these FV-induced tumor cells was directed against specificities associated with both the $H\text{-}2D^b$ molecule and a viral molecule, presumably gp70. The $H\text{-}2K^b$ molecule, on the other hand, appears not to be involved in this phenomenon.

Non-random incorporation of *H-2* molecules into progeny FV virions from infected hosts. Since both viral molecules and cellular $H\text{-}2D^b$ molecules seemed to be involved in the antigenic specificities recognized by T lymphocytes on FV-induced tumor cells, a physical association between these two molecules might be the basic phenomenon giving rise to our observations. If such an association exists on the plasma membrane at the site of virus assembly, then the *H-2* molecule involved might be incorporated into the virion during the budding process. To detect the possible incorporation of *H-2* antigen into virus particles, we examined preparations of NP-40-solubilized virions for their capacity to inhibit the cytotoxicity of monospecific *H-2* antisera. Virus particles form the serum of FV-infected (BALB.B x BALB/c)F_1 mice ($H\text{-}2^b/H\text{-}2^d$) were analyzed in this manner. As demonstrated in Fig. 2, virus preparations from these mice failed to inhibit the complement-dependent cytotoxicity of anti-$H\text{-}2K^b$, anti-$H\text{-}2D^d$ or anti-$H\text{-}2K^d$

antisera tested on normal lymphoid target cells. Anti-*H-2D^b^*-mediated lysis however, was markedly inhibited in the presence of solubilized virus preparations but not in the presence of control preparations. Of the four host *H-2* specificities which might have been expected to be incorporated into progeny virus, only *H-2D^b^* was detected. In control experiments (not

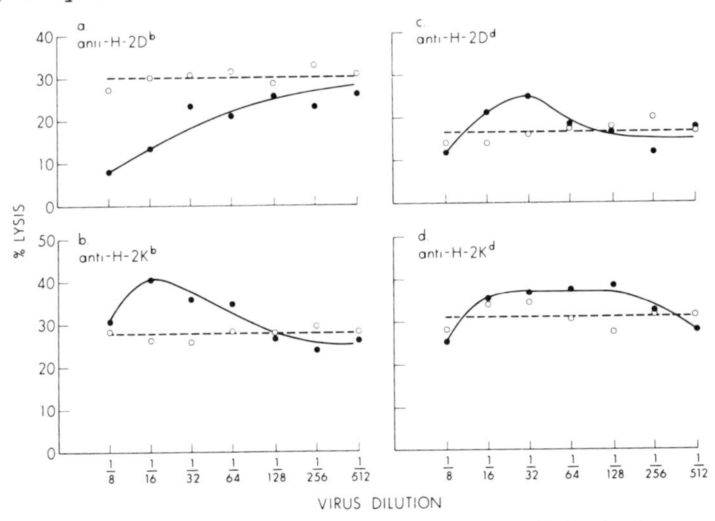

Fig. 2 Anti-H-2 antibody mediated target cell lysis in the presence of NP-40 solubilized Friend virus from the serum of infected (BALB.B x BALB/c)F_1 mice. BALB.B ($H\text{-}2^b$) or BALB/c ($H\text{-}2^d$) mesenteric lymph node cells were lysed by anti-H-2 sera monospecific for K- or D-end antigens in the presence of serially diluted virus (-●-) or control (--o--) preparations. For details see reference 7. Figure reprinted with permission from Nature.

shown), the *H-2D^b^* antigen was detected only in preparations containing virus from the serum of mouse strains possessing the *H-2D^b^* allele. Similarly, virus preparations prepared from infected BALB.B and BALB.G mice inhibited the lytic activity of anti-$H\text{-}2D^b$ but not anti-$H\text{-}2K^b$ or anti-$H\text{-}2K^d$ whereas preparations from FV-infected BALB/c mice failed to inhibit the activity of either anti-$H\text{-}2K^d$ or anti-$H\text{-}2D^d$ antisera.

Adoptive transfer of immunity to primary FV infection by T cells. Since the studies involving the *H-2* restriction of T-cell killing were done with cultured tumor cells, whereas the incorporation of *H-2* molecules into virus particles was demonstrated in vivo, we were concerned with correlating data derived from these two systems. We attempted, therefore, to demonstrate that the antigenic specificities on the cultured tumor cells recognized by immune T cells were the same as those on primary infected spleen cells by performing adoptive transfer experiments. Immune spleen cells from donor mice which had rejected syngeneic HFL tumor cells were injected into non-immune syngeneic recipient mice which then received a high dose of FV. Recipients of non-immune cells developed splenomegaly characteristic of FV-induced disease, whereas recipients of immune cells were refractory to FV (Table 2). This resistance to primary FV infection was mediated by T cells from immune donor mice since pretreatment of the spleen cells with anti-Thy-1.2 and complement abolished their capacity to transfer immunity.

Table 2. Adoptive Transfer of Immunity to FV by Syngeneic Spleen Cells From Mice Immunized With Syngeneic HFL Tumor Cells

Irradiated recipient+	Mean spleen weights (gm ± S.D.)		
	A*	B	C
BALB.B	0.15 ± .01	2.31 ± .27	
BALB.G	0.10 ± .01	1.41 ± .21	2.33 ± .007

+ Recipient mice were lethally irradiated (850R) and reconstituted with syngeneic spleen cells: 24 hrs later the mice received 3000 SFFU of FV iv.

* Group A received immune syngeneic spleen cells pretreated with NMS plus complement; Group B were given immune syngeneic spleen cells pretreated with anti Thy-1.2 serum plus complement; Group C received syngeneic spleen cells from unimmunized mice.

Survival of mice injected with syngeneic FV-induced tumor cells. In an attempt to correlate the physical *H-2*/virus molecule interaction implied by the selective incorporation of *H-2* molecules into virus with the presence of antigenic specificities on FV-induced tumor cells which would cause graft rejection, we inoculated BALB.B, BALB.G and BALB/c mice with syngeneic HFL tumor cells in order to observe their ability to reject these tumor grafts. As shown in Table 3, 100% of BALB.B and 76% of BALB.G mice given syngeneic tumor cells survived these grafts. BALB/c mice inoculated with syngeneic HFL/d cells, however, showed little capacity to reject this tumor, since only about 8% of the animals survived. Thus a strong antigen capable of inducing syngeneic graft rejection appears to exist on tumor cells carrying the $H\text{-}2D^b$ molecule, and this finding correlates with the selective incorporation of $H\text{-}2D^b$ molecules into virus particles.

Table 3. Survival of Mice Challenged With Syngeneic Tumor Cells

Recipient* mice	Tumor cells	Cell dose	No. survivors/ No. inoculated
BALB.B	HFL/b	5×10^6	57/57
BALB.G	HFL/g	5×10^6	16/21
BALB/c	HFL/d	5×10^6	1/12
		5×10^5	2/14

* Mice were inoculated ip with syngeneic tumor cells and observed for 8 weeks.

DISCUSSION

The correlation between *H-2* type and resistance to virus-induced diseases has previously been thought to be the result of the phenotypic expression of immune response genes. The evidence presented in this paper suggests an alternative explanation for this *H-2*/disease association, although it does not exclude a role for immune response genes. We have demonstrated that,in FV-infected mice expressing certain *H-2* haplotypes, *H-2* molecules are selectively incorporated into progeny virions. This incorporation is presumably the result of an interaction between the cellular *H-2* molecule and a major structural component of the virus during assembly of the completed virus particle by budding from the cell surface. However, it appears that not all *H-2* molecules have the same ability to interact with virus molecules. As we have demonstrated, this affinity evidently varies between *H-2K* and *H-2D* molecules as well as between different *H-2D* molecules.

T-cell killing is probably not the only component of the immune response to FV which plays a role in resistance or susceptibility to the FV disease. Nevertheless, our studies suggest that, within the *H-2*-congenic BALB/c family of strains, this response is the decisive factor. HFL cells possessing the $H\text{-}2D^b$ but not the $H\text{-}2D^d$ genotype readily induce cytotoxic T cells when used to inoculate syngeneic mice. The basis for this immunogenicity appears to rise from the property of $H\text{-}2D^b$ but not $H\text{-}2D^d$ molecules to associate with viral molecules (perhaps gp70) in the cell membrane. This molecular association has two consequences: (1) *H-2D*^{b} molecules are selectively included into assembling virions, whereas *H-2D*^{d} molecules are excluded, and (2) molecules associated with *H-2D*^{b} molecules acquire the capacity to elicit T-cell cytotoxicity, a capacity which they lack in the presence of *H-2D*^{d} molecules with which no or only a weak association occurs. Neither the *H-2K*^{b} nor the H-2K^d molecule appears to show an analogous capacity to form molecular associations with viral gene products, but preliminary evidence from our laboratory suggests that, in congenic *H-2*^{k} mice, the *H-2K*^{k} molecule may process an affinity for FV molecules analogous to that of *H-2D*^{b} molecules.

REFERENCES

1. Lilly, F. (1966) *Genetics* 53, 529.
2. Lilly, F. (1968) *J. Exp. Med.* 127, 465.
3. Oldstone, M.B.A., Dixon, F.J., Mitchell, G.F., and McDevitt, H.O. (1973) *J. Exp. Med.* 137, 1201.
4. Lilly, F., and Pincus, T. (1973) *Adv. Cancer Res.*17, 231.
5. Doherty, P.C. and Zinkernagel, R.M. (1974) *Nature* 251, 547.
6. Zinkernagel, R.M., and Doherty, P.C. (1974) *Nature* 248, 701.
7. Bubbers, J.E., and Lilly, F. (1977) *Nature* 266, in press.
8. Blank, K.J., Freedman, H.A., and Lilly, F. (1976) *Nature* 250, 260.
9. Freedman, H.A., and Lilly, F. (1975) *J. Exp. Med.*142, 212.
10. Schrader, J.W., and Edelman, G.M. (1976) *J. Exp. Med.* 143, 601.

IN VIVO INDUCTION OF H-2 RESTRICTED CYTOTOXIC EFFECTOR CELLS IS EITHER NOT H-2 RESTRICTED OR OCCURS VIA HOST PROCESSED ANTIGEN

Michael J. Bevan* and Polly Matzinger†

*Center for Cancer Research, Massachusetts Institute of Technology, Cambridge, Massachusetts 02139, and
†Department of Biology, University of California, San Diego, California 92093.

ABSTRACT. F_1(BALB/c x BALB.B)(H-$2^{d/b}$) mice were immunized in vivo against cells bearing minor histocompatibility differences and the specificity of the induced cytotoxic T cells was studied in vitro in short term ^{51}Cr-release assays. The immunizing cells were from either B10(H-2^b) or B10.D2(H-2^d) mice. These mice carry the same genetic minor H differences from the F_1 responder but in combination with either the responder's paternal (B10) or maternal (B10.D2) H-2 haplotype. We show here that mice primed and boosted in vivo with B10.D2 cells generate an effector population capable of lysing both B10 and B10.D2 targets. Cold target competition experiments, in which the lysis of ^{51}Cr-labelled cells was inhibited by the addition of unlabelled cells, established that separate clones of cytotoxic T cells lysed either target, i.e. the effector cells were H-2 restricted recognizing either B10 minors-plus-H-2^d or B10 minors-plus-H-2^b. Therefore it seems that the B10.D2 immunogen can generate cytotoxic cells in vivo capable of lysing B10 targets even though B10.D2 cells show no detectable binding to these effectors. The results are discussed in terms of the "one receptor" (altered self) and "two receptor" models of H-2 restriction.

INTRODUCTION

Cytotoxic effector T cells directed against membrane antigens which have been induced by viruses (1), chemical modification (2) or are due to minor histocompatibility (H) differences between the responder and stimulator (3,4) are specific for the minor antigen and for membrane structures coded in the major H complex (H-2K and H-2D in mice). This is the finding referred to as "H-2 restriction of T cell function." It means that cytotoxic T cells of BALB/c(H-2^d) mice immunized against the multitude of non-H-2 coded minor H differences carried by B10.D2(H-2^d) cells (5,6) can lyse B10.D2 targets but do not lyse B10(H-2^b) or B10.BR(H-2^k) targets (3). It has been shown that following immunization with antigen X, an H-2 heterozygous mouse has two pools of cytotoxic effector cells; one pool is specific for X-plus-maternal-H-2 and the other is specific for X-plus-paternal-H-2 (1,2,3). For example, F_1(BALB/c x BALB.B), (F_1(C x C.B),H-$2^{d/b}$) mice immunized against F_1(B10 x B10.D2)(H-$2^{b/d}$) cells have two pools of effector cells one of which lyses B10 but not B10.D2 targets, and the other recognizes B10.D2 but not B10 targets.

If such an F_1 is primed by an in vivo injection of B10 cells and the spleen cells boosted in tissue culture with B10 again then the killer cells lyse B10 targets but not B10.D2 targets (3). That is, when the secondary (boosting) immunization is done in vitro, then the only pool of effector cells induced is the one which recognizes the minor antigen plus the H-2 type of the boosting cells. In contrast to this H-2 restricted boosting in vitro, the priming done in vivo is not restricted to the H-2 type of the injected cells (7,8). For example, injecting B10 cells into an F_1(C x C.B) primes the animal for an augmented secondary response to B10.D2 as well as to B10 (7). In this system it has been shown that an F_1 animal primed with B10 has expanded memory pools of cytotoxic cells specific for B10.D2 and B10 (9).

In the work presented here we have done all the immunizations in vivo and assayed the cytotoxic effector cells harvested from the peritoneal cavity and spleen. We show that, unlike in vitro immunizations, the induction of effector cells in vivo is not restricted to the H-2 type of the immunizing cells.

RESULTS

F_1(C x C.B) mice which had been primed by an injection of B10.D2 or B10 spleen cells some months previously were boosted intraperitoneally (i.p.) with B10.D2 spleen cells. Four days later the spleen and peritoneal exudate (PE) were assayed for lysis of various target cells (Table 1). Either group of mice boosted i.p. with B10.D2 spleen cells contained cytotoxic T cells which lysed B10 and B10.D2 targets. The effector cells did not lyse F_1(C x C.B) targets (nor C and C.B targets, Matzinger and Bevan, submitted for publication) showing that lysis of B10 targets was not due to a graft-vs.-host reaction by the injected B10.D2 cells against the H-2^b antigens of the host. Also, B10.BR(H-2^k) targets were not lysed. (B10.BR is congenic with B10 and B10.D2 and therefore carries genetically the same non-H-2 coded minor H antigens).

TABLE I

F_1(C x C.B) CYTOTOXIC T CELLS AGAINST MINOR H ANTIGENS INDUCED IN VIVO[1]

Primed with	Boosted with	Killer cells	% lysis of ^{51}Cr-targets F_1(C x C.B)	B10	B10.D2	B10.BR
B10.D2	B10.D2[2]	Spleen	-8	14	23	-4
B10.D2	B10.D2[2]	PE	-4	51	44	-2
B10	B10.D2[3]	Spleen	N.D.	34	20	-4
B10	B10.D2[3]	PE	N.D.	61	36	-3

[1]All injections of immunizing cells were done i.p. with viable spleen cells suspended in balanced salt solution. The priming injection was of 1-2 x 10^7 cells and the boosting injection 3-4 x 10^7 cells. Responder mice were sacrificed 4 days after the boost and spleen and PE cells assayed for lysis of targets for 4 hours.

[2]The interval between priming and boosting was 8 months. Ratio of aggressors: targets was for spleen 340:1, for PE 60:1.

[3]The interval between priming and boosting was 7 months. Ratio of aggressors: targets was 40:1 for spleen and PE.

Many experiments of this kind have been carried out with various combinations of homozygous priming and boosting cells. The conclusion from these data is that the in vivo induction of cytotoxic T cells is not restricted by the H-2 type of the immunogenic cells but is restricted to targets carrying either H-2 type of the F_1 responder. This is so because B10.BR targets bearing the correct minors with a foreign H-2 are not lysed.

The question as to whether the effector cells lysing B10 targets were the same as, or different from those lysing B10.D2 targets was left open by the data in Table 1. The experiment shown in Fig. 1 answers this question. It proves that even under these conditions of double in vivo immunization the individual effector cells lyse either B10 or B10.D2 targets and not both. F_1 mice which had been primed by injecting B10 cells were boosted 10 weeks later by i.p. injection of B10.D2 cells. Four days later a pool

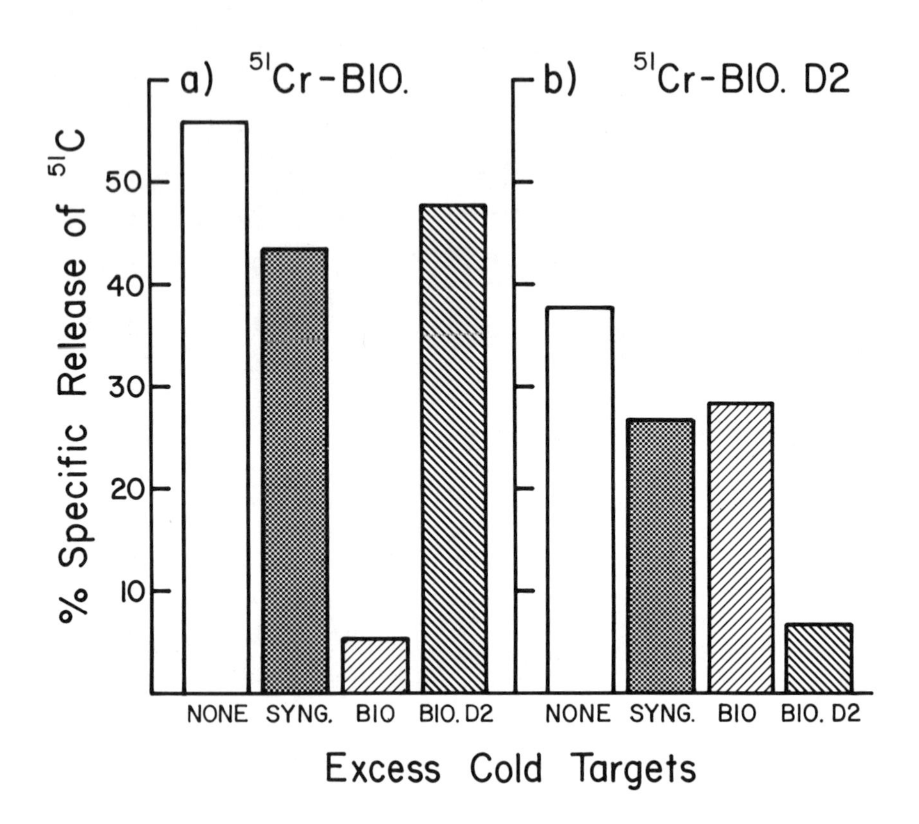

Fig. 1. Cold target competition experiment demonstrating that minor H specific cytotoxic T cells which have been induced in vivo are H-2 restricted. F_1(C x C.B)(H-$2^{d/b}$) mice which had been primed i.p. with B10(H-2^b) cells were boosted i.p. 10 weeks later with B10.D2(H-2^d) cells. A pool of spleen and PE cells was obtained 4 days later and assayed for lysis of Con A blasts from a). B10 and b). B10.D2. Lysis of 3 x 10^4 ^{51}Cr-targets by 7 x 10^6 effectors was studied with no added unlabelled cells (NONE), or in the presence of 9 x 10^5 unlabelled Con A blasts from F_1(C x C.B) (SYNG), B10, or B10.D2.

of spleen and PE cells from these animals lysed B10 and B10.D2 targets (Fig. 1) but not F_1(C x C.B) or B10.BR targets (data not shown). Lysis of labelled B10 or B10.D2 targets was studied in the presence of an excess of unlabelled

F_1(C x C.B), B10 or B10.D2 targets (i.e. a cold target competition experiment). Fig. 1a shows clearly that the rate of lysis of ^{51}Cr-B10 targets was specifically inhibited by cold B10 cells but not by cold B10.D2. Conversely, Fig. 1b shows that with the same population of killer cells the lysis of ^{51}Cr-B10.D2 was inhibited by cold B10.D2 but not by cold B10. Therefore, within this population of effector cells induced by boosting with B10.D2, there are separate pools of effectors lysing B10 or B10.D2, i.e. the effectors are absolutely H-2 restricted.

DISCUSSION

Genetically at least the minor H differences which B10 presents to an F_1(C x C.B) are the same as those presented by B10.D2. The F_1 cytotoxic cells which lyse B10 targets, however, are a separate population from those which lyse B10.D2. H-2 restriction of T cell function has had two main explanations: 1) the T cell has two clonally restricted receptors, one of which binds one of the self H-2 coded structures and the other binds foreign antigen, 2) the one receptor, or "altered self" hypothesis predicts that killer cells have one receptor which is specific for a complex of H-2 and the foreign antigen, i.e. for the F_1(C x C.B) responder B10 minor-plus-H-2^b is a different complex antigen from B10 minor-plus-H-2^d. The finding that in vivo immunization with B10 generates effector cells specific for B10.D2 which do not detectably bind B10 cells can be explained in two ways. Both explanations have important implications for cytotoxic T cell induction.

The first explanation is that the antigen binding requirements for T cell induction are different from those for effector function (target cell lysis). That is, B10 cells as immunogen may be able to bind to and induce anti-B10.D2 cytotoxic precursors though they are not lysed by anti-B10.D2 effector cells. This explanation would fit better with the two-receptor model of H-2 restriction than with the one-receptor model, i.e. a B10 minor H structure binding to one receptor of the cytotoxic precursor is sufficient for induction but not for lysis. (It could also fit with the one-receptor model if sub-sites of the T cell receptor exist which have some affinity for component parts of the complex antigen (10).) We do not favor this interpretation because it does not explain the following: (a) boosting spleen cells in vitro generates effector cells specific only for the H-2 type of the stimulating cells, i.e. F_1 spleen cells primed in vivo with B10 then boosted in vitro with B10.D2 generate effector cells capable of lysing B10.D2 but no effectors specific for B10 (7); (b) Zinkernagel and Doherty (11) also observed H-2 restricted induction of virus specific killer cells when they transferred F_1(A X B) spleen cells into irradiated "multiplier" mice infected with virus. Transfer into an infected irradiated A host gave rise to F_1 effectors which lysed infected A but not infected B targets; (c) injection of H-2^d homozygous tumor cells into the F_1(H-$2^{d/b}$) primes effector cells specific for minor antigens-plus-H-2^d but does not cross-prime effectors for the minor antigens-plus-H-2^b (7,9). Under these three conditions of immunization it is not obvious why the proposed type of non-H-2 restricted induction should not occur.

The second explanation of how an i.p. injection of B10 cells can induce anti-B10.D2 effector cells depends on antigen-presenting cells of host origin being involved. The injected B10 cells may become disrupted and the minor H structures processed and presented on the surface of host (F_1) cells. Now they appear in combination with H-2^b and H-2^d and in this latter form can induce anti-B10.D2 killer cells. This explanation is the one we favor. It allows the antigen binding requirements of induction and effector function to be the same. It also fits equally well with the one-receptor and two-receptor models of H-2 restriction. The failure to cross-induce cytotoxic precursors in vitro or after injection into "multiplier" hosts would be accounted for if the specialized presenting system does not function after the spleen has been teased into single cell suspension. Growing tumor cells may either bypass or inhibit the function of the host antigen presenting system and thereby fail to cross-prime.

In the system of in vivo immunization we have used here cross-boosting of effector cells certainly occurs efficiently. We would like to discover how to reproduce it in tissue culture.

ACKNOWLEDGEMENTS

We thank R.E. Langman, M. Cohn and R.W. Dutton for discussions. Supported by NIH grants to M.J.B., M. Cohn, and R.W. Dutton.

REFERENCES

1. Doherty, P.C., Blanden, R.V. and Zinkernagel, R.M. (1976) Transplant. Rev. 29, 89.

2. Shearer, G.M., Rehn, T.G. and Schmitt-Verhulst, A.M. (1976) Transplant. Rev. 29, 222.

3. Bevan M.J. (1975) J. Exp. Med. 142, 1349.

4. Gordon, R.D., Simpson, E. and Samelson, L.E. (1975) J. Exp. Med. 142, 1108.

5. Snell, G.D., Dausset, J. and Nathenson, S. (1976) Histocompatibility. Academic Press, New York.

6. Graaf, R.J. and Bailey, D.W. (1973) Transplant. Rev. 15, 26.

7. Bevan, M.J. (1976) J. Exp. Med. 143, 1283.

8. Gordon, R.D., Mathieson, B.J., Samelson, L.E., Boyse, E.A. and Simpson, E. (1976) J. Exp. Med. 144, 810.

9. Bevan, M.J. (1976) J. Immunol. 117, 2233.

10. Janeway, C.A. (1976) Transplant. Rev. 29, 164.

11. Zinkernagel, R.M. and Doherty, P.C. (1975) J. Exp. Med. 141, 1427.

IMMUNE RESPONSE TO HISTOCOMPATIBILITY ANTIGENS: *H-2* CONTROL OF *IN VIVO* AND *IN VITRO* EFFECTOR:TARGET INTERACTIONS

Peter J. Wettstein*§, Geoffrey Haughton†, Jeffrey A. Frelinger*

*Department of Microbiology, University of Southern California School of Medicine, Los Angeles, California 90033

†Department of Bacteriology and Immunology, University of North Carolina, Chapel Hill, North Carolina 27514

§Leukemia Society of America Fellow

ABSTRACT. Genes in the *H-2* complex affect both the immunogenicity of the H-7.1 alloantigen and the interactions between H-7.1-incompatible targets and specific effectors *in vivo* and *in vitro*. A gene(s) in the *H-2D* region regulates the rejectability of H-7.1-incompatible skin grafts, the stimulatory efficiency of H-7.1-incompatible lymphocytes in MLC, and the susceptibility of H-7.1-incompatible lymphoblast targets to specific lysis in CML. H-7.1-specific cytotoxic activity *in vitro* was *H-2* restricted. However, cross-priming *in vivo* with H-7.1-incompatible skin grafts bearing different *H-2* haplotypes was not *H-2* restricted, suggesting that *H-2* restriction is not a significant phenomenon *in vivo*.

The cellular *in vivo* response to non-H-2 histocompatibility antigens, including H-Y (1,2), H-4 (3), and H-7 (4), is regulated by *H-2*-linked *Ir* genes. However, the role of *H-2*-linked genes in controlling allograft rejection is not limited to control of recipient responsiveness. Recent reports suggest that genes in the *K* end of the *H-2* complex regulate the rejectability of H-Y-incompatible skin grafts, presumably mediated through quantitative control of H-Y antigen expression (5,6). Further, studies of the *in vitro* response to non-H-2 histocompatibility antigens have demonstrated that, despite evidence substantiating *in vivo* cross-priming (7), the cytotoxic activity of *in vitro* generated cytotoxic effectors specific for non-H-2 antigens is *H-2* restricted. The observed H-2 restriction has been presumed to be of major significance to *in vivo* allograft rejection. We have performed experiments designed to 1) elucidate the role of the *H-2* complex in determining relative rejectability of H-7.1-incompatible skin grafts, and 2) determine the ability of H-7.1-incompatible skin grafts homozygous for different *H-2* haplotypes to reciprocally cross-prime.

We have previously reported that the *in vivo* cellular response to H-7.1-incompatible skin grafts is under *H-2*-linked *Ir* gene control (4). Fast responsiveness is inherited as a dominant trait (10). B10.C-$H\text{-}7^b$ [$H\text{-}2^b$] hosts reject B10 [$H\text{-}2^b H\text{-}7^a$] grafts with a median survival time (MST) of 3.2 weeks, whereas B10-$H\text{-}2^a H\text{-}7^b$ recipients do not reject B10.A [$H\text{-}2^a H\text{-}7^a$] grafts in less than 20 weeks. In order to investigate the possible effects of the $H\text{-}2^b$ and $H\text{-}2^a$ haplotypes on relative rejectability of H-7.1-incompatible skin grafts, (B10.C-$H\text{-}7^b$ x B10-$H\text{-}2^a H\text{-}7^b$)$F_1$ recipients were grafted with either B10, B10.A, or (B10.A x B10)F_1 skin. Tail skin was transplanted orthotopically according to the technique described previously (11). The results of this experiment are presented in Table 1. The key observation was that B10 and

Table 1. *H-2* Determination of Relative Rejectability of H-7.1-Incompatible Skin Grafts on (B10.C-$H\text{-}7^b$ x B10-$H\text{-}2^a H\text{-}7^b$)$F_1$ Hosts.

Primary graft	MST*	Secondary graft	MST*
B10	3.7	B10	1.4
		B10.A	1.5
B10.A	8.0	B10	1.5
		B10.A	3.1
(B10.A x B10)F_1	3.5	B10	1.3
		B10.A	1.5

*Median survival time in weeks

(B10.A x B10)F_1 skin grafts were rejected more rapidly than B10.A grafts. This situation is the reverse of that observed for H-Y in which $H\text{-}2^a$ and not $H\text{-}2^b$ determines relatively high rejectability (6). To determine if primary B10 and B10.A grafts were able to reciprocally cross-prime *H-2* heterozygous recipients, *H-2* heterozygous mice received secondary B10 and B10.A skin grafts 14 days after their rejection of primary grafts. The results of these transplants are included in Table 1. Contrary to restriction dogma, both B10 and B10.A primed for the accelerated rejection of both secondary B10.A and B10 grafts. Primary B10, B10.A, and (B10.A x B10)F_1 grafts were equally effective in accelerating the rejection of secondary B10 grafts. However, primary B10 and (B10.A x B10)F_1 grafts were significantly more efficient in priming for the accelerated rejection of secondary B10.A grafts than were primary B10.A grafts themselves. Therefore, we conclude that cross-priming *in vivo* with H-7.1-incompatible skin grafts is extensive and results in the generation of cytotoxic effectors capable of rejecting H-7.1-incompatible grafts with *H-2* haplotypes other than that of the priming graft donor. The efficiency of cross-priming is ultimately dependent upon the relative rejectability determined by the *H-2* genotype of the primary and secondary graft donors. *H-2* identity of the primary and secondary donors is not required for effective cross-priming.

The intra-*H-2* map position of the gene controlling rejectability of H-7.1-incompatible grafts was determined with the use of intra-*H-2* recombinant haplotypes. (B10.C-$H\text{-}7^b$ x B10-$H\text{-}2^aH\text{-}7^b$)F_1 recipients of primary and secondary H-7.1-incompatible grafts received grafts from B10.A(1R) [$H\text{-}2^{h1}$], B10.A(4R) [$H\text{-}2^{h4}$], B10.A(5R) [$H\text{-}2^{i5}$], B10, and B10.A donors 20-27 weeks after secondary grafting. The data presented in Table 2 represent the results of tertiary grafting of only (B10.C-$H\text{-}7^b$ x B10-$H\text{-}2^aH\text{-}7^b$)F_1 recipients of primary B10.A grafts.

Table 2. Mapping of the Graft Donor Genes Responsible for Rapid Rejection of H-7.1-Incompatible Grafts.

Tertiary graft	Major *H-2* regions				MST*
	K	*I*	*S*	*D*	
B10	b	b	b	b	< 1.5
B10.A	k	k/d	d	d	2.1
B10.A(1R)	k	k/d	d	b	< 1.5
B10.A(4R)	k	k/b	b	b	< 1.5
B10.A(5R)	b	b/d	d	d	2.8

*Median survival time in weeks

Recipients of primary B10.A grafts rejected secondary B10 and B10.A grafts with different MST's, dependent upon the secondary graft rejectability. Tertiary B10, B10.A(1R), and B10.A(4R) grafts were rejected significantly more rapidly than tertiary B10.A and B10.A(5R) grafts. These results are consistent with mapping of the rejectability determining gene telomeric to the $H\text{-}2^{h1}$ recombination site. (B10.C-$H\text{-}7^b$ x B10-$H\text{-}2^aH\text{-}7^b$)F_1 recipients of primary (B10.A x B10)F_1 grafts were also grafted with quaternary grafts from B10, B10.A, B10.A(1R), B10.A(2R) [$H\text{-}2^{h2}$:$K^kI^{k/d}S^dD^b$], B10.A(5R), B10.A(18R) [$H\text{-}2^{i18}$:$K^bI^bS^bD^d$], and B6-Tla^a [$H\text{-}2^b$] donors. B10, B10.A(1R),

B10.A(2R), and B6-Tla^a grafts were rejected significantly more rapidly than B10.A, B10.A(5R), and B10.A(18R) grafts. Therefore, the gene controlling H-7.1-incompatible graft rejectability maps within the *H-2D* region as it is defined by the $H\text{-}2^{h1}$ $H\text{-}2^{h2}$, and $H\text{-}2^{i18}$ recombination sites on the centromeric side and the site of recombination between *H-2* and *Tla* in the selection of B6-Tla^a on the telomeric side. Therefore, the rejectability of H-Y- and H-7.1-incompatible grafts is determined by different genes mapping in the *H-2K* and *H-2D* ends, respectively, of the *H-2* complex. Studies are presently underway to investigate possible effects of these genes on rejectability of H-4- and H-3-incompatible skin grafts in order to determine the antigen specificity of rejectability control.

The *H-2*-linked effects on rejectability led us to believe that the same genes might effect the ability of H-7.1-incompatible cells to stimulate in secondary MLC. Culture conditions were similar to those described by Peck and Bach (12). Responder cells were nylon wool purified splenic T cells obtained from the (B10.C-$H\text{-}7^b$ x B10-$H\text{-}2^aH\text{-}7^b$)F_1 recipients of primary through quaternary H-7.1-incompatible grafts described above. Stimulator cells were Mitomycin C-treated spleen cells obtained from the donors listed in Table 3. ^{3}H-thymidine uptake was measured 3-5 days after initiation of culture. The mean specific uptake (allogeneic combination - syngeneic combination) of ^{3}H-thymidine in each combination at 4 and 5 days is reported in Table 3. B10, (B10.A x B10)F_1, B10.A(1R), B10.A(4R), and B6-Tla^a stimulators were the most effective stimulators, whereas B10.A and B10.A(5R) stimulators were marginal (but significant) stimulators. Therefore, the relative efficiency of stimulation in MLC *in vitro* correlated with the relative rejectability *in vivo*: high immunogenicity of H-7.1 in MLC stimulators requires possession of the *H-2D* region of $H\text{-}2^b$. The one exception to this correlation was B10.A(18R). Although B10.A(18R) grafts were rejected relatively slowly comparable to B10.A, B10.A(18R) stimulators were "intermediate" in stimulating efficiency in MLC. We are presently typing B10.A(18R) mice for serologically detectable public and private specificities coded for by $H\text{-}2D^d$ and possible additional specificities (Ia-like?) which are coded for by genes mapping in the interval between *S* and *D* regions associated with $H\text{-}2^b$. The efficiency of H-7.1-incompatible cells to stimulate generation of H-7.1-specific cytotoxic effectors was evaluated. Nylon wool splenic T cells from (B10.C-$H\text{-}7^b$ x B10-$H\text{-}2^aH\text{-}7^b$)F_1 hosts, grafted with primary-quaternary H-7.1-incompatible grafts as described above, were mixed in MLC with stimulators from B10, B10.A, (B10.A x B10)F_1, B10.A(1R), B10.A(5R), and B10.C-$H\text{-}7^b$ donors. After five days in culture, surviving cells were mixed with ^{51}Cr-labelled lymphoblast targets at a

Table 3. *H-2D* Dependence of Capacity of H-7.1-Incompatible Lymphocytes to Stimulate Proliferation of Primed (B10.C-H-7^b x B10-H-2^aH-7^b)F_1 T Cells.

Stimulator Cell Donor	Specific ^{3}H-Uptake (cpm)*	
	4 days	5 days
C57BL/10	17,319	28,505
B10.A	1,328	2,186
(B10.A x B10)F_1	12,378	19,090
B10.A(1R)	14,296	16,905
B10.A(4R)	14,513	21,604
B10.A(5R)	722	1,109
B10.A(18R)	5,687	11,853
B6-Tla^a	16,468	31,230
B10-H-2^aH-7^b	ns	ns
B10.C-H-7^b	--	--

*Cited when mean uptake in allogeneic combination is significantly different from the mean uptake in the syngeneic combination employing B10.C-H-7^b stimulators at $p < .001$. "ns" indicates lack of stimulation.

20:1 effector:target ratio as described previously (8). The results of this assay are shown in Table 4. There were three important observations made in this experiment. First, H-$2D^b$ H-7.1-incompatible stimulators were more effective than H-$2D^d$ stimulators in stimulating generation of H-7.1-specific effectors. Second, only H-$2D^b$ H-7.1-incompatible lymphoblasts were susceptible to lysis at the 20:1 effector:target ratio. Third, since only H-$2D^b$ cells were susceptible to lysis, and only H-$2D^b$ cells stimulated, H-7.1-specific cytotoxic activity appeared *H-2*-restricted as predicted by restriction theorems.

Although *H-2* restriction of *in vitro* cytolysis of chemically modified targets (13), virus-infected targets (14), and non-*H-2*-incompatible targets (8) is extensively documented, there is presently no evidence for its relevance to the *in vivo* immune response. In this report we have confirmed the observations of Bevan (8) that *H-7*-specific cytotoxic cells generated and assayed *in vitro* appear *H-2*-restricted in their cytotoxic activity. Whether this restriction is simply due

Table 4. *H-2* Determination of Capacity of H-7.1-Incompatible Stimulators to Boost for Restricted Lysis of H-7.1-Incompatible Targets.

Boosting Stimulator cells	Mean % lysis of ^{51}Cr lymphoblast targets						
	C57BL/10	B10.A	$(B10.A \times B10)F_1$	B10.A(1R)	B10.A(5R)	B6-*Tla*^{a}	B10.C-*H-7*^{b}
C57BL/10	39	-20	46	70	9	45	- 1
B10.A	- 3	-16	23	41	5	20	- 4
$(B10.A \times B10)F_1$	45	-14	52	67	10	41	0
B10.A(1R)	45	-15	55	61	7	44	- 4
B10.A(5R)	-31	-18	0	3	8	8	- 7
B10.C-*H-7*^{b}	-27	-10	-12	- 1	6	7	0

to differential stimulatory ability of discrete non-H-2 alloantigens with different *H-2* haplotypes is not clear. This explanation cannot be ruled out for the currently available data on restriction of cytotoxicity to non-H-2 histocompatibility antigens. However, we have conclusively shown that the *in vivo* rejection of H-7-incompatible skin grafts by the very same donors of H-7-specific effectors assayed *in vitro* is not *H-2*-restricted.

ACKNOWLEDGEMENTS

The authors thank Ms. Kathy Mohr for her assistance in production and maintenance of the mice used in this study. This research was supported by National Institutes of Health Grant CA-16246 and American Cancer Society Grant IM-90.

REFERENCES

1. Bailey, D.W. and Hoste, J. (1971) *Transplantation 11,* 404.
2. Gasser, D.L., and Silvers, W.K. (1971) *Transplantation 12,* 412.
3. Wettstein, P.J., and Haughton, G. (1977) *Immunogenetics,* in press.
4. Wettstein, P.J., and Haughton, G. (1977) *Immunogenetics,* in press.
5. Wachtel, S.S., Gasser, D.L., and Silvers, W.K. (1973) *Science 181,* 862.
6. Kralova, J., and Demant, P. (1976) *Immunogenetics 3,* 583.
7. Bevan, M.J. (1976) *J. Exp. Med. 143,* 1283.
8. Bevan, M.J. (1975a) *Nature 256,* 419.
9. Bevan, M.J. (1975b) *J. Exp. Med. 142,* 1349.
10. Wettstein, P.J. (1976) Ph.D. Thesis, Univ. of N. Carolina, Chapel Hill.
11. Bailey, D.W., and Usama, B. (1960) *Transpl. Bull. 7,* 424.
12. Peck, A.B., and Bach, F.H. (1973) *J. Immunol. 3,* 147.
13. Shearer, G.M. (1975) *Transpl. Proc. 7,* 109.
14. Zinkernagel, R.M., and Doherty, D.C. (1974) *Nature 248,* 701.

T LYMPHOCYTE ACTIVATION BY MAJOR HISTOCOMPATIBILITY ANTIGENS: THE ALLOGRAFT REACTION AS A MODEL FOR ALTERED-SELF

Fritz H. Bach and Barbara J. Alter

Immunobiology Research Center and
Departments of Medical Genetics and Surgery
The University of Wisconsin
Madison, Wisconsin 53706

The complexity of T lymphocyte responses can be studied from at least two different perspectives: first, the major histocompatibility complex (MHC) determined antigens that elicit the reactions and second, the functionally different T cell subpopulations, including helper, cytotoxic and suppressor T lymphocytes, that respond (1). In this particular area, the first approach assumes a particularly meaningful role in that one could argue that the T cell system arose in evolution to provide a surveillance mechanism against altered-self antigens. Further, altered-self has recently been shown by the work of several groups to involve modified antigens (no molecular or mechanistic model is intended by this terminology) of the MHC (2,3). Since antigens of the MHC are so very effective in stimulating different populations of T lymphocytes, it is not unlikely that many immunogens are presented to effector lymphocytes in association with MHC antigens (4).

Our present understanding of the T lymphocyte response to MHC antigens has evolved largely from *in vitro* studies using the mixed leukocyte culture (MLC) and cell mediated lympholysis (CML) assays (1). In addition, and more recently, some of these models have been tested *in vivo* (5). It would seem that the very close correlations of findings in allograft systems, *in vitro* and *in vivo*, and altered-self models, involving both viral and chemical modification of cell surfaces, encourages speculation that additional information from the altered-self systems will also parallel the presently available findings in the allograft system.

In Vitro Studies

Cytotoxic T lymphocytes (CTLs) generated against H-2 antigens in an allogeneic combination use target antigens determined by genes of the *K* and *D* regions as their prime target determinants. We refer to these determinants as MHC CD antigens. It is clear, but needs emphasis, that in addition there are CD antigens associated with the *I* region

(6,7,8,9) although in the combinations tested to date these appear to be markedly weaker than the K and D region CD antigens.

Maximal generation of the cytotoxic response against a K or D region CD antigen in most, if not all, cases requires simultaneous stimulation of the responding cells with that CD difference plus an LD difference (1). The strongest LD differences are associated with the I region. It was on this basis that LD-CD collaboration was demonstrated, i.e. that simultaneous activation of the responding cells with LD + CD led to higher anti-CD CML than stimulation with CD alone. (Stimulation with an I region LD stimulus led to no CML against K or D region determined CD antigens.)

At the cellular level, the primary cell apparently responding to an I region LD difference is the helper T cell [Ly 1+ (10), monolayer non-adherent (11)] and CML against CD antigens is effected by the CTL [Ly 2+ (10), monolayer adherent (11)]. It is not clear whether the helper cell can recognize CD determinants or whether the CTL receptor can effectively combine with LD.

One interesting finding in this regard has been that a significant cytotoxic response can be generated against a K or D region difference alone. Although this could simply reflect a response of CTLs to the CD antigens encoded by genes in the K or D region, we suggested, on the basis of the significant proliferative response that was evoked in such combinations, that there was an LD-like locus in the K region as well as one near H-2D.

Since, in addition to the cytotoxicity that was generated in an MLC where responder and stimulator differ by only a K or D region, significant proliferation was also observed and since it was known that CTLs divide, it was not clear whether the division associated with a K or D region difference reflected division of CTLs or whether the Ly 1+ cells which we associate with a response to LD were dividing.

Presented in table 1 are results of an experiment in which this question was probed. Untreated or Ly antisera plus complement treated AQR cells are stimulated with cells from B10.A (a K region difference), B10.M (an entire H-2 difference) or B10.T(6R) (an I + S region difference); mean CPM of tritiated thymidine incorporated on day 4 are given. As previously demonstrated, the great majority of cells responding proliferatively to either an entire H-2 difference or to an I + S region difference are Ly 1+ helper cells. Treatment with anti-Ly 2 serum, which eliminates the CTLs, has little or no effect on the amount of proliferation elicited. In the K region different combination, however, there is a highly significant decrease in proliferation caused by pretreatment of responding cells with either antiserum.

TABLE 1

MLC RESPONSE OF LY ANTISERA TREATED CELLS

	Stimulating Cells			
AQR Responding cells	AQR_x Control	$B10.A_x$ K Region	$6R_x$ I Region	$B10.M_x$ Entire H-2 Region
untreated	8238	16069	28168	46923
αLy1	4545	5223	7783	11202
αLy2	6927	7181	35749	41669

These results suggest that there is at least one gene in the *K* region that codes for an LD-like determinant. Whether this gene is the same one coding for the *K* region CD antigen and thus LD and CD determinants are present on the same molecule or whether they are on separate molecules but both encoded by genes of the *K* region is, of course, not established. One might argue that the response of Ly 1+ cells to a *K* region reflects recognition by those cells of the CD antigens. This cannot critically be ruled out but seems unlikely in view of the findings using ultraviolet light treated stimulating cells which alone fail to elicit proliferation but which in three cell protocols present functional CD determinants to CTLs (1). It is also possible that the cells responding to a *K* or *D* region are Ly 1+2+3+ cells.

A further complication that has arisen in our understanding of T lymphocyte response to MHC antigens is the finding that alloantigens in an MLC stimulate cells to express a suppressor cell phenotype. Although no fine structure analysis is yet complete to analyze which regions and subregions of H-2 are most effective in stimulating this suppressor cell generation, the findings of Sondel et al. (12,13) suggest that it may be the CD antigens to which the suppressor cells primarily respond. If responding cells are exposed to heat or uv-light treated stimulating cells, under which conditions there are no detectable proliferative or cytotoxic responses after five to six days when the normal proliferative responses peak, suppressor cells have nevertheless been generated (13). These findings suggest that suppressor cells can be generated in the absence of help (at least help of the type which allows the development of a maximal cytotoxic response). Since LD is not functional on heat or uv treated cells, we cannot conclude from these studies alone whether LD differences themselves could lead to generation of suppressor cells. Further, we cannot be sure whether the CD determinants on the uv or heat treated cells are those that stimulate suppressor cell generation or whether other determinants expressed on such cells are activating suppressors.

There are some preliminary data which are consistent with the suggestion that suppressor cells generated in MLC are T lymphocytes. If this is substantiated, then a certain "eliciting antigen-responding cell" parallelism would exist between suppressors and CTLs. Suppressor T lymphocytes are Ly 2+ as are CTLs; CTLs respond to CD antigens and so would the suppressor cells. If CD antigens are responsible for suppressor stimulation, it has yet to be critically established whether simultaneous stimulation with LD will simply increase the effectiveness of suppressor generation, or whether increased help will lead to an increased CTL response with a consequent decreased suppressor response. We have recently discussed this problem in greater detail (14).

In Vivo Studies

Over the past several years a number of different techniques have been used in an attempt to correlate the *in vitro* findings discussed above with *in vivo* phenomena. The most direct correlate of the proliferative events seen in response to LD antigens in mixed leukocyte culture has been the proliferative graft versus host as measured by either splenomegaly or lymph node enlargement (15,16,17). Just as with the *in vitro* studies, the strongest stimulus to a proliferative GvH is an *I* region (presumably LD) difference. In contrast, by these *in vivo* assays *K* and *D* region differences in some cases do not lead to a significant response, which may be interpreted to reflect insensitivity of the method at the doses of cells tested. In addition to the proliferative graft versus host reaction, a correlation has been established between the extent of CML *in vitro* and the incidence of fatal GvH disease (18).

A model that we have recently used to test *in vivo* LD-CD collaboration utilizes transplantation of a lobe of the thyroid under the kidney capsule (19). In order to relate the results obtained using this model to LD and CD responses *in vitro*, we assume that thyroid lobe rejection is largely a T lymphocyte mediated phenomenon; further, that rejection is based on the action of CTLs directed at the CD antigens on the thyroid tissue. We recognize, however, that other cellular and humoral mechanisms may play a role in the rejection of a lobe of the thyroid.

If a thyroid is transplanted from a K region different donor that thyroid is relatively slowly rejected. Rejection can be assayed either by determining the amount of radioactive iodine incorporated into the transplanted thyroid and expressing this as a ratio of iodine incorporated as compared to the contralateral control kidney or by histological studies. The speed of rejection of that K region different thyroid can be markedly increased by the intraperitoneal administration to the recipient, at the time of thyroid transplantation, of lymphoid cells from a donor who is I region different from the recipient (5). The lymphoid cells presumably provide the recipient with an LD stimulus which activates recipient helper T cells and thus helps generate a stronger CTL attack against the thyroid and hence speeds up rejection. Most dramatic of the results obtained using this model is the finding that whereas the administration of I region LD different lymphoid cells intraperitoneally at the time of thyroid transplantation leads to rejection of the thyroid at a time not dissimilar from rejection of a thyroid differing by an entire H-2 complex, administration of K region different lymphoid cells from a donor syngeneic to the K region different thyroid donor has a much less marked effect on the speed of rejection. These results would argue strongly for the importance of presenting an LD as well as a CD stimulus to the recipient in terms of determining the strength of the rejection response which takes place.

We would thus interpret these results as being in concert with the model that the strength of a cytotoxic T lymphocyte response *in vivo* as well as *in vitro* depends not only upon the presentation of the foreign CD antigens to those cells but also on the magnitude of the helper cell response. Further, that the magnitude of the helper cell response will depend in large measure on the amount of LD disparity between donor and recipient.

Allogeneic and Altered-Self Induced Responses

Several striking similarities exist in the data obtained from allogeneic and altered-self systems. The altered-self systems most extensively studied are those in which virally transformed cells (2) or cells in which the cell surface has been chemically modified (3) have been used to stimulate responses.

Certainly the most striking of the analogies between the allogeneic and the altered-self model is the pre-eminent importance of the K and D region determined target antigens which play a major role in serving as targets for cytotoxic T lymphocytes in both systems (2,3). More recently, at least using the chemical modification approach, it has been demonstrated that modified *I* region antigens are those that are most important in eliciting a proliferative anti-self response (20). One might well argue that this is analogous to the allogeneic proliferative response of helper T cells to the *I* region LD antigens.

Several other aspects of the allogeneic system remain to be tested in the anti-self response; others have been tested and, at first glance, appear to differ.

It has been pointed out by several authors that there is a cytotoxic response aimed at *I* region determined CD antigens in the allogeneic situation which, while much weaker than the cytotoxic response to *K* or *D* region CD antigens, is certainly significant. Yet, in the anti-self response, no such *I* region modification of target antigens has been noted. This, it would seem to us, has a ready explanation from the development of the findings in the allogeneic system. First, in the initial experiments testing for a cytotoxic response against H-2 determined CD antigens, only those of the *K* and *D* region were detected. Subsequently we and several other laboratories found a significant cytotoxic response against the *I* region. It may thus be that this is simply a matter of developing more sensitive techniques in the altered-self models. Second, and perhaps more pertinent to this comparison, the allogeneic anti-*I* region cytotoxic response is most easily elicited or detected if sensitizing and responding cells differ by only the *I* and *S* regions and are identical for *K* and *D*. In the anti-self system, obviously, this is a difficult end to achieve and it may thus be that the very great strength of the *K* and *D* region determined altered-self antigens has clouded the response to altered-*I* region CD antigens.

A second area in which possible parallelisms should be sought concerns the findings presented in this paper, i.e. that there may well be LD-like antigens in the *K* and *D* regions in addition to the CD determinants. Does virus infection or chemical modification modify not only the CD but also the LD determinants of the *K* and *D* regions in order to allow effective presentation of these determinants in the anti-self system? Perhaps most important from the *in vivo* perspective, does collaboration exist between altered- *I*-region-LD and altered-*K* or *D* region-CD in a manner analogous to LD-CD collaboration in the allogeneic system?

Summary

We have reviewed some of the findings regarding the lymphocyte response to the major histocompatibility complex alloantigens. Included are recent data from attempts to further analyze helper and cytotoxic responses aimed at K and D region antigens. We have emphasized the parallelisms between the findings in the allogeneic system and those in the response to altered-self as well as to raise questions for the altered-self systems based on the allogeneic situation.

References

1. Bach, F.H., Bach, M.L. and Sondel, P.M. (1976) *Nature* 259, 273.
2. Zinkernagel, R.M. and Doherty, P.C. (1977) *Contemporary Topics in Immunobiology*, in press.
3. Shearer, G.M., Rehn, T.G. and Garbarino, C.A. (1975) *J. Exp. Med.* 141, 1348.
4. Thomas, D.W., Yamashita, U. and Shevach, E.M. (1977) *Immunological Reviews*, in press.
5. Sollinger, H.W. and Bach, F.H. (1976) *Nature* 259, 487.
6. Peck, A.B. and Bach, F.H. (1975) *Scand. J. Immunol.* 4, 53.
7. Schendel, D.J. and Bach, F.H. (1975) *Eur. J. Immunol.* 5, 880.
8. Röllinghoff, M. and Wagner, H. (1975) *Eur. J. Immunol.* 5, 875.
9. Nabholz, M., Young, H., Rynbeek, A., Boccardo, R., David, C.S., Meo, T., Miggiano, V. and Shreffler, D.C. (1975) *Eur. J. Immunol.* 5, 594.
10. Cantor, H. and Boyse, E.A. (1975) *J. Exp. Med.* 141, 1376.
11. Bach, F.H., Segall, M., Zier, K.S., Sondel, P.M., Alter, B.J. and Bach, M.L. (1973) *Science* 180, 403.
12. Sondel, P.M., Jacobson, M.W. and Bach, F.H. (1975) *J. Exp. Med.* 142, 1606.
13. Sondel, P.M., Jacobson, M.W. and Bach, F.H. (1977) *Eur. J. Immunol.* 7, 38.
14. Bach, F.H., Grillot-Courvalin, C., Kuperman, O.J., Sollinger, H.W., Hayes, C., Sondel, P.M., Alter, B.J. and Bach, M.L. (1977) *Immunological Reviews*, in press.
15. Livnat, S., Klein, J. and Bach, F.H. (1973) *Nature New Biology* 243, 42.
16. Klein, J. and Park, J.M. (1973) *J. Exp. Med.* 137, 1213.
17. Elkins, W.L., Kavathas, P. and Bach, F.H. (1973) *Transpl. Proc.* V, 1759.
18. Elkins, W.L. (1976) *Transpl. Proc.* VIII, 343.
19. Lafferty, K.J., Cooley, M.A., Woolnough, J. and Walker, K.Z. (1975) *Science* 188, 259.
20. Schmitt-Verhulst, A-M., Garbarino, C.A. and Shearer, G.M. (1977) *J. of Immunol.*, in press.

T CELL RECOGNITION IN CELL-MEDIATED IMMUNITY
I. ANTIGEN RECOGNITION IN THE SYNGENEIC SJL TUMOR SYSTEM*

Janet Roman, Marilyn H. Owens and Benjamin Bonavida

Department of Microbiology and Immunology
School of Medicine, University of California, Los Angeles

INTRODUCTION

SJL mice succumb to a reticulum cell sarcoma (RCS) between the ages of 6-12 months. This tumor is characterized by a mixed cellularity and the nature of the actual tumor cell as well as the etiology of this disease are unknown. We have therefore established *in vitro* cell lines of these tumors. These cell lines have now been in culture for nearly a year and presumably represent pure populations of neoplastic cells. These *in vitro* lines have been compared with *in vivo* passaged tumors in attempts to elucidate the presence of any altered or tumor specific antigens capable of stimulating syngeneic immune responses.

RESULTS AND DISCUSSION

Three SJL RCS lines LA1, LA6, and LA8 passaged both *in vivo* and *in vitro* were examined for ability to stimulate syngeneic proliferative responses (as measured by H^3Thymidine uptake) as well as syngeneic cytotoxic responses (as measured by Cr^{51} release from target cells). A summary of our results is seen in Table I.

Each of the RCS lines, regardless of whether it is grown *in vivo* or *in vitro*, is capable of stimulating a syngeneic proliferative response to some degree. The nature of the stimulus for this syngeneic proliferation has not been determined, and is complicated by a consistently high background in unstimulated SJL lymphocytes as well as the curious finding that mitomycin C treated SJL splenocytes are capable of stimulating autologous splenocytes. In contrast to the proliferative response, only those RCS cells passaged *in vitro* are able to stimulate a syngeneic cytotoxic response or serve as a target for this response. We feel this is not merely an artifact of *in vitro* culture, such as a fetal calf serum antigen, since two of our lines have now lost their antigenicity while a third, simultaneously cultured line has not. Interestingly, coincident with loss of antigenicity of these lines is loss of their ability to form tumors in mice. The failure of *in vivo* lines to stimulate cytotoxicity could not be attributed to the presence of suppressor cells as determined by mixing experiments.

TABLE I

Tumor	Passaged		Stimulates syngeneic		Serves as target for syngeneic CML
	in vivo	in vitro	MLR[a]	CML[b]	
RCS-LA1	+		+	-	-
RCS-LA1		+	++	±	-
RCS-LA6	+		±	-	-
RCS-LA6		+	±	++	++
RCS-LA8	+		±	-	-
RCS-LA8		+	±	++	++
Spontaneous RCS			±	-	-

[a]MLR = mixed lymphocyte reaction, measured by uptake of H^3Thymidine.

[b]CML = cell mediated lympholysis, measured by release of ^{51}Cr from target cells.

Perhaps the majority of cells in SJL "tumors" are non-tumor in nature, and thus do not bear the altered tumor antigens which serve to stimulate a cytotoxic response. Such a proposal is consistent with the facts that: 1) SJL RCS "tumors" are quite difficult to passage in vivo (large inoculum must be used, and the percentage of takes may often be low), and 2) SJL "tumors" do not grow in irradiated mice (1).

The nature of the antigen(s) which stimulate this syngeneic cytotoxic response is suggested by the following findings. SJL spleen cells sensitized to normal spleen cells from C57BL/6 and BALB/c mice are capable of lysing syngeneic RCS cells but not SJL blasts. B10.P, B10.IIIR, and B10.BR cells, on the other hand, do not stimulate a significant syngeneic killing (data not shown). This suggestion of antigens shared between the SJL $H\text{-}2^s$ tumor and $H\text{-}2^d$ and $H\text{-}2^b$ haplotypes is supported by serological data (Table II). Both B10 anti B10.A and C57BL/6 anti BALB/c antisera will cause complement dependent lysis of SJL RCS cells. This anti-tumor activity is absorbed out by BALB/c and LA6 cells but not by SJL cells. The finding of inappropriate H-2 antigens on tumor cells (2,3) raises extremely intriguing questions concerning genetic regulation of histocompatibility antigens as well as the nature of "tumor specific antigens."

TABLE II

Antisera	Absorptions	% Dead Cells (Trypan Blue) RCS-LA6*	SJL spleen
-	-	4	3
B10 α B10.A	-	100	24
"	C57BL/6	92	8
"	BALB/c	6	4
"	SJL	88	4
"	LA6	5	3
C57BL/6 α BALB/c	-	100	22
"	C57Bl/6	100	2
"	BALB/c	8	3
"	SJL	88	6
"	LA6	6	4

*10^6 cells were incubated with antisera at a 1:20 dilution in the presence of C' for one hour, at which time % dead cells was determined by Trypan blue uptake.

REFERENCES

1. Carswell, E.A., Lerman, S.P. and Thorbecke, G.J. (1976) Cellular Immunol. 23, 39.

2. Garrido, F., Scheirmacher, V. and Festenstein, H. (1976) Nature 259, 228.

3. Invernizzi, G. and Parmiani, G. (1975) Nature, 254, 713.

*Supported by NIH CA19753.

T CELL RECOGNITION IN CELL-MEDIATED IMMUNITY II. NON-SPECIFIC STIMULATION OF ALLOSENSITIZED MEMORY LYMPHOCYTES INTO CYTOTOXIC LYMPHOCYTES*

Benjamin Bonavida

Department of Microbiology and Immunology
School of Medicine, University of California, Los Angeles

ABSTRACT. *In vivo* allosensitized and primed splenocytes can be activated *in vitro* into specific secondary cytotoxic lymphocytes by stimulating cells of different H-2 haplotypes than those used for priming. Preliminary studies suggest that stimulating antigens map close to the D end region of the MHC.

INTRODUCTION

Activation of alloantigen primed lymphocytes into specific secondary cytotoxic T lymphocytes (CTL) may be achieved by (a) exposure to the original stimulating antigens (1), (b) LD antigens and third party stimulating alloantigens (2, 3), and (c) by concanavalin A (3, 4). The present report extends these findings and examines what region of the MHC of third party stimulating alloantigens is responsible for activation of primed lymphocytes into secondary CTL.

MATERIALS AND METHODS

In vivo immunization, *in vitro* sensitization, and cytotoxic assays have been done as previously described (4).

RESULTS AND DISCUSSION

Allosensitized BALB/c (H-2^d) anti-EL-4 (H-2^b) splenocytes can be activated into secondary CTL with specificity to C57Bl/6 target antigens by C57Bl/6 (H-2^b), SJL/J (H-2^s), C3H (H-2^k), and CBA (H-2^k) stimulating alloantigens. These results demonstrated that third party stimulating antigens of H-2 haplotypes unrelated to either the responder or stimulating haplotypes are endowed with the capacity to trigger non-specifically memory lymphocytes into CTL specific for the priming antigens. The nature of the non-specific activating antigen(s) was examined by using stimulating antigens from recombinant mice. A representative experiment is shown in Table I. Group 1 shows that various non-H-2 antigens present on B10.D_2 or DBA/2 are not good stimulators. In group 2, although B10.BR can activate memory lymphocytes, B10.A does

Table I. Genetic Mapping of Antigens Involved in the Activation of Secondary Cytotoxic Lymphocytes

Responder	Group	Stimulator	H-2 Complex								CMC ^{51}Cr C57B1/6 blasts E:T (5:1)
			K	A	B	J	C	S	G	D	
BALB/c(H-2^{d} anti EL-4 (H-2^{b}) (50d)	1	B10	b	b	b	b	b	b	b	b	67.4 ± 1.7
		DBA/2	d	d	d	d	d	d	d	d	-2.1 ± 2.0
		B10.D2	d	d	d	d	d	d	d	d	-3.7 ± 1.0
	2	B10.BR	k	k	k	k	k	k	k	k	66.7 ± 1.6
		B10.A	k	k	k	k	d	d	d	d	-4.2 ± 1.4
		B10.A(2R)	k	k	k	?	d	d	d	d	63.8 ± 2.5
		B10.HTT	s	s	s	s	k	k	k	d	2.6 ± 1.0
	3	B10.S	s	s	s	s	s	s	s	s	16.8 ± 0.9
		B10.S(7R)	s	s	s	s	s	s	s	d	54.9 ± 1.1
		B10.S(9R)	s	s	?	k	d	d	d	d	3.4 ± 0.2
		B10.HTT	s	s	s	s	k	k	k	d	2.6 ± 1.0
		---									1.2 ± 0.9
Normal BALB/c		B10	b	b	b	b	b	b	b	b	34.8 ± 0.5
		B10.S	s	s	s	s	s	s	s	s	2.4 ± 0.8

BALB/c mice were primed *in vivo* with EL-4 and 50 days later, spleen cells were cultured *in vitro* with the spleen cells from various congenic strains. The cultures were harvested after 3 days of incubation and tested for cytotoxicity against ^{51}Cr-C57B1/6 blasts.

not, thus ruling out the K, IA, IB, and IJ regions in stimulation. Since B10.HTT is a poor stimulator, the IC, S, and G regions shared with B10.BR can also be ruled out. Likewise, by comparing the non-stimulator B10.A with stimulator B10.A(2R), all regions to the left of the D end can be ruled out. Thus, it appears that a region in the proximity of the D end is involved in stimulation.

Results shown in group 3 supports the role of the D end region. The K, IA, IB, and IJ regions are ruled out by comparing the results of B10.S and B10.HTT. Since the IC, S and G regions are not involved in the H-2^k haplotype, we can extrapolate that these regions are also not involved in the H-2^s haplotype. The findings that B10.S(7R) is a good stimulator, while B10.A is not, demonstrates that the D region itself is not responsible for the activation of memory lymphocytes. The results suggest that the stimulating antigens map either at the left hand or the right hand side of the D region. B10.S(1R) and B10.S(15R) stimulated equally well, thus ruling out any involvement of the Hh1 region in stimulation (unpublished). Preliminary experiments using other responder memory lymphocytes of the H-2^b and H-2^s haplotypes and various stimulating antigens from recombinant mice also show that regions near the D end of the MHC are involved in the triggering events.

Our preliminary results offer two alternative explanations:

Firstly, the third party stimulating alloantigens cross react with the primary sensitizing antigens and share antigenic regions closely associated with the D end region. Cold target cell inhibition experiments should verify this if the cross-reacting antigens involved in the triggering step are also expressed on the target used for cytotoxicity. Preliminary experiments demonstrate that the third party stimulating alloantigens cannot compete completely for the cytotoxicity directed against the antigens used for priming. Thus, although cross-reaction at the cytotoxic level is poor, cross-reactivity at the stimulating phase, however, cannot be ruled out.

Secondly, it may be assumed that memory lymphocytes respond to non-specific signals. These non-specific signals may be mediated by Con A or by functionally similar entities present on the stimulating antigens. These entities may not be involved in the cytotoxic event and, thus, do not serve as target antigens. Such non-specific stimulating antigens which map close to the D end region trigger memory lymphocytes via an antigen specific membrane receptor. In addition to these receptors, the lymphocyte may carry a specific target antigen receptor which serves as a focusing device. It can also be inferred from this hypothesis that the non-specific stimulating antigens are polyclonal with greater affinity for memory lymphocytes than non-primed cells. Thus, selective triggering of the expanded memory clone would explain the specific secondary CTL obtained. The attractiveness of the second interpretation lies in the demonstration that activation of secondary lymphocytes is mediated via a one non-specific signal and is reminiscent of Coutinho and Möller's model of B cell responses by mitogens (5). It remains, however, to be shown which cell type(s) is receiving the non-specific triggering.

Our results cannot discern between the two interpretations and also cannot determine which cell type in the responder population is involved in activation. The purification and enrichment of memory lymphocytes and the use of Ly antisera will be helpful to answer these questions.

ACKNOWLEDGMENTS

We wish to thank Ed Clark and Dale Kipp for valuable discussion and for providing several of the recombinant mice used in the studies.

REFERENCES

1. Cerottini, J.-C., Engers, H.D., McDonald, H.R. and Brunner, K.T. (1974) J. Exp. Med. 140, 703.
2. Alter, B.J., Grillot-Courvalin, C., Bach, M.L., Zier, K., Sondel, P.M. and Bach, F.H. (1976) J. Exp. Med. 143, 1005.
3. Heininger, D., Touton, M., Chakrabarty, A. and Clark, W.J. (1976) J. Immunol. 117, 2175.
4. Bonavida, B. (1977) J. Exp. Med. 145, 293.
5. Coutinho, A., Gronowicz, E. and Moller, G. (1974) Progress in Immunol. II, vol. 2, L. Brent and J. Holborow, eds., North-Holland Publishing Co., p. 167.

*Supported by NIH CA12800.

POSITIVE AND NEGATIVE REGULATORY EVENTS CONTROL THE GENERATION OF CYTOTOXIC T CELLS

Linda M. Pilarski, Abdul R. Al-Adra and Linda L. Baum

Department of Immunology, University of Alberta
Edmonton, Alberta, Canada T6G 2H7

ABSTRACT. The generation of alloantigen-specific cytotoxic T cells *in vitro* is controlled in both a positive and a negative manner by regulatory T cells which are antigen-specific. Helper T cells are required for the generation of cytotoxicity from thymocyte precursors. These helper cells are present in populations of normal spleen cells and precursors of the helpers can be primed *in vitro* to yield an efficient population of T cells whose function is antigen-specific and radio-resistant. A second system for the study of negative regulatory events involves the suppression of cytotoxic responses by *in vitro* primed suppressor cells. Low doses of anti-Balb/c suppressor cells are able to completely suppress the cytotoxic response of CBA lymph node to Balb/c spleen. These suppressors are antigen-specific, radio-resistant T cells, the precursors of which are cortisone-sensitive. The suppressors appear to be functionally distinct from cytotoxic T cells.

INTRODUCTION

Although a great deal of data has accumulated implicating both positive and negative regulatory mechanisms in the control of humoral immunity, relatively little has been done concerning the regulation of cell-mediated immunity. We have been developing an all-purpose experimental system in this laboratory which will hopefully allow us to dissect the cytotoxic T cell response to alloantigens into its various regulatory components. Here we report the development of *in vitro* systems for generating both T cells which help (positive control) and T cells which suppress (negative control) a cytotoxic T cell response.

Previous work (1) indicated that helper T cells were an essential requirement for the generation of cytotoxicity. This was shown in a system where no cytotoxicity was generated in the absence of helper T cells which are resident in normal spleen cell populations. CBA thymus cells do not respond to irradiated Balb/c stimulator spleen cells; in the presence of irradiated CBA spleen cells as well, however, a substantial degree of cytotoxicity is generated. The collaborating activity in the irradiated CBA spleen was removed by treatment with anti-theta serum plus complement prior to culture which indicated that this is a T-dependent function. The helper activity was also found to be antigen-specific. This was determined in the following manner. It was reasoned that since normal spleen cells contained helper activity, the cell population might be expected to contain helpers of any given specificity (i.e. the cell population would be multiclonal). Therefore, a spleen population tolerant of Balb/c antigens should be deficient in anti-Balb helper T cells, but possess normal numbers of anti-C57Bl helper T cells. Spleen cells from irradiated Balb/c mice reconstituted with CBA bone marrow cells two months prior to use were the source of CBA helper cells tolerant to Balb/c antigens. These cells were unable to help a CBA thymocyte response to Balb/c stimulator cells but, in contrast, were able to efficiently help an anti-Balb/c response to (Balb x C57Bl)F1 stimulator cells which possess C57 determinants for helper recognition. In addition, these helper cells were able to efficiently help across allogeneic strain barriers; that is, they are _not_ strain-specific.

In the work to be discussed here, we extend these observations by utilizing a system in which helper cell precursors are primed _in vitro_ to alloantigens to yield an enriched population of helper cells which are theta-bearing, antigen-specific and radio-resistant. Similarly, we have been able to prime suppressor cell precursors _in vitro_, producing a highly efficient population of alloantigen-specific radio-resistant inhibitory T cells.

METHODS

Priming and assay of helper T cells: First step cultures consist of 8-10 x 10^6 CBA spleen cells cultured with 1500 rad irradiated stimulator spleen cells (8-10 x 10^6) in acrylamide rafts (2) for two days. These cells are then irradiated. The helper cell activity is measured by adding various doses of first step cells to a second step culture consisting of 2-5 x 10^5 CBA thymocytes plus 2 x 10^6 irradiated stimulator spleen cells and incubating in v-bottom microtiter trays for five days. The cells in each microtiter well are then assessed for their ability to lyse P815 or EL4 ^{51}Cr-labelled target cells in a four hour assay.

In vitro generation and assay of suppressor cells: First step cultures consist of 3 x 10^6 CBA spleen cells cultured with 3 x 10^6 irradiated Balb/c spleen cells for three days in acrylamide rafts (2). The cells are then irradiated, assayed for cytotoxic activity, and added at various doses to second step cultures to assay suppressor activity. Second step cultures contain 5 x 10^5 CBA lymph node cells plus 16 x 10^6

irradiated stimulator cells, +/- suppressor cells, cultured in acrylamide rafts for five days. Cytotoxicity is assayed as previously described (2). Cortisone-treated CBA mice received 5 mg of hydrocortisone three days prior to use of their spleen cells in first step cultures.

RESULTS AND DISCUSSION

Priming of helper cells: Culturing CBA spleen cells with irradiated Balb spleen cells yielded a population of cells which could efficiently cooperate with thymocyte killer precursors to yield a cytotoxic response. The efficiency of these helper cells was ten or more fold greater than the efficiency of the helper cells resident in normal unprimed spleen. The kinetics of helper cell generation were rapid with a sharp peak of helper cell activity at day 2 of culture. The helpers are radio-resistant cells which were eliminated by treatment with anti-theta serum plus complement (Table 1, lines 2 and 3). The activity of the helpers is specific for the antigen to which they were primed *in vitro*. Helper cells primed to Balb/c antigens do not help a response to C57Bl/6 (H-2^b) (Table 1, line 2). This is a specific enrichment since irradiated normal spleen contains helpers able to cooperate in both anti-Balb and anti-C57Bl/6 responses (Table 1, line 1).

TABLE 1 HELPER CELLS PRIMED *IN VITRO* BEAR THETA ANTIGENS AND ARE ANTIGEN-SPECIFIC.

IRRADIATED HELPER CELL	TREATMENT	IRRADIATED STIMULATOR CELL	PERCENT SPECIFIC CYTOTOXICITY P815	EL4
1. NORMAL CBA SPLEEN	NONE	BALB/c	14.6%	-
		(CBA x C57BL/6)F1	-	18.1%
2. CBA SPLEEN PRIMED TO BALB/c *IN VITRO*	NORMAL MOUSE SERUM + C´	BALB/c	25.4%	-
		(CBA x C57BL/6)F1	-	1.4%
3. CBA SPLEEN PRIMED TO BALB/c *IN VITRO*	ANTI-THETA SERUM + C´	BALB/c	5.6%	-
		(CBA x C57BL/6)F1	-	6.9%
4. NONE	NONE	BALB/c	5.2%	-
		(CBA x C57BL/6)F1	-	5.1%

5 x 10^5 CBA THYMUS (H-2^k) CELLS WERE CULTURED IN MICROTITER TRAYS FOR FIVE DAYS WITH OR WITHOUT 5 x 10^5 *IN VITRO* PRIMED HELPER CELLS OR 1 x 10^6 NORMAL SPLEEN HELPER CELLS WITH EITHER 2 x 10^6 BALB/c SPLEEN CELLS OR 2 x 10^6 (CBA x C57BL/6) SPLEEN CELLS AS STIMULATORS. CYTOTOXICITY VALUES REPRESENT THE MEAN OF FOUR REPLICATE CULTURES.

A second set of experiments demonstrated that the helper cell precursor which is stimulated to divide and differentiate in the first step cultures is also antigen-specific (Figure 1). These were based on the theory that spleen cells tolerant to Balb/c antigens should be unable to produce anti-Balb helper cells when stimulated in vitro but should be capable of being stimulated by C57Bl/6 antigens to yield anti-H-2^b specific helper cells. (CBA x Balb/Fl spleen cells provide a population of helper precursors deficient in anti-Balb specificities for self-tolerance reasons. First step cultures containing (CBA x Balb)Fl spleen cells with either (Balb x C57Bl/6)Fl stimulators or C3H-SwSn stimulators (H-2^b) were assayed for their ability to help a second step anti-Balb cytotoxic response of thymocytes to Balb or (Balb x C57Bl/6)Fl stimulator cells (Figure 1, b and c). Both helper populations were able to efficiently help an anti-Balb response if H-2^b antigens were also present on the second step stimulator cell (50-52% lysis). They were unable to help a response to Balb/c spleen cells; in contrast, CBA anti-Balb helpers allowed thymocytes to increase in cytotoxicity from no cytotoxicity without help (Fig. 1, d) to 21% lysis in the presence of help (Fig. 1, a). Studies are currently in progress to determine

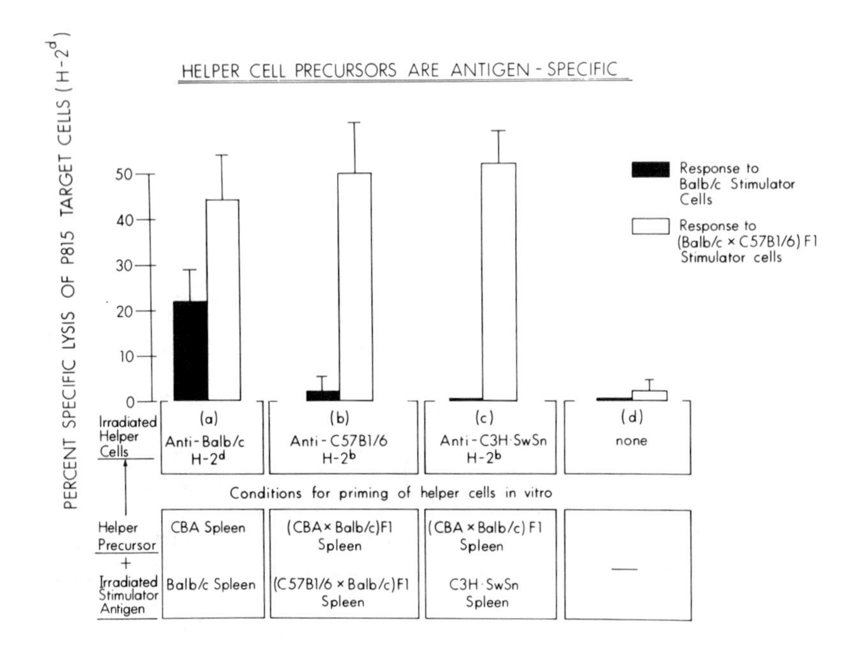

Figure 1

the mechanism by which these helpers collaborate with cytotoxic T cell precursors. Preliminary evidence indicates that the helper and the killer precursor collaborate via associative recognition of linked determinants and do not act via recognition of antigenic determinants on third party cells.

In vitro generation of suppressor cells. Normal CBA spleen cells which have been cultured for three days with irradiated Balb spleen cells yield progeny cells capable of efficient inhibition of cytotoxic T cell responses. A representative experiment is presented in Figure 3. CBA lymph node responds to Balb/c spleen to yield 71% cytotoxicity. Addition of 10^6 irradiated first step cells to this culture inhibits the response by ten fold (to 8% lysis). Lower doses (10^4) of these irradiated first step cells reproducibly enhance the cytotoxic response in second step cultures. The inhibitory activity generated in first step cultures is antigen-specific; anti-Balb (H-2^d) suppressor cells do not inhibit the response of CBA lymph node to C3H-SwSn stimulator cells (H-2^b) (Figure 3b). The enhancing effect of low first step cell doses applies to both anti-H-2^d and anti-H-2^b responses and could be due to cross-reactive anti-H-2^d helper cells since such a cross-reaction has been observed (unpublished). Further studies indicated that the suppressor activity was completely

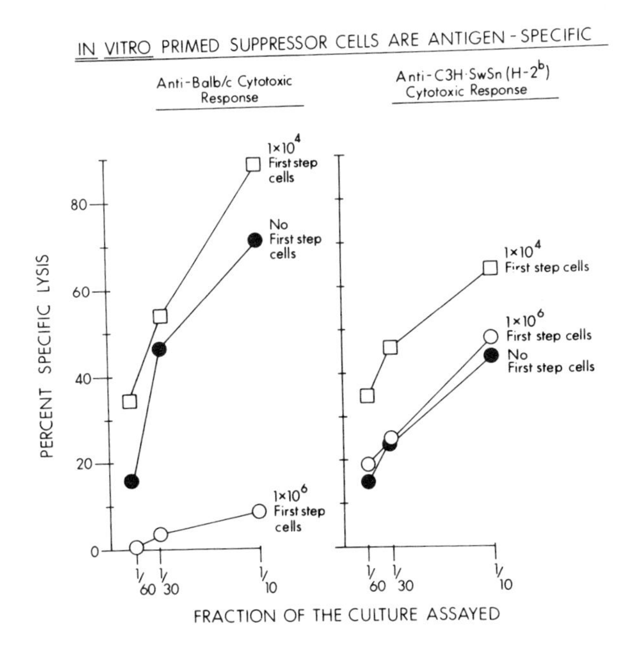

Figure 2

inhibited by treatment with anti-theta serum plus complement but was unaffected by treatment with normal mouse serum plus complement. In addition, the inhibition was not strain-specific; CBA anti-Balb suppressor cells were able to completely inhibit the cytotoxic response of C3H-SwSn lymph node to Balb/c spleen.

Since three day first step cells did contain cytotoxic T cells, it was important to establish whether or not the suppression was due to an inhibitory feedback effect of the cytotoxic cells. Other workers have found suppressive effects which appear to be attributable to such a mechanism in suppressor cell populations generated over longer culture periods (3,4). Treatment of CBA mice with hydrocortisone *in vivo* followed by stimulation of their spleen cells in first step cultures produced a population of cells which had cytotoxic activity after a three day culture period but, unlike untreated cells, did not yield suppressor cells (Figure 4). Figure 4a illustrates the cytotoxic activity of first step cultures containing either untreated responder cells or cortisone-treated responder cells. Normal spleen cells yield 8-10 times more cytotoxicity than do cortisone-treated spleen cells at day three of culture with irradiated Balb spleen. In Figure 4B, the suppressive effect of these same two cell populations is measured. A wide range of irradiated first step cell doses (1-20 x 10^5 per culture) derived from normal spleen cells were able to completely inhibit the response of second step cultures. In contrast, the first step cells derived from cortisone-treated spleen were unable to suppress the response at any cell dose (1-20 x 10^5/culture). In fact, the cortisone-treated cells appeared to generate helper cells which were most effective at the lowest cell doses used (1 x 10^5 per culture) (Figure 4b). Since 1 x 10^5 first step cells from normal cells suppressed the respond one hundred per cent, while 20 x 10^5 first-step cells from cortisone-treated mice did not suppress at all and yielded increasing amounts of help on dilution, we tentatively conclude that the suppression is not due to the cytotoxic activity of the first step cultures. Since other interpretations of this data are still possible, we are currently extending this work in an attempt to unequivocally eliminate or implicate the cytotoxic T cell as the suppressor cell.

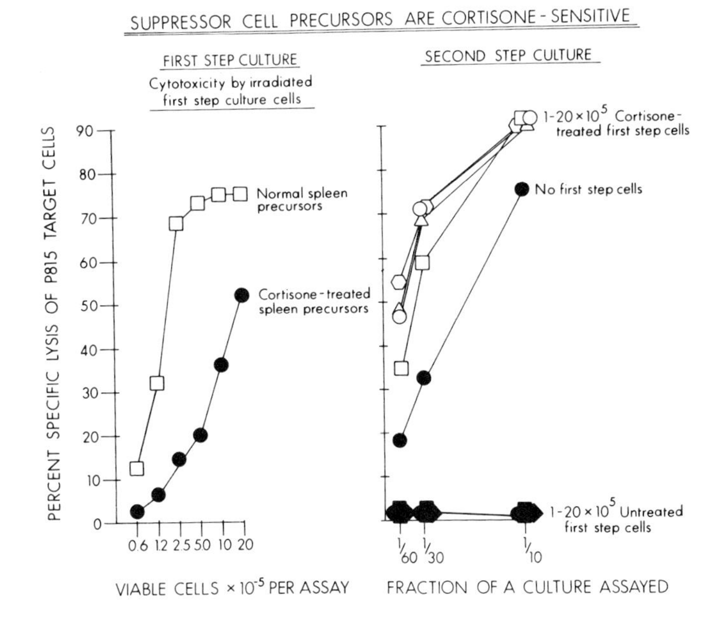

Figure 3

ACKNOWLEDGEMENTS

We are grateful to Ludmilla V. Borshevsky for her dedicated, highly skilled technical assistance. We also wish to acknowledge many thought-provoking discussions with Dr. Peter Bretscher. This work was supported by the Banting Research Foundation and the Medical Research Council of Canada.

REFERENCES

1. Pilarski, L.M. (1977) J. Exp. Med. 145:709.

2. Pilarski, L.M. and Borshevsky, L.V. (1976) Eur. J. Immunol. 12:906.

3. Fitch, F., Engers, H., Cerottini, J.C., Brunner, T. (1976) J. Immunol. 116:716.

4. Sinclair, N., Lees, R., Wheeler, M., Vichos, E., Funy, F. (1976) Cell Immunol. 27:153.

Workshop: Multiplicity of T Cell Receptors

H. Wigzell, Uppsala
M. Katz, Los Angeles
D.B. Wilson, Philadelphia

A list of seven "facts" concerning the immunological specificity of T cells served as the principal focus for discussion in this workshop. In the following account, each of the items on the list is presented, along with the principal reasons for considering an item as fact, and the reaction that was provoked among the workshop discussants.

(1) All immunocompetent T cells have clonally distributed receptors for MHC (self-and allo-) gene products. This statement is based on the findings of (a) the disproportionately high frequency of alloreactive T cells and (b) the restrictive regulation by MHC gene products in the induction and expression of T cell responses to conventional antigens, (i.e., cytotoxicity, DTH, proliferation, and helper functions).

While there seemed to be general agreement with the principle involved in this statement, it did provoke some discussion as to the word "all"; because the premise had not been tested, it could not be accepted as fact. Most agreed, however, that the "majority" of T cells would seem to express such receptors.

(2) T cells recognize "conventional" antigens (i.e., non-MHC) with discriminative fine specificity similar to B cell recognition of antigen. This is based on studies of the fine specificity of T cell populations seen in the discrimination of related hapten conjugates in DTH reactions, in cytolysis of hapten modified target cells, in the specificity of helper function, as well as antigen induced lymphocyte proliferation and tolerance. There was general agreement on this point.

(3) The specificity of T cells for conventional antigens is physiologically adapted by the MHC environment in which antigen is presented. This is based on recent findings that T cell specificity for conventional antigens can be regulated by allogeneic MHC gene products in a manner apparently not different from self MHC restriction (i.e., helper effects for allogeneic B cells with T cells from double chimeras, generation of CTL to hapten-modified allogeneic target cells in T cell populations negatively selected for target cell MHC

alloantigens).

Again, this seemed to be an "acceptable" fact, although there was some resistance to the implication of facts (1) and (3) that cells reactive to allogeneic MHC determinants do not represent a subpopulation distinct from those reactive to conventional antigens. The basis for dissent was essentially due to the continued reaction to the word, "all", in fact (1).

(4) The frequency of T cells in a normal animal reactive to conventional antigens is significantly less than that of MHC reactive T cells. This "fact" is based on the findings that T cells functionally responsive to MHC alloantigens in GVH and MLC reactions normally amount to several percent in normal T cell populations, and specific anti-MHC surface receptors on these cells can also be detected in high frequency with anti-idiotypic sera. This is in contrast to the significantly lower frequency of idiotype positive T cells with specificity for conventional antigens in non-immune T cell populations. Other than points of clarification, this item of the list provoked no resistance.

(5) Some T cells recognize self Ig molecules. This "fact" follows from the demonstration of T cell mediated allotype suppression. Ig recognition by helper T cells appears to be a normal component of this system, although no evidence is available that suppressor T cells also recognize Ig.

(6) Receptors for both MHC alloantigens and conventional antigens use V_H gene products. This was based on recent demonstrations from genetic studies that expression of MHC and conventional antigen specific T cell receptors is linked to genes which determine C_H allotypes (mouse, rat). Also, structural studies by the KÖLN group have demonstrated that specific T cell receptors isolated from haptenated nylon nets bear idiotypes specified by V_H region genes (mouse) and can be removed on anti-V_H allotype columns (rabbit). This statement was accepted without dissent by the workshop participants, and it was pointed out that available evidence does not rule out the possible involvement of V_L gene products in the specificity of the T cell receptor.

(7) (In the mouse), Ly lymphocyte surface markers functionally determine, at least in part, T cell specificity for MHC alloantigens. This statement is based on observations of the binding and activation of lymphocytes of different Ly

phenotypes to allogeneic target cells; Ly 1+ cells bind to MHC I region gene products, whereas Ly 23+ T cells bind to MHC K and D region gene products on target cells. Such a statement can also be inferred indirectly from the finding that helper proliferative responses for conventional antigens by Ly 1+ T lymphocytes are restricted by MHC I region genes, and cytotoxic T cell functions of Ly 23+ cells are restricted by MHC K and D region genes. Although the presence of a given Ly structure appears to be functionally linked with the fine anti-allo or self-MHC specificity of that cell, it is not known whether the Ly gene product is directly involved in the receptor site. No evidence is available for direct interference of antigen binding receptors on T cells by antisera directed to Ly surface molecules. Whether or not Ly phenotypes dictate fine specificity for helper and suppressor functions to conventional T dependent antigens is not known.

It was clear from the ensuing general discussion that it was not yet possible from available evidence to distinguish between a one receptor-"altered self" model and a dual - or even multi-receptor model to account for the current phenomenology of T cell specificity. Several groups are currently involved in experiments designed to determine whether cells positively selected to specific MHC determinants can demonstrate immune reactivity to determinants of the conventional antigenic universe. Other approaches include the generation of killer T cells to hapten modified allogeneic targets in negatively selected lymphocyte populations specificially depleted of responsiveness to target MHC alloantigens.

It also seemed clear that the technology currently available does permit a resolution of this issue, at least in terms of whether or not T cells have more than one antigen specific binding site. If they do express more than one, however, whether these reside on different molecules or represent multiple combining sites on a single molecule may have to await the application of sequence technology.

Workshop on "Self-Recognition"

A. Cunningham and O. Stutman, Chairmen

A central question in immunology concerns the ability of the immune system to react with virtually any foreign antigen and yet not react harmfully against self. This workshop dealt with self-recognition: those reactions which sometimes do occur against self and against antigens closely related to self, and with the mechanisms of self-tolerance. It was suggested that a revolution may be taking place in our views on this topic: the older point of view, that anti-self-reactive cells are normally eliminated, appears likely to be replaced by the idea that self-reactive-immunocompetent cells are common in the body, but are normally held in check by complex control mechanisms. This "revolution" has important implications for clinical immunological problems such as autoimmune disease, cancer and homograft rejection.

1) Existence of self-reactive cells

A.J. Cunningham (Ontario Cancer Institute) described an example of self-tolerance in mice which seems to be maintained by active suppressor mechanisms. Mouse red blood cells contain "internal" antigens that can be revealed by treatment with proteolytic enzymes such as bromelain. Normal, uninjected adult mice of all strains tested have larger numbers of antibody plaque-forming cells (PFC) against these autoantigens (about 5,000 PFC per spleen and varying numbers in other tissues). Mice appear to be naturally highly primed against these autoantigens, since exposure of spleen cells to a B cell mitogen *E. coli* LPS, either *in vivo* or *in vitro*, causes a rapid and large increase in the numbers of PFC. Two lines of evidence suggest that the normal level of these autoreactive PFC is maintained as a balance between constant antigenic stimulation, and some active suppressor mechanism: 1) injecting bromelain-treated erythrocytes into a syngeneic mouse produces little or no elevation of PFC levels (although such treated erythrocytes provoke a strong response in other species); 2) injection of anti-lymphocyte serum causes a 20-fold enhancement in the number of these PFC.

C. Havele (Univ. of Alberta) put forward evidence, from another experimental system, which favored deletion rather than suppression as a tolerance mechanism. She prepared tolerant chickens (between FP and SC strains) either by paraboisis at 12 days of embryonation, or by injecting spleen cells from one to the other at 15 days of embryonation. The lymphoid cells of these chimeras were tolerant of one another by two criteria: lack of GVH (splenomegaly) reactions when chimeric cells were injected into either of the "parental" strains, and greatly reduced numbers of pocks in test on "parental" chorioallantoic membranes, compared with non-chimeric controls.

The mutual tolerance did not appear to be maintained by serum blocking factors, nor was it reversed by trypsin or PHA treatment of cells from the chimeras. "Mixture" experiments, adding (FP x SC) chimera cells to lymphoid cells of either strain of normal chickens did not affect the ability of the latter to make GVH reactions. In discussion of this work, the experiments of Elkins' were cited in which suppressor mechanisms in rat chimeras could be demonstrated, but only after tolerance was first broken.

Y. McHugh (U.C.L.A.) described the generation of large numbers of autoreactive PFC against syngeneic thymus cells in LPS-stimulated tissue cultures. Within 2 days of culture, spleen cells from several strains of mice (including nude mice) developed 2000-3000 PFC, detected as plaques of trypan blue stained thymocytes in agarose. Similar numbers of PFC were detectable on syngeneic, allogeneic or mixed thymocytes, probably indicating that the antibody responsible was reacting with common antigens on these cells.

The experiments of McHugh and of Cunningham agree with reports from others (e.g. Moller, Weigle), that anti-self-reactive B cells are common in the body. The report of Havele shows, however, that in a model system which probably closely resembles self-tolerance, "Self"-reactive T cells are absent or very difficult to detect.

2. Natural reactivity (against "self" and/or endogenous virus)

Two main areas were discussed: 1) natural cytotoxic cells capable of destroying mouse T lymphomas (Wigzell, Uppsala Univ., Sweden) of chemically induced fibrosarcomas (Stutman, Sloan-Kettering Institute) and some related phenomena such as hybrid resistance (Carlson, Univ. of Alberta; Shearer, N.I.H.) and 2) the natural reactivity directed against "modified self" (Doherty, Wistar Institute) endogenous mouse leukemia virus (Nowinski, Fred Hutchinson Cancer Center) and antigens coupled to self (Levy, Univ. British Columbia). The natural killer cell was defined in both systems (Wigzell, Stutman) as being neither T, B nor macrophage and lacking readily detectable Fc receptors. Both effector cells are highly radioresistant although they are derived from radio-sensitive bone marrow precursors (reactivity abolished by treatment of the cell donors with the bone-seeking isotope 89Sr; this last aspect being common to other non-conventional recognition responses such as hybrid resistance, resistance to xenogeneic hemopoietic tissue, resistance to focus formation by Friend virus and early resistance to *Listeria monocytogenes*). Both effector cells have a trypsin-sensitive structure which regenerates rapidly. The natural killer cells in the T lymphoma system can be agglutinated by Helix Pomatia agglutinin. Although both systems show similarities, some differences were noted: a) the natural killer (NK) cell against T lymphomas is not adherent (not retained in nylon columns) while the NK cell against solid tumors is partially retained; b) incubation at 37°C for relatively short periods of time decreases NK activity against lymphomas and does not affect NK cells against solid tumors; c) NK activity against T lymphomas appears related to murine leukemia (MuLV) antigens while susceptibility of solid tumors to NK cytotoxicity is independent of MuLV antigen expression; d) NK activity against T lymphomas appears at 3-4 weeks of age and declines with age while NK activity against solid tumors appears at birth and is maintained throughout life. In both systems there is variable susceptibility of targets to NK cytotoxicity and there are high and low NK mouse strains, high being dominant in the F1 hybrids (in the T lymphoma system the genetic predisposition is controlled by several genes, one linked to H2). In radiation chimeras, resistance to transplanted leukemias correlates with high NK donor strains. In the solid tumor system there is variability in susceptibility of different targets to NK activity: some targets are highly susceptible and killed by all strains tested while others show a restricted pattern of susceptibility. Both systems show no H2 restriction and even human T lymphomas are susceptible to mouse NK. Some sort of recognition is observed in both systems: in the T lymphoma system, demonstrated by inhibition with cold target; in the solid tumor system, by depletion of NK cells after adsorption unto susceptible targets (and not on resistant targets). The question of the _in vivo_ relevance of these observations was brought up by _Moller_ (Karolinska Institute, Sweden). In the T lymphoma system there is _in vivo_ evidence of increased resistance to transplanted lymphomas in the NK-high strains. In the solid tumor system there is evidence of inhibition of tumor growth using mixtures of tumor cells and spleen cells enriched for NK cells in a Winn assay. G. Carlson presented data on the rapid "rejection" of parental cells by irradiated F1 hybrids using the clearance of labelled cells on endpoints: the response appeared within 24 hours. The possibility of an anti-idiotypic response was suggested by Simonsen, however, the response seems too rapid for a conventional immune response, thus most probably related to hybrid resistance mediated by an NK-like cell (Stutman and Shearer, indicated some differences between NK and effector cells in hybrid resistance, mainly the sensitivity of the later to treatment with silica or cyclophosphamide, while NK activity is resistant to such agents). The possibility that NK cells would represent a form of natural surveillance was proposed by

Shearer, who also asked if the 89Sr-treated animals (depleted of NK cells) had increased susceptibility to tumor development. Such experiments have not been performed (Stutman), while it is well accepted that T deprived mice (nude, etc.) are not more susceptible to tumor development, questioning the T-dependency of surveillance (as originally proposed by Burnet). A version of surveillance was discussed by Doherty. Natural killer cells may be a back-up device for the T surveillance, mediated by recirculating T cells which would mainly detect K or D alterations. A definition was proposed in which the aim of surveillance would be to monitor cell surface integrity. T cells would recognize K or D modifications as altered-self. While the recognition step seems well documented, it was not clear how this surveillance was implemented (perhaps at a local level and not as a generalized phenomena). The natural response to endogenous viruses was discussed by Nowinski. Using sensitive assay for antibody and different high and low leukemia mouse strains, three main points were stressed which regulated the actual levels of the antibody response: 1) the production of endogenous virus by the different strains; 2) Ir genes (some linked to H2) which control the response and 3) the T dependency of the antibody response. It became apparent that these different modalities of natural reactivity may represent a possible surveillance device relevant to tumor development (and perhaps autoimmunity).

J. Levy (Univ. of British Columbia) described the immunization of mice with antigen (Rabbit Fab' fragments) attached to syngeneic, allogeneic or semiallogeneic F1 irradiated spleen cells. Specific anti-rabbit Fab plaque-forming cell numbers were measured. The response to antigen on allogeneic cells (maximum about 25,000 PFC/spleen at 3 weeks) was up to 5 times higher than the response against the same antigen presented on syngeneic cells. Two observations suggested possible active suppression in the latter case: 1) the response was also low when coupled F1 cells were used as antigen in spite of the presence of foreign ("carrier") antigen on such cells; 2) after _in vitro_ culture of immune spleen cells with antigen, it was found that those spleen cells exhibiting lowered PFC numbers _in vivo_ (i.e. those mice immunized with antigen on syngeneic or F1 cell surfaces) showed a response equal to or greater than that of cells from mice originally immunized with Fab on allogeneic cells.

3. Reactions to alloantigens

K. H. Lafferty (Australian National University) cultured mouse thyroids *in vitro* for 28 days and showed that this tissue, which before culture would have been rapidly rejected, was now permanently accepted as a functioning implanted graft under the kidney capsule of an allogeneic mouse. The culturing stage probably has its effect by removing "stimulator" cells, notably graft lymphocytes. Dr. Lafferty expressed the view that the way in which graft alloantigens are "presented" to the host is critical: presentation by graft lymphocytes induces reactive host lymphocytes which destroy the graft. Also described were some recent experiments showing that induction of alloreactive cytotoxic T cells requires 2 "signals", the first being provided by antigen, while the second, which normally comes from a lymphocyte associated with the alloantigen, may be replaced by a soluble factor secreted by Con-A stimulated spleen cells.

It was agreed that there is a host response to its own growing tumor, which appears once the tumor is established (Prehn, Jackson Laboratories and Stutman). It was felt that the evidence for a strictly T-dependent surveillance mechanism capable of eliminating nacient tumors) was far from proven. However, some of the non-T dependent mechanisms may perhaps serve as surveillance mechanisms in the "Burnetian" sense.

THYMUS DEPENDENT CELL-MEDIATED CYTOTOXICITY (CMC)

B. Bonavida, H. Wagner, and E. Grimm

Reports pertaining to three separate aspects of CMC were presented during this workshop: 1. The first part of the workshop concerned histocompatibility restriction in the generation and expression of cytotoxic T lymphocytes (CTL), since other sessions were devoted to the mechanisms of dual recognition and altered self theories in syngeneic models, this workshop was directed to concentrate on the concept itself, particularly generalizations and exceptions; 2. the second area was a discussion of several methods to enumerate precursor and functional CTL frequencies; and 3. the final area included work on characteristics of the actual CMC lytic interaction, including CTL cell phenotypes, and biochemical parameters of the lethal hit interaction.

1. Billings reported on studies showing the generation of CTL against I region differences using recombinant strains of mice. The CML was blocked by anti-I serum but not by anti H-2 K or D serum. Thus, he concluded that it was possible to get killing specific for I region independent of K or D regions. Other laboratories (Klein, Wagner), have also reported similar findings. The specificity of the cytotoxic activity against I region was not fully resolved.

Burakoff showed that stimulation of lymphocytes with TNP modified cells results in cytotoxic cells capable of killing not only the allogeneic target but also TNP-modified syngeneic cells. He suggested that T cell receptors recognize modified self and any reactivity to allogeneic cells is due to cross-reactivity.

Matzinger reported on lack of restriction during the induction of CTL against minor antigens. Restriction, however, was required for boost *in vitro*. *In vivo* cross-priming was not due to cross-reactivity. A mechanism was proposed whereby a macrophage presents antigens shed from stimulators, and thus the H-2 restriction was maintained during the sensitization phase.

Marshak studies in the rat suggested that MHC restriction of CTL activity might not be a general phenomenon; T cells from Ag-B rats preimmunized with tissue from rats thought to be Ag-B (MHC) compatible differentiated into cells able to recognize determinants expressed by rat strains of six different Ag-B haplotypes. She suggested that recognition by CTL in some systems is not subject to the same restrictions shown for other minor determinants.

Wagner showed that insertion of Sendai virus material into membrane by incubation of target and virus for 15-20 minutes, made the cells susceptible to killing by anti-virus cytotoxic cells. Cytotoxicity was still H-2 restricted. These results implied that physical modification by virus during the protein synthesis of H-2 is not necessary for syngeneic target lysis.

Doherty presented a report on the fine specificity of the cytotoxic response to influenza virus, and showed that cross-reactivity was due to determinants from the viral internal structural protein.

Forman showed that spleen cells treated with TNBS radio-iodinated with lactoperoxidase and followed by detergent lysis and immune precipitation with anti-TNP, no H-2 antigenic activity remained in the supernatant. These results demonstrated that H-2 molecules were derivatized with TNP. Several experiments were done which support the altered self hypothesis: (a) Anti-TNP serum blocked the action of anti-TNP cytotoxic effector cells, (b) there was a direct correlation between the extent of derivatization of H-2 antigens and the ability of such cells to act as stimulator cells, (c) similar concentration ranges of TNBS were required to create antigenic determinants on the target cells as well as immunogeneic determinants on the stimulator cells that can be recognized by cytotoxic T cells.

2. Teh described culture conditions to estimate the frequency of cytotoxic lymphocyte precursor (CLP) to alloantigens. The frequency of precursors were calculated by fitting the proportion of non-responding cultures to the zero order term of the Poisson distribution; e.g., frequency of precursor in the RNC nu/+ LN population, responding to the H-2 haplotype was 1/776. The B6 nu/+ average clone size for precursor after 7 days in culture was ≃1040. Most clones derived from single precursors of CL were specific.

Lindahl employed a sensitive procedure of limiting dilution of activated lymphocytes and ^{51}Cr release from highly labeled target cells to estimate the frequency of CTL present in a population of mouse splenocytes activated to alloantigens. The frequency ranged from 0.7 to 1.2%. The frequency of CLP was also estimated. For example, 4-15 CLP/10^4 LN of DBA/2, or 1-2% of all CLP are responsive to any given H-2 haplotype difference if assume that 5-10% of peripheral T cells (Ly 2^+3^+) have potential to develop into cytotoxic T lymphocytes.

Grimm presented a single-cell CMC method employing effector-target conjugates that resulted in CTL frequencies to tumor allografts. Since all killer cells are believed to conjugate to their target cell prior to the "lethal hit," enumeration of this binding was used as an estimate of CTL frequency. The absolute frequency was determined by performing a single-cell CMC assay on the conjugates of each population to determine the % killers, a value which averaged slightly over 70%. CTL frequencies in PEL populations, for example, were 14-17%, since an average of 20-21% conjugated and 70-80% of these were shown to be CTL by the single-cell assay.

3. Stutman reported that the expression of cytotoxicity against syngeneic tumor cells by mice immunized with MTV was dependent on both Ly 2^+ and Ly 1^+ lymphocytes in a 24 hour CMC assay. The lytic curve was shown to be bimodal, with the early peak requiring only Ly 2^+3^+ cells. The later peak was dependent on Ly 1^+ cells, since treatment with anti Ly 1^+ sera eliminated its appearance.

Truitt reported suppression of the induction of CTL from normal lymphocytes by spleen cells derived from allogeneically stimulated mice. The suppression was not specific to the antigen used for generating the suppressor cell. The suppressor cell was characterized as T cell.

Bonavida reported Bradley's work that cytotoxic T cells inactivated each other provided Con A or PHA was present. The autologous killing reaction argued for non-specificity of the lytic step and suggested that antigenic recognition is not required for manifestation of lysis but other receptors are involved.

One approach toward understanding the mechanism of CMC was presented by Golstein who showed evidence supporting the theory that the lethal hit was a "metabolic complex" state. Golstein described a Ca^{++} pulse method in which effector and target populations are mixed for 40 min in the presence of Mg^{++} to allow binding. Then Ca^{++} is added which allows the lethal hit to occur. Addition of a variety of agents prior to the Ca^{++} prohibited the manifestation of the lethal hit. These agents included DMSO, cytocholasin B, phenol, and sodium azide. Martz presented results on the mechanism of CMC, distinguishing the binding phase from the lethal hit by lowering the temperature to 15°C which allows only binding to occur. If cytocholasin B was added to this mixture, and the temperature raised to 37°C, this lysis is observed. Martz concluded that cytocholasin B does not inhibit the lethal hit, while other data he presented showed that it was effective in blocking adhesions. The discussion subsequent to these two presentations revealed that commercially available cytocholasin B is not standardized, and should be considered when interpreting such data.

Other participants present included B. Alter, F. Bach, M. Bevan, J. Frelinger, K. Lafferty, L. Pilarski, R. Rich, J. Roman, G. Shearer, B. Terry and J. Thoma.

MEMBRANE EVENTS IN CELL SIGNALLING

Martin C. Raff and Durward Lawson

Medical Research Council Neuroimmunology Project,
Zoology Department, University College London, London, WC1E 6BT

ABSTRACT. The mast cell is an unusually accessible cell for studying the early events in signal transmission by cell surface receptors and for visualising the molecular events in membrane fusion occurring during exocytosis. The available evidence is consistent with the following view of mast cell activation and histamine secretion: when multivalent ligands cross-link IgE receptors on the mast cell surface, Ca^{2+} channels are indirectly, but locally, opened in the plasma membrane; this allows a transient influx of Ca^{2+} which somehow induces a multifocal, lateral displacement of cytoplasmic components and proteins in the plasma membrane and underlying granule membrane, permitting the two membranes to fuse in these regions; the fused lipid bilayers, being depleted of protein, bleb from the cell, vesiculate and break off, opening the granule contents to the extracellular space; this enables histamine to escape by cation exchange and allows the cell to dispose of excess lipid while conserving the membrane proteins.

INTRODUCTION

Cells communicate with each other in two principal ways. One involves cell contact and the formation of small channels called 'gap junctions', which traverse the interacting plasma membranes and allow the intercellular passage of small molecules ($\leq \sim$ 1200 Daltons) such as ions and nucleotides (1). Gap junctions are now known to be the structural basis for electric and metabolic coupling between cells. Although the cells of many tissues in the body are connected by gap junctions, the role of this type of coupling is largely unknown. The other way that cells communicate with each other is by the secretion of chemical signals, which belong to two operationally-defined classes: (i) those that are hydrophobic and dissolve through any plasma membrane and bind to specific receptors in the cytoplasm; the ligand-receptor complex then passes into the nucleus where it binds to DNA and regulates transcripton (2); the steroid and thyroid hormones belong to this class. (ii) those that bind to specific cell surface receptors, which is the case for the majority of chemical signals, including all of the known neurotransmitters, peptide hormones and growth factors.

Using mast cells as a model system, we shall discuss the membrane events occurring during two aspects of cell signalling: the transduction of extracellular signals into intracellular signals by cell surface receptors and the secretion of chemical signals by exocytosis.

MEMBRANE RECEPTORS AS TRANSDUCERS

Membrane receptors have two functions: they bind specific ligands and communicate this fact to the cell interior. It is useful to divide this transduction process into three operationally-defined steps: (i) the binding of ligand (the 'first message'); (ii) the change in the receptor and/or the ligand that initiates the next event; and (iii) the generation of an intracellular signal - the 'second message'. In most cases, one ligand binds to one receptor and the interaction probably induces a conformational change in the receptor and/or the ligand, although in no case has this been directly demonstrated. In the immune system, the stimulation of basophils, mast cells and possibly lymphocytes by antigen, antibody or lectins requires that the receptors be cross-linked by multivalent ligands; the reason for this requirement remains a mystery. There are only two known ways of generating intracellular signals following the binding of ligands to surface receptors: (i) increasing or decreasing the activity of a plasma membrane enzyme (such as adenyl cyclase) or (ii) changing the permeability of the membrane to small molecules (such as ions) - both of which lead to a change in the intracellular concentration of biologically important molecules which, in turn, serve as the second messages.

Complexity of lymphocyte responses

Although lymphocyte activation has been a popular model for studying the early steps in cell triggering, the systems have been too complex to resolve the molecular events occurring at the level of individual receptors. The responses usually involve cell differentiation and/or proliferation and require complex cell interactions. In order to activate a sufficiently large proportion of the cells, various polyclonal activators are used, most of which bind to a variety of cell surface molecules, only some of which are probably involved in the activation process.

The problem can be illustrated by considering the complexity of cell interactions involved in the most commonly studied lymphocyte model - the proliferative response of T lymphocytes induced by the lectins phytohaemagglutinin (PHA) and Concanavalin A (Con A). There is increasing evidence in guinea pigs that these responses are absolutely dependent on macrophages (3,4). One of us (MCR), in collaboration with

Dr S. Habu, has recently shown that the same is true for at least some of these responses in mice (5); the rigorous removal of adherent and phagocytic cells from a suspension of lymph node lymphocytes produced a population of >98% Ly-1^+,2^- T lymphocytes which did not synthesise DNA in response to Con A stimulation and responded very poorly to PHA. These responses were completely reconstituted by adding back peritoneal cells or macrophages to a final concentration of 1%. The reconstituting cells were Thy-1^-,Ia^+ and were found in normal numbers in nude mice; thus they were not a subpopulation of T cells. Since it has been suggested by others that at least two subpopulations of T lymphocytes (Fc^+,Ia^+ and Fc^-,Ia^-) (6,7) are involved in the proliferative response to Con A in mice, it is probable that at least three different cell types are normally interacting in this response. For this reason alone it is difficult to study the detailed molecular events in receptor signalling.

Mast cells: An accessible model of cell signalling

IgE antibodies, secreted by activated B lymphocytes, bind with high affinity to the surface of mast cells and basophils by means of Fc receptors that are specific for IgE. When specific antigen, anti-IgE antibodies or Con A bind to and cross-link the cytophilic IgE, these cells release their histamine by exocytosis, a process referred to as 'degranulation'. Such ligand-induced degranulation is energy dependent (8), requires extracellular Ca^{2+} (9), and does not kill the cell (10).

Rat peritoneal mast cells are a particularly attractive model for studying receptor transducing events. The cells make up 2-5% of the cells in a peritoneal wash-out and are easily purified by density gradient centrifugation. The stimulating antigens or anti-IgE antibodies bind only to the IgE receptors and can be readily labelled, so that their fate can be followed during the signalling process. Degranulation will occur in water, with sucrose to maintain osmolarity, and in the presence of Ca^{2+} (11); no other ions or proteins are required. Thus, it is a one ligand-one receptor-one cell response. Histamine secretion begins within seconds and is over within a minute; it can be readily assayed by a variety of techniques, and the degranulation is easily visualised by light and electron microscopy. Not only can the mast cell be activated by antigen, antibody or lectin - but the same ligands, in too high or too low a concentration (12), or binding in the absence of Ca^{2+} (13), can 'desensitise' the cells (see below).

The requirement for receptor cross-linking: Mast cells and basophils are the prototype systems where a requirement for receptor cross-linking has been unequivocally demonstrated:

monovalent antigens (14-16) and anti-IgE antibodies (17-19) bind but do not activate. In these systems it has been possible to directly determine whether the requirement for cross-linking reflects a requirement for receptor redistribution in cell activation. In both basophils (18) and mast cells (19), studies using florescent and ferritin-coupled ligands have demonstrated that capping, pinocytosis and gross aggregation of receptors are not required for stimulation; in mast cells, it has been shown that if receptor aggregation is required at all, only a small number of IgE molecules (≤ 6) are probably involved (19). These studies, and those of Siraganian et al. (16), suggest that the formation of receptor dimers is sufficient to signal the cell. Whether the cross-linking requirement reflects a need to bring the few Fc receptors together so that they can interact in some way,or indicates that the IgE molecules must be 'tweeked' in order to induce a conformational change in the Fc receptors is still unclear.

The role of Ca^{2+}: There is increasing evidence to suggest that it is an influx of Ca^{2+} that initiates the exocytosis process in mast cells: Ca^{2+} is a required (9) and sufficient ion for degranulation (11), and the artificial introduction of Ca^{2+} into mast cells by microinjection (20) or by the use of the divalent cation ionophore A23187 (21) induces energy-dependent degranulation. Moreover, sensitised mast cells stimulated by antigen show an increased permeability to ^{45}Ca (21,22). There is also evidence that the increased permeability to Ca^{2+} following activation is transient, lasting a few minutes; this is suggested by the fact that mast cells exposed to antigen in the absence of Ca^{2+} rapidly lose their sensitivity to the stimulating effect of added Ca^{2+} (13), a phenomenon known as desensitisation. Recently it has been directly demonstrated that the increased permeability to ^{45}Ca of activated mast cells decays rapidly after the first minute (22). The simplest interpretation of these results is that gated Ca^{2+} channels open following ligand binding (allowing Ca^{2+} to enter the cell down the steep concentration gradient that exists between the extracellular fluid and cytosol in all cells), and then close. Similar gated Ca^{2+} channels have been demonstrated in mouse T (23) and B (24) lymphocytes - although their role in lymphocyte activation remains to be demonstrated.

It has been shown that various manipulations which increase the concentration of cyclic AMP within mast cells and basophils inhibit histamine secretion (25). The fact that this inhibition can be bypassed by moving Ca^{2+} into the inhibited cells with the ionophore A23187 (26), suggests that elevated cyclic AMP levels may interfere with the triggered Ca^{2+} influx. Direct evidence for cyclic AMP inhibition of a gated Ca^{2+} pathway has been obtained in lymphocytes (23) and, more recently, in mast cells (22). It is conceivable that cyclic AMP is involved in the normal closing of the Ca^{2+} channels in mast cells and basophils.

Initiating and modulating receptors

A differentiated cell is phenotypically committed to a particular function, and the various chemical signals that play on its surface initiate or modulate that function. Thus, one can operationally divide cell surface receptors into initiating receptors and modulating receptors: the latter can be inhibitory (increasing the threshold for activating the cell) or enhancing (decreasing the threshold for cell activation). In mast cells and basophils, IgE molecules attached to Fc receptors serve as initiating receptors for histamine secretion. In basophils, H-2 histamine and β adrenergic receptors are both functionally linked to adenyl cyclase and thus the binding of agonists to these receptors leads to an increase in intracellular cyclic AMP and inhibition of histamine secretion (25). On the other hand, the binding of acetylcholine to muscarinic cholinergic receptors on basophils lowers the threshold of stimulation but, on its own, does not initiate histamine secretion (27). Thus, information transfer at the level of individual receptors is usually simple - of the 'start-stop' or 'speed-slow' type, and, in principle, cells need relatively few types of transducing mechanisms.

MEMBRANE EVENTS DURING EXOCYTOSIS

At the other end of the cell signalling process is the secretion of the chemical signals. In most secretory cells, the triggered release of specific materials stored in secretory granules occurs by exocytosis, which involves the fusion of granule membrane with plasma membrane (and subsequently, granule membranes with each other) and results in the exposure of the granule contents to the extracellular space (28). In mast cells, this results in histamine being released from the anionic binding sites on heparin and protein by exchange with extracellular cations - largely Na^{+} (11), which leads to readily detectable alterations in the ultrastructural appearance of the granule, including swelling and reduced electron density of the granule contents.

Very little is known about the molecular events in exocytosis, or in membrane fusion in general. For example, one would like to know how Ca^{2+} initiates the exocytosis process, and what the molecular basis is for the specificity of fusion that allows granule membranes to fuse with the plasma membrane (and eventually with each other) but not with the nuclear or mitochondrial membranes, and what happens to the membrane lipids and proteins during and after membrane fusion. In collaboration with Drs N.B. Gilula, C. Fewtrell and B.D. Gomperts, we have used ferritin-coupled ligands together with thin section and freeze-fracture electron microscopy to study membrane changes occurring during degranulation of rat mast cells (29). The extensive membrane interactions occurring all around the circumference and inside degranulating mast cells make these cells particularly attractive for studying these membrane events.

Membrane protein displacement during membrane fusion (29)

When degranulating mast cells were labelled with ferritin-coupled lectins (Con A and PHA) and anti-immunoglobulin antibodies, we found that these ligands did not bind to the plasma membrane in areas where it was interacting with underlying granule membrane, or to regions of exposed granule membranes where they were interacting with other granule membranes. This was always the case when the interacting membranes had fused to form a pentilaminar structure; where interacting membranes were not fused and some intervening cytoplasm remained, ligand binding was often reduced, while in places where no intervening cytoplasm was seen, no label was present, even though the membranes were often quite far apart. Monovalent Fab fragments of anti-Ig antibodies gave similar results, and pre-fixing the cells with gluteraldehyde before labelling with the ferritin-coupled lectins did not alter the observations; thus ligand-induced redistribution of membrane proteins could not be responsible for the displacement of proteins from the interacting membranes. Another class of membrane proteins was also found to be displaced during membrane fusion; these proteins were those that are visualised as intramembrane particles in freeze-fracture electron microscopy, which are probably different from the glycoproteins labelled by lectins and antibodies (30). Intramembrane particles were largely absent from both protoplasmic and external fracture faces of plasma and granule membranes in regions where these membranes appeared to be interacting. Both the externally applied ligands and intramembrane particles were sometimes seen concentrated at the edges of fusion sites, suggesting that membrane proteins were displaced laterally into adjacent membrane regions during fusion. The frequent decrease or absence of ligand binding to regions where two membranes were interacting but not yet fused suggested that protein displacement preceded fusion. In fact, membrane protein displacement appeared to occur hand-in-hand with the displacement of visible cytoplasm, suggesting the possibility that displacement of cytoplasm and membrane proteins allowed the spontaneous fusion of the two lipid bilayers.

Interestingly, when degranulating mast cells were labelled with cholera toxin, which has been shown to bind to the ganglioside, GM_1 (31), there was increased labelling in regions of fused membranes (32). This suggests that at least some of the glycolipids are not displaced during membrane fusion and that not all of the gangliosides are associated with protein.

Bleb and whorl formation: Physiological or artifactual? (29)

In regions of plasma and granule membrane interaction in degranulating mast cells, we sometimes observed multi-vesiculated membranous structures blebbing from the cell surface. These blebs were seen by thin section and freeze-fracture electron microscopy and were always unlabelled by the lectins and antibodies and were free of intramembrane particles. Although they were sometimes seen in unstimulated mast cells and in other cell types, they were much more frequent in degranulating mast cells. They were usually seen overlying altered granules. Although blebs were not seen in regions where granule membranes were interacting with each other, multilamellar membrane whorls (myelin figures) were sometimes seen in these areas within cavities containing altered granules. Blebs and myelin figures have usually been considered fixation artifacts and we have been unable to exclude this possibility. On the other hand, they would provide the cell with a simple mechanism for disposing of excess lipid while conserving membrane proteins; in addition, the breaking off of blebs could be the mechanism by which the granules are opened to the extracellular space following membrane fusion.

MAST CELL ACTIVATION: A PROPOSED SEQUENCE OF EVENTS

It is very likely that the cross-linking of IgE receptors leads to the opening of gate Ca^{2+} channels in the mast cell membrane. In preliminary experiments we have not been able to demonstrate a relationship between IgE receptors and intramembrane particles (33), and since an ion channel would be expected to have sufficient mass in the lipid bilayer to be visualised as an intramembrane particle in freeze-fracture studies (30), it seems unlikely that the Fc receptors are physically associated with the putative Ca^{2+} channels. Thus, we would suggest that the Ca^{2+} channels open indirectly. Recently, we have shown that Con A, covalently linked to Sepharose 4B beads, induces Ca^{2+}-dependent localised degranulation along the region of contact between the bead and adherent mast cells (34). This indicates that whatever the mechanism of Ca^{2+} channel opening is, it operates locally in the region of the cross-linked Fc receptors. The mechanism of Ca^{2+}

channel closing, which appears to occur within minutes of opening, is also unknown.

Since the introduction of Ca^{2+} into the cells by means of the ionophore A23187 gives results that are morphologically indistinguishable from those seen when mast cells are stimulated by cross-linking ligands, it is likely that, in each case, it is an influx of Ca^{2+} that initiates the sequence of events that leads to membrane fusion and exocytosis. How Ca^{2+} acts inside the cell in this respect is unknown; since Mg^{2+} cannot substitute for Ca^{2+} (9) and since ATP is required following Ca^{2+} influx (21), it is probable that the mechanism is more complex than simple charge neutralisation or lipid segregation, which may explain the role of Ca^{2+} in the fusion of negatively charged liposomes (35). The fact that degranulation is localised when stimulated by Con A on Sepharose 4B beads (see above) indicates that Ca^{2+} acts locally to initiate membrane fusion, rather than activating a process which is transmitted to the whole cell.

Although it is difficult to determine from static pictures the sequence of events in a rapidly evolving process such as exocytosis in mast cells, we feel that our observations are most consistent with the following hypothetical sequence of events (29): the influx of Ca^{2+} leads to a multifocal displacement of cytoplasmic components and proteins in plasma and underlying granule membranes (possibly mediated by contractile elements, such as microfilaments) which allows the two membranes to fuse; the fused lipid bilayers, being depleted of protein and thus, perhaps, unstable, bulge from the cell to form a bleb which vesiculates and breaks off, opening the granule contents to the extracellular space so that histamine is released by cation exchange. In this way, membrane lipid is lost (and thus cell size remains unchanged) and the membrane proteins are conserved and possibly re-used in forming new granule membranes. It seems likely that at least some of these postulated events occur in other examples of exocytosis and that membrane protein displacement is a common feature of transient membrane fusion.

REFERENCES

1. Gilula, N.B. (1974) In: *Cell Communication* R.P. Cox, ed. John Wiley and Sons, Inc., New York, p.1.
2. Gorski, J. and Gannon, F. (1976) *A Rev. Physiol.* 38,425.
3. Ellner, J.J., Lypsky, P.E. and Rosenthal, A.S. (1976) *J. Immunol.* 116,868.
4. Rosenstreich, D.L., Farrar, J.J. and Dougherty, S. (1976) *J. Immunol.* 116,131.
5. Habu, S. and Raff, M.C. (1977) (submitted for publication).

6. Niederhuber, J.E., Frelinger, J.A., Dine, M.S., Shoffner, P., Dugan, E. and Shreffler, D.C. (1976) *J. Exp. Med.* 143,372.
7. Stout, R.D. and Herzenberg, L.A. (1975) *J. Exp. Med.* 142,1041.
8. Perera, B.A.V. and Mongar, J.L. (1965) *Immunology* 8,519.
9. Foreman, J.C. and Mongar, J.L. (1972) *J. Physiol.* 244,753.
10. Johnson, A.R. and Moran, N.C. (1969) *Amer. J. Physiol.* 216,453.
11. Uvnas, B. (1974) *Fed. Proc.* 33,2172.
12. Norn, S. and Stahlskov, P. (1974) *Clin. Exp. Immunol.* 18,431.
13. Foreman, J.C. and Garland, L.G. (1974) *J. Physiol.* 239, 381.
14. Levine, B.B. (1965) *J. Immunol.* 94,111.
15. Magro, A.M. and Alexander, A. (1974) *J. Immunol.* 112,1757.
16. Siraganian, R.P., Hook, W.A. and Levine, B.B. (1975) *Immunochemistry* 12,149.
17. Ishizaka, K. and Ishizaka, T. (1968) *J. Immunol.* 103,588.
18. Becker, K.E., Ishizaka, T., Metzger, H., Ishizaka, K. and Grimeley, P.M. (1973) *J. Exp. Med.* 138,394.
19. Lawson, D., Fewtrell, C., Gomperts, B. and Raff, M.C. (1975) *J. Exp. Med.* 142,391.
20. Kanno, T.D., Cochrane, D.E. and Douglas, W.W. (1973) *J. Physiol. Pharmacol.* 51,1001.
21. Foreman, J.C., Mongar, J.L. and Gomperts, B.D. (1973) *Nature* 245,249.
22. Foreman, J.C., Hallett, M. and Mongar, J.L. (1977) (submitted to *J. Physiol.*)
23. Freedman, M.H., Raff, M.C. and Gomperts, B.D. (1975) *Nature* 255,378.
24. Freedman, M.H. and Gelfand, E.W. (1976) *Immunol. Comm.* 5,517.
25. Bourne, H.R., Lichtenstein, L.M., Melmon, K.L., Henney, C., Weinstein, Y. and Shearer, G.M. (1974) *Science* 184,19.
26. Foreman, J.C., Mongar, J.L., Gomperts, B.D. and Garland, L.G. (1975) *Biochem. Pharmacol.* 24,538.
27. Kaliner, M., Orange, R.P. and Austen, K.F. (1972) *J. Exp. Med.* 136,556.
28. Palada, G. (1975) *Science* 189,347.
29. Lawson, D., Raff, M.C., Gomperts, B., Fewtrell, C. and Gilula, N.B. (1977) *J. Cell Biol.* 72,242.
30. Bretscher, M.S. and Raff, M.C. (1975) *Nature* 258,43.
31. Cuatracasas, P. (1973) *Biochemistry* 12,3558.
32. Lawson, D. and Herschman, H. unpublished observations.
33. Lawson, D., Gilula, N.B., Fewtrell, C., Gomperts, B.D. and Raff, M.C. unpublished observations.

34. Lawson, D., Field, P.M., Fewtrell, C., Gomperts, B.D. and Raff, M.C. unpublished observations.
35. Papahadjopoulos, D., Poste, G., Schaeffer, B.E. and Vail, W.J. (1974) *Biochim. Biophys. Acta* 352,10.

EARLY MOLECULAR EVENTS IN IMMUNOGLOBULIN E MEDIATED MAST CELL EXOCYTOSIS

Henry Metzger

Section on Chemical Immunology,
Arthritis and Rheumatism Branch,
National Institute of Arthritis, Metabolism
and Digestive Diseases
National Institutes of Health
Bethesda, Maryland 20014

ABSTRACT. Immunoglobulin E (IgE) serves as the specific antigen receptor on mast cells. It in turn interacts with a cell surface component which is the presumptive receptor for IgE-mediated exocytosis. The receptor-IgE interaction has been elucidated and structural studies on the receptor have been begun. Recent data on the distribution, valence, mobility and requirements for triggering now permit one to suggest specific models for the early molecular events in stimulation. Progress in the isolation of the receptor as well as the discovery of a stimulatable tumor suggest that it may soon be possible to test some of these models.

INTRODUCTION

Immunoglobulins play a central role in both the "recognition" and "response" aspects of the immune response (1). However, despite considerable information about the structure of immunoglobulins, and in particular about their antigen combining sites, relatively little is known at a molecular level about how they function. In part this is because most studies have limited themselves to either the structure of the antibody molecule itself or at most to the structure of antibody-antigen complexes. Very few investigations have been performed on the critical ternary complex of antigen-antibody-receptor*. Only recently have studies been reported on one such system - the classical complement system in which Clq plays the role of the receptor; studies at the molecular level of antigen-antibody-cell receptor complexes are almost non-existant.

* I use the word receptor in the sense of a specific macromolecule, changes in which initiate a physiological change elsewhere in a system in which it serves as a specific recognition unit.

A number of years ago my laboratory undertook a long-range project to examine such a system and for this purpose we chose to look at antigen-induced IgE-mediated mast cell degranulation. There were a number of attractive features about this system, the prinicipal ones being that the essential components were few and had been defined, and the response was rapid and easily measured. In this paper I will describe the progress we and others have made in describing this system; I will then present some speculations and finally future plans designed to test these ideas.

THE IgE-MAST CELL SYSTEM

The system can be briefly described as follows. Antibodies of the IgE class are produced by plasma cells. The IgE has the unique property of interacting strongly with mast cells via the Fc region of the molecule. Such mast cells (or related basophils) thus can become "sensitized" with antibodies of diverse specificity. In the natural course of events the cells degranulate when they are exposed to the appropriate antigen - that is to antigen against which the cell bound IgE molecules are directed. In the laboratory this reaction can be mimicked by adding anti-IgE antibodies (2,3).

One aspect of this system which has played a very important role in directing our studies is the evidence that bridging or cross-linking of the IgE molecules plays an essential role in the response. Evidence for this comes from a variety of experiments (4) and I shall review some new data on this point here. It is worth noting that whereas cross-linking appears to be required, gross redistribution of the IgE does not appear to be essential (5,6). Indeed, under conditions of extreme cross-linking, where patching and/or capping of the surface IgE occurs, exocytosis is inhibited (5,7). The essential question to which we have directed our work is: How does the cell 'know' that antigen has reacted with the cell bound IgE antibody; what are the initial molecular events in the triggering of exocytosis? The approach we have taken to answer this question has so far been limited to trying to define the component on the cell which binds IgE. We assume (for lack of other candidates) that this is the receptor in the critical antigen-IgE-receptor complex. We have tried to define its interaction with IgE, its structure, its distribution and properties _in situ_, and are beginning to consider experiments to define its function. I shall now review what we know about these matters.

IgE-RECEPTOR INTERACTION

Monomeric IgE binds to the receptors by a simple reversible reaction (8). The forward bimolecular reaction is unremarkable but the unimolecular back reaction is exceedingly slow leading to an extraordinarily high binding constant. Although others have claimed somewhat lower values (9), our own studies suggest K_A values $\geq 10^{10}$ M^{-1} for the cell bound receptor and even higher values ($\sim 10^{12}$) when the solubilized receptor is studied (10). Quantitative analysis has confirmed older data that the reaction is highly species specific, is dependent upon an undenatured (by reducing agents, heat or pH) IgE and cannot be easily blocked by even high concentrations of other immunoglobulins.

RECEPTOR STRUCTURE

Composition. Froese and his colleagues were the first to show that the receptor could be solubilized by non-ionic detergents, that it could be iodinated and in its soluble state was sensitive to proteases (11,12). Kulczycki et al (13) confirmed these studies (as we have) and in addition reported that radioactive aminoacids and sugars could be incorporated into the receptor. More indirect studies from our own laboratory suggested that the molecule could be denatured by moderate temperature or pH variations (14). All these results as well as the receptor's distribution (below) suggest that it is a glycoprotein.

Size. Based on its mobility in polyacrylamide gel electrophoresis in the presence of sodium dodecyl sulfate the receptor's mass appears to be 50-60,000 daltons (12,13). We recently attempted to determine it size by sedimentation-diffusion analysis in non-ionic detergent and obtained a value (Table 1) of $\sim$130,000 for the receptor-detergent complex (15). Based on a variety of (somewhat gratuitous) assumptions we suggested a molecular weight of 80,000 for the detergent-free receptor. This approach involves a variety of methodological problems (15,16) and we concluded that other approaches, e.g. the use of internal cross-linking reagents, would be necessary to define more precisely the polypeptide multiplicity.

RECEPTOR DISTRIBUTION, VALENCE, MOBILITY

Fluorescence- and electronmicroscopy utilizing labeled IgE or anti-IgE have demonstrated that the receptors for IgE

TABLE 1
MOLECULAR PROPERTIES OF RECEPTOR FOR IgE*

Specimen	$S_{20,w}$	$D_{20,w}$	$\bar{v}$	MW (x 10^{-5})
IgE	7.0	3.3	0.74	2.0
IgE-Receptor Complex	7.3	2.5	0.78	3.1
Receptor	3.3	3.2	0.81	1.3†

* All determinations performed in the presence of 0.05% Nonidct P40.

† On the basis of an assumed $\bar{v}$ of 0.72 for the detergent -free receptor, the molecular weight of the latter was calculated to be ∿80,000.

Adapted from Ref. 15.

are more or less randomly distributed on the surface of the cell. Depending somewhat on the natural history of the cell (8,17), there are several hundred thousand to more than a million sites available for binding. We recently compared the numbers of IgE binding sites on intact *versus* detergent solubilized cells and found them to be identical (10). We have also investigated the question of the receptor valence and the possibility that receptors occurred in functional clusters (18). To do this we exposed cells to a mixture of rhodamine-labeled and fluorescein-labeled IgE and then reacted the cells with anti-fluorescein. There was no evidence of comigration of the rhodamine-labeled IgE when the fluorescein-labeled IgE became patched and/or capped. This experiment unambiguously demonstrates that the receptors are effectively univalent and independently mobile. (The molecular weight studies referred to above, included a comparison of the weight of the empty receptors *versus* the weight of the receptor-IgE complexes. The difference, 180,000 was also consistent with the univalence of the receptor since IgE has a molecular weight of about 200,000.)

To define more precisely the mobility characteristics of the receptor we collaborated with J. Schlessinger et al., in a study which utilized the newly developed method of photobleaching of fluorescence (19). With this method one can focally irradiate a portion of the surface of a cell thereby irreversibly quenching the fluorescence of fluorochrome-labeled surface components. The rate and extent of fluorescence recovery as well as its sensitivity to a variety of cell poisons may provide clues about the interaction of cell surface components with other cell structures. Labeled IgE on normal rat mast cells (and related tumor cells) were examined using this approach. The results were consistent with partial immobilization of the receptors. Surprisingly immobilization was enhanced in the presence of cytochalasin B, a reagent one of whose properties is to depolymerize microfilaments. Moderate cross-linking of the fluorescein-labeled IgE with anti-fluorescein caused no detectable changes but more profound cross-linking abrogated all IgE movement (Table 2). The independent mobility of rhodamine-conjugated IgE under these conditions was again confirmed.

FUNCTIONAL ASPECTS

The results so far presented allow us to sketch a possible sequence of events. Monomer IgE molecules of diverse specificity become bound to monovalent surface receptors. When exposed to antigen the IgE molecules bind to it. Because of the mobility of the receptors several IgE molecules can become attached to the same antigen and this apposition of IgE-receptor complexes can presumably lead to some type of signal.

MINIMAL UNIT SIGNAL

We have just completed a study (20) whose aim was to determine how large such an IgE-receptor complex has to be in order to trigger a unit signal; the summation of such signals leading to a detectable response. IgE was artificially cross-linked with dimethyl subarimidate and the derivatized molecules fractionated into monomeric, dimeric, trimeric, etc., IgE by gel filtration. These fractions were then assayed *in vivo* by passive cutaneous anaphylaxis and *in vitro* by sensitizing mast cells and monitoring for histamine (21) or serotonin release (22). Both the *in vivo* and *in vitro* assays gave consistent and completely unambiguous results: monomeric IgE did nothing, while dimeric IgE gave excellent responses (Figure 1). Higher oligomers were not significantly more effective than the dimers. The latter result effectively rules out that (undetected) high molecular weight contaminants were responsible for the effectiveness of the dimers. The ineffectiveness of the monomers

TABLE 2
PHOTOBLEACHING OF FLUORESCENCE STUDY OF RECEPTOR-IgE COMPLEXES ON NORMAL MAST CELLS*

Additive	Conjugate†	Distribution	D(x 1C^{10}) cm^2 sec^{-1}	% Fluorescence Recovery
1. Buffer	R-IgE	diffuse	1.2±0.6	60-80
	F-IgE	diffuse	1.6±0.3	50-80
2. Horse anti-fluorescein	R-IgE	diffuse	1.2±0.6	50-70
	F-IgE	microaggregates	1.5±0.5	30-80
3. Horse anti-fluorescein, then rabbit anti-horse IgG	R-IgE	diffuse	1.1±0.7	40-80
	F-IgE	patches	<.06	∿10%

* Adapted from Ref. 19.

† R-IgE, rhodamine-labeled IgE; F-IgE, fluorescein labeled IgE

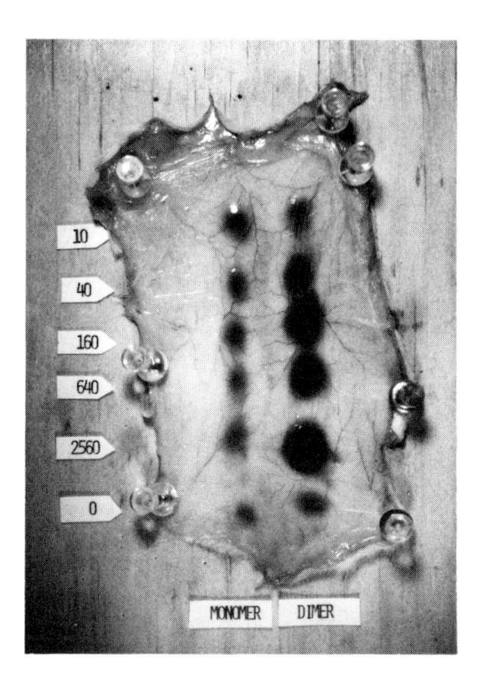

Fig. 1: Passive cutaneous anaphylaxis analysis of monomeric and dimeric rat IgE. IgE was treated with dimethyl subarimidate and fractionated by gel filtration. The peak tubes of monomeric and dimeric molecules were diluted to the same concentrations. A ~300g Sprague Dawley rat was injected intravenously with 2.5 mg Evans Blue dye and then with 0.05ml of the diluted fractions intradermally. After 25 min. the internal surface of the skin was displayed. The numbers on the left side refer to the nanograms of IgE injected at each site. The diluent was phosphate buffered saline, pH 7.2 containing 100 μg/ml bovine serum albumen. Segal, D., Taurog, J.D., and Metzger, H., Unpublished work and (20).

over a wide range of doses, demonstrates that conformational isomers due to the derivatization were also not the cause. These findings strongly suggest that pairs of apposed IgE-receptor complexes trigger the exocytosis event. Data from Lichtenstein's laboratory suggest that the number of such signals required to obtain a maximal response is likely to be a few thousand or less (23). That is, less than 1% of the total number of available receptors may be required.

WHAT IS THE MESSAGE?

While we know more about how the signal may be generated the nature of the message is still unknown. I would like to discuss briefly some possibilities and then discuss how they may be explored.

Three observations may be interestingly related. It has been observed that during the initial stages of mast cell exocytosis fusions occur between the perigranular membranes of the superficial granules and the plasma membrane (24). Concomittantly the fused areaa become depleted of membrane proteins (25,26). That this process may be a relatively simple phenomenon (in the sense that it may not require complicated cytoskeletal element contractions) is suggested by the observation that isolated chromaffin granules exposed to Ca^{++} show similar fusion phenomena (27). The final observation is that certain proteins are capable of forming ion channels in their polymerized form whereas they fail to do so in the monomeric state (28). Thus a possible mechanism for antigen-induced IgE-mediated exocytosis would involve the dimerization of surface IgE-receptor complexes leading to the formation of a Ca^{++} ionophore which can then initiate the reaction. Other possibilities can of course be envisioned in view of our almost total ignorance. It is possibile that the aggregation of the receptor leads to a conformational change which converts the receptor into an enzyme. One is reminded of the protein Clq which, when reacting with antigen antibody complexes, appears to undergo such a conformational change (in this case by a process of intra-molecular aggregation) (29). It is noteworthy in this regard that there is evidence for involvement of an "activatable esterase" in the rodent mast cell degranulation (3). Could it be related to the receptor for IgE?

FUTURE WORK

We are pursuing three major kinds of work: 1) We and others are attempting to produce antibodies specific for the receptor (30-33). On the basis of the considerations reviewed above one would predict that such antibodies should be capable of initiating a reaction providing they are capable of dimerizing the receptors. We feel strongly that until monospecific anti-receptor antibodies are available experiments of this type can only lead to ambiguous results. Providing a positive response could be elicited it would demonstrate that the component which binds IgE to the cell surface was a true receptor (see Footnote page 1).

2) A second approach involves purification and large scale production of the receptor. The approach we are taking towards purification (34) appears to be working. We haptenate IgE, react it with receptor-containing cell extracts, absorb the mixture on an anti-hapten column and elute with hapten. The process can be repeated and preparations of approximately 75% purity have been achieved.

Freeing receptors from the isolated IgE-receptor complexes presents a different problem with which we think we are also having some success. As far as large scale production the statistics are as follows: One can easily culture rat basophilic leukemia cells to 1×10^6 cells/ml bearing 1×10^6 receptors/cell (8,17). Assuming a molecular weight of 8×10^4 for the receptor (15) and a 100% yield (!) one can obtain 120 μg receptor per liter. Correcting for more realistic yields one can anticipate obtaining 10-20 μg pure material. We have preliminary data which suggest that working with solid tumors of these cells grown on new-born rats, may be a more efficient and economic approach. These data suggest that one can realistically anticipate obtaining 50-150 μg/cage. Since the tumors appear to be storable substantial amounts of purified receptor suitable for detailed structural studies are forseeable.

3) At present one cannot distinguish between receptors from normal rat mast cells and those from the rat basophilic leukemia cells. However, the failure of the latter to respond normally to stimuli (35), gives one concern that they may not be the appropriate source to use for studying functional receptors. Ideally one would like a functional tumor. Recent studies show that such cell lines do occur (36). A representative experiment which utilized cultured cells of the AB-CBF_1-MCT-1 mastocytoma is shown in Fig. 2. While we are not certain that the particular mouse mastocytoma we are investigating will be completely appropriate its discovery makes us optimistic that similar tumors will become available in the future.

CONCLUDING REMARKS

I have reviewed the present status of our knowledge about one relatively simple system which appears to involve as a critical component an antigen-antibody-receptor complex. The system resembles in several significant ways the antigen-antibody-complement reaction. My own feeling is that more detailed exploration of both these systems may yield insights into related phenomena. One of the most central of these is of course the antigen mediated activation of B cells. In that system also, a variety of experiments have implicated polyvalent interactions as playing a critical role (37,38). The relationship between such other antigen-induced antibody-mediated cell triggering reactions and the IgE system are certainly worth exploration.

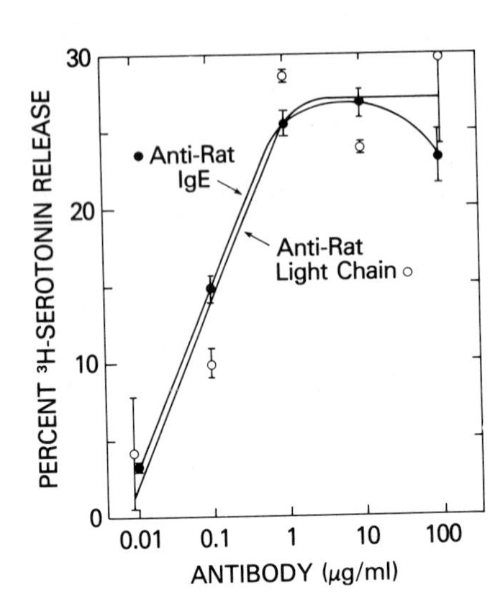

Fig. 2: Washed mastocytoma cells were incubated with ^{3}H-serotonin and IgE. The cells were rewashed and then resuspended with varying concentrations of rabbit anti-IgE or anti-light chain. Supernatants were tested 45 min. later (J.D. Taurog, G.A. Mendoza, W.A. Hook, R.P. Siraganian and H. Metzger In preparation).

ACKNOWLEDGEMENTS

While I assume responsibility for this paper it should be clear from the citations that the results described from my laboratory were obtained by the work of many associates and collaborators. In addition to those cited, Drs. J. Halper and D. Carson made valuable contributions and Mr. G. Poy and J. Rivera provided valued technical assistance.

REFERENCES

1. Metzger, H. In, '*Proc. of the 1974 ICN-UCLA Symposium on Molecular Biology*,' Academic Press, N.Y. p.1 (1974).
2. Ishizaka, K., and Ishizaka, T. (1971) *Ann. N.Y. Acad. Sci.* 190, 443.
3. Becker, E.L., and Henson, P.M. (1973) *Adv. in Immunol.* 17, 94.
4. Metzger, H. In, '*Receptors and Recognition*,' Chapman and Hall Ltd., London (In press).
5. Becker, K.E., Ishizaka, T., Metzger, H., Ishizaka, K., and Grimley, P.M. (1973) *J. Exp. Med.* 138, 394.

6. Lawson, D., Fewtrell, C., Gomperts, B., and Raff. M.C. (1975) *J. Exp. Med.* 142, 391.
7. Magro, A.M., and Alexander, A. (1974) *J. Immunol.* 112, 1762.
8. Kulczycki, Jr., A., and Metzger, H. (1974) *J. Exp. Med.* 140, 1676.
9. Conrad, D.H., Bazin, H., Sehon, A.H., and Froese, A. (1975) *J. Immunol.* 114, 1688.
10. Rossi, G., Newman, S.A., and Metzger, H. (1977) *J. Biol. Chem.* 252, 704.
11. Conrad, D.H., Berczi, I., and Froese, A. (1976) *Immunochem.* 13, 329.
12. Conrad, D.H., and Froese, A. (1976) *J. Immunol.* 116, 319.
13. Kulczycki, Jr., A., McNearney, T.A., and Parker, C.W. (1976) *J. Immunol.* 117, 661.
14. Metzger, H., Budman, D., and Lucky, P. (1976) *Immunochem.* 13, 417.
15. Newman, S.A., Rossi, G., and Metzger, H. (1977) *Proc. Nat'l. Acad. Sci.* (In press).
16. Tanford, C., and Reynolds, J.A. (1976) *Bioch. Biophys. Acta.* 457, 133.
17. Isersky, C., Metzger, H., and Buell, D. (1975) *J. Exp. Med.* 141, 1147.
18. Mendoza, G.R., and Metzger, H. (1976) *Nature* 264, 548.
19. Schlessinger, J., Webb, W.W., Elson, E.L., and Metzger, H. (1976) *Nature* 264, 550.
20. Segal, D.M., Taurog, J.D., and Metzger, H. (1977) *Proc. Nat'l. Acad. Sci., U.S.A.* (In press).
21. Siraganian, R.P. (1974) *Anal. Biochem.* 57, 383.
22. Morrison, D.C., Rosen, J.F., Henson, P.M., and Cochrane, C.G. (1974) *J. Immunol.* 112, 573.
23. Lichtenstein, L. In, '*Proc. IX International Congress Allergy*,' Buenos Aires, *Excerpta Medica*, Amsterdam (In press).
24. Uvnas, B. In, '*Mechanisms in Allergy*,' Marcel Dekker, Inc., New York p. 369 (1973).
25. Chi, E.Y., Lagunoff, D., and Koehler, J.K. (1976) *Proc. Nat'l. Acad. Sci.* 73, 2823.
26. Lawson, D., Raff, M.C., Gomperts, B., Fewtrell, C., and Gilula, N.B. (1977) *J. Cell. Biol.* 72, 242.
27. Schober, R., Nitsch, C., Rinne, U., and Morris, S.J. (1977) *Science* 195, 494.
28. Rosenstreich, D.L., and Blumenthal, R. (1977) *J. Immunol.* 118, 129.
29. Muller-Eberhard, H.J. (1975) *Ann. Rev. Biochemistry* 44, 697.
30. Froese, A. (1976) *Immunol. Communications* 5, 437.

31. Isersky, C., Mendoza, G.R., and Metzger, H. (1977) *Fed. Proc.* 36, 1217.
32. Conrad, D.H., Yiu, S.H., and Froese, A. (1977) *Fed. Proc.* 36, 1217.
33. Ishizaka, T., Chang, T.H., Taggart, M., and Ishizaka, K. (1977) *Fed. Proc.* 36, 1217.
34. Rossi, G., Kanellopoulos, J., and Metzger, H. (In preparation).
35. Siraganian, R.P., Kulczkycki, Jr., A., Mendoza, G.R., and Metzger, H. (1975) *J. Immunol.* 115, 1599.
36. Taurog, J.D., Hook, W.A., Siraganian, R.P., and Metzger, H. (1977) *Fed. Proc.* 36, 1215.
37. Feldman, M. (1972) *J. Exp. Med.* 135, 735.
38. Dintzis, H.M., Dintzis, R.Z., and Vogelstein, B. (1976) *Proc. Nat'l. Acad. Sci., U.S.A.* 73, 3671.

Membrane Phase Transitions
and the Regulation of
B Lymphocyte Activation

K.A. Krolick, B. Wisnieski
and E.E. Sercarz
University of California
Department of Bacteriology
and the
Molecular Biology Institute
Los Angeles, California 90024

INTRODUCTION

During the last several years, attention has been directed toward examining the structural and functional roles played by the lipid bilayer with regard to events occurring at the surface of animal cells. It is becoming apparent that the lipid components of cell membranes can dramatically influence the behavior of membrane-associated active moieties (membrane-bound enzymes, receptors, etc.) (1,2). In addition, the character of the association between surface receptors and the first cytoplasmic components of the cell activation machinery may determine the triggerability of that cell by specific stimuli.

To date, most studies into the mechanical aspects of anti-immunoglobulin (Ig) induced redistribution of cell surface Ig (sIg) on B lymphocytes have examined sIg movement primarily as a function of time at a few select temperatures (3). We have employed an alternate approach, namely, the investigation of membrane surface behavior as a function of temperature over a constant increment of time (4). This tactic has been successfully exploited by a number of investigators for correlating changes in membrane function with changes in membrane physical state (5). To determine the extent to which sIg modulation (patching, capping) is dependent on the physical state of the B cell membrane, studies were initiated to compare and correlate anti-Ig induced redistribution of sIg with changes in the physical state of each monolayer of the plasma membrane.

Changes in the physical state of membranes as a function of temperature can be monitored by electron spin resonance spectroscopy (ESR), specifically by measuring the partitioning of ESR spin labels between the aqueous and hydrocarbon (i.e., membrane) milieu of a membrane suspension. Arrhenius plots of partitioning ratios of ESR probe in hydrocarbon to ESR probe in water versus 1/degree K show discontinuities at characteristic temperatures where changes in state (i.e., melting or freezing) occur (6). These discontinuities arise because the probes employed are more soluble in fluid membrane hydrocarbon than in frozen hydrocarbon. While a pure lipid system undergoes a change in state (frozen → fluid) over a very discrete and very narrow temperature range, observations on binary lipid systems show that "melting" occurs over a wider range of temperatures. The boundaries of this process, termed lateral phase separation, are defined as "characteristic temperatures," t_l and t_h, which mark the onset and completion of lipid melting:

$$\text{frozen} \xrightarrow{t_l} \begin{array}{c}\text{frozen and}\\ \text{fluid domains}\end{array} \xrightarrow{t_h} \text{fluid}$$

Recently, ESR studies have shown that the inner and outer monolayers of animal cell membrane preparations display two non-identical sets of characteristic phase separation temperatures due to their asymmetric distribution of lipids and proteins across the bilayer (7).

EXPERIMENTAL

ESR analysis of mouse B cell membranes

If a homogeneous population of animal cells were examined by ESR, a total of four phase transition temperatures would be expected: one pair associated with the onset and completion of the phase transition process in each monolayer. However, ESR analysis of mouse splenic B lymphocytes (90% sIg^+) revealed 7-8 characteristic temperatures. In addition, when a probe which preferentially associates with the outer membrane monolayer was employed, four transition temperatures were observed rather than the two normally seen with intact membrane systems.

Membrane fluidity and the mobility of receptors on B cells

Since the number of discontinuities observed was twice the expected number, the data suggested that the sIg^+ cells examined consisted of at least two distinct

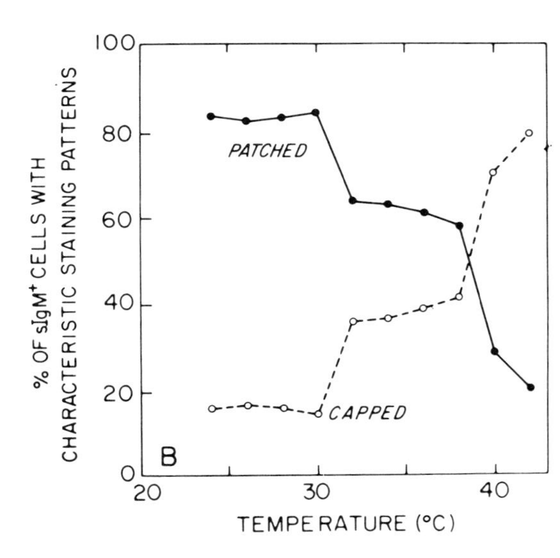

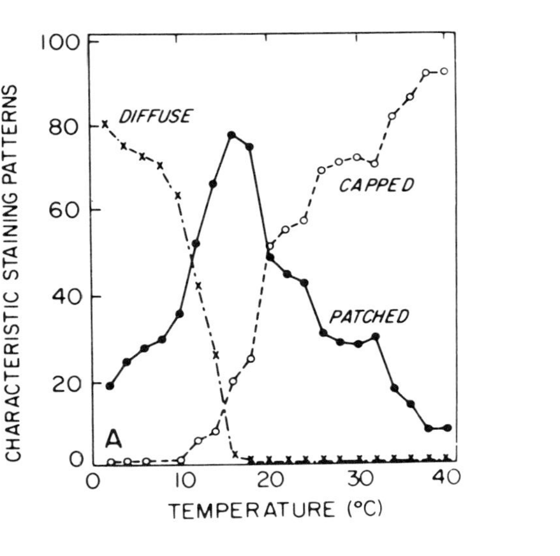

Extent of splenocyte receptor movement induced at various temperatures (2-40°C) during 15 minute (A) or 5 minute (B) exposures to 50 μg/ml/10^7 cells FITC-goat-anti-mouse-IgM. Incubations were terminated by addition of a 10x volume of 4% paraformaldehyde. 100-200 sIgM positive cells were scored as diffuse, patched, or capped.

cell populations. The sIgM and sIgG-bearing cells seemed a reasonable starting point to examine this possibility since they are two major B cell types.

When the temperature dependence of sIgM movement was examined (Fig 1), we found that upon exposure to FITC-GAMIgM the rate of receptor redistribution increased as temperature was increased. This rate increase is reflected by the increased extent of receptor movement from a diffuse distribution to a patched and finally capped distribution on the cell surface. The increasing rate of receptor movement is a complex function of temperature showing discontinuities at very specific points. For sIgM movement, there are discontinuities in such a curve at 18, 24, 32, and 38 degrees C. These characteristic temperatures associated with changes in receptor behavior correlate nicely with four of the eight transition temperatures revealed by ESR. When the same receptor redistribution experiment is performed using anti-IgG2 specific antibodies, the discontinuities observed are different from those seen with the anti-IgM generated data, but correlate exquisitely with the four remaining temperatures of the eight revealed by ESR.

This suggests that the two major populations we are observing are an IgM-bearing and a non-IgM-bearing, IgG-bearing population.

Results summarized in Table I indicate three main points. First, IgM- and IgG-bearing cells demonstrate changes in receptor behavior at different temperatures. For $sIgM^+$ cells, characteristic temperatures are found at 18, 24, 32, and 38 degrees C; for $sIgG2^+$ cells, characteristic temperatures are found at 14, 22, 28, and 36 degrees C. Second, changes in receptor behavior observed at these temperatures are a function of membrane phase transitions. This implies that membrane melting is occurring at different temperatures in the two B cell populations; at any given temperature the membranes of sIgG2-bearing lymphocytes will be intrinsically more fluid than the membranes of sIgM-bearing cells and thereby allow freer receptor movement than the sIgM positive cells.

Table 1.

Inner and Outer monolayer temperature assignments related to discontinuities in sIg behavior

Transition Boundaries[a]	Native		LPS Treated	
	$sIgM^+$	$sIgG2^+$	$sIgM^+$	$sIgG2^+$
$t_l \rightarrow t_h$ (inner monolayer)	24→38C	22→36C	24→36C	28→40C
$t_l \rightarrow t_h$ (outer monolayer)	18→32C	14→28C	18→30C	22→30C

[a] t_l and t_h refer to the lower and upper temperature boundaries, respectively, of phase transitions which occur in each of the plasma membrane monolayers of $sIgM^+$ or $sIgG2^+$ CBA/J spleen cells. Inner versus outer monolayer temperature assignments were made by monitoring phase transitions with two ESR probes, one which freely associates with both monolayers and another which is restricted to the outer monolayer (4).

Effect of LPS on B cell receptor mobility

To pursue the relationship between B cell surface activity and membrane physical state, we investigated how the physical state of the plasma membranes of differentiated subpopulations of B lymphocytes may affect the outcome of interactions with lipophilic effector molecules. One system we have examined was the differential reactivity of *E. coli* lipopolysaccharide (LPS) with IgM-bearing versus IgG-bearing cells. LPS is a B cell mitogen which activates IgM^+ cells preferentially over IgG^+ cells (8). It is of interest that despite the failure of LPS to trigger IgG-bearing B cells, short-term exposure (10 minute) of spleen cells to LPS at 37 degrees C had a profound effect on the temperature-dependent receptor mobility of IgG2-bearing cells.

Upon exposure to LPS, sIgM-bearing cells demonstrate a consistent enhancement of receptor mobility at temperatures above 24 degrees C (Table I). However, the same short-term exposure to LPS dramatically shifts the curve of sIgG2 receptor mobility to temperatures even higher than those characteristic of native sIgM mobility.

This indicates that LPS exerts a membrane rigidifying effect on sIgG2$^+$ cells and promotes the opposite effect on sIgM$^+$ cells. Such LPS-induced differences (membrane-fluidizing versus membrane-rigidifying) may be interpreted via a model of lipid A intercalation into the bilayer which either disrupts the bilayer structure as in the case of the sIgM$^+$ cells, or stabilizes the bilayer as in the case of the sIgG2$^+$ cells. Each effect is predetermined by the physical state of the membrane at 37 degrees C. These direct and opposite membrane effects may explain the preferential activation of sIgM-bearing (versus sIgG-bearing) lymphocytes into mitosis and polyclonal antibody synthesis.

Membrane fluidity and B cell development

In a more general scheme of development, changes in membrane fluidity may occur analogous to those observed when sIgM-bearing cells are compared to the more mature sIgG-bearing cells. Evidence from this laboratory (Krolick and Sercarz, manuscript in preparation) summarized in Figure 2 suggests that a general increase in B cell

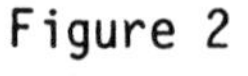

Figure 2

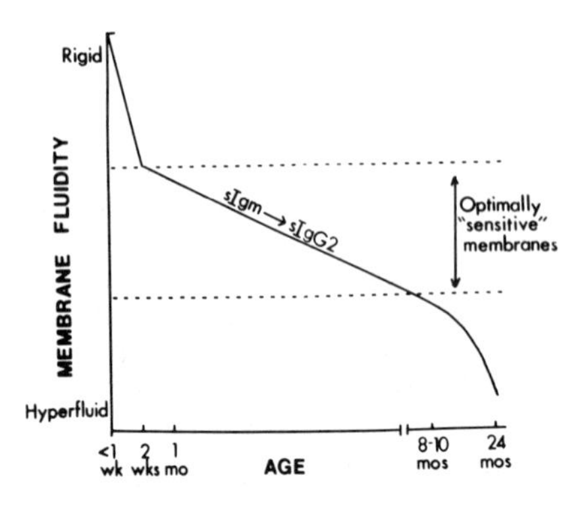

Representation of age-related changes of the fluid nature of mouse B lymphocyte membranes. During the ontogeny of CBA/J mice, changes in membrane fluidity are reflected by changes in the temperature-dependent behavior of immunoglobulin receptors in response to anti-Ig.

membrane fluidity occurs during early life, followed by a stable period in which there are no dramatic changes in this property, but then followed again by increasing membrane fluidity in the aged animal. Such developmental changes in membrane fluidity could serve as a coarse regulatory mechanism in controlling signals at crucial stages of B cell activation. The optimally "sensitive" membrane, present on lymphocytes in the young adult would efficiently transduce signals from external antigens to the cell activation machinery. A very rigid fetal cell membrane may provide a basis for the efficient neonatal tolerance induction reported from various laboratories (9,10). Similarly, if cells from aged animals contained deregulated, hyperfluidized plasma membranes, immune regulation might be difficult. We postulate that a crucial element in B lymphocyte activation is a membrane of appropriate consistency to facilitate and support the proper contacts between receptors for antigens, mitogens, T cell products, etc., and the next component of the activation mechanism of the B cell. The "next component" may well be some structural element of the cytoskeletal network underlaying the membrane. Or, perhaps, some membrane-associated enzyme (for example, adenyl cyclase) depends on bilayer support for determining the availability of its active site (open or closed) for binding with its substrate. Subsequently, the allosteric influence of a closely associated ligand-bound receptor on the activity of this membrane-bound enzyme would be contingent on the physical state of hydrocarbon components which surround the receptor-enzyme "contact points."

Regulation of such molecular interactions on a coarse level could be attained through alterations in membrane lipid composition, thereby shifting the transition temperatures found in the physiological ranges (36 degrees C for IgG2-bearing cells and 38 degrees C for IgM-bearing cells) and consequently the fluidity of those membranes. It is interesting to note that the transitions occurring around 37 degrees C in the lymphocyte populations examined, are associated with the inner monolayers of their plasma membranes. Therefore, slight changes in membrane composition could be efficiently produced by inserting, metabolizing or deleting components of this monolayer, consequently affecting changes in membrane fluidity. Since our data indicate that Ig receptors for antigen are associated with both membrane monolayers, subtle changes in inner monolayer fluidity could influence the interactions of external receptor molecules with their cytoplasmic partners in the B cell activation machinery.

ACKNOWLEDGEMENTS

Research was supported by NIH grants AI-11183 & GM-22240. K.A. Krolick is a predoctoral trainee of USPHS Training Grant AI-00431.

REFERENCES

1. deKruyff, B., P.W.M. vanDyck, R.W. Goldbach, R.A. Demel and L.L.M. vanDeenen. 1973. in Bioch. Bioph. Acta 330:269.

2. Rittenhouse, H.G., R.E. Williams, B.J. Wisnieski and C.F. Fox. 1974. Bioch. Bioph. Res. Commun. 58:222.

3. Schreiner, G.F. and E.R. Unanue. 1977. Adv. in Immunol. 24:37.

4. Krolick, K.A., B.J. Wisnieski and E.E. Sercarz. 1977. Submitted to Proc. Nat. Acad. Sci.

5. Wisnieski, B.J., J.G. Parkes, Y.O. Huang and C.F. Fox. 1974. Proc. Nat. Acad. Sci. (U.S.A.) 71:4381.

6. Shimshick, E.J. and H.M. McConnell. 1973. Bioch. 12:2351.

7. Wisnieski, B.J. and K.K. Iwata. 1977. Bioch. 16(7):1321.

8. Melchers, F. and J. Andersson. 1974. Eur. J. Immunol. 4:181.

9. Nossal, G.J.V. and B.L. Pike. 1975. J. Exptl. Med. 141:904.

10. Metcalf, E.S. and N.R. Klinman. 1976. J. Exptl. Med. 143:1327.

FUNCTIONAL CORRELATES OF SURFACE Ig EXPRESSION FOR T-INDEPENDENT ANTIGEN TRIGGERING OF B CELLS

D.E. Mosier, J.J. Mond, I. Zitron, I. Scher, and W.E. Paul

Laboratory of Immunology, NIAID, NIH, Bethesda, Md. 20014

ABSTRACT

Experiments which have analyzed the correlation between the response of mouse spleen cells to T-independent (T-I) antigens and the surface Ig phenotype of B cells have suggested that: 1) two classes of T-I antigens exist, those which can trigger neonatal or CBA/N defective B cells (designated T-I 1 antigens), and those which can trigger only either more mature B cells or different B cells (T-I 2 antigens); 2) the ability to respond to T-I 2 antigens is correlated with a reduction in total sIg density and acquisition of sIg-D; and 3) although most adult B cells can respond to both classes of T-I antigens, only T-I 2 antigens require the participation of sIg-D for antigen triggering to occur.

INTRODUCTION

During the course of B cell differentiation there occur many changes in the expression of cell surface molecules which might affect the process of B cell activation by antigen. The first and most obvious of these changes is the acquisition and expression of cell surface immunoglobulin (sIg). Many other molecules (e.g., Ia, C 3 receptor, F_c receptor, "mitogen" receptors, MHC products, etc.) are expressed in the course of B cell differentiation or, alternatively, on one or another subline of B cells (1). In studying the response to several T-I antigens of B cells with different surface phenotypes, we have come to the conclusion that the major determinant of responsiveness to these antigens is the sIg phenotype. This conclusion is based on two approaches. The first is the comparison of antigen-triggering requirements of spleen cells from newborn mice, adult mice, and CBA/N mice, a strain with an X-linked defect that leads to the accumulation of B cells with a relatively immature surface phenotype (2-5). The second approach involves separating normal adult splenic B cells on the basis of their density of sIg. These approaches suggest that B cells expressing relatively little sIg and having a low μ/δ ratio are responsible for the antibody response to most T-I antigens and are absolutely required for the response to one subclass of T-I antigens (T-I 2).

METHODS

The normal mice used in these studies were (C57BL/6N x DBA/2N) F_1 (BDF_1) hybrids raised from birth in our laboratory. Immunologically defective CBA/N or (CBA/N x DBA/2) F_1 males or normal F_1 females were supplied by the Small Animal Section, NIH. Spleen cells were cultured in microplates as previously described (5) and generation of trinitrophenyl (TNP-specific antibody forming cells assayed by local hemolysis of TNP-erythrocytes in agarose (6). Phosphorylcholine (PC)-specific plaque-forming cell (PFC) responses were assayed against C II substance-conjugated erythrocytes. The antigens employed were TNP-*Brucella abortus*, TNP-lipopolysaccharide (TNP-LPS), TNP-aminoethylcarbamylmethyl-Ficoll (TNP-Ficoll), TNP-aminoethylcarbamylmethyl-dextran (TNP-dextran), and PC-*B. abortus*. Cells were stained for sIg with a fluorescein-conjugated purified $F(ab')_2$ fragment of goat anti-mouse immunoglobulin (7). Anti-immunoglobulin columns (8) or plates (9) were prepared by attaching purified goat anti-mouse Ig reagents to Sephadex G-200 or Petri dishes. Cells adherent to anti-Ig columns were eluted by incubation with normal mouse Ig or purified myeloma proteins. Cells adherent to anti-Ig plates were eluted by vigorous washing. Cells recovered from such fractionation were characterized as to their sIg expression by staining with fluorescein-anti-Ig either immediately after separation or following 24 hr. of culture to allow recovery from any modulation of sIg caused by the separation procedures. Staining profiles were analyzed on the fluorescence-activated cell sorter (FACS).

RESULTS

Analysis of sIg density by staining intensity on the FACS has shown (7,10,11) that neonatal and CBA/N B cells have a higher mean sIg density than normal adult B cells. The increase in sIg density seems to be due mainly to an increase in sIg-M. By contrast, adult B cells have less intense staining and a significant portion of the sIg is sIgD (7). The major change that occurs, therefore, in sIg expression during B cell development (maturation and/or appearance of sublines) is the relative reduction of sIg-M and the relative increase in sIg-D. If the failure of the defective CBA/N mice to respond to some T-I antigens (e.g., TNP-Ficoll) were linked to its relatively low expression of sIg-D, then we would expect that the time of appearance of TNP-Ficoll responsiveness should correlate with the acquisition of δ-bearing cells. Conversely, sIg-D+ B cells should not be required for the T-I responses which CBA/N mice can make (e.g., to TNP-*B. abortus*) should appear prior to the development of significant numbers of sIg-D+ cells, which occurs at about 7 days of age (12).

Spleen cells from neonatal BDF_1 mice were stimulated with the panel of antigens listed in Table 1 and their direct PFC responses measured after 2-4 days of culture. The first responses to TNP-*B. abortus* and TNP-LPS were detected within one day of birth and adult levels of responsiveness were attained by 7-10 days of age (13). Antibody responses to TNP-Ficoll

TABLE 1: Correlation of sIg Phenotype with T-I Responses

sIg:	μ+++, δ±	μ+, δ++
Appearance in Ontogeny:	Early	Late
antigen responses:[1]		
1. TNP-LPS	+	+
TNP-*B. abortus*	+	+
2. TNP-Ficoll	-	+
TNP-dextran	-	+
SSS III	-	+
PC-*B. abortus*	-	+
differentiation antigen:		
Ly b 5	-	+
susceptibility to antigen-Ig inhibition:		
anti-μ	↑	↓
anti-δ	-	↑

[1]T-I antigen designation: T-I 1^+, 2^- T-I 1^+, 2^+

and the remaining T-I antigens first appeared at 5-7 days of age and did not reach adult levels until about 21 days of age. This was the case with PC-*B. abortus* as well (in contrast to TNP on the same carrier), which may be due to the relatively late appearance of the PC specificity (14) or to an intrinsic difference between PC and TNP as ligands. The ability to respond to the "late" T-I antigens correlates with the acquisition of sIg-D although other maturational events are occurring at the same time. As a further test of the hypothesis that the change in sIg phenotype allowed antibody triggering by late T-I antigens, cells from adult BDF_1 spleens were fractionated on anti-Ig substrates. Spleen cells with low sIg density were non-adherent by such techniques. The fraction of sIg^+ cells in the nonadherent low sIg subpopulation was only 1-4 percent of unfractionated spleen cells. The majority of B cells were adherent to anti-Ig adsorbents and had a relatively high density of sIg by FACS analysis. These techniques thus yielded populations of B cells with relatively low sIg density (non-adherent), relatively high sIg density (adherent), and unfractionated cells which included intermediate sIg density representatives. Each fraction was stimulated _in vitro_ with TNP-_B. abortus_ or TNP-Ficoll and the TNP-specific direct PFC response measured 4 days later. Table 2 presents data that demonstrate that the low sIg density fraction generated many more PFC per sIg^+ cell than either unfractionated or high sIg density fractions.

TABLE 2: Separation of Adult B Cells on Anti-Ig Substrates

Technique:	Relative sIg Density[1]		
	High (adherent)	Normal (Unfractionated)	Low (Nonadherent)
1. Anti-Ig dish[2]			
percent sIg^+	91	45	2
TNP-_B. abortus_ PFC response[3]	646	1,040	488
PFC/sIg^+ cell ($\times 10^{-4}$)	7	33	344
2. Anti-Ig column[4]			
percent sIg^+	91	45	7
TNP-_B. abortus_ PFC response	526	433	638
PFC/sIg^+ cell ($\times 10^{-4}$)	6	10	91
TNP-Ficoll PFC response	146	280	256
PFC/sIg^+ cell ($\times 10^{-4}$)	2	7	36

[1]Relative sIg density was confirmed by staining of cells with a polyvalent anti-mouse Ig reagent and analysis on the FACS.

[2]Performed in collaboration with Drs. M. Mage and T. Rothstein.

[3]Numbers are direct PFC/10^6 cells cultured. Data are from 1 experiment in number 1 and from 8 replicate experiments in 2.

[4]Performed in collaboration with Drs. H. Herrod.

Although both TNP-*B. abortus* and TNP-Ficoll are T-independent, the possibility that the high number of T cells in the nonadherent fraction was contributing to the magnitude of the B cell response was considered and eliminated by the following two observations. First, treatment of spleen cells fractionated by sIg density with anti-Thy 1.2 serum and complement failed to alter the pattern of responsiveness; and second, dilution of unfractionated splenic B cells or adherent B cells into T cells to achieve the same ratio present in the nonadherent fraction did not result in enhanced B cell PFC responses. Even though the PFC response to TNP-*B. abortus* in the newborn does not require the participation of low sIg density cells, it seems clear from these data that much of the response in the adult is derived from cells in this category.

A direct demonstration that sIg-D is required for triggering of some T-I responses (like TNP-Ficoll) would be provided if anti-δ reagents preferentially inhibited such responses. This evidence has been obtained recently and will be presented elsewhere (I. Zitron *et al.*, manuscript in preparation). We have shown also that those T-I responses made by neonatal BDF_1 spleen cells or CBA/N cells are highly susceptible to inhibition with anti-μ reagents. These findings contribute to the model that $\mu^{+++}/\delta^{\pm}$ and μ^{+}/δ^{++} cells are involved in different classes of T-I responses. The final marker of the late-appearing low sIg density subpopulation is the Ly b 5 allo-antigen which is present on about 50 percent of normal adult splenic B cells (15). Preliminary results indicate that anti-Ly b 5 antisera will block preferentially TNP-Ficoll responses *in vitro* (B. Subbarao, A. Ahmed, I. Scher, and D. Mosier, unpublished results). These properties are summarized schematically in Table 1. Taken together, these data strongly suggest a heterogeneity among B cells and sIg receptors involved in antibody responses to different T-I antigens.

DISCUSSION

This cursory review of data concerning the influence of sIg phenotype on the ability of mouse splenic B cells to respond to different T-I antigens suggests the following conclusions: 1) relatively early appearing B cells, which predominate in neonatal and CBA/N mice, are restricted in the type of T-I antigens to which they can respond; 2) an overall reduction in the density of sIg, decrease in sIg-M expression and a relative increase in sIg-D occurs at the same time as the acquisition of responsiveness to a late-appearing subset of T-I antigens; 3) among adult splenic B cells, those with the least relative amount of sIg are the most reactive to all T-I antigens tested; and 4), the PFC response to the "late"-appearing subset of T-I antigens (TNP-Ficoll, etc) is apparently much more susceptible to inhibition with anti-δ reagents or anti-Ly b5 than are responses to the "early" subset of T-I antigens. If we call those T-I antigens which can stimulate neonatal and CBA/N B cells T-I 1 (positive in column 1, Table 1) and the remaining T-I antigens which fail to stimulate neonatal of CBA/N cells T-I 2 (positive in column 2, Table 1), then the B cells corresponding to these reactivities can be designated as T-I 1^+, 2^- and T-I 1^+, 2^+. These designations, introduced for ease of discussion, are meant to distinguish two subpopulations of functionally different B cells. It is not clear whether these two subpopulations arise sequentially in a linear maturation scheme, or, alternatively, develop as independent and parallel lines of differentiation. The sIg phenotype of the T-I 1^+, 2^- cell is high sIg, μ^{+++}, $\delta^{\pm}$, the T-I 1^+, 2^+ cell has a low sIg, μ^+, high δ^{++} phenotype. Except for the Ly b5 alloantigen, we do not yet know what other cell surface markers correlate with this subset designation. On the basis of their sensitivity to anti-Ly b5, a significant fraction of B cells which respond to T-I antigens in adult spleen appear to belong to the T-I 1^+, 2^+ subset. Their reactivity to T-I 2 antigens appears to require involvement of sIg-D, in contrast to T-I 1 responses which seem not to be sensitive to anti-δ inhibition. These findings lead us to postulate that a critical difference between T-I 1^+, 2^- and T-I 1^+, 2^+ B cells is the acquisition of sIg-D. Other differences which may exist, with the possible exception of the expression of Ly b5, would appear to be of lesser importance and are, perhaps, related to regulation rather than initiation of immune responses.

Two additional observations can be adduced for this model of B cell heterogeneity. First, T-I 1 antigens should trigger more B cells (two subsets) in the adult than T-I 2 antigens (1 subset). This prediction has been confirmed by limiting dilution analysis (Mosier et al., manuscript in preparation). Secondly, the amount of sIg-D expressed per B cell might be expected to correlate with the responsiveness to a T-I 2 antigen. In several replicate experiments, the relative amount of δ-like material (identified by surface iodination, immunoprecipitation, and polyacrylamide gel electrophoresis) on spleen cells from different mice has shown a near perfect correlation with the magnitude of the TNP-Ficoll response of the same cells.

Two final points regarding our observations and the model we have proposed may clarify the inability of other workers to find a role for sIg-D: 1) the number of B cells activated by T-I antigens is drawn disproportionately from that relatively small fraction expressing very low amounts of sIg, and 2), only T-I 2 type responses would be expected to be sensitive to inhibition with anti-δ reagents. The failures (16) to inhibit primary T-dependent responses with anti-δ would suggest that they more closely resemble the T-I 1 subset of antigens in their triggering requirements.

ACKNOWLEDGMENT

This paper presents views derived from experiments performed, and discussions with, the following people: Drs. Donna Sieckmann, Afted Ahmed, Bondada Subbarao, Kim Bottomly, Michael Mage, Henry Herrod, Steve Kessler, and Richard Asofsky. Drs. John Inman and Ian Zitron provided some of the antigen preparations, and Ms. Sue Sharrow was invaluable in providing the FACS analysis. The technical assistance of Barbara Johnson and Sue Pickeral is gratefully acknowledged.

REFERENCES

1. Hammerling, U., Chin, A., Abbot, J. (1976) Proc. Nat. Acad. Sci., USA 73, 2008.
2. Amsbaugh, D.F., Hansen, C., Prescott, B., Stashak, P., Barthold, D. and Baker, P. (1972) J. Exp. Med. 136: 931.
3. Scher, I., Ahmed, A., Strong, D., Steinberg, A. and Paul, W.E. (1975) J. Exp. Med. 141: 788.
4. Cohen, P.L., Scher, I., and Mosier, D.E. (1976) J. Immunol. 116: 301.
5. Mosier, D.E., Scher, I. and Paul, W.E. (1976) J. Immunol. 117: 1363.
6. Jerne, N.K. and Nordin, A. (1963) Science (Wash.D.C.) 140: 405.
7. Scher, I., Sharrow, S., Wistar, R.,Jr., Asofsky, R.A. and Paul, W.E. (1976) J. Exp. Med. 144: 494.

8. Chess, L., MacDermott, R.P., Sondel, P.M. and Schlossman,S.F. S.F. (1974) Prog. Immunol. II, 3: 125.
9. Mond, J.J., Mosier, D.E., Herrod, H, Mage, M., Rothstein, T. and Asofsky, R.A. Manuscript in preparation.
10. Finkelman, F.D., Smith, A., Scher, I. and Paul, W.E. (1975) J. Exp. Med. 142: 1316.
11. Scher, I., Sharrow, S. and Paul, W.E. (1976) J.Exp.Med. 144: 507.
12. Vitetta, E.S., Melcher, U., McWilliams, M., Lamm, M.E., Phillips-Quagliatta, J.M. and Uhr, J.W. (1975) J. Exp. Med. 141: 206.
13. Mosier, D.E., et al. (1977) J. Infect. Diseases, in press.
14. Klinman, N.R. and Press, J.L. (1975) Fed. Proc. 34: 47.
15. Ahmed, A., Scher, I., Sharrow, S., SMith, A., Paul, W.E., Sachs, D.H. and Sell, K.W. (1977) J. Exp. Med. 145: 101.
16. Parkhouse, M. Personal communication.

DEFECTIVE RECEPTOR CAPPING AND REGENERATION IN THE ANTIGEN BINDING CELLS OF TOLERANT MICE

Robert F. Ashman and David Naor*

Departments of Microbiology/Immunology and Medicine
UCLA School of Medicine, Los Angeles, California 90024

ABSTRACT. Administration of trinitrobenzene sulfonic acid to mice rendered them specifically tolerant to the trinitrophenyl (TNP) determinant. Since Fidler and Golub (1,2) had demonstrated that this form of tolerance may be induced and expressed in the B lymphocyte in the absence of T cells, an opportunity was provided to contrast the behavior of the tolerized antigen binding cells' antigen receptors to that of potentially responsive cells. The TNP specificity of the cells binding TNP-donkey red cells (TNP-D) was demonstrated by hapten inhibition. Counts of TNP binding cells in tolerized/challenged mice approximated the level seen in unimmunized mice, whereas immunization with TNP sheep red cells (TNP-S) caused TNP binding cells to increase about 2-fold. TNP-binding cells from immunized and unimmunized mice capped their receptors under the influence of bound TNP-D with kinetics similar to that shown by sheep erythrocyte binding cells. But although cells from TNP-tolerized and tolerized/challenged mice capped bound sheep erythrocytes normally, they could not cap TNP-D. Nor did the tolerized/challenged cells show the partial recovery of TNP binding by 2 hours after pronase treatment characteristic of immunized cells. Thus in this form of tolerance, a specific inability of the cells with receptors for TNP to mature to antibody secretion is associated with a specific inability to cap bound antigen or replace receptors.

INTRODUCTION

The antigen induced capping of receptors on the antigen binding lymphocyte requires metabolic energy, and therefore is an early visible sign that a signal has been generated by the crosslinking of membrane molecules. In some circumstances this signal may not be sufficient or necessary (3) for lymphocyte activation, so the question of its relevance to the later fate of the cell remains interesting and controversial. In the case of surface Ig on the easily tolerized neonatal B

*Present address: Lautenberg Center for General and Tumor Immunology, Hebrew University, Hadassah Medical School, Jerusalem, Israel

cell (4) or receptors for DNP coupled to the tolerogenic d-GL copolymer (5), capping is incomplete and regeneration of new receptors fails to occur. In contrasting reports polymerized flagellin in tolerogenic concentrations fails to cap altogether (6), whereas DNP-flagellin caps all the surface Ig on the antigen binding cell, and it is the immunogenic concentrations which produce partial capping (7).

To clarify the relationship of capping to the B cells' decision between tolerance and activation we have examined the behavior of antigen receptors in a system where the tolerance has been shown by appropriate transfer experiments to be manifested in the B cells themselves, rather than in a lack of T cell help or an excess of T cell suppression. Such a system is the specific tolerance to the TNP determinant described by Fidler and Golub (1,2).

METHODS

Four groups of CBA mice were examined: Immunized mice received intravenously 0.2 ml of 2% v/v sheep erythrocytes derivatized with 60 mg/ml TNBS (TNP_{60}-S) as previously described (8). Unimmunized mice received no treatment. Tolerant mice received 5 mg of TNBS intraperitoneally. Tolerant/challenged mice received TNBS 1 hour before immunization with TNP_{60}-S. Four days after this treatment, spleen non-adherent antigen binding cells with specificity for the TNP determinant were assayed by rosetting with non-crossreacting TNP-D. Alternatively, lymphocytes were rosetted with TNP-S after cells binding sheep erythrocytes alone had been removed by sedimentation on Ficoll.

As with sheep erythrocyte rosettes (9), merely incubating preformed rosettes at 37° allows bound TNP red cells to induce active receptor movements. Cells which shift bound red cells into 270° of the lymphocyte's circumference or less are scored as having shown such movements.

To investigate the ability of tolerized antigen binding cells to replace receptors, our first approach was to pretreat the cells with pronase (2 mg/ml for 1 hour, 37°) to reduce the number of cells which can bind antigen by 80-90% (10). Then cells were cultured to allow recovery of antigen binding.

RESULTS AND DISCUSSION

Table 1 confirms the observation of Fidler and Golub (1) in that TNBS treatment selectively prohibited the maturation of anti-TNP-secreting cells following TNP-S immunization, but permitted normal development of cells secreting antibody to sheep red cell determinants. This selective inhibition of

TABLE 1: PLAQUE FORMING CELLS PER SPLEEN (DIRECT PLUS INDIRECT) IN TNP-TOLERANT AND NON-TOLERANT ANIMALS

	Experiment 1			Experiment 2		
	TNP-D	S	D	TNP-D	S	D
Unimmunized	1737	675	0	583	167	0
Immunized	58537	28875	0	29800	43600	83
Tolerized/ challenged	2212	41250	125	760	64700	25
Tolerized	+	+	+	325	215	50

*S refers to sheep erythrocytes, D to donkey erythrocytes. For immunization, S are derivatized with TNBS at 60 mg/ml for 30 min at 23°, whereas cells for rosetting are derivatized with 20 mg/ml TNBS.

+Not done.

the response to TNP was also manifested at the antigen binding level. The frequencies of antigen binding cells to sheep in both immunized and tolerized/challenged animals were about 4 times the levels in unimmunized animals. The frequency ranges of TNP-D rosettes per thousand lymphocytes were 0.7 to 1.9 for unimmunized, 0.6 to 0.9 for tolerized, 1.1 to 1.6 for tolerized/challenged and 1.7 to 4.1 for immunized animals. The contribution of cells binding donkey red cells to the TNP-D rosette population ranged from less than 5% in TNP-S immunized animals to 5-25% in unimmunized or tolerant animals. By indirect immunofluorescence with anti-Ig and anti-T cell sera, the TNP-D rosettes ranged from 83-88% Ig^+ cells and 10-17% T cells in unimmunized, immunized, and tolerized/challenged animals, in contrast to the bulk of spleen lymphocytes which appeared to be 40-45% Ig^+ and about 50% T cells.

The specificity of TNP-D rosettes for TNP was assessed by inhibition of rosette formation using ε-2,4,6-trinitrophenyl caproic acid (TNP-EAC) and α,ε-bis-2,4-dinitrophenyl-l-lysine (DNP_2-lys). At concentrations in excess of 5 x 10^{-4} M, both haptens inhibited 95% of TNP-D binding in immunized cells and in tolerized animals inhibition is in the 80-90% range. Sheep rosettes are not inhibited. TNP-S rosette forming cells out-numbered TNP-D rosettes 2-3 fold, and were less readily inhibitable by TNP-EAC than the TNP-D rosettes (70-90%) probably because they included many cells whose receptors are directed to determinants including both TNP and groups adjacent to TNP on the red cell surface.

Shifting receptors to which bulky red cells are attached should require a great deal more energy than moving protein molecules like anti-Ig, and indeed only 30-40% of rosettes could perform this feat. If the tolerized cell is rendered less effective at moving membrane molecules, moving red cells

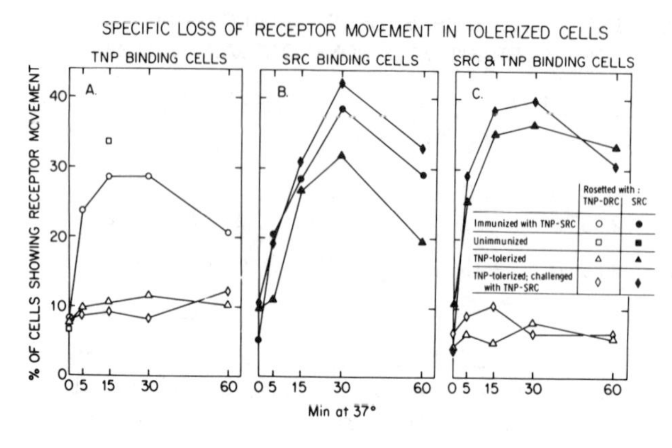

FIGURE 1: Effect of tolerance on the kinetics and extent of antigen induced receptor movements.

Experiment A: Capping kinetics of TNP-D binding cells from TNP-tolerant and non-tolerant animals.

Experiment B: Capping kinetics of sheep erythrocyte binding cells from TNP-tolerant and non-tolerant animals.

Experiment C: Capping kinetics of sheep erythrocyte binding cells and TNP-D binding cells from the same animals.

---

should become impossible, and Fig. 1 confirms this prediction. Panel A shows that immunized and unimmunized cells capped normally, but that tolerized and tolerized/challenged cells did not. However, tolerized cells still capped sheep erythrocyte receptors (B & C) so the defect in receptor movement was confined to cells recognizing TNP and did not represent a general toxic effect on cell membranes.

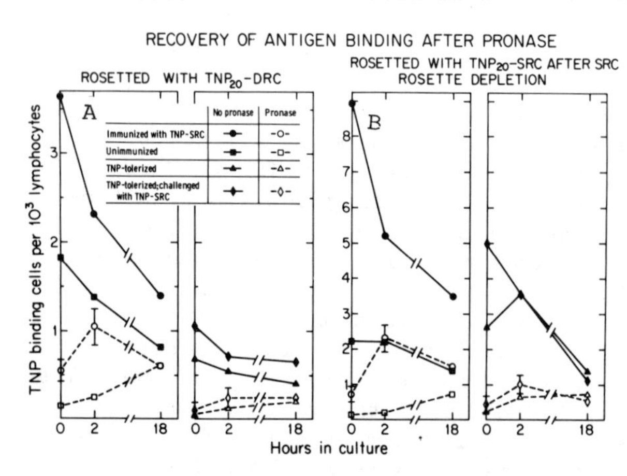

FIGURE 2: Recovery of TNP-D and TNP-S binding after pronase treatment.

Cells from immunized, unimmunized, tolerized and tolerized/challenged animals were incubated with (or without 2 mg/ml pronase for 1 hr at 37°C. After assessment of the proportion of TNP-D (A) or TNP-S (B) binding cells (cells binding underivatized sheep red cells were removed first), pronase treated and untreated cells were then cultured for 2 or 18 hours and reexamined for recovery of ability to bind antigen. Only 0.1 per 10^3 cells rosetted with donkey erythrocytes alone. Brackets indicate the 95% confidence intervals by standard error of proportions.

After pronase treatment, significant (though incomplete) recovery of TNP-D or TNP-S binding occurred by 2 hours in immunized cells (Fig. 2) and there was considerable recovery by unimmunized cells by 18 hours. Recovery was much less evident in the tolerized cells. The slight recovery in the tolerant TNP-S binding cells at 2 hours may be attributed to residual sheep cell binders.

In summary, though tolerant mice have TNP binding cells, this population fails to cap its receptors in response to bound antigen, fails to replace receptors removed by enzyme, fails to undergo expansion after TNP-S immunization, and fails to mature to antibody production, even though cells in the same spleen recognizing the sheep red cell determinants on the TNP-S can perform these functions normally. This data cannot be construed as evidence that receptor capping on the B cell plays a direct role in activation. Rather, it suggests that in this form of B cell tolerance, the "tolerizing event" occurs at a step that follows antigen binding and is necessary for both capping and activation. Alternatively, there may be more than one site of action for TNBS, which is able to derivatize a variety of self proteins; but specificity of the defect for the cells recognizing TNP suggests the B cell antigen receptor mediates the negative signal. However, our data does not rule out an intermediary role for non-B cells in inducing membrane dysfunction in TNP specific B cells.

ACKNOWLEDGEMENTS

The technical assistance of Heinrich Kolbel and Gina Robinson is gratefully acknowledged. Supported by NIH grant CA12800. We are especially grateful to Dr. Irving Weissman for his gift of rabbit anti-mouse T cell serum.

REFERENCES

1. Fidler, J.M. and Golub, E.S. (1973) *J. Immunol.* 111,317.
2. Fidler, J.M. and Golub, E.S. (1974) *J. Immunol.* 112, 1891.
3. Coutinho, A. and Moller, G. (1974) *Scand. J. Immunol.* 3,133.
4. Sidman, C.L. and Unanue, E.R. (1975) *Nature (Lond)* 257,149.
5. Ault, K.A., Unanue, E.R., Katz, D. and Benacerraf, G. (1974) *Proc. Nat. Acad. Sci. USA* 71, 3111.
6. Diener, E., Kraft, N., Lu, K.C. and Shiozawa, C. (1976) *J. Exp. Med.* 143, 805.
7. Nossal, G.J.V. and Layton, J.E. (1976) *J. Exp. Med.* 143,511.
8. Naor, D., Morecki, S. and Mitchell, G.F. (1974) *Eur. J. Immunol.* 4, 311.
9. Ashman, R.F. (1973) *J. Immunol.* 111, 212
10. Ashman, R.F. (1975) *Scand. J. Immunol.* 4, 337.

TRIGGERING AN *IN VITRO* ANTIHAPTEN IgG RESPONSE WITHOUT RECEPTOR Ig-ANTIGEN INTERACTION

S. Cammisuli and L. Wofsy

Department of Bacteriology and Immunology
University of California
Berkeley, California 94720

ABSTRACT. We report experiments which indicate that Ig receptor-antigen interaction is not essential for generating an hapten specific IgG response. Keyhole limpet hemocyanin (KLH) has been bound to the $H2^d$, Ia.7, or Ig L chain surface antigens of spleen cells from Balb/c mice primed against the antigen azophenyl-β-lactoside-KLH (lac-KLH). Selective attachment of KLH was achieved by a hapten-sandwich technique: cells were labeled with benzenearsonate (ars)-coupled anti $H2^d$, anti Ia.7, or anti L chain antibodies, followed by a conjugate of anti ars antibody with KLH (KLH-anti ars). The KLH labeled cells were cultured with spleen cells from mice primed against azophenylglucoside-KLH (azo KLH) as a source of KLH specific T helper cells. In each case, we detected a highly significant, T dependent IgG anti lac plaque forming response.

INTRODUCTION

While interaction between B cell receptor immunoglobulin and antigen determines the specificity of the immune response, it is not clear whether this is a necessary requirement for the differentiation of resting B lymphocytes into immunoglobulin secreting plasma cells. Divergent hypotheses have been argued (1). It has been proposed that Ig receptor-antigen interaction provides a unique signal resulting in tolerance, unless it is coupled to a T cell "second signal" which produces immune induction (2). A contrasting proposal is that receptor Ig serves passively to focus antigen on the lymphocyte membrane so that an activation signal may be elicited at other (non-Ig) mitogenic sites (3). Others postulate a dichotomy between proliferation and differentiation, with Ig receptor-antigen binding per se initiating only proliferation (4,5). Finally, it has been suggested that an optimal redistribution of receptor Ig molecules on the lymphocyte membrane is sufficient stimulus for B cell activation (6,7).

Experimental support for the above models has been obtained generally with complex probes. Evidence that Ig receptor-antigen interaction is not required for B cell activation has come essentially from studies with mitogens (3), which bind to a wide spectrum of largely undefined surface receptors, perhaps in some cases including Ig itself. On the other hand, it is also difficult to dissect the processes leading to B cell activation when antigen is itself the probe, since it may not be feasible to distinguish clearly between contributions from specific B cell receptor binding of antigen, T cell interactions, and possible non-specific mitogenic properties of a particular antigen.

EXPERIMENTAL DESIGN

We have devised an experimental *in vitro* system in which it is possible to provide B memory cells with T help in essentially normal conditions while excluding antigen binding by the B cell receptor Ig. We have investigated whether Ig receptor-antigen interaction is required in the generation of an *in vitro* secondary response to the hapten, phenyl β-lactoside (lac). In the lac *in vitro* system (8), spleen cells from Balb/c mice immunized with lac-KLH give a predominantly IgG PFC response when cultured with lac-KLH. The response is T dependent and requires macrophages. The response of T depleted lac-KLH primed spleen cells can be restored completely by addition of spleen cells primed against KLH or azo-phenylglucoside-KLH (azo KLH). It is thus possible to manipulate separate sources of lac primed spleen cells ("B memory cells") and KLH primed spleen cells ("T helper cells").

Our experimental scheme (Fig. 1) is (a) to attach KLH selectively to various membrane antigens on spleen cells primed against lac-KLH, (b) to culture the labeled cells with additional KLH primed spleen cells in the absence of the lac-KLH antigen, and (c) to assay after five days for an anti-lac PFC response.

For selective binding of KLH, we have used an application of the hapten-sandwich labeling technique (9). Azobenzenearsonate (ars) hapten groups were coupled to anti cell surface antibodies (anti $H2^d$, anti Ia.7, or anti Ig L chains) with a bifunctional amidinating reagent (9), a procedure which yields extensive modification without inactivation of antibody. Cells treated with an ars-coupled anti cell surface antibody were labeled with a glutaraldehyde conjugate of anti ars antibody with KLH (KLH-anti ars). Thus, lac-KLH primed cells, labeled with KLH at either H2, Ia, or Ig L surface antigens, could be investigated for their ability to give an *in vitro* anti lac response with KLH-specific T cell help in the absence of the lac-KLH antigen.

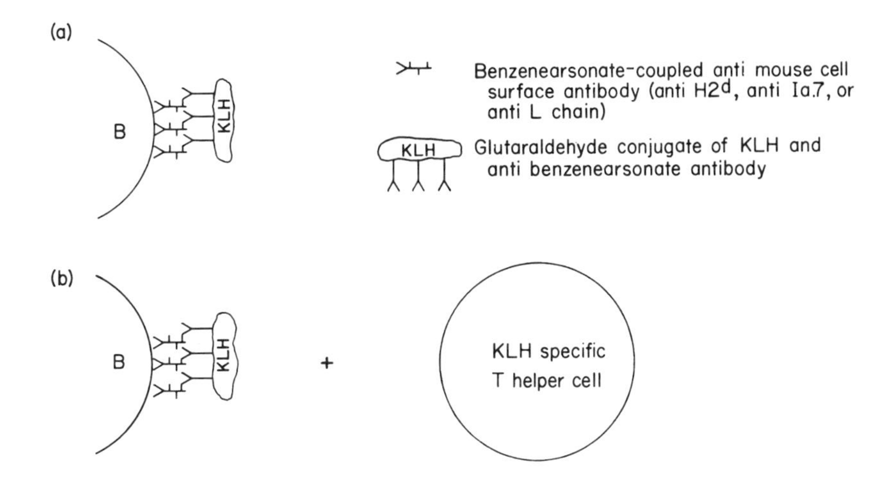

FIG. I. (a) Selective binding of KLH to various B cell membrane antigens (b) Specific T cell help to B cells in absence of Ig receptor–antigen interaction.

ANTIGEN INDEPENDENT DIFFERENTIATION OF B MEMORY CELLS

Table 1 records results of experiments in which B memory cell populations were labeled with KLH and added at a 1:2 ratio to a source of KLH primed T helper cells. Attachment of KLH to H2, Ia, or Ig L surface antigens was similarly effective in stimulating a secondary anti lac IgG response. In each case, we achieved a significant fraction of the response elicited by the lac-KLH antigen itself. Cultures in which B memory cells were treated with ars-anti $H2^d$ or ars-anti L chains without KLH-anti ars gave negligible responses. Responses were absent also when the lac primed cells were treated with the modified anti cell surface antibodies followed by anti ars antibody to simulate cross-linking induced by the KLH-anti ars conjugate. No response was observed when ars-modified normal rabbit gamma globulin (RGG) was substituted for a specific anti cell surface antibody.

Table 1 suggests that hapten-sandwich attachment of KLH to B memory cells in these experiments resulted in

TABLE 1

TREATMENT OF LAC-KLH PRIMED SPLEEN CELLS	LAC INDIRECT PFC RESPONSE PER CULTURE	% OF LAC-KLH TREATED CULTURE
EXP. 1		
ARS-ANTI $H2^D$ + KLH-ANTI ARS	1348	14.6
ARS-ANTI $H2^D$	< 10	< 0.1
ARS-ANTI L CHAIN + KLH-ANTI ARS	861	9.4
ARS-ANTI L CHAIN	< 10	< 0.1
ARS-RGG + KLH-ANTI ARS	< 10	< 0.1
NO TREATMENT	35	0.4
LAC-KLH ADDED IN CULTURE	9207	100
EXP. 2		
ARS-ANTI IA.7 + KLH-ANTI ARS	324	7.7
ARS-ANTI IA.7 + ANTI ARS	33	0.8
ARS-ANTI $H2^D$ + KLH-ANTI ARS	602	14.4
ARS-ANTI $H2^D$ + ANTI ARS	76	1.8
ARS-ANTI L CHAIN + KLH-ANTI ARS	413	9.8
ARS-ANTI L CHAIN + ANTI ARS	64	1.5
KLH-ANTI ARS	< 10	< 0.1
NO TREATMENT	< 10	< 0.1
LAC-KLH ADDED IN CULTURE	4186	100

ACTIVATION OF LAC SPECIFIC IGG PFC PRECURSORS IN THE ABSENCE OF IG RECEPTOR-ANTIGEN INTERACTION. LAC PRIMED SPLENOCYTES WERE TREATED AS INDICATED, MIXED AT A 1:2 RATIO WITH AZO KLH PRIMED SPLENOCYTES AND CULTURED FOR FIVE DAYS UNDER MISHELL-DUTTON CONDITIONS.

stimulation of only a fraction of the anti lac IgG precursors. We reasoned that this could be due to competition for KLH-specific T helper cells. The number of these helper cells is necessarily limiting, since all lymphocytes in the B memory population, regardless of specificity, have bound KLH and can interact with KLH-specific T cells. To verify this assumption, we have varied the ratio of B memory cells to T helper cells in a series of experiments. With a constant final cell density of 1.5 x 10^7 cells per ml culture, B and T primed spleen cell populations were mixed at ratios of 1:2, 1;9, and 1:30. The results in Table 2 show that as the B:T ratio is reduced, a higher percent of the potential response is obtained. At the 1:30 ratio, more than one-third of the potential lac-specific PFC are generated.

The differentiation from memory to plasma cell observed in our experiments is shown to be T dependent by the results

TABLE 2

TREATMENT OF LAC-KLH PRIMED SPLEEN CELLS	RATIO LAC PRIMED: AZO KLH PRIMED SPLEEN CELLS 1:2	1:9	1:30
	LAC INDIRECT PFC/CULTURE	LAC INDIRECT PFC/CULTURE	LAC INDIRECT PFC/CULTURE
ARS-ANTI $H2^D$ + KLH-ANTI ARS	1348 (14.6)	467 (18.9)	332 (38.5)
ARS-ANTI $H2^D$	<10 (<0.1)		
ARS-ANTI L CHAIN + KLH-ANTI ARS	861 (9.4)	432 (17.5)	313 (36.3)
ARS-ANTI L CHAIN	<10 (<0.1)		
NO TREATMENT; LAC-KLH IN CULTURE	9207 (100)	2475 (100)	863 (100)

EFFECT OF VARYING THE RATIO OF LAC B MEMORY CELLS TO AZO KLH T HELPER CELLS, LAC PRIMED SPLEEN CELLS WERE TREATED AS INDICATED, MIXED AT EITHER 1:2, 1:9, OR 1:30 WITH AZO KLH PRIMED SPLEEN CELLS AND CULTURED FOR FIVE DAYS UNDER MISHELL-DUTTON CONDITIONS. NUMBERS IN PARENTHESIS INDICATE THE PERCENTAGE OF THE CONTROL RESPONSE IN PRESENCE OF LAC-KLH.

in Table 3. As expected (8), lac-primed cells treated with anti mouse brain (anti MBr) serum and complement to eliminate T cells failed to respond to lac-KLH. Such T depleted populations, labeled with ars-anti $H2^d$ followed by KLH-anti ars, gave an anti lac response only when cultured with added azo KLH primed T helper cells. The response was not induced by addition of normal spleen cells.

TABLE 3

TREATMENT OF T DEPLETED LAC-KLH PRIMED SPLEEN CELLS	SPLEEN CELLS ADDED IN CULTURE	LAC INDIRECT PFC PER CULTURE	% OF LAC-KLH TREATED CULTURE
ARS-ANTI $H2^D$ + KLH-ANTI ARS	AZO KLH PRIMED	1166	21.2
ARS-ANTI $H2^D$ + ANTI ARS	AZO KLH PRIMED	30	0.6
ARS-ANTI $H2^D$ + KLH-ANTI ARS	NORMAL	<10	< 0.2
LAC KLH ADDED IN CULTURE	AZO KLH PRIMED	5489	100
LAC KLH ADDED IN CULTURE	NORMAL	82	1.5

REQUIREMENT OF T HELP FOR ACTIVATION OF LAC-SPECIFIC MEMORY CELLS. LAC PRIMED SPLEEN CELLS WERE DEPLETED OF T LYMPHOCYTES WITH ANTI-MBR + C', TREATED AS INDICATED, AND MIXED AT 1:2 RATIO WITH EITHER AZO KLH PRIMED OR NORMAL SPLEEN CELLS. CELLS WERE CULTURED FOR FIVE DAYS UNDER MISHELL-DUTTON CONDITIONS.

We conclude that Ig receptors do not deliver a unique signal for the differentiation of memory B cells to plasma cells. We do not exclude their role in delivery of a proliferative signal. Our data are restricted to the generation of a secondary response and do not rule out an Ig receptor signal in the induction of virgin B cells to tolerance or immunity. We are presently investigating whether there is a requirement for Ig receptor-antigen interaction in virgin lymphocyte activation.

As indicated by this study, hapten bridging procedures should permit a variety of investigations in which cell surface interactions are induced selectively between different receptors on the same or on different cells. Hapten-sandwich labeling was developed primarily to improve the sensitivity with which alloantigens may be visualized on cell surfaces. It also provides the basis for manipulating subpopulations of lymphocytes in various *in vitro* procedures, including some improved methods for specific cell fractionation as reported at this meeting (Workshop 9: Separation and Characterization of Subsets of Cells).

ACKNOWLEDGEMENTS

We appreciate the advice and help of Claudia Henry and John Kimura. This work was supported by USPHS Grant AI-06610.

REFERENCES

1. Transplant. Rev. (1975) *23*.
2. Bretscher, P.A. and Cohn, M. (1968) Nature (Lond.) *220*, 444.
3. Coutinho, A. and Möller, G. (1974) Scand. J. Immunol. *3*, 133.
4. Hünig, T., Schimpl, A. and Wecker, E. (1974) J. Exp. Med. *139*, 754.
5. Dutton, R.W. (1974) *in*: The Immune System: Genes, Receptors, Signals, eds. Sercarz, E.E., Williamson, A.R. and Fox, C.F.
6. Diener, E. and Feldmann, M. (1972) Transplant. Rev. *8*, 76.
7. Dintzis, H.M., Dintzis, R.Z. and Vogelstein, B. (1976) Proc. Natl. Acad. Sci. USA *73*, 3671.
8. Henry, C. (1975) J. Cell. Immunol. *19*, 117.
9. Cammisuli, S. and Wofsy, L. (1976) J. Immunol. *117*, 1695.

MOLECULAR EVENTS IN LYMPHOCYTE DIFFERENTIATION: KINETICS OF NONHISTONE NUCLEAR PROTEIN SYNTHESIS IN RABBIT PERIPHERAL BLOOD LYMPHOCYTES STIMULATED BY ANTI-IMMUNOGLOBULIN

Janet M. Decker and John J. Marchalonis

Basic Research Program, Frederick Cancer Research Center, Frederick, Maryland 21701

ABSTRACT. Changes in nuclear protein synthesis were investigated in rabbit peripheral blood lymphocytes at early times after stimulation with heterologous anti-immunoglobulin serum (anti-Ig), which reacts with 7S IgM surface immunoglobulin. By 4 hours following stimulation with anti-Ig, a striking increase in the synthesis of a particular size class of nonhistone chromatin protein (apparent molecular weight 30-40,000 daltons) was observed. This increased synthesis persisted through at least 8 hours post anti-Ig addition, by which time the synthesis of other size classes of nuclear proteins was also elevated by anti-Ig treatment.

INTRODUCTION

Stimulation of cell division by binding of anti-immunoglobulin (anti-Ig) reagents (1) can be used as a model for the study of biochemical events resulting from combination of antigen with the lymphocyte surface antigen-specific receptor (2-4). One of the earliest of the changes reported following lymphocyte activation is the synthesis of nonhistone nuclear proteins (5,6), which have been implicated in the regulation of DNA transcription (7). In this paper, we will describe quantitative changes in the synthesis of a particular size class of nonhistone nuclear proteins that occur during the first 8 hours following anti-Ig stimulation of rabbit peripheral blood lymphocytes (PBL).

MATERIALS AND METHODS

Cell preparation and culture. PBL from the blood of adult outbred rabbits were prepared and cultured in RPMI 1640 (Gibco, San Francisco, CA), supplemented with HEPES, 2-mercaptoethanol, fetal calf serum (Commonwealth Serum Labs, Melbourne, Australia) and antibiotics, as previously described (4). A 1:100 dilution of commercially-prepared goat antiserum to rabbit Ig (shown to give optimal stimulation of ^{125}I-deoxyuridine incorporation) (4) was added at time 0. At 2-hour intervals following anti-Ig addition, the cultures were washed, resuspended in Dulbecco's modified Eagle's medium minus leucine, and pulsed for 2 hours with ^{3}H-leucine (50-100 μCi of >270 mCi/mmol, Amersham, Bucks, England). Parallel control cultures given phosphate-buffered saline (PBS) or normal goat serum were pulsed for the same period with ^{14}C-leucine (5-10 μCi of >270 mCi/mmol, Amersham). At the end of the leucine pulse, the control and experimental cell pellets were combined and the nuclei were prepared by extraction with 0.1% Nonidet-P40 in PBS, followed by centrifugation through 0.32M sucrose. The pellet, which by phase microscopy was seen to contain intact nuclei without visable cytoplasmic contamination, was dissolved in tris buffer containing sodium dodecyl sulfate (SDS) and 2-mercaptoethanol and analyzed by SDS-polyacrylamide gel electrophoresis (PAGE) as described by Laemmli (8). The gels were cut into 2-mm slices, solubilized overnight at room temperature in 0.5 ml Soluene 350 (Packard Instrument Co., Downers Grove, Illinois), and counted in a Packard Tri-Carb liquid scintillation counter.

RESULTS

Nature of the surface molecule involved in anti-Ig stimulation of rabbit PBL. To support the validity of using anti-Ig stimulation of lymphocytes as a model for antigen-antigen receptor (Ig) stimulation, it was important to insure that the surface molecules involved in the stimulation were the same in both cases. Lactoperoxidase-catalyzed radioiodination was used to label PBL surface proteins, which were then extracted and precipitated with the commercial anti-Ig as previously described (9). The precipitate was reduced and analyzed on PAGE as described elsewhere (9). Figure 1 shows the results of this analysis. Anti-Ig precipitated only molecules with the mobilities of immunoglobulin μ and light chains from the population of labelled PBL surface proteins. In combination with experiments such as absorption of mitogenic activity on a column of Sepharose conjugated with rabbit IgG, lack of stimulation by normal goat serum, and stimulation with allotype antisera, the results demonstrate that the mitogenic factor in the anti-Ig serum is antibody to surface Ig.

Nuclear protein synthesis: double label experiments. We have previously described the increased synthesis of a particular size class of nonhistone nuclear proteins at 4 hours following anti-Ig stimulation of rabbit PBL (4). Similar results were obtained when the leucine isotopes were reversed, and when the chromatin was isolated before PAGE analysis (4).

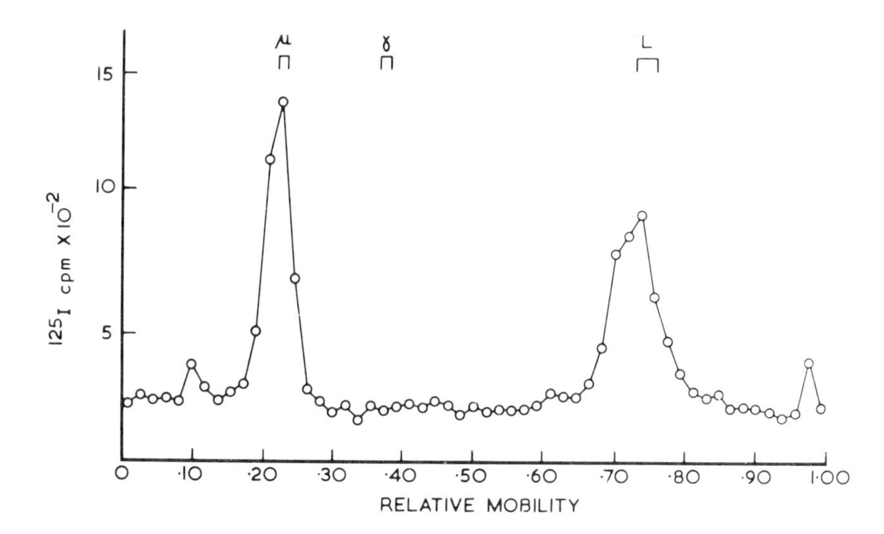

Fig. 1 - PAGE analysis of lactoperoxidase radioiodinated surface protein from rabbit PBL precipitated by goat anti-rabbit Ig serum. 10% acrylamide, reducing conditions. Positions of Ig standards are noted.

Figure 2 illustrates the results from a single experiment in which cells were pulsed with radioactive leucine for 2-hour intervals at several times following saline or anti-Ig stimulation (the mitogen remained in the culture during the leucine pulse). The control and experimental cultures were pooled and analyzed as described in Materials and Methods, and the ratio of ^{3}H (experimental) / ^{14}C (control) radioactivity in each gel slice was calculated and plotted vs relative mobility. Absolute counts for each isotope were normalized, with the highest peak (excluding the first and last 3 slices) taken as 100%. During the first 6 hours of culture, the peak slice was found at approximately 0.55 relative mobility, with a $^{3}H/^{14}C$ ratio constant within a given experiment and equal to the ratio in a control situation where no mitogen was added to either culture. Figure 2 A-C illustrates the results during the first 6 hours following anti-Ig stimulation. At 2 hours (2A), the $^{3}H/^{14}C$ ratio was nearly constant along the length of the gel and equal to the control ratio (no mitogen added, data not shown). By 4 hours (2B), an increase in the $^{3}H/^{14}C$ ratio was quite distinct in the portion of the gel where proteins of apparent molecular weight 30-40,000 daltons would migrate, indicating that more protein in this size class was being synthesized in the stimulated than in the control culture. The increased synthesis in this region of the gels became more pronounced during a 4- to 6-hour post anti-Ig leucine pulse (2C).

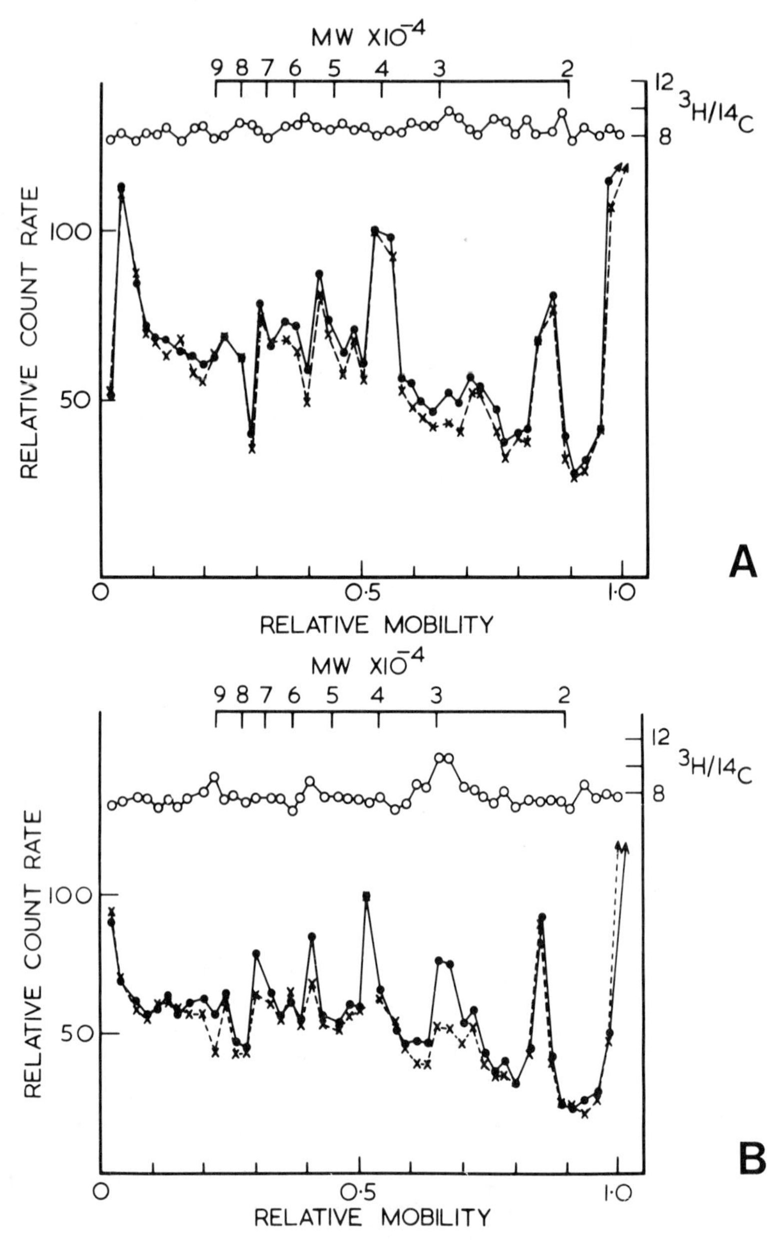

Fig. 2-PAGE analysis of nuclear proteins from rabbit PBL stimulated with saline or anti-Ig for various periods of time and pulsed for 2 hours with ^{14}C-leucine (x) or ^{3}H-leucine (•) respectively. (o) ratio of ^{3}H (experimental)/^{14}C (control) radioactivity. 10% acrylamide, reducing conditions. Leucine pulse 0-2 hours (A) and 2-4 hours (B) of culture.

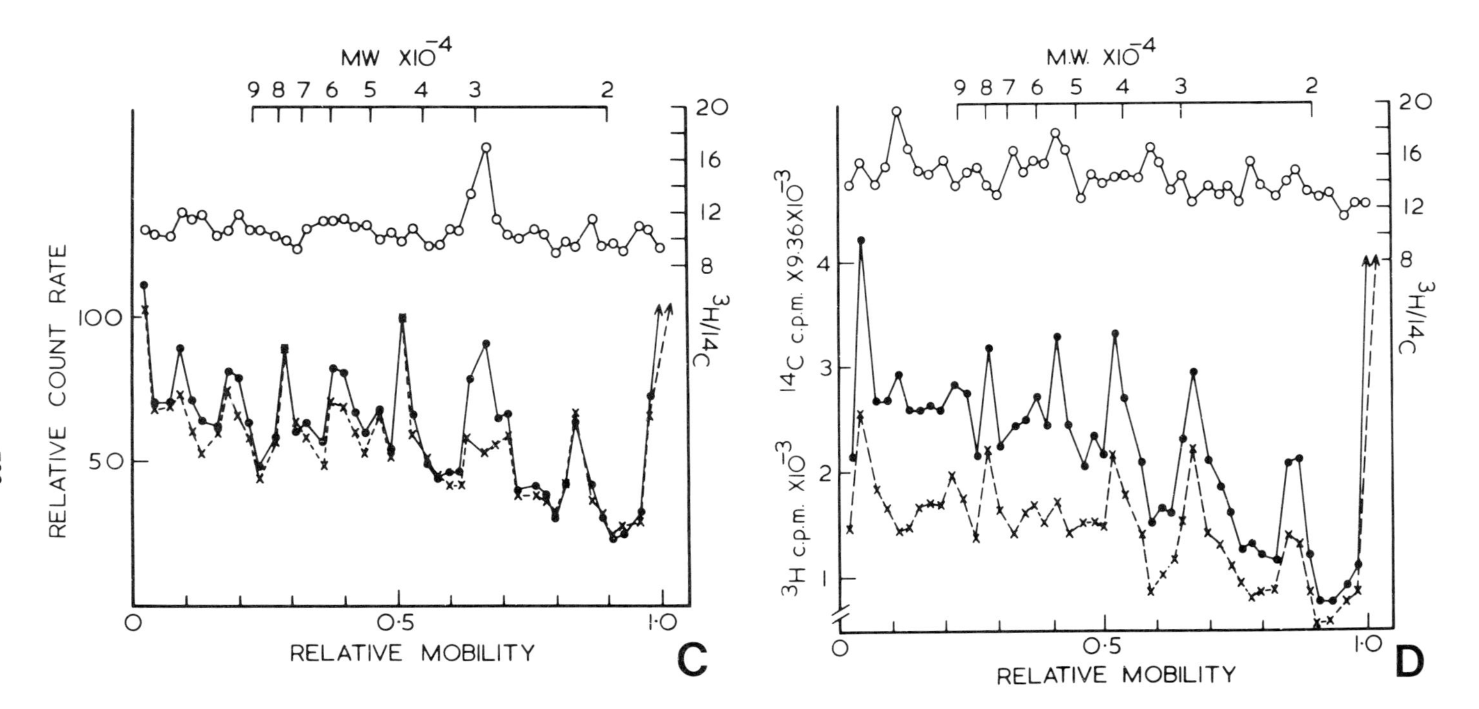

Fig. 2 - Cont. - Leucine pulse 4-6 hours (C) and 6-8 hours (D) of culture. See text for normalization procedure. The factor of 9.36 used to normalize the ^{14}C data in 2D was the ratio of $^{3}H/^{14}C$ radioactivity at 0.53 relative mobility in the saline *vs* saline control at 6-8 hours.

The results of a pulse from 6-8 hours of culture differ from the preceding in that the $^3H/^{14}C$ ratio was higher all along the gel than it was either at earlier time points or in the no-mitogen control. Figure 2D shows that the lowest ratio in any slice at this time point was 11 (compared with 8 in 2C); although the ratio in the area corresponding to 30,000 daltons no longer appeared impressive compared with the rest of the gel, it was still elevated over the control values. We conclude from these data that by 8 hours after mitogen stimulation, the synthesis of nuclear proteins in a wide range of size classes has been stimulated, especially those greater than 50,000 daltons.

DISCUSSION

We have demonstrated that following interaction of anti-Ig with rabbit PBL surface Ig, a significant elevation in the synthesis of a particular size class of nonhistone nuclear proteins occurs relative to the level in unstimulated cells. Because changes in the nonhistone nuclear proteins are observed following hormone contact, neoplastic transformation, and at various stages in the cell cycle (7,10), as well as following mitogen stimulation of lymphocytes (4-6), it has been suggested that these proteins are regulators of gene expression (7). Direct demonstrations of the ability of nonhistone nuclear proteins to affect gene expression are now appearing in the literature (11). Experiments are now in progress in this laboratory to determine whether the protein synthesis we observe at 4 hours is obligatory for the blastogenic transformation and what its function might be in lymphocyte differentiation.

ACKNOWLEDGEMENTS

We wish to thank Ms. Kerry Haynes for expert technical assistance and Ms. Pat Smith for radioiodination and precipitation of rabbit PBL surface Ig. This work was performed at the Walter and Eliza Hall Institute of Medical Research, Melbourne, Australia, supported (in part) by a postdoctoral fellowship from the National Multiple Sclerosis Society (JMD) and by grant CA-20085 from the USPHS and AHA 75-877 from the American Heart Association.

REFERENCES

1. Sell, S. and Gell, P.G.H. (1965) *J. Exp. Med.* 122, 423.
2. Marchalonis, J.J. (1975) *Science* 190, 20.

3. Marchalonis, J.J., Warr, G.W., Moseley, J.M. and Decker, J.M. (1977) *Handbook of Cancer and Immunology*, in press.
4. Decker, J.M. and Marchalonis, J.J. (1977) *Biochem. Biophys. Res. Comm.* 74, 589.
5. Levy, R., Levy, S., Rosenberg, S.A. and Simpson, R.T. (1973) *Biochemistry* 12, 224.
6. Johnson, E.M., Karn, J. and Allfrey, V.G. (1974) *J. Biol. Chem.* 249, 4990.
7. Stein, G.S. Spelsberg, T.C. and Kleinsmith, L.J. (1974) *Science* 183, 817.
8. Laemmli, U.K. (1970) *Nature (London)* 227, 680.
9. Cone, R.E. and Marchalonis, J.J. (1974) *Biochem. J.* 140, 345.
10. Busch, G.I., Yeoman, L.C., Taylor, C.W. and Busch, H. (1974) *Physiol. Chem. Phys.* 6, 1.
11. Stein, G., Park, W., Thrall, C., Mans, R. and Stein, J. (1975) *Nature (London)* 257, 764.
12. Wang, J.L., Gunther, G.R. and Edelman, G.M. (1975) *J. Cell Biol.* 66, 128.

THE ROLE OF J CHAIN IN B CELL ACTIVATION

Elizabeth L. Mather and Marian E. Koshland

Department of Bacteriology and Immunology
University of California, Berkeley, Ca. 94720

ABSTRACT. The relationship between J chain synthesis and B cell differentiation was investigated by fractionating rabbit spleen cell populations and analyzing their J chain and IgM content by radioimmunoassay. When small cell fractions were separated by 1 x g velocity sedimentation, they were found to contain 4 to 10 fold less J chain than the unfractionated population. Since the extent of J chain depletion correlated with the removal of activated B cells, these data indicated that J chain was not being synthesized in the precursor B cell. The small cell fractions were then cultured in the presence of pokeweed mitogen and the amounts of intracellular J chain, monomer IgM, and secreted pentamer were followed as a function of time. New J chain synthesis was detected at 18 hours and the intracellular concentration increased 22 fold over 3-1/2 days. In contrast, a significant change in the intracellular IgM concentration was not detected until 1-1/2 days and the amount present increased only 2.3 fold over 3-1/2 days. These results indicated that a) J chain synthesis is initiated as a result of mitogen stimulation and b) once its synthesis is turned on, J chain is produced in such large excess that it is not the rate limiting step in the amount of pentamer secreted.

INTRODUCTION

In the precursor B cell monomer IgM is synthesized and deposited in the membrane as a surface antigen receptor (1). Once the cell comes in contact with antigen, however, IgM synthesis is increased and the monomers are assembled into pentamer IgM which is rapidly secreted (2). The intracellular events in this differentiation process are not known, but studies of *in vitro* IgM polymerization (3) suggest that the J chain plays an important role in initiating polymer assembly and controlling pentamer secretion. The present studies were undertaken to examine these possibilities.

RESULTS AND DISCUSSION

J chain content of rabbit lymphocytes. The synthesis of J chain by B cells was assessed by use of a radioimmunoassay (RIA) which could detect nanogram levels of free J chain.

Rabbit lymphoid cell populations were lysed with detergent and then examined for their capacity to inhibit the binding of radiolabeled J chain to its specific antiserum. Typical results obtained with normal thymus and spleen cell populations are shown in Fig. 1. It can be seen that the thymus

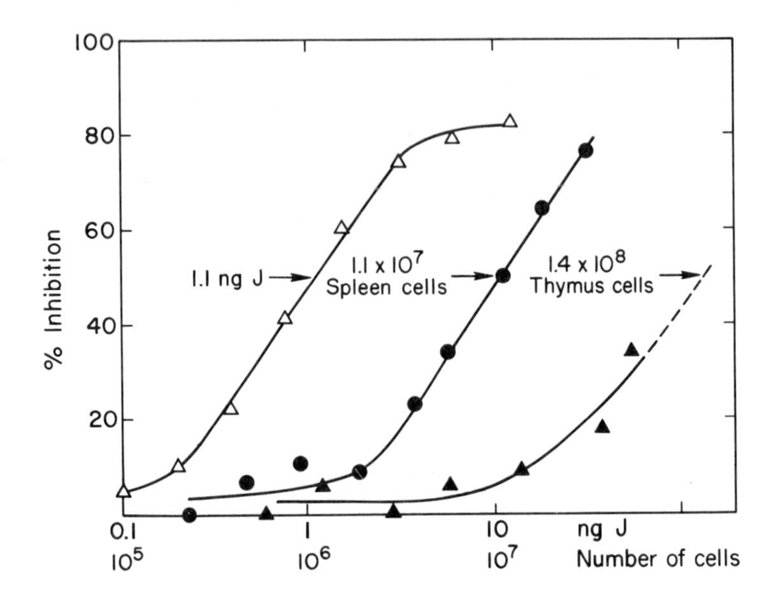

Fig. 1. J chain RIA of rabbit lymphoid cells. Spleen and thymus cells were lysed with 1% (w/v) Nonidet P-40 and then centrifuged at 43,000 x g for 30 min. The supernatants were diluted appropriately and incubated with goat anti-rabbit J chain for 4 hrs at 37°. One ng of lactoperoxidase labeled ^{125}I - J chain was added and the incubation was continued for another 4 hrs at 37°. Sufficient carrier goat IgG and rabbit anti-goat IgG were added to quantitatively precipitate the goat anti-rabbit J chain. The precipitates were washed one time before being counted in a Searle gamma counter, model 1195.

cell lysates contained very little J chain, the equivalent of 1.4 x 10^8 cells being required for 50% inhibition. By comparing the 50% endpoint of the thymus cell lysate with that observed for the standard unlabeled J chain inhibitor, it was calculated that 10^7 thymus cells contained 0.08 ng or 320 molecules of J chain per cell. These small amounts of J chain may have been contributed by B cells from thymic lymph nodes since no attempt was made to separate nodes from the thymic tissue. In contrast, J chain was found to be present in substantial quantity in normal spleen cells. In this experiment the lysate from 10^7 cells contained 1.1 ng of J chain and the values in seven other experiments ranged from 1 to 2.8 ng/10^7 cells or 4,000 to 11,000 molecules per cell.

Since normal spleen cell populations contain a small percentage of activated B lymphocytes (1), it was necessary to remove these cells in order to measure J chain synthesis by the precursor population. Advantage was taken of the increase in cell size that accompanies activation. Spleen cell populations were subjected to 1 x g velocity sedimentation according to the methods of Miller and Phillips (4) and the small cell fractions were then examined in the J chain RIA. As the results in Table 1 show, the J chain content ranged from 0.2 to 0.6 ng/10^7 small cells, 4 to 10 fold less than that of the unfractionated spleen cells. This decrease was not due to a general depletion of B cells because fluorescent staining for surface Ig indicated that the small cells contained the same percentage of Ig positive cells as the original populations. The decrease did, however, correlate with sedimentation rate (Table 1), the smaller the size of the cell fraction examined, the lower the J chain content. Moreover, in other experiments with Sephadex G-10

TABLE 1

J CHAIN CONTENT OF THE SMALL CELL FRACTION FROM RABBIT SPLEEN CELLS

Exp. No.	Small Cells		Unfractionated Cells
	S value (mm/hr)	J chain (ng/10^7 cells)	J chain (ng/10^7 cells)
1	2.3-2.5	0.2	2.1
	2.6-2.9	0.4	
2	2.0-2.8	0.2	-
3	2.8-3.3	0.3	1.9
4	2.8-3.4	0.35	-
5	2.0-3.5	0.5	2.8
6	2.3-3.7	0.6	2.1

separated spleen cells (5), the extent of J chain depletion could be correlated with the depletion in plaque forming cells. Thus it appears likely that the small amounts of J chain detected in the precursor enriched fractions represented the product of a few activated cells that were not removed by the separation methods employed. However, the possibility that precursor B cells synthesize a small amount of J chain cannot be ruled out.

Mitogen stimulation of J chain and IgM synthesis. The finding that little or no J chain was present in precursor B cells indicated that its synthesis must be initiated during B cell activation to effect pentamer assembly and secretion. Evidence to support this conclusion was obtained by examining the effects of mitogen stimulation. Small spleen cells with sedimentation rates of 2.0 to 3.5 mm/hr were cultured with pokeweed mitogen and the changes in intra- and extracellular J chain were monitored as a function of time. Radioimmunoassays of the cultured cell lysates (left hand portion of Figure 2) showed that the levels of intracellular J chain began to increase soon after mitogen addition and essentially doubled every succeeding 24 hours. In 108 hours there was a 44 fold increase in J chain content, the ng/culture rising from 0.14 to 6.2 (Table 2). Radioimmunoassays of the corresponding culture supernatants showed that only covalently bound J chain was secreted since no J chain could be detected unless the supernatant protein was reduced and alkylated. It can be seen from Figure 2 that covalently bound J chain appeared in the supernatant shortly after the increase in J chain synthesis and rapidly accumulated to a value of 25 ng/culture by 108 hours. Although more analyses are needed at the early times after mitogen stimulation, the kinetics of the J chain response observed in these experiments suggested that J chain synthesis preceeds IgM polymerization and secretion. This observation is in agreement with the proposed mechanism of polymerization (3) in which J chain is assigned the function of initiating IgM assembly.

The rate of secretion of covalently bound J chain was estimated from the differences in the amount of supernatant J chain between time points and the assumption that the secretion rate was constant over that time period. The results of these calculations are given in Table 2 along with the intracellular J chain content at corresponding times. It is evident from an inspection of the data that the rate of J chain secretion was slow compared to the amounts of intracellular polypeptide apparently available for polymerization. This was true even at the beginning of the time period when the calculated average rate of secretion represented an overestimate. Moreover, intracellular J chain was present in such large excess relative to its

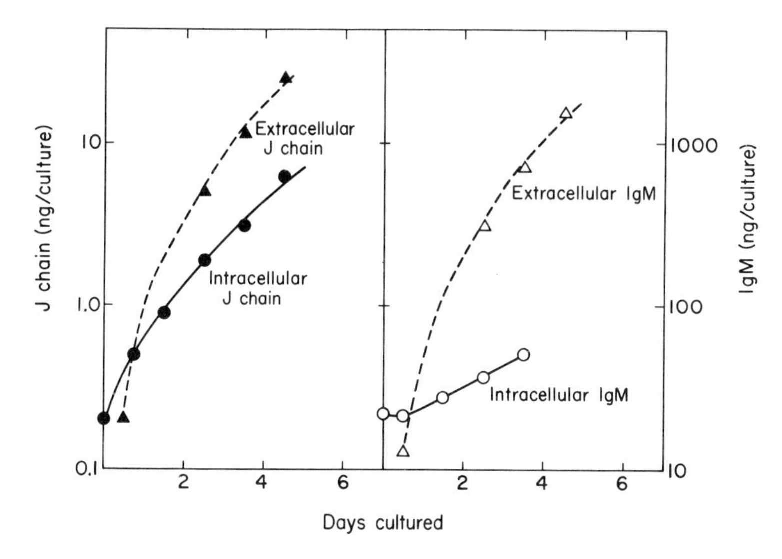

Fig. 2. Synthesis of J chain and IgM by rabbit spleen cells stimulated with PWM. Velocity sedimented small spleen cells (s = 2.0 to 3.5) were cultured in RPMI 1640 - 10% fetal calf serum at 4 x 10^6 cells/ml with 10 µl PWM (GIBCO) per 1 ml culture. Harvested cells were assayed for intracellular J chain as described in Fig. 1. For the determination of intracellular IgM, cells were lysed with 0.4% sodium dodecyl sulfate, sonicated, and centrifuged at 43,000 x g for 30 min. Dilutions of the cell lysates were incubated with goat anti-rabbit µ chain antibody for 4 hrs at 37°. Two ng of ^{125}I-labeled IgM were added and the incubation was continued for another 4 hrs. Facilitation and measurements of radioactivity were carried out as described in Fig. 1. For the determination of extracellular J chain and IgM, culture supernatants were concentrated by filtration, precipitated with 37% $(NH_4)_2SO_4$, completely reduced and alkylated, and then assayed for J chain.

secretion rate that the possibility some of the J chain was contributed by cells synthesizing isotypes other than IgM was not a significant factor. These results indicated that the synthesis of J chain is not, as previously predicted (3, 6), the rate limiting step in IgM secretion. On the contrary, the large amounts found suggested that J chain is produced in excess to help drive the polymerization reaction to completion.

The regulatory role of monomer IgM in pentamer secretion was investigated by following the changes in intracellular IgM

TABLE 2

COMPARISON OF INTRACELLULAR LEVELS OF J AND IgM WITH THE RATE OF PENTAMER SECRETION

Time (hrs)	Intracellular J chain (ng/culture)	Secretion of IgM-bound J chain (ng/culture/hr)
0	0.14	0.02
12	-	
18	0.5	
36	0.9	0.10
60	1.9	
84	3.1	0.26
108	6.2	0.56

Time (hrs)	Intracellular IgM (ng/culture	Secretion rate of IgM (ng/culture/hr)
0	22	1.3
12	20	
18	-	
36	28	6.3
60	37	
84	51	16.4
108	-	35.3

after pokeweed mitogen stimulation. The results of these radioimmunoassays are shown in the right hand portion of Figure 2 and, for comparison, the values for extracellular J chain were converted to ng of pentamer and included in the graph. In contrast to the rapid increase observed in J chain content, the intracellular IgM level did not change significantly until 36 hours after stimulation and increased only 2.3 fold in 84 hours. When these data were compared with the estimated rate of pentamer secretion (Table 2), the intracellular pool of monomer after the first 12 hours was found to exceed the amount secreted per hour by relatively small factors of 3 to 5. Since it was not clear how much of the intracellular IgM was membrane bound and thus unavailable for export, these results could mean that the synthesis of monomer IgM, rather than the synthesis of J chain, is rate limiting in pentamer secretion. However, other intermediates in polymerization, for example, disulfide interchange enzyme, or the process of secretion itself, could be the determining factor. Additional experiments will be required to resolve these alternatives.

ACKNOWLEDGEMENTS

This investigation was support by Research Grant AI 07079 from the National Institute of Allergy and Infectious Diseases, United States Public Health Service and by Grant Number CA 9179, awarded by the National Cancer Institute, DHEW. The authors thank Mrs. Joan Fujita for her excellent technical assistance.

REFERENCES

1. Andersson, J., Lafleur, L., and Melchers, F. (1974) *Eur. J. Immunol.* 4, 170.
2. Melchers, F., and Andersson, J. (1974) *Eur. J. Immunol.* 4, 181.
3. Della Corte, E., and Parkhouse, R.M.E. (1974) *Biochem. J.* 136, 597. Chapuis, R.M., and Koshland, M.E. (1974) *Proc. Nat. Acad. Sci. U.S.* 71, 657. Koshland, M.E. (1975). *Advan. Immunol.* 10, 51.
4. Miller, R.G., and Phillips, R.A. (1969) *J. Cell Physiol.* 73, 191.
5. Ly, I.A., and Mishell, R.I. (1974) *J. Immunol. Methods* 5, 239.
6. Parkhouse, R.M.E., and Della Corte, E. (1973) *Biochem. J.* 136, 607.

ROLE OF CONTRACTILE PROTEINS IN PHAGOCYTOSIS

Thomas P. Stossel, John H. Hartwig, Wayne A. Davies
Ellen C. Jantzen and Stanley G. Pugsley

Medical Oncology Unit, Massachusetts General Hospital,
Department of Medicine, Harvard Medical Scho-l
Boston, Massachusetts 02114

ABSTRACT. Phagocytosis is an example of surface-to-cytoplasm communication. Contact of an appropriate particle with the plasmalemma of phagocytes elicits pseudopod formation, endocytosis and secretion of lysosomal enzymes into phagosomes and the extracellular medium. A molecular explanation for these events is emerging from the systematic study of contractile proteins in macrophages and granulocytes.

A high molecular weight actin-binding protein (ABP) promotes the temperature-dependent gelation of filamentous (F)-actin which comprises ca 7% of the total cell protein. Myosin, in the presence of Mg^{2+}-ATP and a protein cofactor contracts actin-ABP gels. We hypothesize: a) A submembrane ABP-actin gel stabilizes the membrane. b) Particle contact alters ABP-membrane association. c) Activated ABP promotes increased focal gelation of F-actin. c) Since the myosin: actin ratio of phagocytes is low compared to muscle, ABP enhances the efficiency of contraction by cross-linking actin filaments. Contraction of ABP-actin gels forms pseudopods. e) Sustained contraction locally disrupts the F-actin-ABP gel, permitting lysosome:membrane fusion as well as blebbing due to membrane destabilization.

Tests support this theory of a gelation-contraction-disruption cycle: a) The extractibility of ABP from macrophage membranes is greater in phagocytosing than resting cells. b) $>10^{6}$M cytochalasin B destroys pseudopods, induces surface blebs, inhibits phagocytosis and enhances secretion. $>10^{-7}$M cytochalasin B reversibly dissociates ABP from actin, concomitantly dissolving ABP-actin gels without impairing actomyosin contraction. c) Macrophages exposed to nylon wool secrete lysosomal enzymes and bleb, as a hyaline barrier between lysosomes and plasmalemma dissolves. Cytoplasmic extracts of nylon wool-exposed macrophages have reduced ABP and myosin content and capacity to gel, although their F-actin content is similar to that of resting cells.

The idea that cytoskeletal elements have an important role in the functions of cells active in immune responses is popular. Phagocytosis is one such function. During phagocytosis, organelle-excluding pseudopodia containing microfilaments surround objects. Simultaneously, lysosomes enter the cell periphery (from which they are normally excluded by a zone of microfilaments) between the pseudopodia and fuse with the developing phagocytic vacuole, into which they secrete their contents. In this review we briefly summarize our progress in characterizing the contractile proteins of mammalian phagocytes, the interactions of these proteins, and the relationships we propose between these interactions and the complex events of phagocytosis.

CONTRACTILE PROTEINS IDENTIFIED IN MACROPHAGES AND GRANULOCYTES

1. Actin. Electrophoretic analyses of whole cells indicate that actin is about 7% of the total cell protein (1), and we can now isolate about 10% of the cell actin. The isolated actin is quite homogenous by the criterion of molecular weight as determined by polyacrylamide gel electrophoresis with sodium dodecyl sulfate. Interestingly, macrophage and granulocyte actins homogenous by this criteria have at least two components, one major and one minor, separable by an isoelectric focussing technique (2). The electrophoretic components of macrophage and granulocyte actins comigrate with rabbit skeletal muscle actin in our studies. (Fig. 1).

The purified actins assemble into extended linear filaments 4-6nm in diameter in 0.1M KCl solutions (F-actin), suggesting that the thin microfilaments that are especially prominent in the cortical cytoplasm of phagocytes are polymers of this protein (1). To be emphasized is that thus far no major differences between the rheologic behavior of macrophage, granulocyte or muscle actin have been detectable.

2. Myosin. The estimates of macrophage and granulocyte myosin contents by electrophoretic analysis and by yield during purification roughly agree, indicating that myosin constitutes about 1% of the cell protein. The myosins resemble smooth muscle and most other purified cytoplasmic myosins in subunit structure. The phagocyte myosins bind to actin but do not demonstrate impressive activation of their low Mg^{+}ATPase activities (2,13). This ATPase activity is intimately related to "contraction" of muscle actomyosin (4). The subcellular location of myosin in phagocytes is not clearly known, but we have concluded that much of the protein may be closely associated with the plasmalemma (but easily extracted from it) on the basis of a) immunofluorescent localization and b) subcellular fractionation. (Fig. 2).

3. Actomyosin Cofactor. This "factor", similar to a "cofactor" discovered in *Acanthamoeba* (5), which has not yet

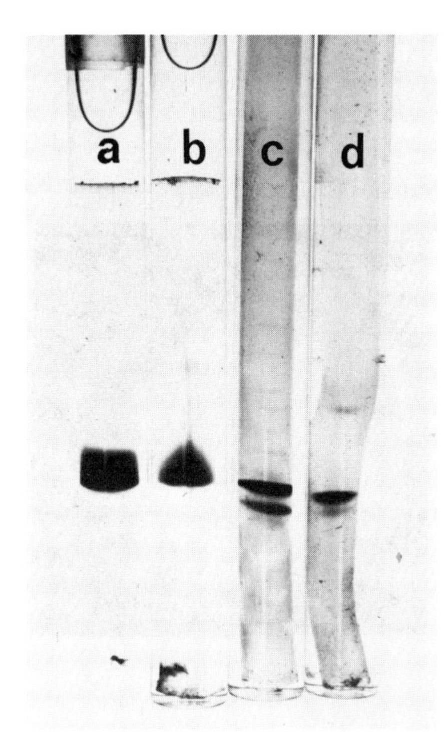

Fig. 1. 5% polyacrylamide gels showing purified rabbit skeletal muscle (a) and rabbit lung macrophage (b) actins after electrophoresis in sodium dodecyl sulfate, or after isoelectric focussing of muscle (c) and macrophage (d) actins with pH ampholytes in 10mM Tris-HCl buffer, pH 8.0, containing 0.1mM ATP, 0.1mM $CaCl_2$, 0.1% 2-mercaptoethanol. The gels were fixed and stained according to reference 36. Muscle actin was purified by the technique of Spudich and Watt (9). Macrophage actin was purified by allowing macrophage cytoplasmic extracts to gel as described (10), followed by dissolving the gel in 0.6M KCl, 0.1mM ATP, 1mM dithiothreitol, 10mM Imidazole-HCl, pH 7.0 and ultracentrifugation to pellet F-actin. The actin in the sediment was purified by ion exchange chromatography on DEAE-sephadex and by gel filtration with Sephadex G-100.

been purified from any cell, has been identified in macrophages (3). Cofactor activates the important Mg^{2+}-ATPase activity of macrophage myosin in the presence of muscle or macrophage actins (3). In light of reports suggesting that the ability of smooth muscle and platelet myosin's Mg^{2+}-ATPase activity to be activated by actin depends on the phosphorylation of myosin light subunits (5-8), we examined whether macrophage myosin could be phosphorylated and whether the phosphorylated protein's Mg^{2+}-ATPase activity was directly activatable by actin. If so, the macrophage cofactor might be a protein kinase. A crude myosin-containing fraction of macrophage homogenates was incubated with MgCl and ATP plus ATP^{32}, or else only with EDTA. Myosin was purified from these mixtures. Myosin from the Mg^{2+}ATP incubation, but not the EDTA incubation, incorporated ^{32}p, 1.6 moles/mole of protein, and the ^{32}p was bound to a 15000 dalton subunit of the myosin. This amount and location of phorphorylation were similar to findings by others studying the phosphorylation of diverse myosins (6-8). However, the Mg^{2+}ATPase activity of neither

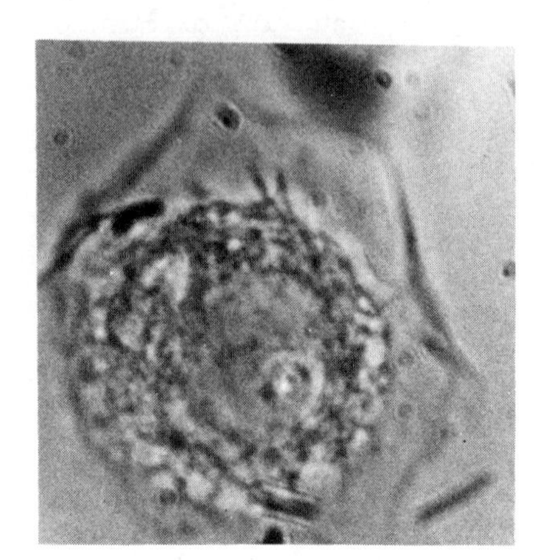

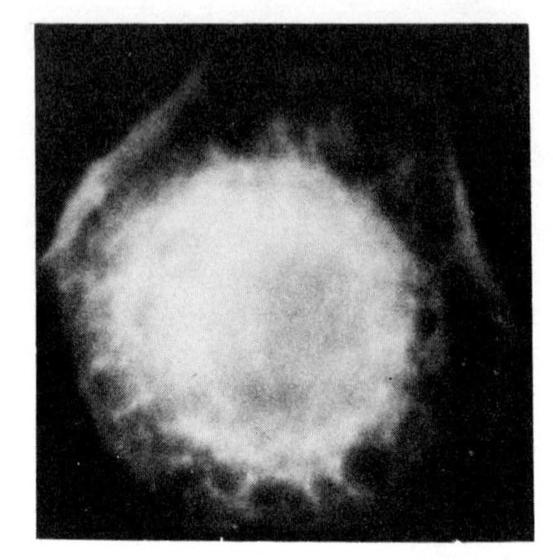

Fig. 2. Phase contrast (left) and fluorescence (right) photomicrographs of a rabbit lung macrophage fixed with acetone at room temperature after having spread on a glass slide in balanced salt solution containing 1% rabbit serum. The fixed cell was reacted for 45 min with rhodamine-labelled anti-macrophage myosin Fab'_2 and then washed with phosphate-buffered 15mM NaCl solution. Anti-macrophage myosin antiserum was generated by a goat immunized with highly purified (1) macrophage myosin and shown to be monospecific by the criterion of immunoelectrophoresis. Conjugation of the IgG purified from the antiserum with tetramethylrhodamine was done by the method of Cebra and Goldstein (11) and Fab'_2 was prepared according to reference 12. Note: a) decreased stain in the nuclear area; b) relatively diffuse staining of the cell body (no "stress fibers" apparent); c) presence of stain in ruffled areas of the spread hyaline veil, suggesting a peripheral location of myosin molecules. This type of staining pattern was diminished by pre-incubation of the fixed cells with unconjugated anti-macrophage myosin IgG prior to reacting it with conjugated antimyosin Fab'_2. Magnification: 1200X.

phosphorylated nor nonphosphorylated macrophage myosins was activated by actin unless cofactor was added. Therefore, phosphorylation of myosin is unlikely to be the mechanism of action of cofactor.

4. Actin-binding Protein (ABP). This high-molecular weight subunit (ca 280,000 daltons) protein comprises about 1% of the total protein of granulocytes and macrophages by electrophoretic analysis (1), but only a third of the protein is extractible and subsequently purifiable from resting cells (10). As amplified below, this protein is the only entity in macrophages and granulocytes that we have found capable of making viscous filamentous actin form a solid gel (10). ABP resembles a protein subsequently isolated from smooth muscle in subunit and amino acid composition and chromatographic behavior, and there is evidence that ABP or very similar proteins exist in diverse cells (Table 1). The subcellular distribution of ABP is also not definitely known but may resemble that of myosin in being associated with plasma membrane (24,25).

INTERACTIONS

1. Actin and Actin-binding Protein. We continue to obtain evidence that ABP causes the gelation of filamentous actin: a) In crude macrophage extracts the gelation of actin is proportional to the concentration of ABP (10). b) Highly purified ABP causes actin gelation in a concentration-dependent manner. As little as 50μg/ml of ABP will gel a 2mg/ml solution of F-actin. c) Only high molecular weight material isolated by chromatography from cytoplasmic extracts of macrophages causes the gelation of actin. d) Antiserum raised in rabbits against human granulocyte ABP inhibited the gelation of crude cytoplasmic extracts (13).

2. Actin and Myosin (+ Cofactor). "Contraction" or "syneresis" of granulocyte or macrophage crude cytoplasmic actin gels occurs in the presence of Mg^{2+}ATP. This phenomenon can be reproduced with a mixture of purified actin, ABP and myosin in the presence of Mg^{2+}ATP, and macrophage cofactor accelerates the reaction (10). The results indicate that this system can generate mechanical force. Antiserum against purified granulocyte myosin inhibits the syneresis of actin gels (13).

Although free calcium ions cause rapid aggregation of crude cytoplasmic extracts (10,13), no good evidence yet exists that calcium regulates the "contractile" activity of phagocytes. In studies so far, we have found: a) that

TABLE 1

HIGH MOLECULAR-WEIGHT ACTIN-BINDING PROTEINS

Name of Protein	Cell of Origin	Apparent Subunit Molecular Weight	Properties	Ref.
ABP	Rabbit Lung Macrophage	$\sim$280,000	Purified protein binds, gels actin.	(1,3,10)
	Human Blood Granulocytes	$\sim$280,000	Purified protein binds, gels actin.	(13)
	Human Platelets	>220,000	Purified protein binds, gels actin.	(14,15)
	Acanthamoeba	$\sim$280,000	Associated with actin gels	(16)
HMW	HeLa Cells	$\sim$280,000	Associated with actin gels	(17)
HMW	Sea Urchin Eggs	>220,000	Associated with actin gels	(18)
HMW	_Amoeba Proteus_	$\sim$280,000	Associated with actin gels	(19)
Filamin	Chicken Gizzard, Smooth Muscle	$\sim$280,000	Binds actin. Antibody to this protein reacted with various cultured cells.	(20,21)
Spectrin	Erythrocytes	$\sim$250,000 230,000	Binds actin.	(22,23)

gelation and contraction of mixtures of purified macrophage contractile proteins occur essentially with equal efficiency in the presence or absence of free calcium (10); and b) that calcium does not affect the Mg^{2+}-ATPase activity of even crude macrophage actomyosin preparations.

CONTRACTILE PROTEINS AND PHAGOCYTOSIS

1. ABP Extractibility Changes With Phagocytosis. The fraction of total cell ABP extractible from macrophages is significantly greater than that of resting cells (10). This finding is consistent with a change in membrane association of ABP during phagocytosis.

2. Cytochalasin B Alters ABP-Actin Interaction and Phagocytosis. The drug, cytochalasin B, among diverse effects on cells, inhibits phagocytosis by macrophages and granulocytes (26-28) and yet enhances the secretion of lysosomal enzymes into the extracellular medium by these cells when challenged with objects that they would phagocytize (29,30). In 0.1M KCl solution, cytochalasin B dissolves crude macrophage cytoplasmic extract actin gels and gels composed of actin plus purified ABP at concentrations $\geq 2\times10^{-7}$M, amounts less than those required to influence the motile functions of intact cells ($\geq 10^{-6}$M) (31). At these concentrations, cytochalasin B does not depolymerize actin filaments or inhibit the Mg^{2+}-ATPase activity of cofactor-activated macrophage actomyosin (31). The results support the idea that actin gelation is important in a) the formation of pseudopodia and b) the control of lysosomal fusion.

3. Neutrophil Actin Dysfunction. The neutrophils of an infant with recurrent pyogenic infections were markedly impaired in locomotion and phagocytosis, yet secreted excessive amounts of lysosomal enzymes into the extracellular medium. Actin from these neutrophils was very deficient in polymerizability (32). Only about half of the actin from neutrophils of both of the infant's parents polymerized under conditions which fully polymerized normal neutrophil actin, suggesting that the disorder is inherited as an autosomal recessive trait. No evidence was found for an inhibitor of polymerization. The non-polymerizable actin was purified and found to have a different isoelectric point than normal neutrophil actin.

4. Redistribution of Contractile Proteins During Phagocytosis. Taken together, the available facts suggest

that actin gelation is necessary for the creation of pseudopodia required for phagocytic vacuole formation, whereas lysosomal secretion depends on local dissolution of the gel which permits the lysosomes to gain access to the plasma membrane. Since the events occur simultaneously, different influences on the state of actin must exist in different regions of the developing phagocytic vacuole: actin gelation predominates in advancing pseudopodia and gel dissolution occurs at the base of the vacuole where lysosomes fuse. Assuming that ABP and myosin molecules, both of which can cross-link actin filaments, regulate the state of actin, two mechanisms are possible to explain how these accessory proteins might control the actin state during phagocytosis. 1) These proteins could be uniformly distributed around the cell periphery and "activated." This activation could propagate with the sequential construction and spreading of the pseudopodia around an object. Inactivation of ABP and myosin at or near the original site of membrane:particle contact would return actin to the sol state. 2) The control proteins could be in asymmetrical distribution throughout the cell periphery and could actually advance with the pseudopodia, abandoning the central region that becomes the site of lysosomal secretion. The location of these proteins would therefore determine the presence or absence of an actin gel state. In addition to deciding between these alternatives, the question as to whether the depolymerization of actin filaments accompanies the solvation of actin gel requires resolution.

To approach these problems, we have taken advantage of the similarity between cell spreading and phagocytosis (33). Phagocytes recognize nylon wool fibers and spread pseudopodia on them in an apparent attempt at ingestion (34), and lysosomes fuse with the plasmalemma at the nylon wool fiber contact site. Since no true phagocytic vacuole can form on the flat nylon wool fiber surface, the lysosomal enzymes are secreted to the extracellular medium (34). Removal of the spread cells from the nylon wool surface by mechanical shear results in the separation of the spread pseudopods from the cell bodies, and these pseudopodia are isolable by differential centrifugation as a pure population of "podosomes" (35). (Fig. 3).

We compared the consistency of extracts prepared from cell bodies of spread and unspread rabbit lung macrophages and correlated the findings with the protein composition of cell bodies and of podosomes.

Cytoplasmic extracts of the cell bodies eluted from nylon fibers contained two-thirds less ABP and myosin and about 30% less actin than extracts of cells sheared in the absence of nylon wool fibers. Nearly all of the actin and

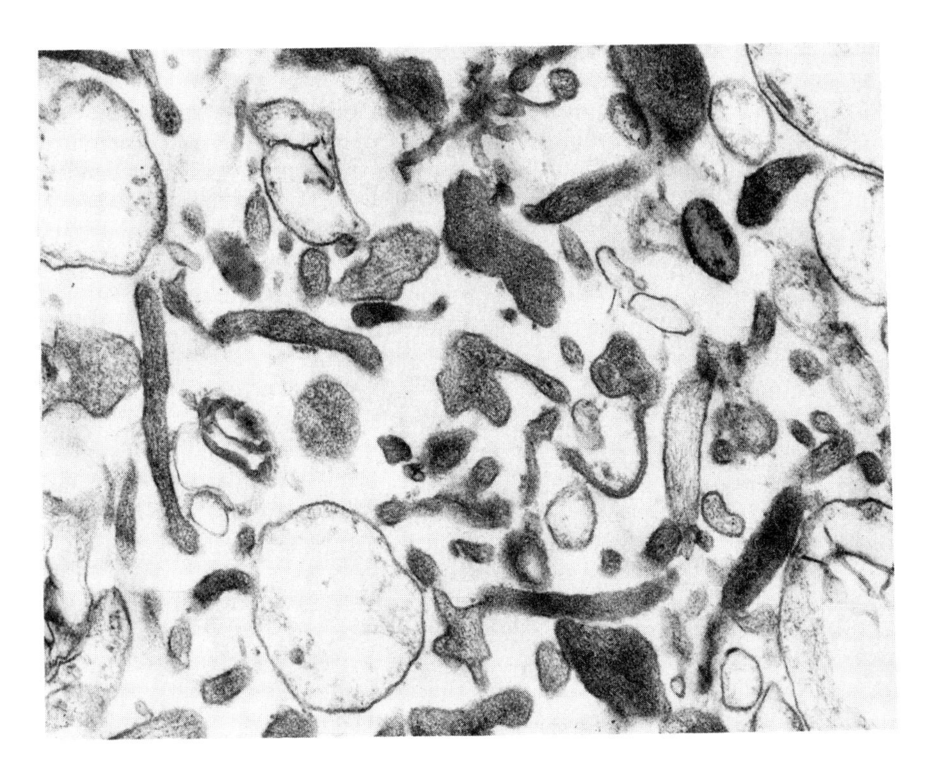

Fig. 3. Electron micrograph of a thin section of podosomes, hyaline blebs isolated from rabbit lung macrophages (35). These structures were enriched 3-fold over whole homogenates in specific adenylate cyclase activity and not enriched in markers for mitochondria (succinic dehydrogenase activity), nuclei (DNA), ribosomes (Trichloroacetic acid-precipitable ^{14}C-uracil), lysosomes (β-glucuronidase activity). The podosomes contained LDH activity and actin and were enriched in myosin and ABP concentrations relative to macrophage cytoplasmic extracts. The findings indicate that podosomes are sacs of plasmalemma surrounding peripheral cytoplasm. Magnification: 52,000X.

two-thirds of the other two proteins were accounted for in podosomes, indicating transfer of the proteins from cell bodies to podosomes. The alterations in protein composition correlated with a marked diminution in the capacity of extracts of nylon wool fiber-treated cell bodies to gel, a property dependent on the interaction between ABP and F-actin. However, the actin in the extracts was fully capable of polymerizing.

We propose that translocation of contractile proteins accounts for the concomitant differences in organelle exclusion that characterize phagocytosis. ABP and myosin move into pseudopodia where prominent gelation and syneresis of actin occur. Actin in the regions from which ABP and myosin move disaggregates without depolymerizing, permitting lysosomes to gain access to the plasmalemma. The signals that are responsible for the activation and movement of these proteins and the precise mechanisms of movement remain to be defined.

REFERENCES

1. Hartwig, J.H. and Stossel, T.P. (1975) J. Biol. Chem. 250,5699.
2. O'Farrell, P.H. (1975) J. Biol. Chem. 250,4007.
3. Stossel, T.P. and Hartwig, J.H. (1975) J. Biol. Chem. 250,5705.
4. Szent-Györgyi, A. (1947) Chemistry of Muscular Contraction Acad. Press, N.Y.
5. Pollard, T.D. and Korn, E.D. (1973) J. Biol. Chem. 248,4691.
6. Adelstein, R.S. and Conti, M.A. (1975) Nature 256, 597.
7. Aksoy, M.O., Williams, D., Sharkey, E.M. and Hartshorne, D.J. (1976) Biochem. Biophys. Res. Commun.69,35.
8. Chacko, S., Conti, M.A. and Adelstein, R.S. (1977) Proc. Nat. Acad. Sci. USA 74,129.
9. Spudich, J.A. and Watt, S. (1971) J. Biol. Chem. 245,4866.
10. Stossel, T.P. and Hartwig, J.H. (1976) J. Cell Biol. 68,602.
11. Cebra, J.J. and Goldstein, G. (1965) J. Immunol. 95,230.
12. Nisonoff, A. (1964) Methods Med. Res. 10,134.
13. Boxer, L.A. and Stossel, T.P. (1976) J. Clin. Invest. 57,964.
14. Schollmeyer, J.E., Rao, G.H.R. and White, J.G. (1976) Circulation 54, suppl 2,197.
15. Lucas, R.C., Gallagher, M. and Stracher, A. (1976) Contractile Systems in Non-muscle Tissues, N. Holland Biomed. Press 133.
16. Pollard, T.D. (1976) J. Cell Biol. 68,579.
17. Weihing, R.R. (1976) Cell Motility, Cold Spring Harb. Press 3,671.
18. Kane, R.E. (1976) J. Cell Biol. 71,704.
19. Taylor, D.L., Rhodes, J.A. and Hammond, S.A. (1976) J.Cell Biol. 70,112.

20. Wang, K., Ash, J.F. and Singer, S.J. (1975) Proc. Nat. Acad. Sci. USA 72,4483.
21. Shizuta, Y., Shizuta, H., Gallo, M., Davies, P., Pastan, I. and Lewis, M. (1976) J. Biol. Chem. 251,6562.
22. Marchesi, V.T. and Steers, E., Jr. (1968) Science 158,203.
23. Tilney, L.G. and Detmers, P. (1975) J. Cell Biol. 66,508.
24. Boxer, L.A., Richardson, S. and Floyd, A. (1976) Nature 263,259.
25. Pincus, S.H. and Stossel, T.P. (1976) Clin. Res. 24,109A.
26. Allison, A.C., Davies, P. and De Petris, S. (1971) Nature New Biol. 232,153.
27. Davis, A.T., Estensen, R. and Quie, P.G. (1970) Proc. Soc. Exp. Biol. Med. 137,161.
28. Malawista, S.E., Gee, J.B.L. and Bensch, K.G. (1971) Yale J. Biol.Med.44,286.
29. Zurier, R.B., Hoffstein, S. and Weissman, G. (1973) Proc. Nat. Acad. Sci. USA. 70,844.
30. Davies, P., Allison, A.C. and Haswell, A.D. (1973) Biochem. J. 134,33.
31. Hartwig, J.H. and Stossel, T.P. (1976) J. Cell Biol. 71,295.
32. Boxer, L.A., Hedley-Whyte, E.T. and Stossel, T.P. (1974) N. Engl. J. Med. 291,1093.
33. North, R.J. (1970) Semin. Hemat. 7,161.
34. Klock, J.C. and Bainton, D.F. (1976) Blood 48,149.
35. Davies, W.A. and Stossel, T.P. (1976) J. Cell Biol. 70,296a.
36. Fairbanks, G., Steck, T.L. and Wallach, D.F.H. (1971) Biochem.10,2606.

WORKSHOP 17

MEMBRANES AND SIGNALLING

F. Melchers* and B. Pernis+, Convenors
K.A. Krolick, Scibe

The workshop first dealt with the role of surface-bound Ig in the regulation of proliferation and maturation of B-cells to Ig-secretion. Anti Ig-antibodies were used in several laboratories to probe surface Ig on B-cells, and two apparently contradictory sets of results were reported. Several laboratories found anti Ig-antibodies to stimulate proliferation, as measured by thymidine uptake, while others did not see such stimulation. Whereever stimulation by anti Ig was seen, it was only to proliferation, but not to Ig-secretion. Experiments using rabbit peripheral lymphocytes and their stimulation by anti-Ig to proliferation were reviewed by *Sell*. Nylon wool passed T-cells, negative for surface Ig, stimulatable by Con A and subject to killing by ATS and C', with demonstrable helper and suppressor functions, do not respond to anti-Ig. B-cells, surviving ATS + C' treatment, surface Ig-positive and unreactive to Con A, on the other hand, can be stimulated. This identifies the prime target of anti-Ig action as B-cells, although some action on T-cells could not be excluded, since Con A stimulated 80% and anti-Ig 70-80% of all peripheral lymphocytes (anti μ 80%, anti γ 70%, anti κ 30%, anti ε none). Anti allotype a and b are stimulatory, antibodies against other allotypes do not stimulate. The anti-Ig antibodies appear endocytosed quickly, go into small vescicles, then reappear on the surface after 24-36 hours. *Kermani* reported stimulation of human lymphocytes to 2-6 fold increased thymidine uptake by anti IgD. *Sieckmann* reported stimulation of mouse lympho-

*Present address: Basel Institute for Immunology
Grenzacherstrasse 487
CH-4005 Basel, Switzerland

+Present address: Department of Microbiology
Columbia University
College of Physicians and Surgeons
701-711 West 168th Street
New York, 10032, N.Y., USA

cytes from spleen, lymph nodes, bone marrow and Peyer's patches to increased thymidine uptake, measured in 64 hrs stimulated cultures as an 18 hrs pulse (anti μ 40 fold, anti γ / κ 10 fold, anti κ 10 fold over background). Nylon wool purified T-cells did not respond, and were not needed for anti-Ig action. This response was first detectable in 4 weeks old spleen, reaching adult levels at 12 weeks of age. CBA/N spleen cells could be stimulated by LPS but not by anti-Ig, C57bl10/Cr could be stimulated by anti-Ig but not by LPS, while F1 spleen cells of crosses of the two strains could be stimulated by both LPS and anti-Ig. This was interpreted as indicating that anti-Ig could stimulate a mature B-cell subpopulation missing in CBA/N mice. *Möller* concluded from these results that Ig could not serve as the triggering molecule in B-cell surface membranes, since CBA/N mice have Ig-positive cells and elicit normal T-cell dependent responses. The positive LPS-response of CBA/N mice was also surprising, since studies by *Huber, Coutinho and Melchers* indicate that this mouse strain has at least 100 times less LPS-reactive B-cells maturing to IgM-secretion. *Melchers* reviewed the experimental evidence obtained with murine lymphocytes that anti-Ig antibodies do not stimulate B-cells to significant increases of either proliferation or of maturation to Ig-secretion. Macrophage-depleted B-cells of nude mice from spleen, lymph node, bone marrow or thoracic duct were used. Rabbit, guinea pig or chicken anti mouse μ, α, κ, or λ antibodies or their pepsin (Fab')2 or papain Fab fragments had no stimulating effects. When, however, B-cell mitogens, such as LPS, PPD, Dextransulfate, Nocardia or fetal calf serum, were added to anti Ig-coated B-cells, inhibition of polyclonal stimulation was observed. This was most evident when the maturation to Ig-secreting PFC was monitored. Proliferation, as measured by thymidine uptake, was either slightly enlarged (factors 2-4), unaltered, or for some antisera, inhibited. Pepsin $(Fab')_2$-fragments were equally inhibitory, indicating that Fc-receptors on B-cells were not involved in the inhibitory action. Papain Fab fragments had 100 fold lower inhibitory capacity. The B-cells could be relieved from anti-Ig-inhibition by proteolysis of the attached anti-Ig or by shedding of the Ig-anti Ig complexes from the surface of B-cells as long as LPS was not added to the anti Ig-coated cells, so that later LPS-triggering resulted in proliferation and maturation to Ig secretion. This suggests that the state of surface Ig, occupied by anti-Ig or not occupied, may modulate the reactivity of B-cells toward mitogens such as LPS, leading to either suppression or stimulation. B-cell mitogens circumvent the binding step to stimulate or suppress B-cells via mitogen-receptor structures directly. A receptor complex between surface Ig and mitogen receptors may exist on B-cells, in which the growth-regulating mitogen receptors can be modulated by the Ig molecule for either stimulation or inhibition. While this inhibition of proliferation and maturation appears to exist in one subpopulation of splenic B-cells, which is also found in abundance in thoracic duct, other subpopulations exist which, upon mitogenic stimulation, are inhibited only for secretion but not for proliferation, or which only proliferate but never mature and which are either inhibited or not inhibited by anti-Ig antibodies.

Parker, also using murine splenic lymphocytes, could not detect stimulation of lymphocytes to proliferation by soluble anti-Ig antibodies, Anti Ig (μ or κ) attached to polyacrylamide beads, however, lead to a 10-20 fold increase in thymidine uptake and in blast transformation. The stimulation was independent of T-cells or macrophages. Soluble anti-Ig competed in the stimulation, inhibiting it. These studies could be extended to human peripheral lymphocytes, which were stimulated by anti-IgM and anti-IgD to increased thymidine uptake. Polyacrylamide beads, by themselves, had no stimulatory effect. The objection that polyacrylamide acts as a mitogen for B-cells, could, however, not be ruled out, since the control using Ig-unrelated antibodies specific for other B-cell surface structures (H2, MBLA) coupled to polyacrylamide beads had not yet been done.

The discrepancies in the results obtained in exposing B-cells to anti-Ig antibodies remained unresolved. Possible sources of experimental variations between different sets of experiments were a different heterogeneity in B-cell subpopulations differently susceptible to stimulation, distinctions of reactions leading to proliferation from those leading to maturation (Ig secretion) and possible contamination of the anti-Ig antibody preparations by B-cell mitogens. The physiological role of the action of anti-Ig antibodies on B-cells remains to be evaluated; it is, however, likely that the immune system would favor suppressive actions (in a possible network of lymphocyte interaction) over stimulatory actions.

Receptors other than Ig must exist on B-cells, which regulate growth and maturation. *Möller* presented evidence that unresponsiveness to the T-cell-independent antigen α-

1,6-dextran results from a defect in the capacity of ACA, A, ATH and ATL-mice to produce Ig-structures recognizing epitopes on α 1,6-dextran. In these mice α 1,6-dextran is recognized normally as a mitogen, probably by mitogen-receptors, leading to polyclonal Ig secretion with specificities for any hapten except for α 1,6-dextran. Suppressor T-cell action could be excluded in these responses. In the same nonresponder mice FITC- α 1,6-dextran was used as a T-cell independent antigen in tolerogenic high doses to induce unresponsiveness to FITC "in vivo". When cells from the tolerant mice were activated "in vitro" with another B-cell mitogen, LPS, activating B-cell subpopulations which are mitogen-noncrossreactive with α 1,6-dextran, a normal FITC-specific response was obtained, while dextran as a mitogen could never break the FITC-dextran-induced tolerance to FITC. The results indicate that different mitogen-reactive B-cell subpopulations exist, which all carry the full repertoire of FITC-specific cells. The tolerance-inducing signal for FITC-specific cells is induced in the α 1,6-dextran-mitogen-reactive B-cell subpopulation by high doses of FITC-dextran through the mitogenic action of dextran, possibly via dextran-specific mitogen (non Ig) receptors. This appears to be a further case for mitogen-mediated suppression of B-cells, in this case in cells binding high numbers of epitopes onto their Ig-surface structures. It may, therefore, depend on the number of surface Ig-molecules occupied by either hapten or anti-Ig molecules, whether the binding cells are subsequently stimulated or suppressed by mitogens.

Cammisuli presented evidence, that binding of the carrier to Ig on B-cells is not needed in the carrier primed-T-cell-dependent activation of memory B-cells induced to IgG-secretion, as long as the carrier (antigen) is bound to the B-cell by other interactions. Lactoside-KLH-primed lymphocytes were treated with haptenated (Ars-coupled) anti H2, anti Ia-7, anti-γ, anti-μ and anti L antibodies, and subsequently coated with KLH by adding KLH covalently linked to anti-ars-antibodies. KLH-specific T-helper cells were added and the response measured as Lac-specific IgG-secreting PFC. Stimulation was T-cell dependent. No suppressor T-cells were present. KLH, bound via structures other than Ig to B-cell surfaces, could, therefore, concentrate the KLH-specific T-cell factors of helper T-cells onto B-cells, including those specific for Lac. The prediction of these results, not yet experimentally tested, would be that these KLH-specific helper factors could induce a polyclonal B-cell response, since KLH is concentrated via anti H2, anti Ia or anti Ig onto all B-cells regardless of the hapten specificity of their surface Ig. The experiments lend strong support to the idea that T-helper-factors can induce B-lymphocytes to proliferate and mature polyclonally to Ig secretion. It should be emphasized here again that concentrating the carrier via anti μ or anti L chain antibodies did not result in an inhibition of the subsequent stimulation by T-cell factors. It remains to be tested whether at possibly higher concentrations of anti Ig presented to the B-cells, inhibitory effects will be seen.

Diener studied the response of unprimed B-cells to TNP-RGG coupled to a backbone of Ficoll in the presence of T-cells. The response, measured as PFC to TNP, appeared T-dependent and essentially independent of the coupling-density of TNP to RGG. It, however, appeared dependent on the density of RGG-carrier determinants on the Ficoll backbone, showing increased responses with increased carrier densities. This suggests that carrier-specific helper factors have to be presented to the B-cells in high enough concentrations to be activating, an effect which is achieved by higher numbers of carrier molecules presented per B-cell surface area. His notion of a T-cell factor having amphipathic properties, by which it can insert itself into the lipophilic portions of the surface membrane of B-cells, was challenged by the remark of *Möller* that mitogen-recognizing structures on the B-cell are molecules, probably proteins, which are encoded by genes and that, according to recent studies by *Melchers, Andersson and Coutinho,* one B-cell appears to be responsive in one strain (C_3H/Tif) to LPS and Nocardia, while in the LPS-nonresponsive C_3H/HeJ strain the same cell still shows mitogen-reactiveness to Nocardia. Stimulatory mechanisms envisaging a general disturbance of the lipid bilayers of the plasma membrane thus appear not likely.

B-cells from tolerant mice were shown by *Ashman* not to cap surface Ig specific for the tolerizing antigen, while other antigen-specific B-cells showed unaltered capping capacities.

Mechanisms of B-cell triggering either via cross-linking of Ig receptors or via allotype changes within the Ig molecule after antigen binding were discussed by *Koshland.* In her experiments Lac-KLH primed lymphocytes were challenged with monomeric or with polymeric Lac-RNAase in the presence of RNAase-primed T-cells. No differences were observed between the monomeric and the polymeric antigen in Lac-specific responses. The conclusion that crosslinking of Ig-receptors is not a requisite for triggering was questioned by *Raff* who remarked that carrier-T-cell-factor-complexes may very well aggregate the monomeric form of the antigen.

The workshop ended with discussions on molecular events in the surface membrane and in the nucleus of lymphocytes which follow the binding of the triggering agents to the lymphocytes. *Streilein* showed that α 2-macroglobulin inhibits mitogenic stimulation of lymphocytes. Since α 2-macroglobulin is known to act as a protease inhibitor, she concluded that protease action may be instrumental in lymphocyte triggering. *Wisnieski* and *Krolick* have probed the fluidity of the inner and the outer layer of the lipid bilayer of the lymphocyte plasma membrane with nitroxides which specifically localize in either of the layers and which are detected by electronspinresonance measurements. Characteristic changes in fluidity are detected with changing temperature at 15^0, 21^0, 30^0 and 37^0C. The outer layer of the membrane melts between $15^0$and 30^0C, the inner layer between 21^0 and 37^0C. The influence of these melting processes on the behaviour of surface Ig in B-cells was probed by capping experiments, using fluorescent anti-Ig-antibodies. The rate of capping showed four changes at 15^0, 21^0, 30^0 and 37^0, indicating that the fluidity of both layers of the plasma membrane may influence the behaviour of membrane-bound Ig molecules. This may be interpreted to mean that surface Ig spans both layers of the plasma membrane. Addition of LPS resulted in small, but significant changes in melting temperatures (1.5-2^0C lower) of the inner and outer layers of the surface membrane. Changes in membrane fluidity induced by lipid-containing mitogens, it was concluded, may be a necessary part of the membrane-mediated activation events in lymphocytes.

Finally, *Decker* reported experiments which show that stimulation of rabbit peripheral blood lymphocytes by either Con A or by anti Ig antibodies resulted in a specific increase of non-histone protein synthesis in the cell nucleus. Since non-histone proteins are thought to act as units regulating DNA synthesis in eukaryotic nuclei, these studies may point to the earliest changes in a lymphocyte which will finally result in DNA-synthesis and cell division.

Author Index

(Article numbers are shown following the names of contributors. Affiliations are listed on the title page of each article)

Subject Index

(Citations are to article number. The article numbers appear in the articles on top of the left hand pages in front of the names of the contributors.)

A

B

C

D

F

G

H

I

J

L

M

N

P

R

A
B 7
C 8
D 9
E 0
F 1
G 2
H 3
I 4
J 5